Genome Mapping and Molecular Breeding in Plants
Volume 3

Series Editor: Chittaranjan Kole

Volumes of the Series
Genome Mapping and Molecular Breeding in Plants

Volume 1
Cereals and Millets

Volume 2
Oilseeds

Volume 3
Pulses, Sugar and Tuber Crops

Volume 4
Fruits and Nuts

Volume 5
Vegetables

Volume 6
Technical Crops

Volume 7
Forest Trees

Chittaranjan Kole (Ed.)

Pulses, Sugar and Tuber Crops

With 45 Illustrations, 8 in Color

Springer

Chittaranjan Kole
Department of Horticulture
316 Tyson Building
The Pennsylvania State University
University Park, PA 16802
USA

e-mail: cuk10@psu.edu

Library of Congress Control Number: 2006930258

ISBN-10 3-540-34515-9 Springer Berlin Heidelberg New York
ISBN-13 978-3-540-34515-2 Springer Berlin Heidelberg New York

Springer is a part of Springer Science+Business Media
springer.com

Editor: Dr. Sabine Schreck, Heidelberg, Germany
Desk Editor: Dr. Jutta Lindenborn, Heidelberg, Germany
Cover design: WMX Design GmbH, Heidelberg, Germany
Typesetting and Production: LE-TEX Jelonek, Schmidt & Vöckler GbR, Leipzig, Germany
39/3100/YL 5 4 3 2 1 0 – Printed on acid-free paper

Preface to the Series

Genome science has emerged unequivocally as the leading discipline of this new millennium. Progress in molecular biology during the last century has provided critical inputs for building a solid foundation for this discipline. However, it has gained fast momentum particularly in the last two decades with the advent of genetic linkage mapping with RFLP markers in humans in 1980. Since then it has been flourishing at a stupendous pace with the development of newly emerging tools and techniques. All these events are due to the concerted global efforts directed at the delineation of genomes and their improvement.

Genetic linkage maps based on molecular markers are now available for almost all plants of significant academic and economic interest, and the list of plants is growing regularly. A large number of economic genes have been mapped, tagged, cloned, sequenced, or characterized for expression and are being used for genetic tailoring of plants through molecular breeding. An array of markers in the arsenal from RFLP to SNP; tools such as BAC, YAC, ESTs, and microarrays; local physical maps of target genomic regions; and the employment of bioinformatics contributing to all the "-omics" disciplines are making the journey more and more enriching. Most naturally, the plants we commonly grow on our farms, forests, orchards, plantations, and labs have attracted emphatic attention, and deservedly so. The two-way shuttling from phenotype to genotype (or gene) and genotypte (gene) to phenotype has made the canvas much vaster. One could have easily compiled the vital information on genome mapping in economic plants within some 50 pages in the 1980s or within 500 pages in the 1990s. In the middle of the first decade of this century, even 5,000 pages would not suffice! Clearly genome mapping is no longer a mere "promising" branch of the life science; it has emerged as a full-fledged subject in its own right with promising branches of its own. Sequencing of the *Arabidopsis* genome was complete in 2000. The early 21st century witnessed the complete genome sequence of rice. Many more plant genomes are waiting in the wings of the national and international genome initiatives on individual plants or families.

The huge volume of information generated on genome analysis and improvement is dispersed mainly throughout the pages of periodicals in the form of review papers or scientific articles. There is a need for a ready reference for students and scientists alike that could provide more than just a glimpse of the present status of genome analysis and its use for genetic improvement. I personally felt the gap sorely when I failed to suggest any reference works to students and colleagues interested in the subject. This is the primary reason I conceived of a series on genome mapping and molecular breeding in plants.

There is not a single organism on earth that has no economic worth or concern for humanity. Information on genomes of lower organisms is abundant and highly useful from academic and applied points of view. Information on higher animals including humans is vast and useful. However, we first thought to concentrate only on the plants relevant to our daily lives, the agronomic, horticultural and technical crops, and forest trees, in the present series. We will come up soon with commentaries on food and fiber animals, wildlife and companion animals, laboratory animals, fishes and aquatic animals, beneficial and harmful insects, plant- and animal-associated microbes, and primates including humans in our next "genome series" dedicated to animals and microbes. In this series, 82 chapters devoted to plants or their groups have been included. We tried to include most of the plants in which significant progress has been made. We have also included preliminary works on some so-called minor and orphan crops in this series. We would be happy to include reviews on more such crops that deserve immediate national and international attention and support. The extent of coverage in terms of the number of pages, however, has nothing to do with the relative importance of a plant or plant group. Nor does the sequence of the chapters have any correlation to the importance of the plants discussed in the volumes. A simple rule of convenience has been followed.

I feel myself fortunate to have received highly positive responses from nearly 300 scientists of some 30-plus countries who contributed the chapters for this series. Scientists actively involved in analyzing and improving particular genomes contributed each and every chapter. I thank them all profoundly. I made a conscientious effort to assemble the best possible team of authors for certain chapters devoted to the important plants. In general, the lead authors of most chapters organized their teams. I extend my gratitude to them all.

The number of plants of economic relevance is enormous. They are classified from various angles. I have presented them using the most conventional approach. The volumes thus include cereals and millets (Volume I), oilseeds (Volume II), pulse, sugar and tuber crops (Volume III), fruits and nuts (Volume IV), vegetables (Volume V), technical crops including fiber and forage crops, ornamentals, plantation crops, and medicinal and aromatic plants (Volume VI), and forest trees (Volume VII).

A significant amount of information might be duplicated across the closely related species or genera, particularly where results of comparative mapping have been discussed. However, some readers would have liked to have had a chapter on a particular plant or plant group complete in itself. I ask all the readers to bear with me for such redundancy.

Obviously the contents and coverage of different chapters will vary depending on the effort expended and progress achieved. Some plants have received more attention for advanced works. We have included only introductory reviews on fundamental aspects on them since reviews in these areas are available elsewhere. On other plants, including the "orphan" crop plants, a substantial amount of information has been included on the basic aspects. This approach will be reflected in the illustrations as well.

It is mainly my research students and professional colleagues who sparked my interest in conceptualizing and pursuing this series. If this series serves its purpose, then the major credit goes to them. I would never have ventured to take up this huge task of editing without their constant support. Working and interacting with many people, particularly at the Laboratory of Molecular Biology and Biotechnology of the Orissa University of Agriculture and Technology, Bhubaneswar, India as its founder principal investigator; the Indo-Russian Center for Biotechnology, Allahabad, India as its first project coordinator; the then-USSR Academy of Sciences in Moscow; the University of Wisconsin at Madison; and The Pennsylvania State University, among institutions, and at EMBO, EUCARPIA, and Plant and Animal Genome meetings among the scientific gatherings have also inspired me and instilled confidence in my ability to accomplish this job.

I feel very fortunate for the inspiration and encouragement I have received from many dignified scientists from around the world, particularly Prof. Arthur

Kornberg, Prof. Franklin W. Stahl, Dr. Norman E. Borlaug, Dr. David V. Goeddel, Prof. Phillip A. Sharp, Prof. Gunter Blobel, and Prof. Lee Hartwell, who kindly opined on the utility of the series for students, academicians, and industry scientists of this and later generations. I express my deep regards and gratitude to them all for providing inspiration and extending generous comments.

I have been especially blessed by God with an affectionate student community and very cordial research students throughout my teaching career. I am thankful to all of them for their regards and feelings for me. I am grateful to all my teachers and colleagues for the blessings, assistance, and affection they showered on me throughout my career at various levels and places. I am equally indebted to the few critics who helped me to become professionally sounder and morally stronger.

My wife Phullara and our two children Sourav and Devleena have been of great help to me, as always, while I was engaged in editing this series. Phullara has taken pains ("pleasure" she would say) all along to assume most of my domestic responsibilities and to allow me to devote maximum possible time to my professional activities, including editing this series. Sourav and Devleena have always shown maturity and patience in allowing me to remain glued to my PC or "printed papers" ("P3" as they would say). For this series, they assisted me with Internet searches, maintenance of all hard and soft copies, and various timely inputs.

Some figures included by the authors in their chapters were published elsewhere previously. The authors have obtained permission from the concerned publishers or authors to use them again for their chapters and expressed due acknowledgement. However, as an editor I record my acknowledgements to all such publishers and authors for their generosity and good will.

I look forward to your valuable criticisms and feedback for further improvement of the series.

Publishing a book series like this requires diligence, patience, and understanding on the part of the publisher, and I am grateful to the people at Springer for having all these qualities in abundance and for their dedication to seeing this series through to completion. Their professionalism and attention to detail throughout the entire process of bringing this series to the reader made them a genuine pleasure to work with. Any enjoyment the reader may derive from this books is due in no small measure to their efforts.

Pennsylvania, Chittaranjan Kole
10 January 2006

Preface to the Volume

The number of "groups" of economic plant species is too many! This caused a serious problem in terms of allocating them under the seven planned volumes of the series with consideration of uniformity of size and inclusion of all relevant groups. Certain groups, for example cereals and millets, oilseeds, fruits, vegetables, and forest trees, have enough economic species those attracted the attention of molecular biologists and biotechnologists. In comparison, the number of pulse crops is too few to deserve an entire volume to itself. We had to accommodate pulse crops, sugar crops, and tuber crops together in volume 3 to maintain a more or less uniform coverage across all volumes. Except for the common bean, pea, and cowpea, most pulse crops are grown mainly in developing countries and have attracted relatively little attention of scientists from developed countries. These include the "orphan crops" such as chickpea, pigeonpea, mungbean, lentil, *Lathyrus*, etc. Thanks to certain labs in the USA and Australia, appreciable work has been done on these crops. There are still some more neglected pulse crops, we could easily ascribe the term "beggar crops" to them, such as urdbean, rice bean, adzuki bean, etc. in which almost no molecular work has been done. We must wait for future editions for their inclusion in this series. However, we have included two not-so-well-known pulse crops, quinoa and bambara groundnut, on which considerable work has been done. This volume can boast of introducing these two crops with comprehensive reviews for the first time.

Pulse crops will play a crucial role in global agriculture in the near future. Their shorter duration, docility for adaptation to several cropping schemes, tolerance to abiotic stresses particularly drought, and the preference by people in developing countries for vegetable protein to animal sources will definitely make an impact sooner rather than later. The research on these crops will fill an entire volume in a year or two!

Sugarcane has been included in this volume as well. This cash crop has generated much interest, particularly for its genomic proximity to other members of the "grass family" that comprises extensively studied crop plants like rice and maize. Sugar beet could be included here as well, but we will deal with it under beets in volume 5, which is dedicated to "vegetables".

The tuber crops included in this volume are potato, sweetpotato, cassava, and yam. Granted, potato could have been included under vegetables in volume 5, but that would have forced us to consider another "subvolume" for vegetable crops!

The contents of the chapters in this volume may appear somewhat contrasting. Crops like common bean, pea, cowpea, potato, and sugarcane contain elaborate deliberations on molecular aspects. For others, fundamental information besides preliminary molecular efforts are also discussed in depth. I hope the reader will appreciate the relative importance attached to the formulation of the contents.

In the last few years, my own research interests and research projects of my students and staff in India have mostly related to pulse crops. This gave me access to the literature accumulated on the pulse crops. For sugarcane and the tuber crops included in this volume, I had to be a student again before being an editor. The first two volumes of this series have been well received by readers. We hope this volume will also earn their appreciation.

If this volume finds favor with readers, credit must go to the authors and the publisher. The mistakes are mine alone, and I will rectify them upon the readers' welcome suggestions for improvement.

Pennsylvania, 3 March 2006 Chittaranjan Kole

Contents

Contributors

Phillip N. Miklas
USDA-ARS
24106 N. Bunn Road
Prosser, WA 99350-9687, USA

Shree P. Singh
Department of Plant, Soil and Entomological Sciences
University of Idaho
3793 North 3600 East
Kimberly, ID 83341-5076, USA
singh@kimberly.uidaho.edu

K. E. McPhee
USDA-ARS
P.O. Box 646434
Pullman, WA 99164-6434, USA
kmcphee@mail.wsu.edu

Michael P. Timko
Department of Biology
University of Virginia
Charlottesville, VA 22904, USA
mpt9g@virginia.edu

Jeff D. Ehlers
Department of Botany and Plant Sciences
University of California
Riverside, CA 92521-0124, USA

Philip A. Roberts
Department of Nematology
University of California
Riverside, CA 92521-0124, USA

Christopher J. Lambrides
School of Land and Food Sciences
Plant Improvement Group
The University of Queensland
St Lucia, QLD 4072, Australia
chris.lambrides@uq.edu.au

Ian D. Godwin
School of Land and Food Sciences
Plant Improvement Group
The University of Queensland
St Lucia, QLD 4072, Australia

Rebecca Ford
BioMarka
Faculty of Land and Food Resources
The University of Melbourne
Victoria 3010, Australia
rebeccaf@unimelb.edu.au

Rubeena
Department of Crop and Soil Sciences
Washington State University
291 Johnson Hall, PO Box 646420
Pullman, WA 99164, USA

R. J. Redden
Australian Temperature Field Crops Collection
Department of Primary Industries
Horshan, Victoria 3401, Australia

M. Materne
Australian Temperature Field Crops Collection
Department of Primary Industries
Horshan, Victoria 3401, Australia

P. W. J. Taylor
BioMarka
Faculty of Land and Food Resources
The University of Melbourne
Victoria 3010, Australia

Beata Skiba
School of Medical Sciences
RMIT University, Bundoora
Victoria 3083, Australia

Allison M. Gurung
Faculty of Land and Food Resources
The University of Melbourne
Parkville, Victoria 3010, Australia
a.gurung@unimelb.edu.au

Edwin C.K. Pang
School of Applied Sciences
RMIT University
Bundoora, Victoria 3083, Australia

Milind B. Ratnaparkhe
Virginia Polytechnique Institute
and State University
Blacksburg, VA 24061, USA
ratnaparkhem@missouri.edu

Vidya S. Gupta
National Chemical Laboratory
Pune 411008, India

Peter J. Maughan
Brigham Young University
Provo, UT 84602, USA
Jeff_Maughan@byu.edu

Alejandro Bonifacio
Fundación PROINPA
Edificio Marconi
Piso 2
La Paz, Bolivia

Craig E. Coleman
Brigham Young University
Provo, UT 84602, USA

Eric N. Jellen
Brigham Young University
Provo, UT 84602, USA

Mikel R. Stevens
Brigham Young University
Provo, UT 84602, USA

Daniel J. Fairbanks
Brigham Young University
Provo, UT 84602, USA

S. Basu
Division of Agricultural
and Environmental Sciences
School of Biosciences
University of Nottingham
Sutton Bonington Campus
Loughborough
Leicestershire, LE12 5RD, UK

J.A. Roberts
Division of Plant Sciences
School of Biosciences
University of Nottingham
Sutton Bonington Campus
Loughborough
Leicestershire, LE12 5RD, UK

S.N. Azam-Ali
Division of Agricultural
and Environmental Sciences
School of Biosciences
University of Nottingham
Sutton Bonington Campus
Loughborough
Leicestershire, LE12 5RD, UK
Sayed.Azam-ali@nottingham.ac.uk

S. Mayes
Division of Agricultural
and Environmental Sciences
School of Biosciences
University of Nottingham
Sutton Bonington Campus
Loughborough
Leicestershire, LE12 5RD, UK

Giovanni Cordeiro
Centre for Plant Conservation Genetics
Southern Cross University
PO Box 157
Lismore, NSW 2480, Australia
gcordeir@scu.edu.au

Ouzi Amouyal
Centre for Plant Conservation Genetics
Southern Cross University
PO Box 157
Lismore, NSW 2480, Australia

Frances Eliott
Centre for Plant Conservation Genetics
Southern Cross University
PO Box 157
Lismore, NSW 2480, Australia

Robert Henry
Centre for Plant Conservation Genetics
Southern Cross University
PO Box 157
Lismore, NSW 2480, Australia

Dan Milbourne
Plant Biotechnology Unit
Teagasc
Crops Research Centre
Oak Park
Carlow, Eire
dmilbourne@oakpark.teagasc.ie

Barnaly Pande
Department of Environmental Horticulture
University of California, Davis
One Shields Avenue
Davis, California 95616, USA

Glenn J. Bryan
Genetics Programme
Scottish Crop Research Institute
Invergowrie
Dundee, DD2 5DA, UK

Umesh K. Reddy
Department of Biology
West Virginia State University
Institute, WV 25112, USA
ureddy@wvstateu.edu

George T. Bates
Alcorn State University
Professor Emeritus 118 Bates Rd.
Natchez, MS 39120, USA
Present address:
1304 E. Pass Road
Gulfport, MS 39507, USA

Janice Ryan-Bohac
U.S. Vegetable Laboratory
USDA, ARS
Charleston, SC 29414, USA

Padma Nimmakayala
Department of Biology
West Virginia State University
Institute, WV 25112, USA

Y. Lokko
International Institute of Tropical Agriculture (IITA)
Oyo Road PMB 5320
Ibadan, Nigeria
Current address:
Plant Breeding & Genetics Section
International Atomic Energy Agency
Wagramer Straße 5
PO Box 200
A-1400 Vienna, Austria
Official.Mail@iaea.org

E. Okogbenin
Centro Internacional de Agricultura Tropical
Apartado Aéreo 6713
Cali, Colombia

C. Mba
Plant Breeding Unit
Joint FAO/IAEA Agriculture and Biotechnology Laboratory
International Atomic Energy Agency Laboratories
A-2444 Seibersdorf, Austria

A. Dixon
International Institute of Tropical Agriculture (IITA)
Oyo Road PMB 5320
Ibadan, Nigeria

A. Raji
International Institute of Tropical Agriculture (IITA)
Oyo Road PMB 5320
Ibadan, Nigeria

M. Fregene
Centro Internacional de Agricultura Tropical
Apartado Aéreo 6713
Cali, Colombia

Hodeba D. Mignouna
Virginia State University
Agricultural Research Station
Box 9061
Petersburg, VA 23806, USA
jmignoun@vsu.edu

Current address:
Technical Operations Manager
African Agricultural Technology
Foundation (AATF)
c/o ILRI, P.O. Box 30709
Nairobi, Kenya
h.mignouna@aatf-africa.org

Mathew M. Abang
International Center for Agricultural
Research in the Dry Areas (ICARDA)
P.O. Box 5466
Aleppo, Syria and
IRAD, Root and Tuber Program
B.P. 2123 Messa
Yaoundé, Cameroon

Robert Asiedu
International Institute of Tropical
Agriculture (IITA)
c/o L.W. Lambourn & Co.
Carolyn House
26 Dingwall Road
Croydon CR9 3EE, UK

Abbreviations

ABA	Abscisic Acid
ABR	Ascochyta Blight Resistance
AFLP	Amplified Fragment Length Polymorphism
ANOVA	Analysis Of Variance
AP-PCR	Arbitrarily Primed PCR
APR	Adult Plant Resistance
ARS	Agriculture Research Service
ASAP	Allele Specific Associated Primer
AUDPC	Area Under the Disease Progress Curve
AYT	Advanced Yield Trial
BAC	Bacterial Artificial Chromosome
BAMFOOD	Increasing the productivity of Bambara Groundnut (*Vigna subterranea* L. Verdc) for sustainable food production in Semi-Arid Africa
BC	Backcross
BCMNV	Bean Common Mosaic Necrosis Virus
BCMV	Bean Common Mosaic Virus
BCTV	Beet Curly Top Virus
BGMV	Bean Golden Mosaic Virus
BGYMV	Bean Golden Yellow Mosaic Virus
BICMV	Blackeye Cowpea Mosaic Virus
BLRV	Bean Leaf Roll Virus
*Bru*1	Brassinosteroid-regulated protein
BSA	Bulked Segregant Analysis
C	Haploid Genome Content
CAPS	Cleaved Amplified Polymorphic Sequences
CC-NBS-LRR	Coiled-Coiled domain-containing NBS-LRR protein
cDNA	Complementary DNA
CIAT	Centro Internacional de Agricultura Tropical (Cali, Colombia)
CID	Carbon Isotope Discrimination
CIM	Composite Interval Mapping
cM	centi-Morgan
CMD	Cassava Mosaic Disease
CMS	Cytoplasmic Male Sterility
CPB	Colorado Potato Beetle
cpDNA	Chloroplast DNA
CRSD	Cowpea Collaborative Research Support Program
cSNP	SNP in coding region
CTCRI	Central Tuber Crops Research Institute (Trivandrum, India)
DAF	DNA Amplification Fingerprinting
DH	Doubled Haploid
DNA	Deoxyribonucleic Acid
DNC	Dry Matter Content
DR	Defense-Related
DS	Drought-Stressed
EBN	Endosperm Balance Number

EM	Expectation Maximization (algorithm)
EMBRAPA	Empresa Brasileira de Pesquisa Agropecuária (Goiana, Brazil)
EST	Expressed Sequence Tag
EU	European Union
FAO	Food and Agricultural Organization
FGS	Fast-Growing Salmon
FISH	Fluorescence In Situ Hybridization
Foc	Fusarium Wilt Resistance
GAT	Zentralinstitut für Genetik und Kulturpflanzenforschung (Gatersleben, Germany)
gDNA	Genomic DNA
GISH	Genomic In Situ hybridisation
GM	Genetically Modified
GMO	Genetically Modified Organism
GMS	Genetic Male Sterility
GP	Gene Pool
GP1	Primary Gene Pool
GP1A	Primary Gene Pool – Domesticated
GP2	Secondary Gene Pool
GP3	Tertiary Gene Pool
GPIB	Primary Gene Pool – Wild
GRIN	Germplasm Resources Information Network
GSIRI	Guangzhou Sugarcane Industry Research Institute (China)
Hs	Average Hardy-Weinberg expected heterozygosity per subpopulation
Ht	Hardy-Weinberg heterozygosity of the total population
ICARDA	International Center for Agriculture Research in the Dryland Areas (Aleppo, Syria)
ICRISAT	International Center of Research For Semi-Arid Tropics (Hyderabad, India)
IDS	Initial Disease Score
IITA	International Institute of Tropical Agriculture (Ibadan, Nigeria)
ILDIS	International Legume Database & Information Service
IM	Interval Mapping
indel	insertion/deletion
IPGRI	The International Plant Genetic Resources Institute (Harare, Zimbabwe)
ISSR	Inter Simple Sequence Repeat
ITS	Internal Transcribed Spacer
JIC	John Innes Center
LD	Linkage Disequilibrium
LG	Linkage Group
LOD	Logarithm Of Odds
LRR	Leucine Rich Repeat
LRS	Likelihood Ratio Statistic
LTR	Long Terminal Repeat
MAS	Marker-Assisted Selection
MDSS	Mean Disease Severity Scores
MIM	Multiple Interval Mapping
mRNA	Messenger Ribonucleic Acid
mtDNA	Mitochondrial DNA

MtGI	*Medicago truncatula* Gene Index
MYMV	Mungbean Yellow Mosaic Virus
NARS	National Agricultural Research System
NBPGR	National Bureau of Plant Genetic Resources (New Delhi, India)
NBS	Nucleotide Binding Site
NIL	Near Isogenic Lines
NOR	Nucleolar Organizer Region
NPGS	National Plant Germplasm System
NRCRI	National Root Crop Research Institute (Umuahia, Nigeria)
NS	Non-Stressed
ODAP	β-N-Oxalyl-L-α,β-DiaminoPropanoic acid
ORSTOM	Institut Français de la Recherché Scientifique pour le Développement en Coopération (now IRD; Montpellier, France)
PCN	Potato Cyst Nematode
PCR	Polymerase Chain Reaction
PEMV	Pea Enation Mosaic Virus
PI	Plant Introduction
PIs	Proteinaceous Inhibitors
PPB	Participatory Plant Breeding
PPD	Post harvest Physiological Deterioration
PPO	PolyPhenol Oxidase
PRINS	Primed In Situ (DNA Labeling)
PSbMV	Pea Seed-borne Mosaic Virus
PVX	Potato Virus X
PYT	Preliminary Yield Trial
QTA	Quantitative Trait Allele
QTL	Quantitative Trait Loci
RAF	Randomly Amplified Fragment
RAPD	Random Amplified Polymorphic DNA
rDNA	Ribosomal DNA
RFLP	Restriction Fragment Length Polymorphism
RGA	Resistant Gene Analog
RI	Recombinant Inbred
RIL	Recombinant Inbred Line
RSD	Ratoon Stunting Disease
SAM	Shoot Apical Meristem
SAT	Semi Arid Tropics
SCAR	Sequence Characterized Amplified Region
SDM	Single Dose Marker
SG	Striga Race
SGG	Slow-Growing Gray
SI	Self Incompatibility
SLA	Specific Leaf Area
SNP	Single Nucleotide Polymorphism
SPLAT	Specific Polymorphic Locus Amplification Test
SR	Specific Resistance
SSD	Single Seed Descent
SSLP	Simple Sequence Length Polymorphism
SSR	Simple Sequence Repeat
STMS	Sequence Tagged Microsatellite Site
STS	Sequence Tagged Sites

SUCEST	Sugarcane EST project
TCA	TriCarboxylic Acid
TIGR	The Institute for Genomic Research
TIR	Toll and Interleukin Receptor
UHD	Ultra-High Density
USAID	United States Agency of International Development
USDA	United States Department of Agriculture
UYT	Uniform Yield Trial
VIGS	Virus Induced Gene Silencing
WASDU	West African Seed Development Union
WHO	World Health Organization
WUE	Water Use Efficiency
YAD	Yam Anthracnose Disease
YMV	*Yam Mosaic Virus*

1 Common Bean

Phillip N. Miklas [1], Shree P. Singh [2]

[1] USDA-ARS, 24106 N. Bunn Road, Prosser, WA 99350-9687, USA
[2] Department of Plant, Soil and Entomological Sciences, University of Idaho, 3793 North 3600 East, Kimberly, ID 83341-5076, USA, *e-mail*: singh@kimberly.uidaho.edu

1.1 Introduction

A book (Singh 1999 a), workshop proceedings (Singh 2000), and two book chapters (Singh 2001b, 2005) on the common bean (*Phaseolus vulgaris* L.) have been published within the last few years. Also, review articles are available on broadening the genetic base of cultivars (Singh 2001 a), development of integrated linkage map (Gepts 1999), and marker-assisted selection (MAS) (Kelly and Miklas 1999). More recently, Kelly et al. (2003) and Miklas et al. (2006) reviewed tagging and mapping of genes and quantitative trait loci (QTL) of economic importance and molecular MAS. Nonetheless, in this chapter we shall briefly describe the history of the crop, botanical description, economic importance, and breeding objectives and achievements of conventional breeding. The remainder of the chapter will be dedicated to the construction of linkage maps, tagging of genes and QTL of economic importance, and progress achieved by MAS.

1.1.1 History of the Crop

The common bean is among the five domesticated *Phaseolus* species that are native to the Americas (Gepts and Debouck 1991). From its origin and domestication regions in the Andean South America, Central America, and Mexico, the common bean has expanded into other parts of the Americas (from about 35°S to >50°N latitude and from sea level to >3000 m altitude) (Gepts et al. 1988; Singh 1992). Subsequently, it was introduced into Africa, Asia, Europe, and Oceania (Gepts and Bliss 1988).

Wild populations of common bean are distributed from northern Mexico (Chihuahua) to northeastern Argentina (San Luis) (Gepts et al. 1986). The common bean is a noncentric crop that had multiple domestications throughout the range of its wild populations (Harlan 1975; Gepts et al. 1986). Hybrids between wild and cultivated beans are fully fertile and no major barriers exist for introgression and exchange of favorable alleles and QTL (Singh et al. 1995; Koinange et al. 1996; Zizumbo-Villarreal et al. 2005).

Through domestication the common bean shifted from extreme indeterminate climbing to determinate bush types; from sensitivity to insensitivity to long photoperiod; from small to large leaves, pods, and seeds; and from a few gray, brown, beige, and cream colored spotted and speckled seeds that mimicked surroundings in wild grassland and oak-pine forest habitats to highly attractive and showy colors except blue and green with solid as well as stripes, spots, speckles, etc. Similarly, the common bean has evolved from having an impermeable to a water-permeable seed coat, and from types that shatter due to highly fibrous and parchmented pod walls to forms with less fiber that are less subject to shattering (Gepts and Debouck 1991; Gepts 1998). Major alleles and QTLs that influenced common bean domestication have been identified and mapped (Koinange et al. 1996; Freyre et al. 1998; Gepts 1999). These traits are growth habit (*fin*), photoperiod insensitivity (*ppd, hr*), pod fiber (*St*), seed dormancy, and seed size, color, and shape. Existence of a considerably larger variation in the evolutionary marker, phaseolin types (Gepts 1988 a), in wild bean populations compared to cultivars suggests that not all wild beans were domesticated and cultivars may have reduced genetic diversity (Koenig et al. 1990; Gepts 1998; Zizumbo-Villarreal et al. 2005). The Andean South American wild and cultivated common beans differ from those of Central America and Mexico. These differences occur in seed size and other mor-

Genome Mapping and Molecular Breeding in Plants, Volume 3
Pulses, Sugar and Tuber Crops
C. Kole (Ed.)

phological (Singh et al. 1991 c), isozyme (Koenig and Gepts 1989; Singh et al. 1991 b), physiological (White et al. 1992), molecular (Becerra-Velásquez and Gepts 1994; Haley et al. 1994 c), and adaptive traits (Singh 1989). Also, there are occasional incompatibilities between the two groups of wild (Koinange and Gepts 1992) and cultivated (Singh and Gutiérrez 1984; Gepts and Bliss 1985) germplasm such that they are considered two distinct gene pools. Singh et al. (1991 a) further divided the Andean and Middle American gene pools into six races: three Andean gene pool (all large-seeded) = Chile, Nueva Granada, and Peru races; and three Middle American gene pool = Durango (medium-seeded semiclimber), Jalisco (medium-seeded climber), and Mesoamerica (all small-seeded) races. Beebe et al. (2000) reported the existence of additional diversity within the Middle American gene pool, especially within a group of Guatemalan climbing bean accessions that were distinct from previously defined races.

1.1.2 Botanical Description

Freytag and Debouck (2002) described in considerable detail the taxonomy, distribution, and ecology of over 25 *Phaseolus* species, including *P. vulgaris*, that are native to North America, Mexico, and Central America. Cultivated and wild *P. vulgaris* (Brücher 1988) and other *Phaseolus* species (Debouck 1999) are also native to Andean South America. The natural habitat of wild common bean ranges between ca. 800 to 2750 m elevations. Indeterminate climbing populations have a perennial tendency in the wild, but when planted in the field they may behave as an annual similar to most cultivated types.

The genus *Phaseolus* belongs to family Leguminosae, subfamily Papilionoideae. *P. vulgaris* belongs to its section Phaseoli. There is continuous variation in growth habit from determinate bush to indeterminate climbing cultivars. Singh (1982), however, classified growth habits into four major classes using the type of terminal bud (vegetative vs. reproductive), stem strength (weak vs. strong), climbing ability (nonclimber vs. strong climber), and fruiting patterns (mostly basal vs. along entire stem length or only in the upper part). These are the Type I = determinate upright bush, Type II = indeterminate upright bush, Type III = indeterminate, prostate, nonclimbing or semiclimbing, and Type IV = indeterminate, strong climbers. Roots are generally fibrous with a marked tap or main root. Under most field conditions, especially in cool subtropical and temperate environments, they may bear nitrogen-fixing nodules from a few weeks after emergence through flowering. The main stem derives from the axis of the seed embryo. The number of branches and branching pattern may vary greatly depending upon the genotype and environment. Often more than 50% of the pods are borne on branches. The two unifoliolate leaves borne above the cotyledonary node are opposite to each other followed by one trifoliolate leaf at each node in an alternate phyllotaxy. The fully developed trifoliolate leaf has a long (> 7 cm) petiole, a small (< 3 cm) petiolule, very small pulvini, and three leaflets of which the central one is often symmetrical and chordate, ovate, or lanceolate. The inflorescence is a pseudoraceme often with several flowers of which only the basal few bear pods; an exception are small-diameter snap bean that bear a profusion of pods. Also, dry bean of outrigger types bearing six or more pods can be rarely found. Papilionaceous flowers can be pink, purple, white, or bicolor with or without stripes at the outer base of a very pronounced standard. Sessile bracteoles often are larger in Middle American compared to Andean genotypes and may be chordate, ovate, or lanceolate. Bilabiate calyx is small (< 5 mm) with the upper two teeth united. The two keels may be coiled up to two times. There is a single vexillary stamen on the upper side and nine stamen united into a long sheath or tube around the style. The introrse stigma tends to extend around the tip of the style. Flowers are cleistogamous and normally are highly self-pollinated (< 1% outcrossing). Nonetheless, Ibarra-Pérez et al. (1997) reported outcrossing rates ranging from 0.0 to 78% for individual families with a mean rate for six dry bean genotypes ranging from 4.4 to 10.2% in California. Anthesis occurs in early morning hours, and crosses are made with or without emasculation of anthers prior to anthesis. Mature pods are straight to slightly curved with five to eight seeds. There is considerable variation in size, shape, and color of pods and seeds. Germination is epigeal with cotyledons dropping off a couple of weeks after emergence.

Common bean is a short-day crop (White and Laing 1989). Cultivars adapted to higher latitudes either have evolved during dissemination from the primary centers of domestication or have been developed by breeding. Mildly cool environments favor growth and development. Thus, under nonstressed environments with 18 to 22 °C mean growing temperatures and about 12-h day-length, most cultivars complete their growing cycle from germination to seed maturity in 70 to 120 d. In the highlands (above 2000 m elevation) of Bolivia, Colombia, Ecuador, and Peru, climbing cultivars often require more than 250 d to mature. In the humid highlands of Guatemala and Mexico and in Principado de Asturias, Spain, climbing cultivars require ca. 150 d to mature.

At higher latitudes in temperate climates, dry bean cultivars of growth habit Types I, II, and III predominate. These are harvested within 90 to 120 d of planting. Cultivars of growth habit Types I, II, and III are grown in monoculture as well as under different relay, strip, and intercropping systems throughout the world (Singh 1992). Type IV cultivars always require support. Thus, these are grown in either association with maize (*Zea mays* L.) and other crops or on trellises or stakes. Although dry bean is grown in a wide range of soil types, light loamy soils with pH 7.0 and rich in organic matter are more suitable for production. A 90- to 120-d crop with a yield of 2500 kg ha^{-1} will usually remove 60 to 80 kg of soil nitrogen and 40 kg of phosphorus.

P. vulgaris and a great majority of other cultivated and wild *Phaseolus* species have $2n=2x=22$ chromosomes. The *P. vulgaris* chromosomes are extremely small, and all 11 chromosomes have been identified (Mok and Mok 1977; Cheng and Bassett 1981). They were also recently assigned to the respective linkage groups (LGs) (Table 1) using the fluorescence in situ hybridization (FISH) (Pedrosa et al. 2003). However, they have been of little or no use in breeding. The common bean has one of the smallest genomes in the legume family with 0.65 pg/haploid genome or 635 mbp (Arumuganathan and Earle 1991).

Table 1. Integration of common bean (*Phaseolus vulgaris* L.) linkage and chromosome maps

Chromosome	Linkage group	
	Florida	Davis (BJ)
1	G	B6
2	H	B1
3	F	B8
4	A	B7
5	C	B3
6	J	B11
7	E	B5
8	I	B10
9	D	B2
10	B	B4
11	K	B9

1.1.3 Economic Importance

The common bean is the most important of over 30 *Phaseolus* species native to the Americas, occupying more than 85% of areas sown to these species worldwide. There are two principal types of common bean: snap and dry. Fully developed green pods of snap bean harvested for fresh-market or processing purposes have reduced fiber in the pod walls and sutures. The USA, Europe, and China are the largest producers of snap bean. Although exact area planted to snap bean is not known, it is estimated to be <3 million ha. For further details on snap bean, the reader should refer to Myers and Baggett (1999) and Myers (2000).

Dry bean is grown in more than 14 million ha in the world. The Americas are the largest dry bean producing regions (6.7 million MT), and Brazil (2.5 million MT) is the largest producer and consumer (Singh 1999b). Asia (2.2 million MT), Africa (2.1 million MT), and Europe (~1 million MT) follow the lead of the Americas in dry bean production in the world. The USA (1.3 million MT) and Mexico (0.98 million MT) follow Brazil as leading dry bean producers. Production has increased substantially in the last 50 years in Argentina, Bolivia, Brazil, Canada, and the USA. Consumer preferences for dry bean size, color, shape, and brilliance vary a great deal (Singh 1992; Voysest 2000). In Latin America, the highest per-capita consumption of dry bean is in Brazil and Mexico (>10 kg per year). In Rwanda and Burundi, per-capita consumption is over 40 kg per year. Dry, green-shelled, and snap bean have high nutritional value, especially in conjunction with cereals and other carbohydrate-rich foods, and can reduce cholesterol and cancer risks (Andersen et al. 1999; Myers 2000). Dry bean (average of 22% protein) dishes range from simply beans boiled in water to more sophisticated preparations of baked beans, cakes, chips, creams, pastes, salads, soups, and stews (Hosfield et al. 2000).

1.1.4 Objectives and Achievements of Classical Breeding

Nearly a century of organized genetics and breeding of common bean has been carried out in the USA and elsewhere in the world. Early efforts emphasized breeding for disease resistance, early maturity, and upright determinate bush growth habit Type I to facilitate mechanical harvest, especially in snap bean. Initially, selection within and between landraces followed by pedigree, mass-pedigree, and recurrent backcrossing were used. Common bean breeding accelerated in the second half of the 20th century in the Americas and Europe. Improved germplasm and cultivars were developed using recurrent backcross (Pompeu 1982), pedigree (Kelly et al. 1994 a), and mass-pedigree (Singh et al. 1989) methods and their modifications. Congruity backcrossing (Mejía-Jiménez et al. 1994; Urrea and Singh 1995), single-seed descent (SSD) (Kelly et al. 1989; Urrea and Singh 1994), recurrent (Kelly and Adams 1987; Beaver and Kelly 1994; Singh et al. 1999), and gamete (Singh 1994; Singh et al. 1998) selection methods have been used more recently.

Favorable alleles and QTLs have been introgressed from the tepary bean (*P. acutifolius* A. Gray) for common bacterial blight [caused by *Xanthomonas campestris* pv. *phaseoli* (Smith) Dye] resistance (Scott and Michaels 1992; Singh and Muñoz 1999), from runner bean (*P. coccineus* L.) for common bacterial blight (Miklas et al. 1994) and white mold [caused by *Sclerotinia sclerotiorum* (Lib.) de Bary] resistance (Miklas et al. 1998 a), and from wild common bean for the bean weevil (*Zabrotes subfasciatus* Boheman) resistance (Cardona et al. 1990). Singh and Muñoz (1999), while introgressing common bacterial blight resistance from the tepary bean (VAX 1 and VAX 2), also pyramided the highest level of common bacterial blight resistance to develop breeding lines VAX 3 to VAX 6. Nonetheless, most breeding has largely utilized favorable alleles and QTLs available between and within cultivated common bean market classes, races, and gene pools. The major breeding achievements in the Americas include introgression of upright growth habit Type II from race Mesoamerica into traditional Type III cultivars of race Durango using phenotypic recurrent selection (Kelly and Adams 1987) and other breeding methods (Coyne et al. 2000; Kelly et al. 2000). Recently developed cultivars also carry resistance to *Bean common mosaic virus* (BCMV, a potyvirus), *Bean common mosaic necrosis virus* (BCMNV, a potyvirus), and rust [caused by *Uromyces appendiculatus* (Pers.) Ung.]. Adams (1982) and Grafton et al. (1993) used Type II cultivars to change Type I growth habit of navy and small white cultivars into more stable high-yielding Type II. Similarly, cream-striped carioca beans (traditionally a Type III) with growth habit Type II and resistance to leafhopper (*Empoasca kraemeri* Ross & Moore) and five diseases were developed using gamete selection (Singh et al. 1998). Seed yield was improved using mass-pedigree (Singh et al. 1993) and recurrent (Singh et al. 1999) selection methods from interracial populations within the Middle American gene pool and from Andean×Middle American intergene pool crosses using recurrent selection (Beaver and Kelly 1994; Singh et al. 1999). Pereira et al. (1993) increased nodule number and weight after three cycles of recurrent selection. Bliss et al. (1989) developed five high N_2-fixing genotypes.

Schneider et al. (1997 a, b) and Rosales-Serna et al. (2000) developed drought-resistant breeding lines from biparental populations using seed yield and/or random amplification of polymorphic DNA (RAPD) markers as selection criteria. Singh (1995) and Terán and Singh (2002) developed drought-resistant breeding lines from double-cross interracial and intergene pool populations using a bulk-pedigree method. Similarly, breeding lines such as A 321, A 445, and A 744 resistant to low soil fertility were developed from interracial populations within Middle American gene pool (Singh et al. 2003 b).

Kelly et al. (1994 b) developed anthracnose [caused by *Colletotrichum lindemuthianum* (Sacc. and Magn.) Bri. & Cav.], BCMV, BCMNV, and rust-resistant black-seeded cultivar Raven, which was then used to develop Phantom with similar resistance (Kelly et al. 2000). Kelly et al. (1999 c) also combined good canning quality, BCMV resistance, and the Andean *Co-1* and Middle American *Co-2* alleles for anthracnose resistance in a large-seeded light red kidney bean Chinook 2000. Both anthracnose resistance alleles were also combined with BCMV and rust resistance in small black cultivar Jaguar (Kelly et al. 2001). Good canning quality and resistance to anthracnose, BCMV, *Beet curly top virus* (BCTV, a leafhopper vectored geminivirus), and rust, either singly or in various combinations, have also been bred into dark and light red kidney, red mottled, white kidney, and cranberry beans for North America (Miklas and Kelly 2002; Miklas et al. 2002 a) and for the tropics and subtropics (Beaver et al. 2003). Singh et al. (2003 a) developed angular leaf spot [caused by *Phaeoisariopsis griseola* (Sacc.) Ferr.], anthracnose, BCMV, rust, and halo blight [caused by *Pseudomonas syringae* pv. *phaseolicola* (Burkh.) Young] resistant breeding lines A 339, MAR 1, MAR 2, and MAR 3 from interracial populations between the three Middle American races.

Acosta-Gallegos et al. (1995), Ibarra-Pérez et al. (2004), and Sanchez-Valdez et al. (2004) combined resistance to angular leaf spot, anthracnose, BCMV, and rust into high-yielding "bayo," black, "flor de mayo," ojo de cabra, pinto, and shiny black bean cultivars for Mexican highlands. Kelly et al. (1999 a, b), Coyne et al. (2000), and Brick et al. (2001), among others (see Brick and Grafton 1999; Singh 2001 a; Beaver et al. 2003), combined BCMV and rust resistance into great northern and pinto beans. Resistance to BCMV and BCMNV and rust resistance were combined into pinto Kodiak (Kelly et al. 1999 a) and great northern UI98-209G (Stewart-Williams et al. 2003). Pinto germplasm 92US-1006 (Silbernagel 1994) and cultivar Quincy (synonymous with USPT-73) (Hang et al. 2005) carry *I* and *bc-2*^{2} resistance alleles for BCMV and BCMNV. The *I* and *bc-3* alleles imparting resistance to all known strains of BCMV and BCMNV were combined with four to six rust resistance alleles into great northern BelMiNeb-RMR-6 to 13 and pinto BelDakMi-RMR-14 to 23 beans (Pastor-Corrales 2003; Pastor-Corrales et al. 2001).

The recessive resistance allele *bgm-1* for leaf chlorosis induced by *Bean golden mosaic virus* (BGMV, a geminivirus) and *Bean golden yellow mosaic virus* (BGYMV, a geminivirus) from the landrace Garrapato (synonymous with G 2402) was inadvertently transferred into breeding line A 429 (Morales and Singh 1993). A 429 was subsequently used to develop highly resistant small red, black, and carioca bean cultivars for Central America and Brazil, using pedigree, mass-pedigree, or gamete-selection methods. Beaver et al. (1999) were the first to develop BGYMV (*bgm-1* allele), BCMV, common bacterial blight, and rust-resistant large-seeded light red kidney bean breeding line PR9443-4 for the Caribbean. Singh et al. (2000) pyramided a high level of BGYMV resistance in different dry beans using direct disease screening that was subsequently verified by the presence of molecular markers. In the tropics and subtropics of Latin America, resistance to bean pod weevil (*Apion godmani* Wagner), angular leaf spot, anthracnose, BCMV, BGYMV, bruchid, common bacterial blight, and leafhopper, and upright plant type in beige, black, cream, cream-striped, and red beans were incorporated (Singh et al. 1998; Beaver et al. 2003). Silva et al. (2003) in Brazil in 1984 released the first cultivar, EMGOPA-Ouro-201 (synonymous with A 295), that combined angular leaf spot, anthracnose, BCMV, common bacterial blight, halo blight, powdery mildew (caused by *Erysiphe polygoni* DC), and rust resistance.

1.2 Genetic Linkage Maps

1.2.1 Linkage Mapping Prior to 1990

Lamprecht (1961) published the first genetic linkage map for common bean, which summarized previous linkage reports. This map comprised 26 naturally occurring traits. Most of the genes controlled the color of the flower or seed or affected pod traits. The Lamprecht map was extended with additional isozymes, seed proteins, and induced mutations (Bassett 1988; Gepts 1988 b; Arndt and Gepts 1989; Koenig and Gepts 1989; Vallejos and Chase 1991 b). These early maps were extensively reviewed by Bassett (1988, 1991) and culminated in a revised linkage map for common bean consisting of 13 LGs and 47 marker genes that included his own mapping of four recessive marker genes and three recessive induced mutants. For example, Nagata and Bassett (1984) mapped dark green savoy leaf (*dgs*), dwarf seed (*ds*), stipelless lanceolate leaf (*sl*), and round leaf (*rnd*) on LG VII, and spindly branch (*sb*), diamond leaf (*dia*), and progressive chlorosis (*prc*) on LG IX. He also discussed some of the less defined associations that had not yet been mapped. For example, Kyle and Dickson (1988) reported a tight linkage or pleiotropy of the dominant *I* allele imparting resistance to all known strains of the BCMV, and resistance to four related potyviruses. The *I* allele was also linked with seed coat (Temple and Morales 1986; Kyle and Dickson 1988) and hilum-region-darkening allele *B* (Park and Tu 1986). Deakin and Dukes (1975) and Dickson and Petzoldt

(1986) observed linkage of colored seed coat with resistance to Pythium and/or Rhizoctonia root rots. Coyne et al. (1973) found association between resistance to common bacterial blight and late flowering. Valladares-Sanchez et al. (1979) reported linkage between late maturity and indeterminate growth habit, Stavely (1984) among genes for rust resistance, and Osborn et al. (1986) between arcelin and lectin genes. Association of isozyme EST-2 with white flower (Weeden and Liang 1985) and white seed coat color (Weeden 1984) was also reported. These early "classical" maps were rudimentary, providing very little genomic coverage and utility for marker-assisted selections (MAS), but they did provide a point of reference for subsequently developed DNA-based linkage maps.

Bassett (1991) noted in revising the classical map that many previously described mutants lacked a seed source and therefore could not be tested. The renewed *Phaseolus* Genetics Committee in 1987 (Gepts 1988 c) advocated for a repository for genetic stocks that Bassett established and is currently maintained by the USDA-ARS, NPGS (National Plant Germplasm System) at the Western Regional Plant Introduction (PI) Station in Pullman, WA, USA (http://www.ars-grin.gov/cgi-bin/npgs/html/desc.pl?83034).

Another problem encountered in establishing the classical maps was the use of different gene symbols by different researchers for the same gene (Bassett 1991). A subcommittee of the *Phaseolus* Genetics Committee addressed this lack of coordination among geneticists for naming genes and formulated guidelines for gene designation and nomenclature (Myers and Weeden 1988; Bassett and Myers 1999). Bassett (2004) updated the list of genes for *P. vulgaris*.

1.2.2 Linkage Mapping After 1990

Availability of DNA-based markers in the mid-1980s aroused great interest and facilitated development of common bean linkage maps within the last 15 years. Vallejos et al. (1992) and Nodari et al. (1993 a) were among the first to develop molecular linkage maps of common bean, which subsequently evolved into the two major bean-mapping populations. For both maps, widely divergent parents were chosen to maximize (i) polymorphism at the nucleotide level, (ii) phenotypic diversity, and (iii) variability for disease resistance and other traits. Vallejos et al. (1992) developed a backcross (BC_1) population (Florida map) from diverse parents XR-235-1-1 (recurrent parent) and DIACOL Calima, representatives of the Middle American and Andean gene pools, respectively (Table 2). The Florida BC_1 map consisted of the pigmentation gene *P*, 224 restriction fragment length polymorphisms (RFLPs) (from *Pst*I genomic clones), 9 seed proteins, and 9 isozyme marker loci, which sorted into 11 LGs labeled in descending order of length from A to K. This map covered 963 cM of the estimated 1200 cM of the bean genome. Gepts et al. (1993), using the same prediction model of Hulbert et al. (1988), estimated the genome length to be 1250 cM. Seven additional markers added 17 cM of coverage (980 cM) (Vallejos 1994). To date the Florida map consists of 294 markers, including the addition of RFLP probes from synteny studies (Boutin et al. 1995), and covers 900 cM (Vallejos et al. 2001), which is less than previous reports because of an increased stringency (LOD >2.0) for placement of markers on the map.

Earlier, the Florida BC_1 population and an F_2 population from the same cross were used to map isozymes, seed proteins, and the *P* locus controlling pigmentation (Vallejos and Chase 1991 b), which resulted in the combination and extension of LGs X and XIII of the classical map (Bassett 1991). In a companion study (Vallejos and Chase 1991 a), a linked pair of isozyme markers *Adh-1* and *Got2* was significantly associated with seed size, which in effect identified a QTL. Sax (1923) was the first to note a linkage association between a morphological marker (seed pigmentation) and quantitative trait (seed size) in bean. Johnson et al. (1996) speculated that the phaseolin locus, *Phs*, was the candidate gene underlying the QTL for seed size, and its linkage with *P* was the morphological locus for seed color that Sax (1923) had identified. This mapping population was later used directly to identify QTL for common bacterial blight resistance (Yu et al. 1998).

The Davis map (Table 2), based on an F_2 population, was obtained from the wide cross between BAT 93 of the Middle American and Jalo EEP558 of Andean gene pool (Nodari et al. 1993 a). The map, with 143 markers, consisted of three genes (*I* for resistance to BCMV, *Cor* for seed color pattern, and an unknown gene for flower color), 108 RFLP (primarily *Pst*I clones), 7 isozyme, 7 RAPD, and 18

Table 2. Common bean (*Phaseolus vulgaris* L.) populations used for tagging and mapping genes and quantitative trait loci between 1992 and 2004.

Parents	Source	Abbre-viation	Map size	Predominant marker	Traits mapped	References
BAT 93/Jalo EEP558 (F_2 and RI)	Davis	BJ, core	1226 cM	RFLP, mix	ALS, anthracnose, CBB, Rhizobium	Nodari et al. 1993a,b; Freyre et al. 1998; Gepts 1999
XR235-1-1/DIACOL Calima (BC)	Florida	XD	900 cM	RFLP, mix	Seed size, CBB	Vallejos et al. 1992; Yu et al. 1998
Corel/Ms8EO2 (BC)	Paris	CE	567 cM	RAPD, mix	Anthracnose	Adam-Blondon et al. 1994 a
DOR 364/G 19833	CIAT	DG	Full	RFLP, SSR	BGYMV, low P, root traits, ALS, Anthracnose	Beebe et al. 1998; Blair et al. 2003; López et al. 2003
Midas/G 12873	Davis	MG		RFLP	Domestication traits	Koinange et al. 1996
BAC 6/HT 7719	NE	BH	545 cM	RAPD	CBB, web blight, rust	Jung et al. 1996
Olathe/Sierra	ID, ND	OS	Partial	RAPD	BCMV, rust	Strausbaugh et al. 1999; Kalavacharla et al. 2000
DOR 364/XAN 176	PR, ARS	DX	930 cM	RAPD	ASB, BGYMV, CBB, rust	Miklas et al. 1996 b, 1998 b, 2000 a
PC-50/XAN 159	NE	PX	404 cM	RAPD	CBB, rust, white mold	Jung et al. 1997; Park et al. 2001
CDRK/Yolano	Davis	CY	1487 cM	AFLP	Seed yield	Johnson and Gepts 1999
A 55/G 122	Harris Moran, Davis	AG	1631 cM	AFLP	Seed yield, white mold, heat tolerance	Johnson and Gepts 1999; Miklas et al. 2001; Porch 2001
Eagle/Puebla 152	WI	EP	825 cM	RAPD	Root rot	Vallejos et al. 2001; Navarro et al. 2003
Jamapa/DIACOL Calima	Florida	JC	950 cM	Mix	RGA	Rivkin et al. 1999; Vallejos et al. 2001
Benton/NY6020-4	ARS	BN	Partial	RAPD	White mold	Miklas et al. 2003 b
OAC Seaforth/OAC 95-4	Guelph	S95	1717 cM	Mix	CBB, agronomic traits	Tar'an et al. 2001, 2002
BelNeb-RR-1/A 55	NE	BA	755 cM	RAPD	BBS, HBB, BCMV, Fusarium wilt	Ariyarathne et al. 1999; Fall et al. 2001; Jung et al. 2003; Fourie et al. 2004
Sierra/Lef-2RB	MI		Partial	RAPD	Drought, Fusarium wilt	Schneider et al. 1997 a, b; Brick et al. 2004
Sierra/AC1028	MI		Partial	RAPD	Drought	Schneider et al. 1997 a, b
Isles/FR266	MI		Partial	RAPD	Fusarium root rot	Schneider et al. 2001
Montcalm/FR266	MI	MF	Partial	RAPD	Fusarium root rot	Schneider et al. 2001
Bunsi/Huron	MI	BH	Partial	AFLP	White mold	Kolkman and Kelly 2003
Bunsi/Newport	MI	BN	Partial	AFLP	White mold	Kolkman and Kelly 2003
Montcalm/CDRK-82	ARS, MI, ND		Partial	RAPD	Canning quality	Posa-Macalincag et al. 2002
Montcalm/CELRK	ARS, MI, ND		Partial	RAPD	Canning quality	Posa-Macalincag et al. 2002
Berna/EMP 419	Guelph		Partial	Mix	Leafhopper, seed size	Murray et al. 2004
Bayo Baranda/G 22837 (F_2)	Mexico		497 cM	AFLP	Seed Ca, Fe, Zn, tannin, mass	Guzmán-Maldonado et al. 2003
Moncayo/Primo	ARS	MP	Partial	RAPD	BCTV	Larsen and Miklas 2004
Minnuette/OSU 5630	OR		Full	Mix	Snap bean traits	Myers et al. 2004
Aztec/ND88-106-04	ND, ARS	AN	Partial	Mix	White mold, rust, zinc	Miklas et al. 2005 a
Bunsi/Raven	MI	BR	Partial	Mix	White mold, BCMV	Ender and Kelly 2005
HR67/OAC 95	Harrow	H95	Partial	Mix	CBB	Yu et al. 2004
Red Hawk/Negro San Luis	MI	RN	Parital	Mix	Root rot	Román-Avilés and Kelly 2005

Abbreviations: ALS: angular leaf spot, ASB: ashy stem blight, BBS: bacterial brown spot, BCMV: bean common mosaic virus, BCTV: beet curly top virus, BGYMV: bean golden yellow mosaic virus, CBB: common bacterial blight, HBB: halo bacterial blight, RGA: resistance gene analog marker loci corresponding to clones of known genes, which assorted across 15 linkage groups labeled in descending order of length from D1 to D15. This map covered 827 cM of the genome and was quickly expanded to include 204 markers assigned to 13 LGs covering 1060 cM of the genome (Gepts et al. 1993). This expansion involved RFLP markers from other mapping populations including the Florida (Vallejos et al. 1992) and Paris (Adam-Blondon et al. 1994a) maps in the first effort to align linkage groups among different maps. The Davis map was used directly to map genes/QTL for *Rhizobium* inoculation and resistance to common bacterial blight (Nodari et al. 1993b) and anthracnose (Geffroy et al. 2000).

The Paris BC_1 map, the next significant molecular map published (Adam-Blondon et al. 1994a), was obtained from the backcross Corel (recurrent parent)×*Ms8*EO_2 and developed for the primary purpose of localizing specific anthracnose resistance genes (Table 3). The map consisted of 157 markers including 100 RAPD, 51 RFLP (19 of which were used to align with the Davis map), 2 sequence characterized amplified region (SCAR), and 4 gene loci: *Co-2* (synonymous with *Are*) and *RVI* for resistance to anthracnose, *Ms-8* for male sterility, and *SGou* for pod shape. The map covered 567 cM of the genome.

Many recombinant inbred (RI) populations developed for genetic analysis and mapping of specific traits soon followed (Table 2). Gepts (1999) listed 11 and Kelly et al. (2003) provided an updated list of 14 common bean populations that have been used to develop linkage maps since 1992. Koinange et al. (1996) used an RI population between an Andean snap bean cultivar Midas and a Middle American wild bean G12873 to map the major alleles and QTLs that differentiate the wild from cultivated beans. The Sierra×Olathe, DOR 364×XAN176, BAC 6×HT 7719, BelNeb-RR-1×A 55, PC-50×XAN 159, Eagle×Puebla 152, OAC Seaforth×OAC 95-4, and many other RI populations were developed to study resistance to specific diseases. A55×G122 and CDRK×Yolano RI populations were constructed to look at yield components in an intergene pool cross (Middle American×Andean) (Johnson and Gepts 1999). Sierra×LEF-2RB and Sierra×AC1028 RI populations were used to study drought resistance (Schneider et al. 1997a).

Another significant mapping population, which in effect represents the base mapping population for CIAT (International Center for Tropical Agriculture, Cali, Colombia), consists of an RI population (87 n) from the cross DOR 364×G 19833 (Beebe et al. 1998), which again capitalizes upon diverse parents from the Middle American and Andean gene pools, respectively, to maximize polymorphism and phenotypic variability. The complete CIAT map has not been published; however, it was recently used to identify resistance gene analog (RGA) associations with disease resistance genes/QTL (López et al. 2003) and to integrate SSR markers into the bean map (Blair et al. 2003). The cross was originally made to study the BGYMV resistance present in DOR 364 and tolerance to low soil fertility present in G 19833.

1.2.3
Map Integration

Use of different sizes and types of populations (e.g., F_2, backcross, RI), marker types [e.g., amplified fragment length polymorphism (AFLP), RFLP, RAPD], and lack of coordination among researchers have hindered and slowed the pace of development of fully saturated integrated linkage maps. Nonetheless, Freyre et al. (1998) integrated the linkage maps developed by Vallejos et al. (1992), Nodari et al. (1993a), Adam-Blondon et al. (1994b), and Jung et al. (1996, 1997) into a core map that has a higher density of markers and expanded length. The Davis F_2 mapping population was advanced to an RI population using SSD. The Davis RI population is referred to as the BJ (**B**AT93/**J**alo EEP558) mapping population. The BJ map consists of 563 markers covering 1226 cM of the bean genome (Freyre et al. 1998). Linkage groups for the BJ map are labeled from B1 to B11. Of the 563 markers, about 200 are RAPD that derive from earlier map integration work with the Nebraska and Wisconsin RAPD maps, BAC6 ×HT 7719 (Jung et al. 1996) and PC-50×XAN 159 (Jung et al. 1997), that was never formally published (Skroch et al. 1996). The BJ map has been updated and reviewed by Gepts (1999) and Kelly et al. (2003).

The resulting integration of the Florida and Paris maps with the BJ map enabled the approximate location of more than 1070 markers (Freyre et al. 1998). RFLP was the primary marker used to align the three maps. Although the Florida map is still widely used for integrating maps, the BJ popu-

Table 3. SCAR (STS) markers linked with disease resistance in common bean (*Phaseolus vulgaris* L.)

SCAR name	Marker of origin	Disease	Size (bp)/ orientation	Sequences of SCAR	Tagged locus	Linkage group	Reference
SAP6	AP6	Common bacterial blight (CBB)	820 cis	F – GTC ACG TCT CCT TAA TAG TA R – GTC ACG TCT CAA TAG GCA AA	Major QTL (GN#1 sel 27)	B10	Miklas et al. 2000 c
BAC6	BC409	CBB	1250 cis	TAG GCG GCG GCG CAC GTT TTG TAG GCG GCG GAA GTG GCG GTG	Major QTL (GN#1 sel 27)	B10	Jung et al. 1999
SU91	U9	CBB	700 cis	CCA CAT CGG TTA ACA TGA GT CCA CAT CGG TGT CAA CGT GA	Major QTL (XAN 159)	B8	Pedraza et al. 1997
LG5 syn. BC420	BC420	CBB	900 cis	GCA GGG TTC GAA GAC ACA CTG G GCA GGG TTC GCC CAA TAA CG	Major QTL (XAN 159)	B6	Yu et al. 2000 a
Nitrate reductase	PV-tttc001 (SSR)	CBB	161 cis codominant	TTT ACG CAC CGC AGC ACC AC CTT TCT GCG CCT CTA TGA GTC CA	Major QTL (XAN 159 & HR67)	B7	Yu et al. 2004
R7313		CBB	700 cis	ATT GTT ATC GTC GAC ACG AAT ATT TCT GAT CAC ACG AG	Major QTL (OAC 88-1)	B8	Bai et al. 1997; Beattie et al. 1998
R4865		CBB	950 cis	TCC AAA GCC ATT CTA GTT CAG CTA CTT TCA AAC TGG G	Major QTL (OAC 88-1)		Bai et al. 1997; Beattie et al. 1998
SB10	B10	Halo bacterial blight (HBB)	525 cis	CTG CTG GGA CAA TCA CCA AGT C CTG CTG GGA CTC TCT TAC	*Pse-1*	B4	Fourie et al. 2004
SW13	W13	Bean common mosaic virus (BCMV) & HBB	690 cis	CAC AGC GAC ATT AAT TTT CCT TTC CAC AGC GAC AGG AGG AGC TTA TTA	*I* *Pse-3*	B2	Haley et al. 1994 a; Melotto et al. 1996; Fourie et al. 2004
ROC11	C11	BCMV	420 trans	CCA ATT CTC TTT CAC TTG TAA CC GCA TGT TCC AGC AAA CC	*bc-3*	B6	Johnson et al. 1997
SBD5	BD5	BCMV	1250 cis	GTG CGG AGA GGC CAT CCA TTG GTG GTG CGG AGA GTT TCA GTG TTG ACA	*bc-1*[2]	B3	Miklas et al. 2000 b
SR2	R2	Bean golden yellow mosaic virus (BGYMV)	530/570 codominant	CAC AGC TGC CCT AAC AAA AT CAC AGC TGC CAC AGG TGG GA	*bgm-1*		Urrea et al. 1996; Beebe (personal communication 1996)
SW12	W12	BGYMV	700 cis	TGG GCA GAA GTT CTA GCA TGT GGC TGG GCA GAA GCA CAG TAT GAT TTG	Major QTL (DOR 364)	B4	Miklas et al. 2000 a; Singh et al. 2000
SAS8	AS08	Beet curly top virus (BCTV)	1550 cis	GGC TGC CAG TAT CTT GTC TAA CAC C GGC TGC CAG TGA CGC AAT TCT GCA G	*Bct*	B7	Larsen and Miklas 2004
SK14	K14	Rust	620 cis	CCC GCT ACA CAC CAA TAC CTG CCC GCT ACA CTT GAT AAA ATG TTA G	*Ur-3*	B11	Haley et al. 1994 d; Nemchinova and Stavely 1998; Miklas et al. 2002 b
SA14	A14	Rust	1079/800 codominant	CTA TCT GCC ATT ATC AAC TCA AAC GTG CTG GGA AAC ATT ACC TAT T	*Ur-4*	B6	Miklas et al. 1993; Mienie et al. 2004; Miklas et al. 2002 b

Table 3 (continued)

SCAR name	Marker of origin	Disease	Size (bp)/ orientation	Sequences of SCAR	Tagged locus	Linkage group	Reference
SI19	I19	Rust	460 cis	AAT GCG GGA GAT ATT AAA AGG AAA G AAT GCG GGA GTT CAA TAG AAA AAC C	*Ur-5*	B4	Haley et al. 1993; Miklas et al. 2000 a
SBC6	BC06	Rust	308 cis	GAA GGC GAG AAG AAA AAG AAA AAT GAA GGC GAG AGC ACC TAG CTG AAG	*Ur-6*	B11	Miklas et al. 2002 b; Park et al. 2004 b
SAD12	AD12	Rust	537 cis	AAG AGG GCG TGA GAT CGT CG AAG AGG GCG TCT TGA AGG TT	*Ur-7*	B11	Park et al. 2003, 2004 a
SAE19	AE19	Rust	890 trans	CAG TCC CTG ACA ACA TAA CAC C CAG TCC CTA AAG TAG TTT GTC CCT A	*Ur-11*	B11	Johnson et al. 1995; Miklas et al. 2002 b; Alzate-Marin et al. 2004; Queiroz et al. 2004 c
UR11-GT2	GT02	Rust	450 cis	CGC ACT TAG GAG CAC AAA TGG TGG GTC CCA TAT TTT G	*Ur-11*	B11	Boone et al. 1999; Miklas et al. 2002 b
SF10	F10	Rust	1072 cis	GGA AGC TTG GTG AGC AAG GA GGA AGC TTG GCT ATG ATG GT	(Ouro Negro)	B4	Corrêa et al. 2000; Faleiro et al. 2000 a; Miklas et al. 2002 b
SBA8	BA8	Rust	530 cis	CCA CAG CCG ACG GAG GAG GCC ATG TTT TTT GTC CCC	(Ouro Negro)	B4	Corrêa et al. 2000; Miklas et al. 2002 b
KB126	E-AAC/M-ACC	Rust	405/430 codominant	GAA TTC AAC CTC GGC CAC TAC C TTA AAC CTT CCG GAG GAT TC	*Ur-13*	B8	Mienie et al. 2005
Phs	Phaseolin 'T' & 'S' alleles	White mold & CBB	Multiple	AGC ATA TTC TAG AGG CCT CC GCT CAG TTC CTC AAT CTG TTC	Major QTL (G 122) & (BAT 93)	B7	Nodari et al. 1993 b; Kami et al. 1995; Miklas et al. 2001
SAU5	AU05	White mold	1350 cis	GAG CTA CCG TCA GTT TAC TAA GAG CTA CCG TGG CTT TTT TCT	QTL (minor) (NY6020-4)	B6	Miklas et al. 2003 a
SS18	S18	White mold	1650 cis	CTG GCG AAC TGT ACA TGC AAC ATA C CTG GCG AAC TGA TTC ATA CAT TTT G	QTL (major) (NY6020-4)	B8	Miklas et al. 2003 a
	SE_{ACT}/M_{CCA}	Anthracnose	codominant		*Co-1*^{2}	B1	Vallejo and Kelly 2002
SCAreoli	H20	Anthracnose	1000	GGG AGA CAT CCA TCA GAC AAC TCC GGG AGA CAT CTT CAT TTG ATA TGC	*Co-2*	B11	Adam-Blondon et al. 1994 b; Geffroy et al. 1998
SY20	Y20	Anthracnose	830 cis	AGC CGT GGA AGG TTG TCA T CCG TGG AAA CAA CAC ACA AT	*Co-4*	B8	Arruda et al. 2000; Kelly et al. 2003; Queiroz et al. 2004 b
SC08	C08	Anthracnose	910 cis	AGA ATG CCT TTA GCT GTT GG CAG AGA GGC TAG GCT TAT CG	*Co-4*	B8	Arruda et al. 2000; Kelly et al. 2003; Queiroz et al. 2004 b

Table 3 (continued)

SCAR name	Marker of origin	Disease	Size (bp)/ orientation	Sequences of SCAR	Tagged locus	Linkage group	Reference
SAS13	AS13	Anthracnose	950 cis	CAC GGA CCG AAT AAG CCA CCA ACA CAC GGA CCG AGG ATA CAG TGA AAG	*Co-4*^{2}	B8	Young et al. 1998; Kelly et al. 2003
SH18	H18	Anthracnose	1100 cis	CCA GAA GGA GCT GAT AGT ACT CCA CAA C GGT AGG CAC ACT GAT GAA TCT CAT GTT GGG	*Co-4*^{2}	B8	Alzate-Marin et al. 2001; Awale and Kelly 2001; *Kelly et al. 2003
SBB14	BB14	Anthracnose	1150/1050 codominant	GTG GGA CCT GTT CAA GAA TAA TAC GTG GGA CCT GGG TAG TGT AGA AAT	*Co-4*^{2}	B8*	Awale and Kelly 2001; Kelly et al. 2003
SAB3	AB-3	Anthracnose	400 cis	TGG CGC ACA CAT AAG TTC TCA CGG TGG CGC ACA CCA TCA AAA AAG GTT	*Co-5*		Vallejo and Kelly 2001
SAZ20	AZ20	Anthracnose	845 cis	ACC CCT CAT GCA GGT TTT TA CAT AAT CCA TTC ATG CTC ACC	*Co-6*	B7	Alzate-Marin et al. 2000; Kelly et al. 2003; Queiroz et al. 2004 b
SZ04	Z04	Anthracnose	567 cis	GGC TGT GCT GAT TAA TTC TGG TGC TCA TTT TAT AAT GGA GAA AAA	*Co-6*	B7	Alzate-Marin et al. 1999; Kelly et al. 2003; Queiroz et al. 2004 b
SB12	B-12	Anthracnose	350 cis	CCT TGA CGC ACC TCC ATG TTG ACG ATGGG TTG GCC	*Co-9*	B4	Mendez de Vigo et al. 2002
SF10	F10	Anthracnose	1072 cis	GGA AGC TTG GTG AGC AAG GA GGA AGC TTG GCT ATG ATG GT	*Co-10*	B4	Corrêa et al. 2000; Faleiro et al. 2000 b; Alzate-Marin et al. 2003
SH13	H13	Angular leaf spot	520 cis	GAC GCC ACA CCC ATT ATG TT GCC ACA CAG ATG GGA GCT TTA	(AND 277) dominant gene *Phg-1*		Carvalho et al. 1998; Queiroz et al. 2004 a
SN02	N02	Angular leaf spot	890 cis	ACC AGG GGC ATT ATG AAC AG ACC AGG GGC AAC ATA CTA TG	(Mexico 54 & C49-242) dominant gene *Phg-2*	B8	Nietsche et al. 2000; Sartorato et al. 2000; Miklas et al. 2006
	E-ACA/M-CTT$_{330}$	Angular leaf spot	280/305 codominant	CTT GTT CTG AGT CAT TTA CCT TGC GAA TTC ACA GTC CAA ACT CTA ATC	(G 10474) dominant gene		Mahuku et al. 2004
SAA19	AA19	Angular leaf spot	650 cis	TGA GGC GTG TCA ATG GAT ATA A GAG GCG TGT TGA TAA TTC TGG	(Ouro Negro) dominant gene		Corrêa et al. 2001; Queiroz et al. 2004 a
SBA16	BA16	Angular leaf spot	560 cis	TTC CAC GTC TAT TTT GCA TCA CAC GCA TCA CGC AGA ACT	(Ouro Negro) dominant gene		Faleiro et al. 2003; Queiroz et al. 2004 a
SM02	M02	Angular leaf spot	460 cis	CAA CGC CTC ATT AAA TTG GA CGC CTC TAA ACG GGA GAA AC	(Ouro Negro) dominant gene		Corrêa et al. 2001; Queiroz et al. 2004 a

lation is widely recognized as the "core mapping population" for common bean (Freyre et al. 1998; Gepts 1999). The BJ RI lines have been distributed worldwide, and the molecular marker data spreadsheet for the BJ core map is readily accessible via the Internet (http://www.agronomy.ucdavis.edu/agronomy/). Also, a BAC library has been developed for BAT 93 that will contribute to additional marker development and positional cloning of genes (Kami and Gepts 2000). Separately, the Florida RFLP map was integrated with the Wisconsin RAPD map Eagle×Puebla 152 via the RI mapping population Jamapa×DIACOL Calima (Vallejos et al. 2001). Together the integration of these three maps placed 814 markers (230 RFLP, 464 RAPD, and 120 other) in the bean genome.

Eventually all maps directly or indirectly integrate with either the Davis (BJ) or the Florida map. Early on, the original RFLP clones for both the Davis and Florida maps were widely distributed and used as anchor markers to integrate and align linkage groups from subsequently developed maps. Researchers used the RFLP anchor markers to generate a framework map in their population. Conversely, researchers have placed polymerase chain reaction (PCR)-based markers (RAPD, SCAR, AFLP) generated from their mapping populations in the Davis or Florida populations to align linkage groups with the original maps. This direct mapping is generally done with individual traits or LGs in an effort to locate them on the core BJ map. Larsen and Miklas (2004) recently placed a SCAR linked with the *Bct* gene conditioning resistance to BCTV on LG B7 by directly mapping the marker in the BJ population. A SCAR linked with the $bc\text{-}1^2$ gene for resistance to BCMV was placed on LG B3 (of the BJ core map) by mapping the marker in the CIAT map, DOR 364×G18933 (Miklas et al. 2001), which is aligned with the BJ core map.

Now instead of cumbersome RFLP, rapid PCR-based markers, namely, SCAR, sequence tagged site (STSs), and simple sequence repeat (SSR), are being used as anchor markers for integrating bean maps. Cumulatively, 115 SSR markers have been developed and characterized (Yu et al. 1999, 2000b; Gaitán-Solis et al. 2002; Blair et al. 2003) and located in either the BJ (37 SSR) or CIAT (78 SSR) maps. SSRs are codominant, easier to generate, and more polymorphic than RFLPs because they are multiallelic. SSRs and other PCR-based markers like STS will quickly replace RFLP as the marker of choice for integrating bean LGs as they have in other crops (Cregan et al. 1999). A few RFLP clones from the Florida map have been converted to STS markers, which may contribute to their use as anchor markers (Murray et al. 2002). Ongoing development and mapping of EST (reviewed by Broughton et al. 2003) in common bean will lead to further identification of SSR markers.

Recently the Florida linkage map (Vallejos et al. 1992) was integrated with the chromosomal map (Moscone et al. 1999) using FISH of pooled RFLP clones from each linkage group (Pedrosa et al. 2003). Integration of the genetic and chromosomal maps will serve to unify genetic linkage group nomenclature in common bean (Table 1).

1.3 Tagging and Mapping Favorable Genes and QTLs of Economic Importance

In common bean, molecular markers have been used to characterize genetic diversity and provide credence to earlier results and conclusions arrived at from the use of crossing compatibilities (Singh and Gutiérrez 1984; Gepts and Bliss 1985), and morphological (Singh et al. 1991b), isozyme (Singh et al. 1991c), seed protein (Gepts et al. 1986), and DNA markers (Khairallah et al. 1990; Becerra-Velásquez and Gepts 1994; Haley et al. 1994c; Sonnante et al. 1994; Beebe et al. 2000; McClean et al. 2004; González-Mejía et al. 2005). They also facilitated evolutionary studies (Gepts 1993, 1998; Sonnante et al. 1994) and could be used for fingerprinting of cultivars and germplasm to avoid identity theft. But the most remarkable use of molecular markers has been to tag and map favorable alleles and QTLs controlling phenotypic traits of economic value to producers and consumers (Fig. 1).

Disease resistance traits tagged with SCAR markers are summarized in Table 3. Primarily RAPD followed by AFLP have been used to tag disease resistance traits in bean. Those resistance-linked RAPD and AFLP found useful for MAS are converted to allele-specific markers (SCAR) to facilitate utilization and portability because RAPD and AFLP are notoriously difficult to reproduce across different laboratories. We find that even some SCAR markers may not amplify in our laboratory without further optimization of PCR protocols.

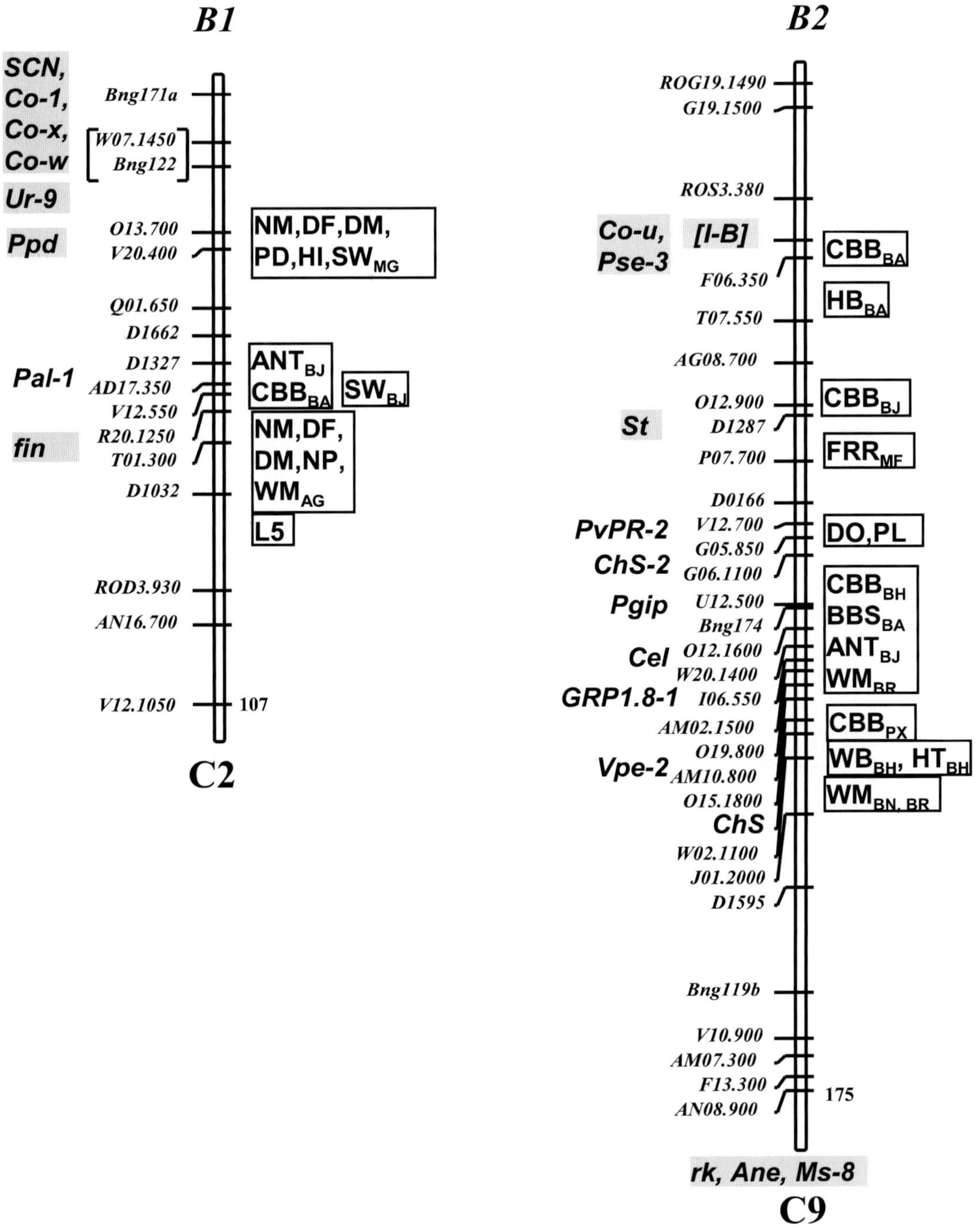

Fig. 1. Genomic map of common bean (*Phaseolus vulgaris* L.) with distribution of genes and QTL. The linkage groups (B1 to B11) correspond to the core map version of Freyre et al. (1998). This map is a compilation of the maps presented by Kelly et al. (2003) and Miklas et al. (2006). Directly to the left of each linkage group are the framework molecular markers (smaller font), biochemical genes (larger font), and phenotypic trait genes (shaded boxes). For explanations on marker and biochemical gene symbols see [Bassett 2004; Tables 4 and 5 in Gepts (1999)], and the original map references (Vallejos et al. 1992; Nodari et al. 1993a; Adam-Blondon et al. 1994a). SCN is the putative location of a soybean cyst nematode resistance gene (Concibido et al. 1996). The *Pse-3* gene replaces *R3* (Fourie et al. 2004). Briefly, disease resistance gene symbols *Co* are anthracnose resistance loci, *Ur* rust resistance loci, *Pse* halo blight resistance loci; *I* and *bc* are dominant and recessive genes, respectively, for resistance to BCMV, *Phg* angular leaf spot resistance locus. *Bct* is a locus for resistance to BCTV (see also Table 3). *Pu-a* is for abaxial leaf pubescence (Jung et al. 1998). [*AI-Arl-Lec*] is the main locus coding for the α-amylase

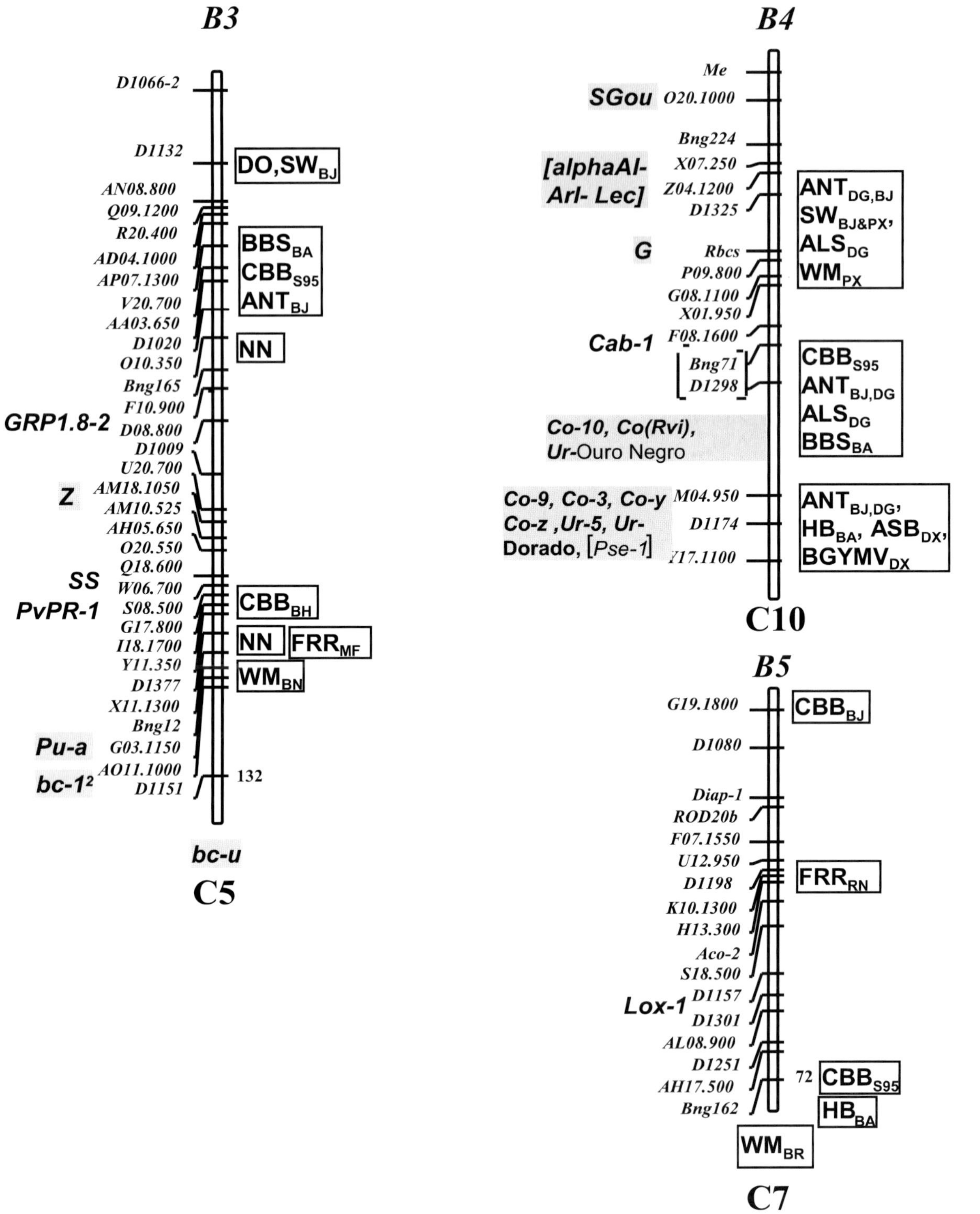

Fig. 1 (continued) inhibitor-arcelin-lectin multigene family. *Fin*, *Ppd*, *St* are genes for determinacy, sensitivity to photoperiod, and pod string formation (Koinange et al. 1996). *B*, *[C-R]*, *G*, *P*, and *rk* are seed color genes and *Ana*, *Ane*, *Bip*, *L*, *T*, and *Z* are seed color pattern genes (McClean et al. 2002). *Asp* is a gene for seedcoat shininess (Arndt and Gepts 1989; Gepts 1999). *Fr* is a male fertility restorer gene (He et al. 1995). *SGou* and *Ms-8* are genes controlling pod cross-section and male sterility (Adam-Blondon et al. 1994 a). To the right (boxed symbols) are QTLs mapped in different populations (Table 2). For disease resistance abbreviations: ALS resistance to angular leaf spot, ANT anthracnose, ASB ashy stem blight, BGYMV bean golden yellow mosaic virus, BBS bacterial brown spot, CBB common bacterial blight, FRR *Fusarium* root rot, HB halo blight,

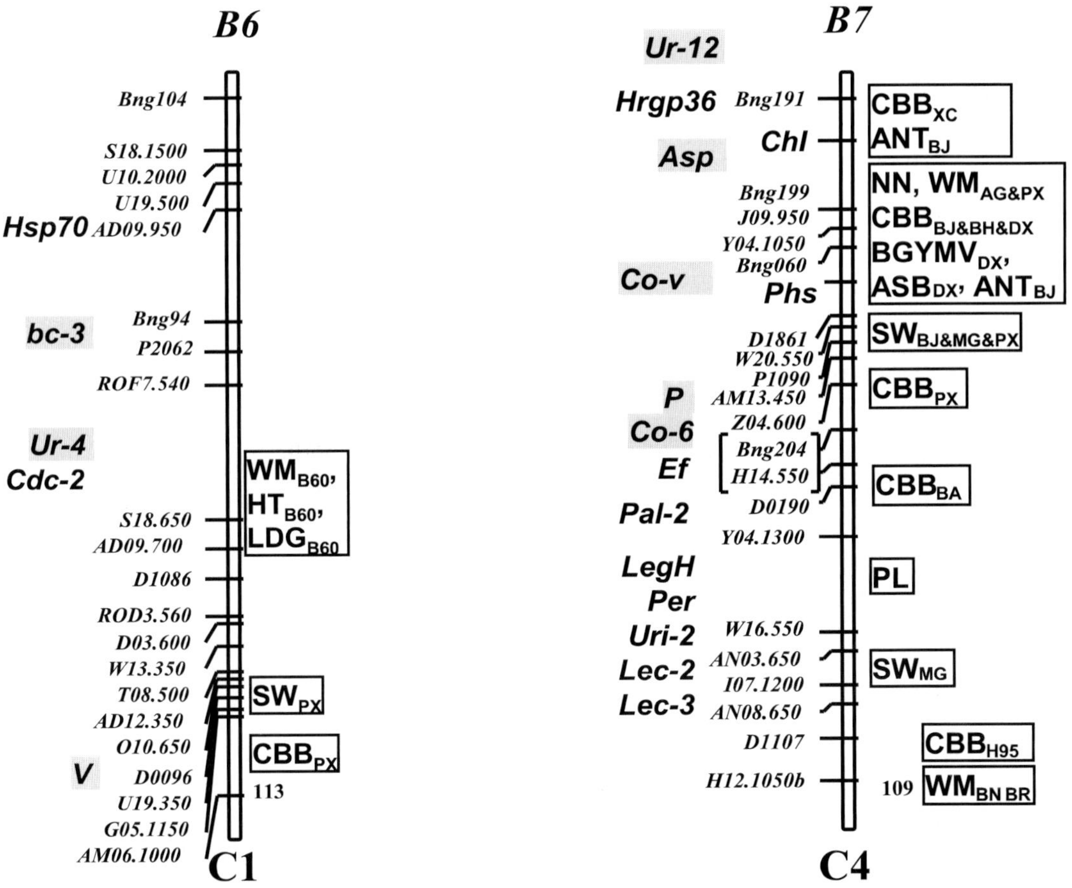

Fig. 1 (continued) WB web blight, and WM white mold resistance. For other abbreviations: APP appearance of canned kidney beans, DF and DM number of days to flowering and to maturity, DO seed dormancy, HI harvest index, HT height, L5 length of the fifth internode, LDG lodging, NM number of nodes on the main stem, NN *Rhizobium* nodulation, NP number of pods, PD photoperiod-induced delay in flowering, PL pod length, and SW seed weight. Abbreviation symbols in subscript represent the source population of the QTL (Table 2)

Bean researchers have generally utilized one or more of the following strategies for tagging genes and QTLs with molecular markers: backcross-derived near-isogeneic lines (NILs) (Miklas et al. 1993), interval mapping (Nodari et al. 1993b), heterogeneous inbred derived NIL (Haley et al. 1994d), bulked segregant analysis (BSA) (Urrea et al. 1996), BSA and NIL (Miklas et al. 1993), BSA and selective mapping (Miklas et al. 1996b), and multitrait bulking and selective mapping (Kolkman and Kelly 2003).

The *Ur-4* gene for rust resistance was the first gene specifically tagged in common bean with a DNA marker (Miklas et al. 1993). The tagging of other genes and QTLs for resistance to bacterial, fungal and viral pathogens, and other traits, as described below, quickly followed.

There has been a recent emphasis on tagging genes conditioning resistance to angular leaf spot, a devastating disease in Africa and the Americas. Ferreira et al. (2000) found a RAPD marker that was linked to a single dominant allele (*Phg-2*) in breeding line MAR2 imparting resistance to race 63.39 of *P. griseola*. Nietsche et al. (2000) also reported RAPD and SCAR (SN02) markers linked to a dominant allele controlling resistance to race 31.17 of *P. griseola* in small back bean Cornell 49-24-2. Moreover, the same marker (SN02) was linked with the resistance found in MAR 2 and Mexico 54. Miklas et al. (2006) mapped the SN02 marker to chromosome 3 of the core BJ map. Caixeta et al. (2003) reported a dominant allele for resistance to race 61.41 of the same pathogen in dry bean breeding line BAT332 that was linked to two RAPD

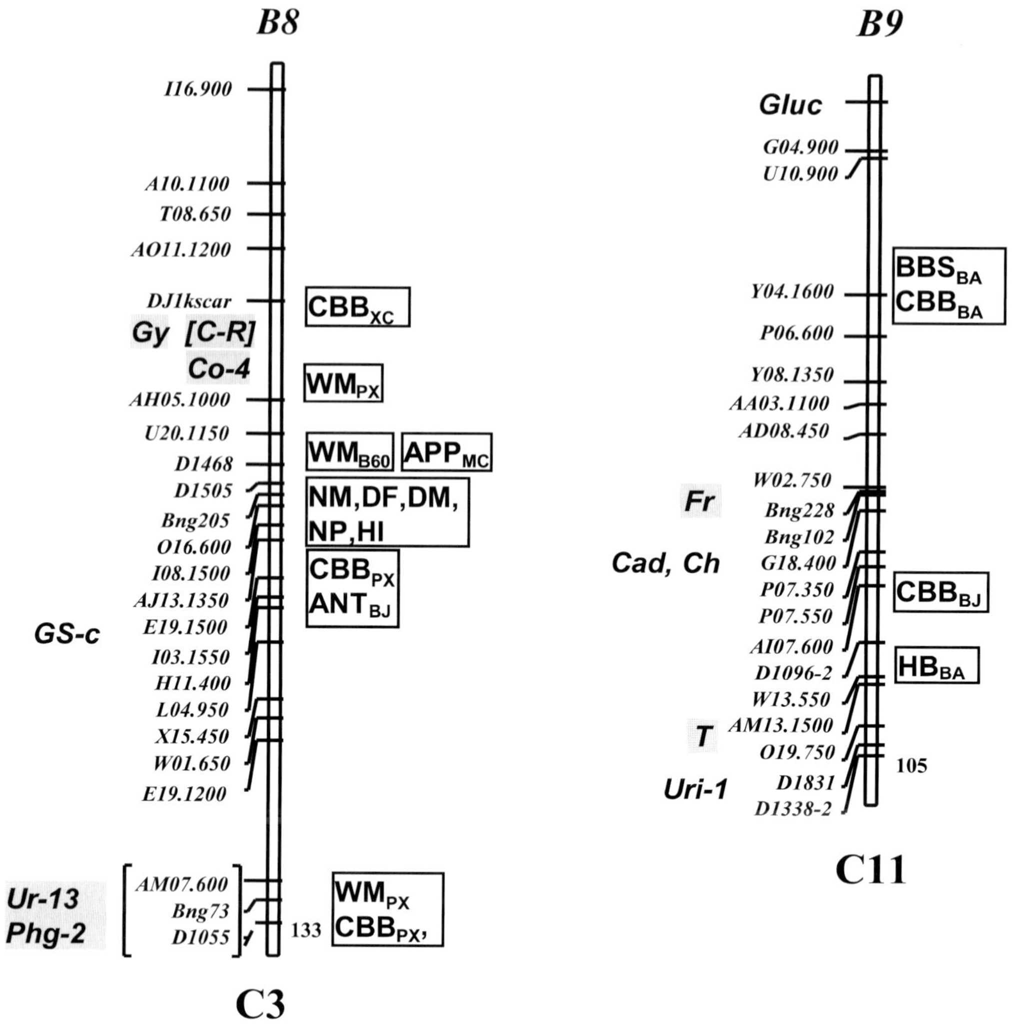

Fig. 1 (continued)

markers in cis-position. The SH13 SCAR linked with putative resistance gene *Phg-1* was identified in AND 277 (Carvalho et al. 1998; Queiroz et al. 2004 b). A SCAR converted from an AFLP marker was linked to a single dominant gene from G 10474 (Mahuku et al. 2004). López et al. (2003) observed five QTLs for resistance to angular leaf spot located on chromosomes 8 and 10. Tests of independence among the tagged genes for resistance to angular leaf spot have not been conducted, and map location of the genes are unknown with the exception of SN02 (*Phg-2*) and the QTL.

As reviewed by Kelly and Vallejo (2004 a), seven of nine major resistance genes for anthracnose have been tagged and mapped (Fig. 1, Table 3). Four of the genes, *Co-1*, *Co-2*, *Co-9*, and *Co-10*, cosegregate with rust resistance and the latter two also with halo blight resistance (Miklas et al. 2002 b, 2005 a; Kelly and Vallejo 2004 a), which is indicative of resistance gene clusters derived from ancestral genes having undergone duplication and divergence events (Michelmore and Myers 1998).

Both recessive and dominant viral resistance genes have been tagged with molecular markers. For BCMV and BCMNV resistance, the *bc-3* (Haley et al. 1994 b; Johnson et al. 1997), $bc\text{-}1^2$ (Miklas et al. 2000 b), and *I* genes (Haley et al. 1994 a; Melotto et al. 1996) were tagged and mapped to chromosomes 1, 5, and 9, respectively (Fig. 1). Urrea et al. (1996) found a RAPD marker (map location unknown) for the *bgm-1* allele imparting resistance to leaf chlorosis induced by BGMV (occurring in Brazil, Argentina, and Bolivia) and BGYMV (found in Central America, Mexico, and the Caribbean). Re-

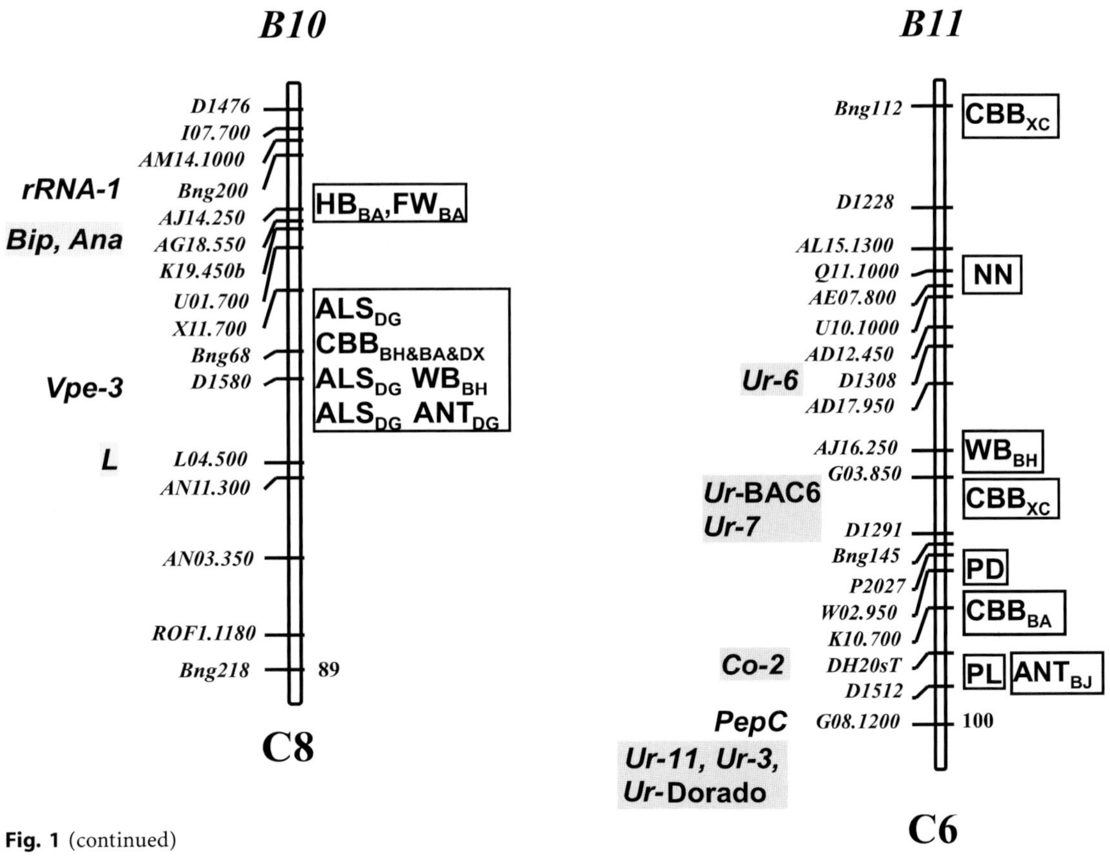

Fig. 1 (continued)

sistance to BGYMV-induced leaf chlorosis found in DOR 364 (synonymous with Dorado) was controlled by two major QTLs (Miklas et al. 1996 b) on chromosomes 4 and 10 (Miklas et al. 2000 a). The QTL on chromosome 10 was tagged with a SCAR marker, SW13. Both QTL for BGYMV resistance appear to reside within gene-rich regions. The QTL on chromosome 10 resides within a cluster of genes conditioning resistance to anthracnose, charcoal rot [synonymous ashy stem blight caused by *Macrophomina phaseolina* (Tassi) Goid.], halo blight, and rust. A SCAR marker for the dominant gene, *Bct*, conferring resistance to BCTV, was generated and mapped to chromosome 4 (Larsen and Miklas 2004). Both *Bct* and the QTL for BGYMV resistance from DOR 364 appear to reside within a cluster of genes conditioning resistance to anthracnose, charcoal rot, common bacterial blight, and white mold.

The dominant alleles *Ur-3* (Haley et al. 1994 d; Nemchinova and Stavely 1998), *Ur-4* (Miklas et al. 1993; Mienie et al. 2004), *Ur-5* (Haley et al. 1993; Melotto and Kelly 1998), *Ur-6* (Park et al. 2004 b), *Ur-7* (Park et al. 2003, 2004 a), *Ur-9* (Jung et al. 1996), *Ur-11* (Queiroz et al. 2004 c; Johnson et al. 1995), and *Ur-13* (Mienie et al. 2005) conferring specific resistance (SR) to rust were tagged using RAPD markers with many eventually converted to SCAR (Table 3) and placed on the linkage map (Fig. 1). Unnamed SR genes for rust include two from DOR 364 (Miklas et al. 2000 a) that map in the vicinity of the *Ur-3* and *Ur-5* loci, one from BAC 6 (Jung et al. 1996) that maps near *Ur-7*, and one from Ouro Negro (Corrêa et al. 2000) located on chromosome 10. Other genes involved with rust resistance have been mapped: *Ur-12* on chromosome 4 conditions adult plant resistance (APR) that is initially expressed at the fourth trifoliolate leaf stage or later (Jung et al. 1998), *Pu*-a on chromosome 5 conditions abaxial leaf pubescence (Jung et al. 1998), and *Crg* (complements resistance gene) on chromosome 3 is necessary for expression of the *Ur-3* gene (Kalavacharla et al. 2000).

Two independent complementary dominant alleles (*Mp-1* and *Mp-2*) controlled resistance to charcoal rot in drought-resistant dry bean breeding line BAT 477, and RAPD markers for one allele in coupling and for the other allele in repulsion phase were identified (Olaya et al. 1996). Mayek-Pérez et al. (2001) also reported a similar inheritance of resistance in BAT 477. However, from a field screening of 119 $F_{5:7}$ RI population from DOR 364/XAN 176 Miklas et al. (1998b) reported a quantitative inheritance (h^2 0.53 and 0.57) for charcoal rot with RAPD markers linked to four QTLs (each controlling 13 to 19% of variation for resistance). Two of the QTLs with major effect across environments (Miklas et al. 2000a) were located within resistance gene clusters on chromosomes 4 and 10 (Fig. 1).

Of over a dozen major and minor QTLs controlling resistance to common bacterial blight, the SU91 (Pedraza et al. 1997) and BC420 (Yu et al. 2000a) SCAR markers are tightly linked with major QTLs that were introgressed from a single tepary bean accession into the common bean breeding line XAN159 (McElroy 1985). A third major QTL for common bacterial blight resistance linked to SCAR SAP6 on chromosome 8 (Miklas et al. 2000c) derives from the great northern landrace Montana No. 5, not tepary bean as previously thought (Miklas et al. 2003a). A fourth major QTL identified on chromosome 4 between the *Asp* locus for seed brilliance and phaseolin seed protein locus *Phs* has not been tagged with any markers (Miklas et al. 2000d; Nodari et al. 1993b). A fifth major resistance QTL present in OAC 95-4, and putatively derived from tepary bean, maps toward the end of chromosome 7 (Tar'an et al. 2001).

Genes and QTLs controlling resistance to other bacterial pathogens have been tagged and mapped. The *Pse-1* gene conditioning resistance to Races 1, 7, and 9; *Pse-3* to Races 3 and 4; and *Pse-4* to Race 5 of *P. syringae* pv. *phaseolicola* were tentatively identified in BelNeb-RR-1/A 55 RI population (Fourie et al. 2004) on chromosomes 1, 9, and 10, respectively. An earlier study with this same population identified QTL that conditioned resistance to halo blight, with map locations that corresponded to the putative location of the *Pse-1*, *Pse-3*, and *Pse-4* genes (Ariyarathne et al. 1999). The *Pse-1* gene and linked SCAR B10 reside within the B4 cluster of genes conditioning anthracnose, rust, ashy stem blight (QTL), and BGYMV (QTL) resistance. A QTL for bacterial brown spot (caused by *P. syringae* pv. *syringae*) resistance (Jung et al. 2003) is located in a nearby gene cluster (Fig. 1), and this QTL on chromosome 10 and four others on chromosomes 4, 5, 9, and 11 were identified using the same BelNeb-RR-1/A 55 RI population. The *Pse-3* gene is completely linked with the *I* gene as no recombinants for these two genes have been observed (Teverson 1991). The *Pse-4* gene may be linked with *Pse-1* on chromosome 10.

Except for Fusarium wilt (caused by *F. oxysporum* f.sp. *phaseoli*), major QTLs controlling resistance to any other root rots have not been found. Fall et al. (2001) identified and mapped a major QTL on chromosome 8 from small black breeding line A 55 that controlled 63.5% of variance for Fusarium wilt resistance. The RAPD markers (except one) associated with resistance to Fusarium root rot in greenhouse tended not to associate with field screening, and individual markers did not explain more than 15% of variation for resistance (Schneider et al. 2001).

Park et al. (2001) reported RAPD markers linked with as many as nine QTLs responsible for white mold resistance in PC-50/XAN 159 common bean population. Most markers linked to field resistance were located on chromosomes 3, 4, 6, and 10, with the QTL on chromosome 4 accounting for the largest (12%) variation in resistance. In a similar region of chromosome 4, Miklas et al. (2001) identified the *Phs* locus linked with a major QTL for the greenhouse straw test (38%) and field resistance (26%) in the A 55/G 122 population. An additional QTL responsible for canopy porosity (34%) and disease avoidance (18%) in the field (mapped on chromosome 2) was also identified and attributable to the *fin* gene for determinant growth habit that mapped to the same location. Kolkman and Kelly (2003) reported white mold resistance QTLs from ICA Bunsi/Newport population that were located on chromosomes 4 and 9. They used RI population from Huron/Newport population to confirm linkage between markers and QTL derived from ICA Bunsi (synonymous with Ex-Rico 23) for white mold resistance across environments. Miklas et al. (2003b) identified two QTLs for white mold resistance in a Benton/NY 6020-4 snap bean population. The QTL on chromosome 1 derived from Benton explained 12% of the variation for resistance, whereas the QTL from NY 6020-4, responsible for 38% variation, was located on chromosome 3. The latter was associated with increased internode length. Thus, QTLs for white mold resistance have been located on 8 of the 11 chromosomes.

Schneider et al. (1997a) reported four QTL-linked RAPD markers in one biparental population and five in another that were consistently associated with drought-stressed (DS) yield, nonstressed (NS) yield, and/or geometric mean yield in DS and NS environments.

Other traits mapped include seed coat color genes and QTLs controlling canning quality and seed size, shape, and mineral content for iron and zinc. McClean et al. (2002) tagged nine genes conditioning seed coat color (*G, Gy, J,* and *V*) or patterning (*C, T, Z, Ana, Bip*) with RAPD markers of which many were later converted to STS markers and placed on the BJ core linkage map (Fig. 1).

Walters et al. (1997) studied canning quality in three navy bean populations across two locations and identified mostly population- and location-specific QTL-marker associations. The QTLs individually had minor effects but taken together explained from 20 to 52% of the variation in visual appeal of the canned product based on regression models using one to six markers. None of the QTLs was located on linkage groups. Two QTLs influencing canning quality in multiple environments were identified by Posa-Macalincag et al. (2002), with one QTL derived from Montcalm dark red kidney possessing desirable characteristics, and the other contributed by parents with undesirable canning performance California Dark Red Kidney 82 in one population and California Early Light Red Kidney in another population. The QTL from Montcalm was located on chromosome 3, had minor effect (<10%), and was expressed in only one population. The second QTL had variable expression of 8% and 20% across populations and was not integrated in the core map.

Five QTLs affecting seed mass (explaining 42% of the phenotypic variance), four tannin content (42%), two each Ca (25%) and Fe (25%) content, and one Zn content (15%) dispersed across four linkage groups (nonintegrated) were identified by Guzmán-Maldonado et al. (2003) from a cultivated×wild cross. For an in-depth look at QTLs conditioning seed size see Johnson et al. (1996). The favorable QTLs for seed accumulation of Ca, Fe, and Zn were derived from the wild parent, joining other favorable traits from wild *P. vulgaris* including arcelin seed protein for combating seed weevils (Cardona et al. 1990) and QTLs for yield improvement. These QTLs were not integrated in the core map. Other research in the area of tagging and mapping genes affecting mineral content of bean is ongoing at CIAT (Blair et al. 2002) and by the Harvest Plus consortium (http://www.harvestplus.org), but detailed results have not been formally published. The research for identifying, characterizing, and improving nutritive attributes of bean for human consumption is a hot topic receiving widespread attention (Welch et al. 2000; House et al. 2002).

Although to date microarrays and other surveys of EST have not been employed to directly identify candidate genes underlying quantitative traits in bean, putative genes underlying QTLs have been identified by map colocalization. The *Phs* seed protein locus is a candidate gene for seed size in bean (Johnson et al. 1996) that maps to the same region on chromosome 4. The *fin* gene is a candidate gene for disease avoidance to white mold (Miklas et al. 2001) that colocalize on chromosome 2. The colocalization of R genes or defense-related genes with disease resistance QTLs is indicative of candidate genes (Kelly and Vallejo 2004b). For instance, anthracnose resistance QTL mapped near defense-related protein *Hrgp36* and enzyme *Pal-2* (Geffroy et al. 2000) on chromosome 4, the *Pv*PR-1 and *Pv*PR-2 (*P. vulgaris* pathogenesis-related) genes were linked with QTLs controlling resistance to Fusarium root rot on chromosomes 5 and 9 (Schneider et al. 2001), and *Pgip* (polygalacturonase-inhibiting protein) is near a QTL for common bacterial blight resistance on chromosome 9 (Fig. 1). R genes and QTLs colocalize for resistance to anthracnose on chromosomes 4 (*Co-v*), 6 (*Co-2*), and 10 (*Co-9*, *Co-10*) (Geffroy et al. 2000; Fig. 1) and similarly for resistance to halo blight on chromosomes 9 (*Pse-3*) and 10 (*Pse-1*) (Fourie et al. 2004). The anthracnose genes on chromosomes 6 and 10 consist of multigene families (Geffroy et al. 1998), so it is hypothesized that a component (or components) of the gene family interacts with certain pathotypes to contribute a quantitative partial resistance.

Mapping of RGA generated from conserved R-gene-sequence motifs, leucine-rich repeats, and nucleotide binding sites has also been a useful approach to identifying putative candidate genes underlying disease resistance QTLs. The most recent efforts are those of López et al. (2003), who mapped RGA near QTLs for resistance to angular leaf spot, anthracnose, and BGYMV.

1.4 Marker-Assisted Selection

Kelly and Miklas (1999) discussed morphological, biochemical, and DNA-based markers for indirect selection in common bean. In brief, among morphological traits of agronomic value, growth habit, plant height, and node and pod number have the most marked effects on and a positive association with the number of days to maturity and seed yield in common bean (Nienhuis and Singh 1986; Tar'an et al. 2002). Change in growth habit from determinate Type I to indeterminate upright Type II increased yield and stability of small white-seeded cultivars (Adams 1982; Grafton et al. 1993).

With regard to biochemical markers, one cycle of indirect selection for specific electrophoretic bands associated with seed protein was as effective as three cycles of direct selection (Delaney and Bliss 1991). Similarly, extremely high levels of resistance to bean weevil (*Z. subfasciatus*) have been successfully transferred from the Mexican wild bean populations to a range of breeding lines using arcelin seed protein as an indirect selection criterion (Cardona et al. 1990). More recently, availability of DNA-based markers has minimized the need for and complemented disease screening in presence of pathogens, especially for traits controlled by genes or QTLs with large effects. The resistance-linked markers have been used in various breeding strategies including marker-assisted backcrossing (Fig. 2), detection and retention of hypostatic genes, gene pyramiding, and direct selection for resistance. There are also many instances where resistance-linked markers have been used to determine genes present in advanced breeding lines and cultivars (e.g., Singh et al. 2000). As an example of marker-assisted breeding (Fig. 2), McMillan et al. (1998) introgressed the *bgm-1* allele for BGMV resistance from dry bean breeding line A 429 into snap bean using a RAPD (Urrea et al. 1996) converted to a SCAR marker (Table 3). The snap bean cultivar Genuine with resistance to BGYMV resulted from this work. Miklas et al. (2003 c), using marker-assisted backcrossing, introgressed the *Co-*4^2 resistance allele into a pinto breeding line USPT-ANT-1. Marker-assisted backcrossing for major-effect QTL linked with SAP6 and SU91 SCAR in combination with intermittent pathogen screening was used by Mutlu et al. (2005 a) and Miklas et al.

Cross

F_1 Resistant parent A 429 (*bgm-1//bgm-1*) × susceptible snap bean cv. (*Bgm-1//Bgm-1*)

BC_1F_1 F_1 (*Bgm-1//bgm-1*) × snap bean cultivar (*Bgm-1//Bgm-1*)

BC_2F_1 BC_1F_1= ½ (*Bgm-1//Bgm-1*): ½ (*Bgm-1//bgm-1*)* × snap bean cv. (*Bgm-1//Bgm-1*)

BC_3F_1 BC_2F_1 = ½ (*Bgm-1//Bgm-1*): ½ (*Bgm-1//bgm-1*)* × snap bean cv. (*Bgm-1//Bgm-1*)

Selfed BC_3F_1= ½ (*Bgm-1//Bgm-1*): ½ (*Bgm-1//bgm-1*)# are selfed to produce BC_3F_2

BC_3F_2=25% resistant (*bgm-1//bgm-1*)§: 50% susceptible (*Bgm-1//bgm-1*): 25% susceptible (*Bgm-1//Bgm-1*)

* indicates BC_nF_1 plants selected by the marker for the next backcross cycle;

# indicates BC_nF_1 plants selected for selfing to produce the BC_nF_2 generation;

§ indicates resistant BC_nF_2 plants selected;

Fig. 2. Marker-assisted backcrossing scheme for incorporating *bgm-1* resistance to BGYMV into snap bean using codominant SCAR marker SR2 (Table 3)

* indicates BC_nF_1 plants selected by marker for next backcross cycle; # indicates BC_nF_1 plants selected by marker for selfing to the BC_nF_2 generation; § indicates resistant BC_nF_2 plants selected by marker.

(2000 b) to develop pinto (Mutlu et al. 2005 b) and kidney bean (Miklas et al. 2005 b) lines with improved resistance to common bacterial blight. Recently, Miklas and Bosak (2004), using a $BC_3F_{2:4}$-derived RI population, showed that marker-assisted backcrossing was effective for moving QTLs for resistance to white mold from G 122 and NY6020-4 into pinto bean. Using a RAPD marker Oliveira et al. (2002) followed the introgression of angular leaf spot resistance alleles into inbred-backcross lines derived from crosses of Ruda with MAR 2 and Mexico 54.

Using markers to detect hypostatic resistance genes in the presence of epistatic genes has been used to retain defeated resistance genes and pyramid genes for more durable resistance to bean rust and BCMV. For development of the BelMiDak-RR-lines (Stavely et al. 1994) with pyramided resistance genes to bean rust, RAPD marker OA14 was used to detect *Ur-4* (Miklas et al. 1993) in the presence of the epistatic *Ur-11* gene. Likewise, Stavely et al. (1998) used a marker to detect *Ur-3* in the presence of *Ur-11* for development of germplasm line Bel-DakMi-RMR-14 with pyramided rust resistance. The retention of genes like *Ur-3* and *Ur-4* that possess resistance to fewer races than the broadly effective *Ur-11* gene is important for attaining durable resistance because these less effective genes have provided resistance against new races that eventually overcame the *Ur-11* gene deployed in rust disease nurseries in Honduras (Mmbaga et al. 1996). Pinto, kidney, and cranberry breeding lines with *I* + *bc-3* combination of resistance genes to BCMV and BCMNV were developed with the assistance of a SCAR marker used to detect the *I* gene in the presence of the epistatic *bc-3* gene (Stavely et al. 1998; Miklas and Kelly 2002; Miklas et al. 2002 a).

Yu et al. (2000 a) used a SCAR marker (BC420) to compare indirect vs. direct selection for common bacterial blight resistance in an Envoy/HR67 population. The cost of selection using the SCAR marker was approximately one third less than that of direct selection. The repulsion AE19 RAPD marker (Johnson et al. 1995) recently converted to a SCAR (Queiroz et al. 2004 c) is being used to obtain the *Ur-11* gene in Carioca market class by selecting against it (Souza et al. 2003). A repulsion marker is used to negatively select for the *bc-3* gene in climbing bean for East Africa (Blair, personal communication, 2003), but this marker cannot be used for selection in multiparent- or backcross-F_1 because all F_1 will possess the repulsion marker and would be selected against. Miklas et al. (1996 a) described a recombination-facilitated MAS strategy to reverse linkage orientation between a resistance gene and linked marker to fit the selection scheme. In this case recombinants with the marker in coupling linkage with the target gene would be identified in a segregating population. The recombinants would then be used to move the recessive gene into the recurrent parent using MAS of F_1 or BCn F_1 that possess the marker.

Schneider et al. (1997 a) used RAPD markers in a biparental population to realize 8% yield gains in nonstressed vs. 11% gains in drought-stressed conditions. Furthermore, they concluded that the effectiveness of MAS for drought resistance was inversely proportional to heritability of yield in drought-stressed conditions. In contrast to the other examples, MAS for drought resistance outside the original mapping population has not been shown.

Although there are considerable numbers of molecular markers available in common bean for both qualitative and quantitative traits, and some are being used routinely in breeding programs (e.g., for BCMV, BCMNV, common bacterial blight, and rust resistance), a great majority of markers remain underutilized or not used at all. Possible reasons are that most QTLs individually have small to moderate effects on traits of interest, that the effectiveness of MAS or simply the presence of markers for a trait has not been confirmed in independent populations, or the effectiveness of markers is gene-pool specific. An example of gene-pool specificity is the OA14 marker for the *Ur-4* gene for rust resistance. OA14 is ubiquitous in the Andean gene pool whether accessions are susceptible or not and thus could only be used for MAS in the Middle American gene pool (Miklas et al. 1993). Similarly, the marker linked to *Bct* is ubiquitous in the Middle American gene pool and can only be used for MAS in Andean germplasm (Larsen and Miklas 2004). The SAP6 marker for common bacterial blight resistance cannot be used effectively in the Mesoamerican race within the Middle American gene pool (Kelly et al. 2003). Other examples of gene-pool specificity of DNA markers linked with useful traits in bean exist (see Kelly and Miklas 1999). Recombination-facilitated MAS (Miklas et al. 1996 a) described above can also be used to overcome gene-pool specificity of markers.

1.5 Future Prospects

Considerable progress has been achieved in improving adaptation, plant type, maturity, seed and pod characteristics, yield, and resistance to major abiotic and biotic stresses in the common bean using conventional genetics and breeding methods and morphological and biochemical markers. The availability of DNA-based markers within the past 20 years has provided new opportunities and challenges to bean researchers. These are being used to characterize genetic variability, fingerprinting, gene function and structure, and MAS. While numerous molecular markers are available, especially for resistance to bacterial, fungal, and viral diseases in common bean, proportionately fewer are routinely used in breeding. Reduced cost, enhanced reliability, and ease of use should increase molecular-marker usage in the future, especially for germplasm enhancement and introgression of a few traits controlled by major genes in otherwise popular cultivars. For improvement of quantitatively inherited traits such as yield and drought resistance, and development of broadly adapted high-yielding cultivars, molecular markers may have to be used in conjunction with conventional genetics and breeding methods.

The availability of saturated high-density linkage maps, cloning of target genes, and reliable transformation systems would be expected to modernize and facilitate common bean germplasm enhancement and cultivar development.

References

Acosta Gallegos JA, Castellanos JZ, Nuñez-Gonzáles S, Ochoa-Márquez R, Rosales-Serna R, Singh SP (1995) Registration of 'Flor de Mayo M38' common bean. Crop Sci 35:941–942

Adam-Blondon AF, Sévignac M, Dron M, Bannerot H (1994a) A genetic map of common bean to localize specific resistance gene against anthracnose. Genome 37:915–924

Adam-Blondon AF, Sevignac M, Bannerot H, Dron M (1994b) SCAR, RAPD, and RFLP markers linked to the dominant gene (*Are*) conferring resistance to anthracnose. Theor Appl Genet 88:865–870

Adams MW (1982) Plant architecture and yield breeding in *Phaseolus vulgaris* L. Iowa State J Res 56:225–254

Alzate-Marin AL, Menarim H, Carvalho GA, Paula Jr TJ, Barros EG, Moreira MA (1999) Improved selection with newly identified RAPD markers linked to resistance gene to four pathotypes of *Colletotrichum lindemuthianum* in common bean. Phytopathology 89:281–285

Alzate-Marin AL, Menarim H, Chagas JM, Barros EG, Moreira MA (2000) Identification of RAPD marker linked to the *Co-6* anthracnose resistant gene in common bean cultivar AB 136. Genet Mol Biol 23:633–637

Alzate-Marin AL, Menarim H, Baía GS, Paula Jr TJ, Souza KA, Costa MR, Barros EG, Moreira MA (2001) Inheritance of anthracnose resistance in the common bean differential cultivar G 2333 and identification of a new molecular marker linked to the *Co-4*^{2} gene. J Phytopathol 149:259–264

Alzate-Marin AL, Costa MR, Arruda KM, Barros EG, Moreira MA (2003) Characterization of the anthracnose resistance gene present in Ouro Negro (Honduras 35) common bean cultivar. Euphytica 133:165–169

Alzate-Marin AL, Souza TLPO, Ragagnin VA, Moreira MA, Barros EG (2004) Allelism tests between the rust resistance gene present in common bean cultivar Ouro Negro and genes *Ur-5* and *Ur-11*. J Phytopathol 152:60–64

Andersen JW, Smith BM, Washnock CS (1999) Cardiovascular and renal benefits of dry bean and soybean intake. Am J Clinical Nutr 70:464S–474S

Ariyarathne HM, Coyne DP, Jung G, Skroch PW, Vidaver AK, Steadman JR, Miklas PN, Bassett MJ (1999) Molecular mapping of disease resistance genes for halo blight, common bacterial blight, and bean common mosaic virus in a segregating population of common bean. J Am Soc Hort Sci 124:654–662

Arndt GC, Gepts P (1989) Segregation and linkage for morphological and biochemical markers in a wide cross in common bean (*Phaseolus vulgaris*). Annu Rep Bean Improv Coop 32:68–69

Arruda MCC, Alzate-Marin AL, Chagas JM, Moreira MA, Barros EG (2000) Identification of RAPD markers linked to the *Co-4* resistance gene to *Colletotrichum lindemuthianum* in common bean. Phytopathology 90:758–761

Arumuganathan K, Earle DE (1991) Nuclear DNA content of some important plant species. Plant Mol Biol Rep 9:208–218

Awale HE, Kelly JD (2001) Development of SCAR markers linked to *Co-4*^{2} gene in common bean. Annu Rep Bean Improv Coop 44:119–120

Bai Y, Michaels TE, Pauls KP (1997) Identification of RAPD markers linked to common bacterial blight resistance genes in *Phaseolus vulgaris* L. Genome 40:544–551

Bassett MJ (1988) Linkage mapping of marker genes in common bean. In: Gepts P (ed) Genetic Resources of *Phaseolus* bean. Kluwer, Dordrecht, pp 329–353

Bassett MJ (1991) A revised linkage map of common bean. HortScience 26:834–836

Bassett MJ (2004) List of genes - *Phaseolus vulgaris* L. Annu Rep Bean Improv Coop 47:1–24

Bassett MJ, Myers JR (1999) Report of BIC genetic committee. Annu Rep Bean Improv Coop 42:vi

Beattie A, Michaels TE, Paul KP (1998) An efficient reliable method to screen for common bacterial blight (CBB) resistance in *Phaseolus vulgaris* L. Annu Rep Bean Improv Coop 41:53–54

Beaver JS, Kelly JD (1994) Comparison of two selection methods for the improvement of dry bean populations derived from crosses between gene pools. Crop Sci 34:34–37

Beaver JS, Rosas JC, Myers J, Acosta J, Kelly JD, Nchimbi-Msolla S, Misangu R, Bokosi J, Temple S, Arnaud-Santana E, Coyne DP (2003) Contributions of the bean/cowpea CRSP to cultivar and germplasm development in common bean. Field Crops Res 82:87–102

Beaver JS, Zapata M, Miklas PN (1999) Registration of PR9443-4 dry bean germplasm resistant to bean golden mosaic, common bacterial blight, and rust. Crop Sci 39:1262

Becerra-Velásquez VL, Gepts P (1994) RFLP diversity of common bean (*Phaseolus vulgaris*) in its centres of origin. Genome 37:256–263

Beebe S, Pedraza F, Rojas M, Gutiérrez J, Tohme J (1998) A genetic map of common bean combining RFLP, RAPD, SCAR, and AFLP markers. Annu Rep Bean Improv Coop 41:95–96

Beebe S, Skroch PW, Tohme J, Duque MC, Pedraza F, Nienhuis J (2000) Structure of genetic diversity among common bean landraces of Middle American origin based on correspondence analysis of RAPD. Crop Sci 40:264–273

Blair MW, Beebe S, Astudillo C, Rengifo J, Giraldo MC, Tohme J, Graham R (2002) Role of nutritional genomics in improving micronutrient content in common beans. In: Proc 1st Intl Conf on Legume Genomics and Genetics: Translation to Crop Improvement 2–6 June 2002, Minneapolis-St Paul, MN

Blair MW, Pedraza F, Buendia HF, Gaitán-Solís E, Beebe SE, Gepts P, Tohme J (2003) Development of a genome-wide anchored microsatellite map for common bean (*Phaseolus vulgaris* L.). Theor Appl Genet 107:1362–1374

Bliss FA, Pereira PAA, Araujo RS, Henson RA, Kmiecik KA, McFerson JR, Teixeira MG, Silva CC (1989) Registration of five high nitrogen-fixing common bean germplasm lines. Crop Sci 29:240–241

Boone WE, Stavely JR, Weeden NF (1999) Development of a sequence-tagged site (STS) marker for *Ur-11*, a gene conferring resistance to the bean rust fungus, *Uromyces appendiculatus*. Annu Rep Bean Improv Coop 42:33–34

Boutin SR, Young ND, Olson TC, Yu ZH, Shoemaker RC, Vallejos CE (1995) Genome conservation among three legume genera detected with DNA markers. Genetics 38:928–937

Brick MA, Grafton KF (1999) Improvement of medium-seeded race Durango cultivars. In: Singh S (ed) Common Bean Improvement in the Twenty-First Century. Kluwer, Dordrecht, pp 223–253

Brick MA, Ogg JB, Schwartz HF, Byrne PF, Kelly JD (2004) Resistance to multiple races of fusarium wilt in common bean. Annu Rep Bean Improv Coop 47:131–132

Brick MA, Schwartz HF, Ogg JB, Johnson JJ, Judson F (2001) Registration of 'Montrose' pinto bean. Crop Sci 41:260

Broughton WJ, Hernández G, Blair M, Beebe S, Gepts P, Vanderleyden J (2003) Beans (*Phaseolus* spp.) - model food legumes. Plant Soil 252:55–128

Brücher H (1988) The wild ancestor of *Phaseolus vulgaris* in South America. In: Gepts P (ed) Genetic Resources of *Phaseolus* Beans. Kluwer, Dordrecht, pp 185–214

Caixeta ET, Borém A, de Azevedo Fagundes S, Niestche S, de Barros EG, Moreira MA (2003) Inheritance of angular leaf spot resistance in common bean line BAT 332 and identification of RAPD markers linked to the resistance gene. Euphytica 134:297–303

Cardona C, Kornegay J, Posso CE, Morales F, Ramírez H (1990) Comparative value of four arcelin variants in the development of dry bean lines resistant to the Mexican bean weevil. Entomol Exp Appl 56:197–206

Carvalho GA, Paula Jr TJ, Alzate-Marin AL, Nietsche S, Barros EG, Moreira MA (1998) Herança da resistência da linhagem AND-277 de feijoeiro-comum à raça 63-23 de *Phaeoisariopsis griseola* e identificação de marcador RAPD ligado ao gene de resistência. Fitopatol Bras 23:482–485

Cheng SS, Bassett MJ (1981) Chromosome morphology in common bean (*Phaseolus vulgaris*) at the diplotene stage of meiosis. Cytology (Tokyo) 46:675–684

Concibido V, Young N, Lange D, Denny R, Danesh D, Orf J (1996) Targeted comparative genome analysis and qualitative mapping of a major partial-resistance gene to the soybean cyst nematode. Theor Appl Genet 93: 234–241

Corrêa RX, Costa MR, Good-God PI, Ragagnin VA, Faleiro FG, Moreira MA, de Barros EG (2000) Sequence characterized amplified regions linked to rust resistance genes in the common bean. Crop Sci 40:804–807

Corrêa RX, Good-God PI, Oliveira MLP, Niestche S, Moreira MA, de Barros EG (2001) Herança da resistência á manchaangular do feijoeiro e identificação de marcadores moleculares flanqueando o loco de resistência. Fitopatol Bras 26:7–32

Coyne DP, Nuland DS, Lindgren DT, Steadman JR, Smith DW, Gonzalcs J, Schild J, Reiser J, Sutton L, Carlson C, Stavely JR, Miklas PN (2000) Weihing great northern disease resistant dry bean. HortScience 35:310–312

Coyne DP, Schuster ML, Hill K (1973) Genetic control of reaction to common blight bacterium in bean (*Phaseolus vulgaris*) as influenced by plant age and bacterial multiplication. J Am Soc Hort Sci 98:94–99

Cregan PB, Jarvik T, Bush AL, Shoemaker RC, Lark KG, Kahler AL, Kaya N, Vantoai TT, Lohnes DG, Chung J, Specht JE (1999) An integrated genetic linkage map of the soybean genome. Crop Sci 39:1464–1490

Deakin MH, Dukes PD (1975) Breeding snap beans for resistance to diseases caused by *Rhizoctonia solani* Kuehn. HortScience 10:269–271

Debouck DG (1999) Diversity in *Phaseolus* species in relation to the common bean. In: Singh SP (ed) Common Bean Improvement in the Twenty-First Century. Kluwer, Dordrecht, pp 25–52

Delaney DE, Bliss FA (1991) Selection for increased percentage protein in common bean. 1. Comparison of selection for seed protein alleles and S_1 family recurrent selection. Theor Appl Genet 81:301–305

Dickson M, Petzoldt R (1986) p gene in beans (*Phaseolus vulgaris* L.): a gene for horizontal mediocrity. In: 22nd Int Hort Congress, Davis, CA. HortScience 21:338 (Abstract 339)

Ender M, Kelly JD (2005) Identification of QTL associated with white mold resistance in common bean. Crop Sci 45:2482–2490

Faleiro FG, Ragagnin VA, Corrêa RX, Vinhadelli WS, Moreira MA, Barros EG (2000 b) Ligação gênica da resistência à ferrugem e à antracnose na variedade de feijão Ouro Negro. Rev Ceres 47:375–382

Faleiro FG, Ragagnin VA, Schuster I, Corrêa RX, Good-God PI, Brommonshenkel SH, Moreira MA, Barros EG (2003) Mapeamento de genes de resistência do feijoeiro à ferrugem, antracnose e mancha-angular usando marcadores RAPD. Fitopatol Brás 28:059–066

Faleiro FG, Vinhadelli WS, Ragagnin VA, Corrêa RX, Moreira MA, Barros EG (2000 a) RAPD markers linked to a block of genes confering rust resistance to the commom bean. Genet Mol Biol 23:399–402

Fall AL, Byrne PF, Jung G, Coyne DP, Brick MA, Schwartz HF (2001) Detection and mapping of a major locus for Fusarium wilt resistance in common bean. Crop Sci 41:1494–1498

Ferreira CF, Borém A, Carvalho GA, Nietsche S, Paula Jr TJ, Barros EG, Moreira MA (2000) Inheritance of angular leaf spot resistance in common bean and identification of a RAPD marker linked to a resistance gene. Crop Sci 40:130–133

Fourie D, Miklas PN, Ariyarathne HM (2004) Genes conditioning halo blight resistance to races 1, 7, and 9 occur in a tight cluster. Annu Rep Bean Improv Coop 47:103–104

Freyre R, Skroch P, Geffroy V, Adam-Blondon AF, Shirmohamadali A, Johnson W, Llaca V, Nodari R, Pereira P, Tsai SM, Tohme J, Dron M, Nienhuis J, Vallejos C, Gepts P (1998) Towards an integrated linkage map of common bean. 4 Development of a core map and alignment of RFLP maps. Theor Appl Genet 97:847–856

Freytag GF, Debouck DG (2002) Taxonomy, distribution, and ecology of the genus *Phaseolus* (Leguminosae-Papilionoideae) in North America, Mexico and Central America. Botanical Research Institute, Fort Worth, TX, USA

Gaitán-Solís E, Duque MC, Edwards KJ, Thome J (2002) Microsatellite repeats in common bean (*Phaseolus vulgaris*): isolation, characterization, and cross-species amplification in *Phaseolus* spp. Crop Sci 42:2128–2136

Geffroy V, Creusot F, Falquet J, Sevignac M, Adam-Blondon AF, Bannerot H, Gepts P, Dron M (1998) A family of LRR sequences at the *Co-2* locus for anthracnose resistance in *Phaseolus vulgaris* and its potential use in marker-assisted selection. Theor Appl Genet 96:494–502

Geffroy V, Sévignac M, De Oliveira J, Fouilloux G, Skroch P, Thoquet P Gepts P, Langin T, Dron M (2000) Inheritance of partial resistance against *Colletotrichum lindemuthianum* in *Phaseolus vulgaris* and co-localization of QTL with genes involved in specific resistance. Mol Plant-Micr Interact 13:287–296

Gepts P (1988 a) Phaseolin as an evolutionary markers. In: Gepts P (ed) Genetic Resources of *Phaseolus* Beans. Kluwer, Dordrecht, pp 215–241

Gepts P (1988 b) Provisional linkage map of common bean. Annu Rep Bean Improv Coop 31:20–25

Gepts P (1988 c) Report of the genetics committee. Annu Rep Bean Improv Coop 31:15

Gepts P (1993) The use of molecular and biochemical markers in crop evolution studies. Evol Biol 27:51–94

Gepts P (1998) Origin and evolution of common bean: past events and recent trends. HortScience 33:1124–1130

Gepts P (1999) Development of an integrated linkage map. In: Singh SP (ed) Common Bean Improvement in the Twenty-First Century. Kluwer, Dordrecht, pp 52–92

Gepts P, Bliss FA (1985) F_1 hybrid weakness in the common bean: differential geographic origin suggests two gene pools in cultivated bean germplasm. J Hered 76:447–450

Gepts P, Bliss FA (1988) Dissemination pathways of common bean (*Phaseolus vulgaris*, Fabaceae) deduced from phaseolin electrophoretic variability. II. Europe and Africa. Econ Bot 42:86–104

Gepts P, Debouck D (1991) Origin, domestication, and evolution of the common bean (*Phaseolus vulgaris* L.). In: Schoonhoven A van, Voysest O (eds) Common Beans: Research for Crop Improvement. CAB, Wallingford, UK and CIAT, Cali, Colombia, pp 7–53

Gepts P, Osborn TC, Rashka K, Bliss FA (1986) Phaseolin protein variability in wild forms and landraces of the common bean (*Phaseolus vulgaris*): evidence for multiple centers of domestication. Econ Bot 40:451–468

Gepts P, Kmiecik K, Pereira P, Bliss FA (1988) Dissemination pathways of common bean (*Phaseolus vulgaris*, Fabaceae) deduced from phaseolin electrophoretic variability. I. The Americas. Econ Bot 42:73–85

Gepts P, Nodari R, Tsai SM, Koinange EMK, Llaca V, Gilbertson R, Guzman P (1993) Linkage mapping in common bean. Annu Rep Bean Improv Coop 36:xxiv-xxxviii

González-Mejía A, Wong A, Delgado-Salinas A, Papa R, Gepts P (2005) Assessment of inter simple sequence repeat markers to differentiate sympatric wild and domesticated populations of common bean (*Phaseolus vulgaris* L.). Crop Sci 45:606–615

Grafton KF, Chang KC, Venette JR, Vander Wal AJ (1993) Registration of 'Norstar' navy bean. Crop Sci 33:1405–1406

Guzmán-Maldonado SH, Martínez O, Acosta JA, Guevara-Lara F, Parades-López O (2003) Putative quantitative trait loci for physical and chemical components of common bean. Crop Sci 43:1029–1035

Haley SD, Afanador L, Kelly JD (1994a) Identification and application of a random amplified polymorphic DNA marker for the *I* gene (potyvirus resistance) in common bean. Phytopathology 84:157–160

Haley SD, Afanador LK, Kelly JD (1994b) Selection for monogenic resistance traits with coupling- and repulsion-phase RAPD markers. Crop Sci 34:1061–1066

Haley SD, Miklas PN, Afanador L, Kelly JD (1994c) Random amplified polymorphic DNA (RAPD) marker variability between and within gene pools of common bean. J Am Soc Hort Sci 119:122–125

Haley SD, Afanador LK, Miklas PN, Stavely JR, Kelly JD (1994d) Heterogeneous inbred populations are useful as sources of near-isogenic lines for RAPD marker localization. Theor Appl Genet 88:337–342

Haley SD, Miklas PN, Stavely JR, Byrum J, Kelly JD (1993) Identification of RAPD markers linked to a major rust resistance gene block in common bean. Theor Appl Genet 86:505–512

Hang AN, Silbernagel MJ, Miklas PN (2005) Release of 'Quincy' pinto dry edible bean. Annu Rep Bean Improv Coop 48:199–200

Harlan JR (1975) Geographic patterns of variation in some cultivated plants. J Hered 66:184–191

He S, Yu Z, Vallejos C, Mackenzie S (1995) Pollen fertility restoration by nuclear gene *Fr* in CMS common bean: an *Fr* linkage map and the mode of *Fr* action. Theor Appl Genet 90:1056–1062

Hosfield GL, Uebersax MA, Occena LG (2000) Technological and genetic improvements in dry bean quality and utilization. In: Singh SP (ed) Bean Reseach, Production and Utilization. Proc Idaho Bean Workshop, University of Idaho, Moscow, ID, pp 135–152

House WA, Welch RM, Beebe S, Cheng Z (2002) Potential for increasing the amounts of bioavailable zinc in dry beans (*Phaseolus vulgaris* L.) through plant breeding. J Sci Food Agric 82:1452–1457

Hulbert SH, Ilott TW, Legg EJ, Lincoln SE, Lander ES, Michelmore RW (1988) Genetic analysis of the fungus, *Bremia lactucae*, using restriction fragment length polymorphisms. Genetics 120:947–958

Ibarra-Pérez FJ, Ehdaie B, Waines JG (1997) Estimation of outcrossing rate in common bean. Crop Sci 37:60–65

Ibarra-Pérez FJ, Acosta-Gallegos JA, Cazares Enriquez B, Castillo-Rosales A, Rosales-Serna R, Kelly JD, Singh SP (2004) Registration of 'Negro Vizcaya' shiny black bean. Crop Sci 44:1866–1867

Johnson WC, Gepts P (1999) Segregation for performance in recombinant inbred populations resulting from intergene pool crosses of common bean (*Phaseolus vulgaris* L). Euphytica 106:45–56

Johnson E, Miklas PN, Stavely JR, Martínez-Cruzado JC (1995) Coupling- and repulsion-phase RAPDs for marker-assisted selection of PI 181996 rust resistance in common bean. Theor Appl Genet 90:659–664

Johnson WC, Menéndez C, Nodari R, Koinange EMK, Magnusson S, Singh SP, Gepts P (1996) Association of a seed weight factor with the phaseolin seed storage protein locus across genotypes, environments, and genomes in *Phaseolus-Vigna* spp.: Sax (1923) revisited. J Agric Genom 2 (previously J Quant Trait Loci): http://www.cabi-publishing.org/jag/papers96/paper596/indexp596.html

Johnson WC, Guzmán P, Mandala D, Mkandawire ABC, Temple S, Gilbertson RL, Gepts P (1997) Molecular tagging of the *bc-3* gene for introgression into Andean common bean. Crop Sci 37:248–254

Jung G, Coyne DP, Skroch P, Nienhuis J, Arnaud-Santana E, Bokosi J, Ariyarathne H, Steadman J, Beaver J, Kaeppler S (1996) Molecular markers associated with plant architecture and resistance to common blight, web blight, and rust in common beans. J Am Soc Hort Sci 121:794–803

Jung G, Skroch P, Coyne DP, Nienhuis J, Ariyarathne H, Kaeppler S, Bassett M (1997) Molecular-marker-based genetic analysis of tepary-bean-derived common bacterial blight resistance in different developmental stages of common bean. J Am Soc Hort Sci 122:329–337

Jung G, Coyne DP, Bokosi JM, Steadman JR, Nienhuis J (1998) Mapping genes for specific and adult plant resistance to rust and abaxial leaf pubescence and their genetic relationship using random amplified polymorphic DNA (RAPD) markers in common bean. J Am Soc Hort Sci 123:859–863

Jung G, Skroch PW, Nienhuis J, Coyne DP, Arnaud-Santana E, Ariyarathne HM, Marita JM (1999) Confirmation of QTL associated with common bacterial blight resistance in four different genetic backgrounds in common bean. Crop Sci 39:1448–1455

Jung G, Ariyarathne HM, Coyne DP, Nienhuis J (2003) Mapping QTL for bacterial brown spot resistance under natural infection in field and seedling stem inoculation in growth chamber in common bean. Crop Sci 43:350–357

Kalavacharla V, Stavely JR, Myers JR, McClean PE (2000) *Crg*, a gene required for *Ur-3*-mediated rust resistance in common bean, maps to a resistance gene analog cluster. Mol Plant-Micr Interact 13:1237–1242

Kami J, Gepts P (2000) Development of a BAC library in common bean genotype BAT93. Annu Rep Bean Improv Coop 43:208–209

Kami J, Velasquez VB, Debouck DG, Gepts P (1995) Identification of presumed ancestral DNA sequences of phaseolin in *Phaseolus vulgaris*. Proc Natl Acad Sci USA 92:1101–1104

Kelly JD, Adams MW (1987) Phenotypic recurrent selection in ideotype breeding of pinto beans. Euphytica 36:69–80

Kelly JD, Miklas PN (1999) Marker-assisted selection. In: Singh SP (ed) Common Bean Improvement in the Twenty-First Century. Kluwer Academic, The Netherlands, pp 93–123

Kelly JD, Vallejo VA (2004 a) A comprehensive review of the major genes conditioning resistance to anthracnose in common bean. HortScience 39:1196–1207

Kelly JD, Vallejo VA (2004 b) QTL analysis of multigenic disease resistance in plant breeding. In: Tuzun S, Bent E (eds) Multigenic and Induced Systemic Resistance in Plants (in press)

Kelly JD, Adams MW, Saettler AW, Hosfield GS, Varner GV, Beaver JS, Uebersax MA, Taylor J (1989) Registration of 'Mayflower' navy bean. Crop Sci 29:1571–1572

Kelly JD, Hosfield GL, Varner GV, Uebersax MA, Brothers ME, Taylor J (1994 a) Registration of 'Huron' navy bean. Crop Sci 34:1408

Kelly JD, Hosfield GL, Varner GV, Uebersax MA, Haley SD, Taylor J (1994 b) Registration of 'Raven' black bean. Crop Sci 34:1406–1407

Kelly JD, Hosfield GL, Varner GV, Uebersax MA, Taylor J (1999 a) Registration of 'Kodiak' pinto bean. Crop Sci 39:292–293

Kelly JD, Hosfield GL, Varner GV, Uebersax MA, Taylor J (1999 b) Registration of 'Matterhorn' great northern bean. Crop Sci 39:589–590

Kelly JD, Hosfield GL, Varner GV, Uebersax MA, Taylor J (1999 c) Registration of 'Chinook 2000' light red kidney bean. Crop Sci 39:293

Kelly JD, Hosfield GL, Varner GV, Uebersax MA, Taylor J (2000) Registration of 'Phantom' black bean. Crop Sci 40:572

Kelly JD, Hosfield GL, Varner GV, Uebersax MA, Taylor J (2001) Registration of 'Jaguar' black bean. Crop Sci 41:1447–1448

Kelly JD, Gepts P, Miklas PN, Coyne DP (2003) Tagging and mapping of genes and QTL and molecular marker-assisted selection for traits of economic importance in bean and cowpea. Field Crop Res 82:135–154

Khairallah MM, Adams MW, Sears BB (1990) Mitochondrial DNA polymorphisms of Malawian bean lines: further evidence for two major gene pools. Theor Appl Genet 80:753–761

Koenig R, Gepts P (1989) Allozyme diversity in wild *Phaseolus vulgaris*: further evidence for two major centers of genetic diversity. Theor Appl Genet 78:809–817

Koenig R, Singh SP, Gepts P (1990) Novel phaseolin types in wild and cultivated common bean (*Phaseolus vulgaris*, Fabaceae). Econ Bot 44:50–60

Koinange EMK, Gepts P (1992) Hybrid weakness in wild *Phaseolus vulgaris* L. J Hered 83:135–139

Koinange EMK, Singh SP, Gepts P (1996) Genetic control of the domestication syndrome in common-bean. Crop Sci 36:1037–1045

Kolkman JM, Kelly JD (2003) QTL conferring resistance and avoidance to white mold in common bean. Crop Sci 43:539–548

Kyle MM, Dickson MH (1988) Linkage of hypersensitivity to five viruses with the *B* locus in *Phaseolus vulgaris* L. J Hered 79:308–311

Lamprecht H (1961) Weitere Kopplungsstudien an *Phaseolus vulgaris* mit einer Übersicht über die Koppelungsgruppen. Agric Hort Genet 19:333–343

Larsen RC, Miklas PN (2004) Generation and molecular mapping of a SCAR marker linked with the *Bct* gene for resistance to *Beet curly top virus* in common bean. Phytopathology 94:320–325

López CE, Acosta IF, Jara C, Pedraza F, Gaitán-Solís E, Gallego G, Beebe S, Tohme J (2003) Identifying resistance gene analogs associated with resistances to different pathogens in common bean. Phytopathology 93:88–95

Mahuku G, Montoya C, Henríquez MA, Jara C, Teran H, Beebe S (2004) Inheritance and characterization of angular leaf spot resistance gene present in common bean accession G 10474 and identification of an AFLP marker linked to the resistance gene. Crop Sci 44:1817–1824

Mayek-Pérez N, López-Castaneda C, López-Salinas E, Acosta-Gallegos JA (2001) Inheritance of genetic resistence to *Macrophomina phaseolina* (Tassi) Goid in common bean. Agrociencia 35:637–648

McClean PE, Lee RK, Otto C, Gepts P, Bassett MJ (2002) Molecular and phenotypic mapping of genes controlling seed coat pattern and color in common bean (*Phaseolus vulgaris* L.). J Hered 93:148–152

McClean PE, Lee RK, Miklas PN (2004) Sequence diversity analysis of dihydroflavonol 4-reductase intron 1 in common bean. Genome 47:266–280

McElroy JB (1985) Breeding for dry beans, *P. vulgaris* L., for common bacterial blight resistance derived from *Phaseolus acutifolius* A. Gray. PhD dissertation, Cornell University, Ithaca, NY

McMillan Jr RT, Davis MJ, McLaughlin HJ, Stavely JR (1998) PCR evaluation of fourteen bean golden mosaic virus (BGMV) resistant snap bean germplasm lines for the presence of the virus. Annu Rep Bean Improv Coop 41:31–32

Mejía-Jiménez A, Muñoz C, Jacobsen HJ, Roca WM, Singh SP (1994) Interspecific hybridization between common and tepary beans: increased hybrid embryo growth, fertility, and efficiency of hybridization through recurrent and congruity backcrossing. Theor Appl Genet 88:324–331

Melotto M, Kelly JD (1998) SCAR markers linked to major disease resistance genes in common bean. Annu Rep Bean Improv Coop 41:64–65

Melotto M, Afanador L, Kelly JD (1996) Development of a SCAR marker linked to the *I* gene in common bean. Genome 39:1216–1219

Mendez de Vigo B, Rodriguez C, Paneda A, Giraldez R, Ferreira JJ (2002) Development of a SCAR marker linked to *Co-9* in common bean. Annu Rep Bean Improv Coop 45:116–117

Michelmore RW, Meyers BC (1998) Clusters of resistance genes in plants evolve by divergent selection and a birth-and-death process. Genome Res 8:1113–1130

Mienie CMS, Liebenberg MM, Pretorius ZA, Miklas PN (2005) SCAR markers linked to the *Phaseolus vulgaris* rust resistance gene *Ur-13*. Theor Appl Genet 111:972–979

Mienie CMS, Naidoo R, Liebenberg MM (2004) Conversion of the RAPD marker for *Ur-4* to a co-dominant SCAR marker $SA14_{1079/800}$. Annu Rep Bean Improv Coop 47:261–262

Miklas PN, Kelly JD (2002) Registration of two cranberry bean germplasm lines resistant to bean common mosaic and necrosis potyviruses: USCR-7 and USCR-9. Crop Sci 42:673–674

Miklas PN, Bosak KM (2004) Marker-assisted backcrossing of QTL for resistance to Sclerotinia white mold in pinto bean. Agron Abstr, Madison, WI

Miklas PN, Stavely JR, Kelly JD (1993) Identification and potential use of a molecular marker for rust resistance in common bean. Theor Appl Genet 85:745–749

Miklas PN, Zapata M, Beaver JS, Grafton KF (1994) Registration of four dry bean germplasm resistant to common bacterial blight: ICB-3, ICB-6, ICB-8, and ICB-10. Crop Sci 39:594

Miklas PN, Afanador L, Kelly JD (1996 a) Recombination-facilitated RAPD marker-assisted selection for disease resistance in common bean. Crop Sci 36:86–90

Miklas PN, Johnson E, Stone V, Beaver JS, Montoya C, Zapata M (1996 b) Selective mapping of QTL conditioning disease resistance in common bean. Crop Sci 36:1344–1351

Miklas PN, Grafton KF, Kelly JD, Steadman JR, Silbernagel MJ (1998 a) Registration of four white mold resistant dry bean germplasm lines: I9365-3, I9365-5, I9365-31, and 92BG-7. Crop Sci 38:1728

Miklas PN, Stone V, Urrea CA, Johnson E, Beaver JS (1998 b) Inheritance and QTL analysis of field resistance to ashy stem blight. Crop Sci 38:916–921

Miklas PN, Delorme R, Stone V, Daly MJ, Stavely JR, Steadman JR, Bassett MJ, Beaver JS (2000 a) Bacterial, fungal, and viral disease resistance loci mapped in a recombinant inbred common bean population ('Dorado/Xan176'). J Am Soc Hort Sci 125:476–481

Miklas PN, Larsen RC, Riley R, Kelly JD (2000 b) Potential marker-assisted selection for *bc-1*^{2} resistance to bean common mosaic potyvirus in common bean. Euphytica 116:211–219

Miklas PN, Smith JR, Riley R, Grafton KF, Singh SP, Jung G, Coyne DP (2000 c) Marker-assisted breeding for pyramided resistance to common bacterial blight in common bean. Annu Rep Bean Improv Coop 43:39–40

Miklas PN, Stone V, Daly MJ, Stavely JR, Steadman JR, Bassett MJ, Delorme R, Beaver JS (2000 d) Bacterial, fungal, and viral disease resistance loci mapped in a recombinant inbred common bean population ('Dorado'/XAN 176). J Am Soc Hort Sci 125:476–481

Miklas PN, Johnson WC, Delorme R, Gepts P (2001) QTL conditioning physiological resistance and avoidance to white mold in dry bean. Crop Sci 41:309–315

Miklas PN, Hang AN, Kelly JD, Strausbaugh CA, Forster RL (2002 a) Registration of three kidney bean germplasm lines resistant to bean common mosaic and necrosis potyviruses: USLK-2 light red kidney, USDK-4 dark red kidney, and USWK-6 white kidney. Crop Sci 42:674–675

Miklas PN, Pastor-Corrales MA, Jung G, Coyne DP, Kelly JD, McClean PE, Gepts P (2002 b) Comprehensive Linkage map of bean rust resistance genes. Annu Rep Bean Improv Coop 45:125–129

Miklas PN, Coyne DP, Grafton KF, Mutlu N, Reiser J, Lindgren DT, Singh SP (2003 a) A major QTL for common bacterial light resistance derives from the common bean great northern landrace cultivar Montana No. 5. Euphytica 131:137–146

Miklas PN, Delorme R, Riley R (2003 b) Identification of QTL conditioning resistance to white mold in snap bean. J Am Soc Hort Sci 128:564–570

Miklas PN, Kelly JD, Singh SP (2003 c) Registration of anthracnose-resistant pinto bean germplasm line USPT-ANT-1. Crop Sci 43:1889

Miklas PN, Grafton KF, Bosak KM (2005 a) Molecular breeding for resistance to white mold in common bean. In: 3rd USDA-ARS, Sclerotinia Initiative Workshop, 19–20 January 2005, Minneapolis, MN

Miklas PN, Kelly JD, Beebe SE, Blair MW (2006) Common bean breeding for resistance against biotic and abiotic stresses: from classical to MAS breeding. Euphytica 147:105–131

Miklas PN, Smith JR, Singh SP (2005 b) Release of USDK-CBB-15 dark red kidney bean germplasm line with improved resistance to common bacterial blight. Annu Rep Bean Improv Coop 48:192–193

Mmbaga M, Steadman JR, Stavely JR (1996) The use of host resistance in disease management of rust in common bean. Integr Pest Mgt Rev 1:191–200

Mok DWS, Mok MC (1977) Monosomics in common bean, *Phaseolus vulgaris*. Theor Appl Genet 49:145–149

Morales FJ, Singh SP (1993) Breeding for resistance to bean golden mosaic virus in an interracial population of *Phaseolus vulgaris* L. Euphytica 67:59–63

Moscone EA, Klein F, Lambrou M, Fuchs J, Schweizer D (1999) Quantitative karyotyping and dual-color FISH mapping of 5S and 18S-25S rDNA probes in the cultivated *Phaseolus* species (Leguminosae). Genome 42:1224–1233

Murray JD, Michaels TE, Cardona C, Schaafsma AW, Pauls KP (2004) Quantitative trait loci for leafhopper (*Empoasca fabae* and *Empoasca kraemeri*) resistance and seed weight in the common bean. Plant Breed 123:474–479

Murray J, Larsen J, Michaels TE, Schaafsma A, Vallejos CE, Pauls KP (2002) Identification of putative genes in bean (*Phaseolus vulgaris*) genomic (Bng) RFLP clones and their conversion to STSs. Genome 45:1013–1024

Mutlu N, Miklas P, Reiser J, Coyne D (2005a) Backcross breeding for improved resistance to common bacterial blight in pinto bean (*Phaseolus vulgaris* L.). Plant Breed 124:282–287

Mutlu N, Miklas PN, Steadman JR, Vidaver AV, Lindgren D, Reiser J, Pastor-Corrales MA (2005b) Registration of common bacterial blight resistant pinto bean germplasm line ABCP-8. Crop Sci 45:806–807

Myers JR (2000) Tomorrow's snap bean cultivars. In: Singh SP (ed) Bean research, production and utilization. In: Proc Idaho Bean Workshop Univ of Idaho, Moscow, ID, pp 39–51

Myers JR, Weeden NF (1988) A proposed revision of guidelines for genetic analysis in *Phaseolus vulgaris* L. Annu Rep Bean Improv Coop 31:16–19

Myers JR, Baggett JR (1999) Improvement of snap bean. In: Singh SP (ed) Common Bean Improvement in the Twenty-First Century. Kluwer Publ, Dordrecht, pp 289–329

Myers JR, Davis J, Yorgey B, Kean D (2004) Mapping quantitative trait loci for green bean traits of horticultural importance. Annu Rep Bean Improv Coop 47:75–76

Nagata RT, Bassett MJ (1984) Linkage relationships of nine induced mutants in common bean. J Am Soc Hort Sci 109:517–519

Navarro F, Sass M, Nienhuis J (2003) Identification and mapping bean root rot resistance in a population of Mesoamerican×Andean origin. Annu Rep Bean Improv Coop 46:213–214

Nemchinova YP, Stavely JR (1998) Development of SCAR primers for the *Ur-3* rust resistance gene in common bean. Phytopathology 88:S67

Nienhuis J, Singh SP (1986) Combining ability analyses and relationships among yield, yield components, and architectural traits in dry bean. Crop Sci 26:21–27

Nietsche S, Borém A, Carvalho GA, Rocha RC, Paula Jr TJ, De Barros EG, Moreira MA (2000) RAPD and SCAR markers linked to a gene conferring resistance to angular leaf spot in common bean. J Phytopathol 148:117–121

Nodari RO, Tsai SM, Gilbertson RL, Gepts P (1993a) Towards an integrated linkage map of common bean. II. Development of an RFLP-based linkage map. Theor Appl Genet 85:513–520

Nodari RO, Tsai SM, Guzmán P, Gilbertson RL, Gepts P (1993b) Towards an integrated linkage map of common bean. 3. Mapping genetic factors controlling host-bacterium interactions. Genetics 134:341–350

Olaya G, Abawi GS, Weeden NF (1996) Inheritance of the resistance to *Macrophomina phaseolina* and identification of RAPD markers linked to the resistance genes in beans. Phytopathology 86:674–679

Oliveira EJ de, Alzate-Marin AL, de Melo CLP, Borém A, de Barros EG, Moreira MA (2002) Backcross assisted by RAPD markers for the introgression of angular leaf spot resistance genes in common bean cultivars. Annu Rep Bean Improv Coop 45:142–143

Osborn TC, Blake T, Gepts P, Bliss FA (1986) Bean arcelin. 2. Genetic variation, inheritance and linkage relationships of a novel seed protein of *Phaseolus vulgaris* L. Theor Appl Genet 71:847–855

Park SJ, Tu JC (1986) Association between BCMV resistant *I* gene and eye color of cv. Steubean. Annu Rep Bean Improv Coop 29:4–5

Park SO, Coyne DP, Steadman JR (2004a) Development of a SCAR marker linked to the *Ur-7* gene in common bean. Annu Rep Bean Improv Coop 47:269–270

Park SO, Coyne DP, Steadman JR, Crosby KM, Brick MA (2004b) RAPD and SCAR markers linked to the *Ur-6* Andean gene controlling specific rust resistance in common bean. Crop Sci 44:1799–1807

Park SO, Coyne DP, Steadman JR, Skroch PW (2001) Mapping of QTL for resistance to white mold diseases in common bean. Crop Sci 41:1253–1262

Park SO, Coyne DP, Steadman JR, Skroch PW (2003) Mapping of the *Ur-7* gene for specific resistance to rust in common bean. Crop Sci 43:1470–1476

Pastor-Corrales MA (2003) Sources, genes for resistance, and pedigree of 52 rust amd mosaic resistant dry bean germplasm lines released by the USDA Beltsville bean project in collaboration with the Michigan, Nebraska and North Dakota Agricultural Experiment Stations. Annu Rep Bean Improv Coop 46:235–241

Pastor-Corrales MA, Stavely JR, Kelly JD, Grafton KF, Steadman JR, Coyne DP, Lindgren DT, Scully BT (2001) Rust and mosaic resistant bean germplasm releases, 1997–1999. Annu Rep Bean Improv Coop 44:101–102

Pedraza F, Gallego G, Beebe S, Tohme J (1997) Marcadores SCAR y RAPD para la resistencia a la bacteriosis comun (CBB). In: Singh SP, Voysest O (eds) Taller de mejoramiento de frijol para el siglo XXI: Bases para una estrategia para America Latina. CIAT, Cali, Colombia, USA, pp 130–134

Pedrosa A, Vallejos CE, Bachmair A (2003) Integration of common bean (*Phaseolus vulgaris* L.) linkage and chromosomal maps. Theor Appl Genet 106:205–212

Pereira PAA, Miranda BD, Attewell JR, Kmiecik KA, Bliss FA (1993) Selection for increased nodule number in common bean (*Phaseolus vulgaris* L.). Plant Soil 148:203–209

Pompeu AS (1982) Catu, Aete-3, Aroana-80, Moruna-80, Carioca-80 e Ayso: novos cultivares de feijoeiro. Bragantia 41:213–218

Porch TG (2001) Genetics and applications of heat tolerance in common bean. PhD dissertation, Cornell University, Ithaca, NY

Posa-Macalincag MCT, Hosfield GL, Grafton KF, Uebersax MA, Kelly JD (2002) Quantitative trait loci (QTL) analysis of canning quality traits in kidney bean (*Phaseolus vulgaris* L.). J Am Soc Hort Sci 127:608–615

Queiroz VT, Sousa CS, Costa MR, Sanglad DA, Arruda KMA, Souza TLPO, Ragagnin VA, Barros EG, Moreira MA (2004 a) Development of SCAR markers linked to common bean anthracnose resistance genes *Co-4* and *Co-6*. Annu Rep Bean Improv Coop 47:249–250

Queiroz VT, Sousa CS, Costa MR, Sanglad DA, Arruda KMA, Souza TLPO, Ragagnin VA, Barros EG, Moreira MA (2004 b) Development of SCAR markers linked to common bean angular leaf spot resistance genes. Annu Rep Bean Improv Coop 47:237–238

Queiroz VT, Sousa CS, Souza TLPO, Costa MR, Sanglad DA, Ragagnin VA, Barros EG, Moreira MA (2004 c) SCAR marker linked to the common bean rust resistance gene *Ur-11*. Annu Rep Bean Improv Coop 47:271–272

Rivkin ML, Vallejos CE, McClean PE (1999) Disease resistance related sequences in common bean. Genome 42:41–47

Román-Avilés B, Kelly JD (2005) Identification of QTL conditioning resistance to Fusarium root rot in *Phaseolus vulgaris* L. J Am Soc Hort Sci (accepted)\CEnote{Please update}

Rosales-Serna R, Ramírez-Vallejo P, Acosta-Gallegos JA, Castillo-González F, Kelly JD (2000) Grain yield and drought tolerance of common bean under field conditions. Agrociencia 34:153–165

Sax K (1923) The association of size differences with seed coat pattern and pigmentation in *Phaseolus vulgaris*. Genetics 8:552–560

Sanchez-Valdez I, Acosta-Gallegos JA, Ibarra-Pérez FJ, Rosales-Serna R, Singh SP (2004) Registration of 'Pinto Saltio' common bean. Crop Sci 44:1865–1866

Sartorato A, Nietsche S, Barros EG, Moreira MA (2000) RAPD and SCAR markers linked to resistance gene to angular leaf spot in common beans. Fitopatol Bras 25:637–642

Schneider K, Brothers M, Kelly JD (1997 a) Marker-assisted selection to improve drought resistance in common bean. Crop Sci 37:51–60

Schneider KA, Rosales-Serna R, Ibarra-Perez F, Cazares-Enriquez B, Acosta-Gallegos JA, Ramírez-Vallejo P, Wassimi N, Kelly JD (1997 b) Improving common bean performance under drought stress. Crop Sci 37:43–50

Schneider KA, Grafton KF, Kelly JD (2001) Genetic and QTL analysis of resistance to Fusarium Root rot in bean. Crop Sci 41:535–542

Scott ME, Michaels TE (1992) *Xanthomonas* resistance of *Phaseolus* interspecific cross selections confirmed by field performance. HortScience 27:348–350

Silbernagel MJ (1994) Release of pinto breeding line 92US-1006 with *I bc2*^{2} resistance to bean common mosaic virus. Annu Rep Bean Improv Coop 37:246

Silva LO, Araujo Moraes E, Aidar H, Thung MDT, Gutiérrez JA, Terán H, Morales FJ, Pastor-Corrales MA, Schwartz HF, Singh SP (2003) Registration of 'EMGOPA 201-Ouro' common bean. Crop Sci 43:1881

Singh SP (1982) A key for identification of different growth habits of frijol *Phaseolus vulgaris* L. Annu Rep Bean Improv Coop 25:92–95

Singh SP (1989) Patterns of variation in cultivated common bean (*Phaseolus vulgaris*, Fabaceae). Econ Bot 43:39–57

Singh SP (1992) Common bean improvement in the tropics. Plant Breed Rev 10:199–269

Singh SP (1994) Gamete selection for simultaneous improvement of multiple traits in common bean. Crop Sci 34:352–355

Singh SP (1995) Selection for water-stress tolerance in interracial populations of common bean. Crop Sci 35:118–124

Singh SP (ed) (1999 a) Common Bean Improvement in the Twenty-First Century. Kluwer, Dordrecht

Singh SP (1999 b) Production and utilization. In: Singh SP (ed) Common Bean Improvement in the Twenty-First Century. Kluwer, Dordrecht, pp 1–24

Singh SP (ed) (2000) Bean research, production, and utilization. In: Proc Idaho Bean Workshop, University of Idaho, Moscow, ID

Singh SP (2001 a) Broadening the genetic base of common bean cultivars: a review. Crop Sci 41:1659–1675

Singh SP (2001 b) The common bean and its genetic improvement. In: Kang MS (ed) Crop Improvement: Challenges in the 21st Century. Food Products Press, New York, pp 161–192

Singh SP (2005) Common Bean (*Phaseolus vulgaris* L.). In: Singh RJ, Jauhar PP (eds) Genetic Resources, Chromosome Engineering, and Crop Improvement Series I – Grain Legumes. CRC, Boca Raton, FL, pp 11–48

Singh SP, Gutiérrez JA (1984) Geographical distribution of *DL1* and *DL2* genes causing hybrid dwarfism in *Phaseolus vulgaris* L., their association with seed size, and their significance to breeding. Euphytica 33:337–345

Singh SP, Muñoz CG (1999) Resistance to common bacterial blight among *Phaseolus* species and common bean improvement. Crop Sci 39:80–89

Singh SP, Cajiao C, Gutiérrez JA, García J, Pastor-Corrales MA, Morales FJ (1989) Selection for seed yield in inter-gene pool crosses of common bean. Crop Sci 29:1126–1131

Singh SP, Gepts P, Debouck DG (1991 a) Races of common bean (*Phaseolus vulgaris*, Fabaceae). Econ Bot 45:379–396

Singh SP, Gutiérrez JA, Molina A, Urrea C, Gepts P (1991 b) Genetic diversity in cultivated common bean. II. Marker-based analysis of morphological and agronomic traits. Crop Sci 31:23–29

Singh SP, Molina A, Urrea CA, Gutiérrez JA (1993) Use of interracial hybridization in breeding the race Durango common bean. Can J Plant Sci 73:785–793

Singh SP, Molina A, Gepts P (1995) Potential of wild common bean for seed yield improvement of cultivars in the tropics. Can J Plant Sci 75:807–813

Singh SP, Cardona C, Morales FJ, Pastor-Corrales MA, Voysest O (1998) Gamete selection for upright carioca bean with resistance to five diseases and a leafhopper. Crop Sci 38:666–672

Singh SP, Morales FJ, Miklas PN, Terán H (2000) Selection for bean golden mosaic resistance in intra- and inter-racial bean populations. Crop Sci 40:1565–1572

Singh SP, Nodari R, Gepts P (1991 c) Genetic diversity in cultivated common bean. I. Allozymes. Crop Sci 31:19–23

Singh SP, Terán H, Muñoz CG, Takegami JC (1999) Two cycles of recurrent selection for seed yield in common bean. Crop Sci 39:391–397

Singh SP, Terán H, Gutiérrez JA, Pastor-Corrales MA, Schwartz HF, Morales FJ (2003 a) Registration of A 339, MAR 1, MAR 2, and MAR 3 angular leaf spot and anthracnose resistant common bean germplasm. Crop Sci 43:1886–1887

Singh SP, Terán H, Muñoz CG, Osorno JM, Takegami JC, Thung MDT (2003 b) Low soil fertility tolerance in landraces and improved common bean genotypes. Crop Sci 43:110–119

Skroch P, Jung G, Nienhuis J, Coyne D (1996) Integration of RAPD marker linkage maps and comparative mapping of QTL for disease resistance in common bean. Annu Rep Bean Improv Coop 39:48–49

Sonnante G, Stockton T, Nodari RO, Becerra Velásquez VL, Gepts P (1994) Evolution of genetic diversity during the domestication of common-bean (*Phaseolus vulgaris* L.). Theor Appl Genet 89:629–635

Souza TLPO, Ragagnin VA, Alzate-Marin AL, Faleiro FG, Moreira MA, Barros EG (2003) Backcross assisted by RAPD markers to develop common bean lines with carioca type grains containing the *Ur-11* rust resistance gene. Annu Rep Bean Improv Coop 46:195–196

Stavely JR (1984) Pathogenic specialization in *Uromyces phaseoli* in the United States and rust resistance in beans. Plant Dis 68:95–99

Stavely JR, Kelly JD, Grafton KF (1994) BelMiDak-rust-resistant navy dry bean germplasm lines. HortScience 29:709–710

Stavely JR, Grafton KF, Kelly JD (1998) Release of BelDakMi-RMR-14 erect, short vine, rust, and mosaic resistant pinto bean germplasm line. USDA-ARS Germplasm Release Notice, Beltsville, MD, p 3

Stewart-Williams KD, Myers JR, Dennis MF, Hayes R, Strausbaugh C, Singh SP (2003) Registration of great northern common bean germplasm UI98-209G. Crop Sci 43:2312–2313

Strausbaugh CA, Myers JR, Forster RL, McClean PE (1999) *Bc-1* and *Bc-u* two loci controlling bean common mosaic virus resistance in common bean are linked. J Am Soc Hort Sci 124:644–648

Tar'an B, Michaels TE, Pauls KP (2001) Mapping genetic factors affecting the reaction to *Xanthomonas axonopodis* pv. *phaseoli* in *Phaseolus vulgaris* L. under field conditions. Genome 44:1046–1056

Tar'an B, Michaels TE, Pauls KP (2002) Genetic mapping of agronomic traits in common bean. Crop Sci 42:544–556

Temple SR, Morales FJ (1986) Linkage of dominant hypersensitive resistance to bean common mosaic virus to seed color in *Phaseolus vulgaris* L. Euphytica 35:331–333

Terán H, Singh SP (2002) Comparison of sources and lines selected for drought resistance in common bean. Crop Sci 42:64–70

Teverson DM (1991) Genetics of pathogenicity and resistance in the halo-blight disease of beans in Africa. PhD dissertation, University of Birmingham, UK

Urrea CA, Singh SP (1994) Comparison of mass, F_2-derived family, and single-seed - descent selection methods in an interracial population of common bean. Can J Plant Sci 74:461–464

Urrea CA, Singh SP (1995) Comparison of recurrent and congruity backcrossing for interracial hybridization in common bean. Euphytica 81:21–26

Urrea CA, Miklas PN, Beaver JS, Riley RH (1996) A codominant randomly amplified polymorphic DNA (RAPD) marker useful for indirect selection of BGMV resistance in common bean. J Am Soc Hort Sci 121:1035–1039

Valladares-Sanchez NE, Coyne DP, Schuster ML (1979) Differential reaction of leaves and pods of *Phaseolus* germplasm to strains of *Xanthomonas phaseoli* and transgressive segregation for tolerance from crosses of susceptible germplasm. J Am Soc Hort Sci 104:648–654

Vallejo V, Kelly JD (2001) Development of a SCAR marker linked to *Co-5* gene in common bean. Annu Rep Bean Improv Coop 44:121–122

Vallejo V, Kelly JD (2002) The use of AFLP analysis to tag the *Co1*^{2} gene conditioning resistance to bean anthracnose. In: Plant & Animal Genome X Conf, San Diego. http://www.pag-intl.org

Vallejos CE (1994) *Phaseolus vulgaris* – The common bean. In: Phillips RL, Vasil IK (eds) DNA-Based Markers in Plants. Kluwer, Dordrecht, pp 261–270

Vallejos CE, Chase CD (1991 a) Linkage between isozyme markers and a locus affecting seed size in *Phaseolus vulgaris* L. Theor Appl Genet 81:413–419

Vallejos CE, Chase CD (1991 b) Extended map for the phaseolin linkage group of *Phaseolus vulgaris* L. Theor Appl Genet 82:353–357

Vallejos CE, Sakiyama NS, Chase CD (1992) A molecular marker-based linkage map of *Phaseolus vulgaris* L. Genetics 131:733–740

Vallejos CE, Skroch PW, Nienhuis J (2001) *Phaseolus vulgaris* – The common bean integration of RFLP and RAPD-based linkage maps. In: Phillips RL, Vasil IK (eds) DNA-Based Markers in Plants, 2nd edn. Kluwer, Dordrecht, pp 301–317

Voysest VO (2000) Mejoramiento genético del frijol (*Phaseolus vulgaris* L.): Legado de variedades de América Latina 1930–1999. CIAT, Cali, Colombia

Walters KJ, Hosfield GL, Uebersaz MA, Kelly JD (1997) Navy bean canning quality: Correlations, heritability estimates, and randomly amplified polymorphic DNA markers associated with component traits. J Am Soc Hort Sci 122:238–343

Weeden NF (1984) Distinguishing among white seeded bean cultivars by means of allozyme genotypes. Euphytica 33:199–208

Weeden NF, Liang CY (1985) Detection of a linkage between white flower color and EST-2 in common bean. Annu Rep Bean Improv Coop 28:87–88

Welch RM, House WA, Beebe S, Cheng Z (2000) Genetic selection for enhanced bioavailable levels of iron in bean (*Phaseolus vulgaris* L.) seeds. J Agric Food Chem 48:3576–3580

White J, Laing DR (1989) Photoperiod response of flowering in diverse genotypes of common bean (*Phaseolus vulgaris*). Field Crops Res 22:113–128

White JW, Kornegay JL, Castillo J, Cajiao CH, Tejada G (1992) Effect of growth habit on yield of large-seeded bush cultivars of common bean. Field Crops Res 29:151–161

Young RA, Melotto M, Nodari RO, Kelly JD (1998) Marker assisted dissection of the oligogenic anthracnose resistance in common bean cultivar, G 2333. Theor Appl Genet 96:87–94

Yu ZH, Stall RE, Vallejos CE (1998) Detection of genes for resistance to common bacterial blight of beans. Crop Sci 38:1290–1296

Yu KF, Park SJ, Poysa V (1999) Abundance and variation of microsatellite DNA sequences in beans (*Phaseolus* and *Vigna*). Genome 42:27–34

Yu K, Park SJ, Poysa V (2000 a) Marker-assisted selection of common beans for resistance to common bacterial blight: efficiency and economics. Plant Breed 119:411–416

Yu K, Park S, Poysa V, Gepts P (2000 b) Integration of simple sequence repeat (SSR) markers into a molecular linkage map of common bean (*Phaseolus vulgaris* L.). J Hered 91:429–434

Yu K, Park SJ, Zhang B, Haffner M, Poysa V (2004) An SSR marker in the nitrate reductase gene of common bean is tightly linked to a major gene conferring resistance to common bacterial blight. Euphytica 138:89–95

Zizumbo-Villarreal D, Colunga-GarcíaMarín P, Payró de la Cruz E, Delgado-Valerio P, Gepts P (2005) Population structure and evolutionary dynamics of wild-weedy-domesticated complexes of common bean in a Mesoamerican region. Crop Sci 45:1073–1083

Fig. 1. Alternative leaf morphology in pea. **a** Conventional (*AfAfStStTlTl*) (*left*) and semileafless (*afafStStTlTl*) (*right*). **b** Intertwining nature of tendrils from adjacent branches providing mutual support

and survive relatively harsh winters. The fall-sown or winter pea is sown in late September to mid-October in northern latitudes and ideally reaches a moderately branched, rosette stage ca. 5 to 7 cm in height prior to entering a winter dormant period. As temperatures rise in the spring the rosette breaks dormancy assuming a vegetative growth pattern followed by a reproductive period and seed fill. Crop maturity is typically 2 to 3 weeks prior to spring-sown types. Improved capacity to take advantage of early spring moisture and avoidance of elevated temperatures during the flowering and seed fill periods allows the fall-sown crop to outyield spring crops by an average of 40%. Market classes of winter pea include the Austrian winter pea (pigmented seed and flowers) and a winter feed pea (with either green or yellow cotyledons) characterized by white flowers and seeds lacking pigmentation in the testa.

Plant type of modern cultivars has changed significantly compared to cultivars of 15 to 20 years previously. Overall plant height has been reduced through selection of short internodes conferred by recessive *le* along with other modifying genes, and leaf morphology has been converted from conventional leaflets to a semileafless type where leaflets are converted to tendrils by recessive *af* (Goldenberg 1965) (Fig. 1). The tendrilled morphology allows neighboring plants to intertwine and provide mutual support for a rigid canopy. Combination of these morphological changes with improved stem strength has resulted in a more upright plant habit and significant resistance to lodging. The upright growth habit reduces foliar disease incidence in the canopy by allowing greater airflow and has improved harvest ease and crop quality through reduced content of foreign material in the harvested seed.

Pisum is a member of family Fabaceae, subfamily Papillionaceae, and tribe Vicieae and comprises two species, *Pisum fulvum* Sibth and Sm. and *Pisum sativum* L. *Pisum sativum* can be further divided into subspecies *P.s.* ssp. *sativum*, *P.s.* ssp. *elatius*, *P.s.* ssp. *humile*, *P.s.* ssp. *arvense*, and *P.s.* ssp. *hortense*. *P. sativum* ssp. *elatius* and ssp. *humile* are considered the progenitors of *P.s.* ssp. *sativum*, the modern cultivated pea. *P.s.* ssp. *arvense* encompasses the field pea and is characterized by colored flowers and pigmented seeds. *P. fulvum* is the wild form of pea and can be crossed with *P. sativum*, but *P. fulvum* must be used as the pollen parent (Muehlbauer 1992; Ben-Ze'ev and Zohary 1973).

Genetic composition of pea is ca. 4800 Mbp spread across $2n=2x=14$ chromosomes. Karyotypes were presented as early as 1931 (Lewitsky 1931, as cited by Blixt 1958). The standard karyotype of pea is based on the type line, L110, following improvements over earlier procedures of Caroli and Blixt (1953) (Blixt 1958; Hall et al. 1997a) (Table 2). Centromere location in chromosomes 1 and 2 is metacentric, while in the remaining five chromosomes it is submetacentric. Chromosomes 4 and 7 contain secondary constrictions and possess satellites. *P. fulvum* contains a third secondary constriction on chromosome 5 (Ben-Ze'ev and Zohary 1973). Chromosomes 2 and 5 have experienced a translocation

Table 2. Relative chromosome length and centromere position of seven pea chromosomes (Hall et al. 1997)

Chromosome	Relative length	Centromere position	Satellites
1	0.125 ± 0.008	Submetacentric	No
2	0.125 ± 0.008	Submetacentric	No
3	0.145 ± 0.007	Acrocentric	No
4	0.138 ± 0.009	Acrocentric	Yes
5	0.172 ± 0.013	Acrocentric	Yes (in *P. fulvum*)
6	0.142 ± 0.008	Acrocentric	No
7	0.151 ± 0.011	Acrocentric	Yes

event, referred to as the Hammarlund translocation (summarized by Kvostova 1973, translated 1983). Numerous other chromosomal rearrangements have also been reported (Kvostova 1973, translated 1983; Lamm and Miravalle 1959; Folkeson 1990 a, b).

Breeding objectives for pea vary depending on production region and end use. Quality of fresh peas is highly refined, and genetics controlling specific traits are controlled through proprietary means. Many of the agronomic, disease, and quality traits for dry peas are common to fresh pea and are listed in Table 3. Primary objectives common to all regions and commodity types include increased yield, multiple disease resistance, and superior quality for the desired end use and agronomic adaptation to local production conditions.

Classical breeding has made significant improvement in crop production and seed quality. The genetic contribution to improved seed yield has come through recurrent selection for adaptation to local production areas, incorporation of multiple disease resistance, improved plant architecture, and resistance to seed shatter. Many common diseases of pea can be controlled by single genes, and plant breeding has been extremely successful in combining many of these genes in improved cultivars or germplasm. Dr. John Kraft (USDA-ARS, retired) developed a number of fresh pea lines with improved resistance to the root rots that are controlled quantitatively and made the germplasm widely available to breeders and geneticists (Kraft and Tuck 1986; Kraft 1989, 1992). Identification of partial genetic resistance to foliar fungal pathogens such as *Mycosphaerella pinodes*, *Phoma medicaginis var. pinodella*, *Sclerotinia sclerotiorum*, *Botrytis cinerea*, and *Pseudomonas syringae* pv. *pisi* (Bevan et al. 1995) has been augmented through avoidance mechanisms of an upright growth habit and a more open canopy resulting from the semileafless morphology. Unfortunately, the open canopy has little effect on reducing the incidence of powdery mildew.

Quality attributes of peas grown for dry seed harvest have focused primarily on the visual appearance of the seed, i.e., uniformity of seed color and intensity as well as shape. Greater attention is being given to the chemical composition of the pea due to its importance in animal feed rations. Pea harvested for fresh products, as either immature seed or whole pods, have been subject to numerous

Table 3. Agronomic, disease and quality traits targeted for improvement in pea breeding programs

Agronomic character	Disease resistance	Quality attribute
	Foliar fungi	*Dry pea*
Seed yield	Mycosphaerella blight	Seed size
Increased biomass	Sclerotinia white mold	Cotyledon color
Lodging resistance	Powdery mildew	Seed shape
Shatter resistance	Downy mildew	Protein content
Multiple podding	*Soil-borne fungi*	Digestibility
	Fusarium wilt race 1 and 2	*Fresh produce pea*
	Fusarium root rot	Blonde peas
	Aphanomyces root rot	Flavor
	Viruses	
	Pea enation mosaic	
	Bean leaf roll	
	Pea seedborne mosaic	
	Clover yellow vein	

and intense selection for myriad of criteria including appearance, flavor, retention of color, size, etc. Effort toward quality improvement in pea directly reflects the importance of satisfying quality expectations whether for dried or fresh product of domestic household consumers, which are far more rigorous than feed markets.

Improvement of peas for resistance to disease and other traits is often limited by inconsistent phenotypic expression or unreliable evaluation and scoring due to significant environmental influence, absence of disease, or inadequate rating scales. This is particularly true for those characters showing quantitative inheritance, yield, quality, and resistance to a number of diseases. Specific examples include resistance to pathogens contributing to the root rot complex (*Aphanomyces euteiches* and *Fusarium solani*) and the ascochyta blight complex (*Mycosphaerella pinodes*, *Phoma medicaginis* var. *pinodella*, and *Ascochyta pisi*). Selection for these traits might be greatly improved through marker-assisted selection (MAS) based on relatively dense genetic maps and known locations of important genes.

2.2 Construction of Genetic Maps

The easy cultivation of pea and its self-pollinating reproductive habit and wealth of genetic variation made pea an easy subject for genetic investigation and served to justify Mendel's (1866) decision to use *Pisum* for his seminal work on genetic inheritance. Mendel studied seven morphological characters in pea (Table 4), and, despite the fact that these seven traits resided in only four linkage groups (LGs), he did not detect any linkage relationships due to the large genetic distance between those traits residing on the same LG. Genetic mapping in pea began in the 1920s with the classical genetic studies of Herbert Lamprecht (summarized in Blixt 1972). As Blixt (1974) summarized, the only two traits for which Mendel may have detected linkage were *v* and *le* on LG III, but as far as is known, Mendel never inspected his data for simultaneous segregation.

Table 4. Gene symbols, linkage group, and description of traits studied by Mendel

Gene	Linkage group	Trait
i	I	Green/yellow cotyledon color
a	II	Pigmented flowers
v	III	Parchmented pods
le	III	Long internodes
fa	III	Fasciated stem
r	V	Round/wrinkled seed
gp	V	Yellow pod coloration

The first genetic linkage maps of pea were established by analyses of numerous two-point crosses (Lamprecht 1948, 1958; Blixt 1972). Figure 2 presents a timeline of marker development and mapping in pea. More than 160 of the nearly 400 morphological mutants reported in these early studies were placed on one of the seven LGs of pea (Blixt 1972). Assignment of genes to individual LGs has required only modest adjustments, and this early framework map made it relatively easy to assign markers derived using modern molecular techniques to the expected seven LGs. Incorrect linkage assignment was due to the large linkage distances between markers and to a limited number of marker loci present in any given population (Weeden et al. 1998).

Following establishment of the LGs efforts were made to assign each LG to one of the seven chro-

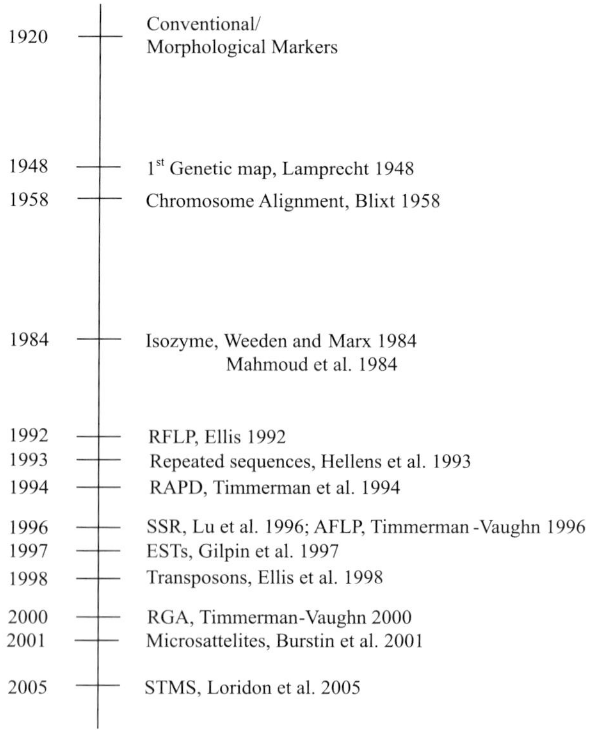

Fig. 2. Timeline following development of genetic maps of *Pisum* and adoption of specific marker technologies

mosomes of pea. Blixt (1958) described the standard karyotype of pea based on Line 110 and subsequently used a set of translocation lines to associate the LGs with individual chromosomes (Blixt 1959). A number of chromosomal rearrangements were found in pea, and translocation tester sets were characterized and used to assign LGs to individual chromosomes (Lamm and Miravalle 1959). Correct assignment of LGs to chromosomes was limited by the inaccurate assignment of genes to LGs.

The 30 years between the mid-1950s and the mid-1980s was characterized by additional classical genetic studies by the late Dr. Jerry Marx and Dr. Ian Murfet (Fig. 2). In the mid-1980s, isozymic markers in *Pisum* were placed on the framework map of Blixt (1974) (Mahmoud et al. 1984; Weeden and Marx 1984, 1987). In the early 1990s, restriction fragment length polymorphism (RFLP) maps of *Pisum* were reported and increased marker density and further defined genetic linkage in pea (Ellis et al. 1992; Timmerman et al. 1993; Dirlewanger et al. 1994).

The advent of polymerase chain reaction (PCR) opened numerous opportunities for a multitude of molecular markers to be added to genetic maps of all crops, including pea. Yu et al. (1995) reported two random amplified polymorphic DNA (RAPD) markers linked to *En*, a gene for resistance to pea enation mosaic virus, and Laucou et al. (1998) reported an integrated map based primarily on RAPD markers (Table 4). Development of amplified fragment length polymorphism (AFLP) technology (Vos et al. 1995) offered the opportunity to rapidly increase marker density (Lu et al. 1996; Timmerman-Vaughn et al. 1996). Repeated sequences in crop genomes are useful sources of polymorphism and were exploited in pea beginning in 1993 (Hellens et al. 1993; Turner and Ellis 1997).

Mapping expressed genes to specific LGs became a reality with the development of expressed sequence tag (EST) and resistance gene analog (RGA) markers. Gilpin et al. (1997) placed 29 ESTs on a *Pisum* map based on RFLP, RAPD, and AFLP markers, and Timmerman-Vaughn et al. (2000) placed eight RGAs on a map based on three different crosses (Table 5). Flavell et al. (1998) reported on the utility of copialike retrotransposon markers (RBIPs) for mapping in pea. Burstin et al. (2001) reported on the variation and polymorphism of microsatellites in pea, and the first comprehensive map based on sequence-tagged microsatellite (STMS) markers was published in 2005 (Loridon et al. 2005).

2.3 Gene Mapping

Gene mapping in pea was initially based on its rich morphological variation, but due to generally low marker density, accurate detection of linkage was inhibited. Lamprecht's (1948) genetic map aligns well with modern maps with the exception that markers assigned to linkage group VII are now considered to be on the long arm of linkage group V and linkage group VII was not represented. The lack of morphological markers on linkage group VII and the significant distance between genes *r* and *cp* on linkage group V explain why these two LGs were assigned to separate linkage groups. The lack of marker genes on linkage group VII was

Table 5. First-generation maps created based on two-point analyses of morphological traits, RFLP, RAPD, and AFLP markers

Map	Population	Marker type	Markers	Population size	Coverage
Lamprecht 1948	Multiple	Morphological	37	–	not reported
Ellis et al. 1992	JI 15×JI 61	RFLP, morphological &	51	38	not reported
	JI 15×JI 1194	others	65	50	
	JI 281×JI 399		153	71	
	JI 813×JI 1201		13	48	
Gilpin et al. 1997	Primo×OSU442-15	ESTs, RFLP, RAPD & AFLP	209	102	1330 cM
Laucou et al. 1998	Terese×K586	RAPD, RFLP	240	139	1139 cM
Timmerman-Vaughn	Primo×OSU442-15	EST, RFLP, RAPD, AFLP	209	102	1330 cM
et al. 2000	JI 1794×Slow	RAPD, RFLP, AFLP	235	51	1289 cM
	Rovar×3176-A26	RFLP, RAPD	84	148	–

noted by Folkeson (1990 a, b) when he assigned the *tl-r-bt* segment to LG V.

The *Pisum* genetic map has evolved over time to become rather dense and well defined. Application of molecular-marker technology contributed significantly to the increased marker density allowing more accurate assignment of genes to LGs and generated a number of closely linked markers, especially disease resistance genes, for use in MAS (Weeden 1996). Consensus maps of *Pisum* encompassing the knowledge of linkage at the time were reported by Weeden et al. (1996, 1998). Although the consensus map summarizes known linkage relations among genes, it is limited by lack of common genes/markers between populations, causing a number of the genes to be placed in the "most likely" position. Establishment of a consensus map based on the codominant and transferable STMS markers and the transfer of these markers to other populations is expected to allow many existing maps to be aligned and will significantly increase the density and definition of the *Pisum* map.

The focus of the first molecular mapping efforts was to locate resistance genes for powdery mildew, ascochyta blight, and a number of viral diseases affecting pea and to identify markers closely linked to these genes to augment selection procedures (Table 6). Physiological traits such as starch synthesis, photoperiod response, plant habit, and symbiotic interactions with *Rhizobium* ssp. have also been the focus of many studies due to their importance in crop production. Recent efforts in genetic mapping

Table 6. Selected individual genes targeted on maps generated since 1993 including population and closest linked marker. More than 300 morphological markers have been mapped via conventional methods prior to 1993 (Lamprecht 1948; Weeden et al. 1998)

Trait	Population	Marker	Reference
rb	JI 281×JI 399	pJC 2-7	Hellens et al. 1993
er-1	Erygel×661	p236	Dirlewanger et al. 1994
	Almota×88V 1.11	pD10$_{650}$ SCAR	Timmerman et al. 1994
		OPU-17 & ScOPD10$_{650}$	Janilla and Sharma 2004
sbm-1	88V 1.11×Almota '88V 1.11×Line 425	GS185	Timmerman et al. 1993
Fw	Erygel×661	p254	Dirlewanger et al. 1994
Mo	Erygel×661	p252	Dirlewanger et al. 1994
Ascochyta pisi race C (*Rap-2*)	Erygel×661	p227	Dirlewanger et al. 1994
En	Alaska×B880-221	P256$_{900}$ & B500$_{400}$	Yu et al. 1995
Rmp1,- 2,- 3 & -4	JI 97×JI 296 JI 1089×JI 296	morphological	Clulow et al. 1991
cyv-1		B302	Yu et al. 1996
sbm-3		B302	Yu et al. 1996
fa	JI 814×Terese	Q11_1380	Rameau et al. 1998
det	JI 2121×Terese	U16_1300, U8_1900	Rameau et al. 1998
sn	HL 59×Terese	AA7_1700	Rameau et al. 1998
dne	Terese×K218	S2_1900	Rameau et al. 1998
Rms2	K524×Terese	A4_430	Rameau et al. 1998
Rms3	WL6042×Terese	V20_1100	Rameau et al. 1998
Rms4	M3T-946×Torsdag	C12_500	Rameau et al. 1998
Fwf	74SN3×A83-22e	*Aatp*	Coyne et al. 2000
Ppi1	JI15×JI 399 & JI 281×JI 399	*Pl*	Hunter et al. 2001
Ppi2	JI15×JI 399 & JI 281×JI 399	*Rrn2*	Hunter et al. 2001
Ppi3	Vinco×Hursts Greenshaft Partridge×Early Onward	OPA-20$_{0.71}$	Hunter et al. 2001
Ppi4	Vinco×Hursts Greenshaft Partridge×Early Onward	OPA-20$_{0.71}$	Hunter et al. 2001
Sym9	P2×JI 281 & P54×JI 281	A5/16	Schneider et al. 2002
Sym10	P56×JI 15	*chs2*	Schneider et al. 2002

in pea have focused on increasing marker density around individual genes with the ultimate goal of using map-based cloning strategies to clone specific genes of interest.

2.4 QTLs Detected

Quantitative trait locus (QTL) analyses allow genomic regions influencing traits highly influenced by the environment and showing continuous variation to be located on the genetic map. Due to the significant environmental effect on traits showing quantitative inheritance, genetically fixed populations that can be evaluated in replicated trials over multiple site-years are desired. Recombinant inbred lines (RILs) derived via single seed descent from individual F_2 plants to the F_6 or later generations are typically used for QTL studies. Other fixed populations include inbred backcross populations. Few analyses are based on F_2- or F_2-derived F_3 families due to constraints of replication and/or multisite studies.

Relatively few QTL analyses in pea have been reported in the literature (Table 7). The first QTL reported was that for seed weight (Timmerman-Vaughn et al. 1996). Other QTL analyses have looked at lodging resistance (Tar'an et al. 2003), plant height (Tar'an et al. 2003), grain yield (Tar'an et al. 2004), seed protein content (Tar'an et al. 2004), maturity (Tar'an et al. 2004), persistence of green seed color (McCallum 1997) tolerance to *Orobanche crenata* (Valderrama et al. 2004), and resistance to *Aphanomyces euteiches* (Pilet-Nayel et al. 2002) and *Mycosphaerella pinodes* (Timmerman-Vaughn et al. 2002; Tar'an et al. 2003; Prioul et al. 2004).

Nearly all QTL studies in pea have used RIL populations as the genetic material with the exception of McCallum (1997) for persistent seed color, Timmerman-Vaughn (1996) for seed size, and Valderrama et al. (2004) for tolerance to *Orobanche crenata*. All three studies listed analyzed data collected on F_2 individuals and/or F_2-derived F_3 family populations. Data analysis and map construction in all studies used MapMaker (Lander et al. 1987; Lincoln et al. 1992), and QTL identification and effect were estimated by composite interval mapping using QTL Cartographer (Basten et al. 2001).

Ascochyta blight, caused by a complex of three pathogens, *Mycosphaerella pinodes*, *Phoma medicaginis*, and *Ascochyta pisi*, is a serious disease of many pea-producing regions worldwide. Three laboratories have investigated resistance mechanisms and reported QTLs for resistance to these pathogens (Timmerman-Vaughn et al. 2002; Tar'an et al. 2003; Prioul et al. 2004). Timmerman-Vaughn (2002) identified 13 QTLs residing on LGs I, II, III, IV, V and VII and Prioul et al. (2004) identified 16 QTLs located on LGs II, III, V, VI and VII while Tar'an et al. (2004) identified three QTLs located on LGs II, IV and VI. The maps could not be compared because of a lack of common markers. However, based on the close association of the CD40 marker and the *a* locus (Weeden et al. 1998) to which the Asc2.1 locus (Timmerman-Vaughn et al. 2002) and the mpII-2 locus (Prioul et al. 2004) are linked, respectively, Asc2.1 and mpII-2 could represent a common QTL. Although few markers surround the QTL on the proximal end of LG II of Tar'an et al. (2004), it could represent a common QTL. It is not likely that Asc3.1 (Timmerman-Vaughn et al. 2002) and mpIII-1 of Prioul et al. (2004), both residing on LG III, represent a common QTL due to the linkage distance between M27 and L109 to which these QTLs are most closely linked, respectively (Weeden et al. 1998). Tar'an et al. (2003) suggest that the QTL they detected on LG IV and that of Timmerman-Vaughn et al. (2002) may be a common QTL based on common linkage with the p628 marker. It is not possible to determine if the QTLs reported on LG VI by Tar'an et al. (2003) and Prioul et al. (2004) are common due to a lack of common markers. Likewise, it is not possible to determine if QTLs located on LG V are common to the maps of Timmerman-Vaughn et al. (2002) and Prioul et al. (2004).

Common root rot, caused by *Aphanomyces euteiches* Drechs. f. sp. *pisi* W. F. Pfender & D. J. Hagedorn, is an important disease in France, and many pea production regions in the USA. Pilet et al. (2002) evaluated a RIL population of 127 lines over 4 site-years and detected seven QTLs, *Aph1* through *Aph7*, conferring partial resistance to aphanomyces root rot. Three QTLs contributed a major effect and explained 47, 32, and 11% of the variation, respectively. *Aph1* is located on LG IVb, *Aph2* is located on LG V near the *r* gene locus controlling seed morphology, and *Aph3* is located on LG Ia near the *af* locus for tendrilled leaf morphology. The four minor QTLs are located on LGs Ib (*Aph4*, *Aph5*), VII (*Aph6*), and an unlinked LG B (*Aph7*).

Table 7. QTLs detected and mapped in pea. Location, name, and closest single marker or marker interval with associated LOD scores and percent of variation explained are also presented

QTL	Population	Location	QTL identification			Marker		References
			Name	LG	Effect (%)	Name	LOD	
Lodging resistance	Carneval×MP1401	11 locations	1	III	47	cacc4	14.5	Tar'an et al. 2003
		in Canada	2	VI	26.4	acct1	3.5	
Plant height	Carneval×MP1401	11 locations	1	III	56.9	cttg7	21.5	Tar'an et al. 2003
		in Canada	2	C	17.2	caag4	3.3	
			3	D	6.5	cagg5	3.1	
Resistance to	Carneval×MP1401	11 locations	1	II	5.0	ccta2	2.9	Tar'an et al. 2003
Mycosphaerella		in Canada	2	IV	19.1	cccc1	3.3	
blight			3	VI	16.8	acct1	3.1	
Resistance to	DP×JI 296	GH and 1 field	mpII-1	II	6 (field, stipule)	AD12-800	3.1	Prioul et al. 2004
Mycosphaerella		location	mpII-2	II	9 (field, stipule)	*a*	4.6	
pinodes					6 (field, stem)		4.3	
			mpIII-1	III	18 (CC, stipule)	E08-980	13	
					20 (CC, stem)		13.9	
					26 (field, stipules)		12.5	
					42 (field, stems)	V03-1200	18	
			mpIII-2	III	7 (CC, stipules)	PSP40SG	4.8	
					9 (CC, stems)	V03-1000	6.1	
			mpIII-3	III	6 (CC, stems)	V03-1000	4.9	
					7 (field, stipules)	PSMPSAA175	3.9	
					6 (field, stems)	F09-1900	3.2	
			mpIII-4	III	29 (field, stems)	PSMPSAA374a	6.8	
			mpIII-5	III	11 (field, stems)	PSMPSAA163.2	5.8	
			mpVa-1	V	10 (CC, stipules)	PSMPSAA163.2	7.2	
					8 (CC, stems)	PSMPAA163.2	5.9	
					7 (field, stems)	T14-2200	3.1	
			mpVa-2	V	16 (field, stems)	G04-950	4.4	
			mpVI-1	VI	15 (CC, stipules)	G04-950	9.3	
					20 (CC, stems)	PSMPSAA399	12.3	
			mpVII-1	VII	5 (CC, stipules)	PSMPSAA399	3.2	
					6 (CC, stems)	PSMPSAA399	3.3	
					9 (field, stipules)	Z17-550	3.0	
			mpVII-2	VII	8 (field, stipules)		4.3	

Table 7 (continued)

QTL	Population	Location	QTL identification			Marker		References
			Name	LG	Effect (%)	Name	LOD	
Seed weight	Primo × OSU442-15	1 location (F_2)	1	III/IV	26	M27/B08_1250	4.6	Timmerman-Vaughn et al. 1996
			2	IV	23	A09_1250/A15_1580	2.2	
			3	V	13	P445/e07_1100	2.1	
	Slow×JI 1794	GH single plants	1	I	–	B526d		
			2	III	–	AFPIx		
			3	III	–	B08-1250		
			4	VII	–	B526c		
Grain yield, seed protein and maturity	Carneval×MP1401	13 environments in Canada	*Grain Yield*					Ta'ran et al. 2004
			1	II	3.4	gccc3	12.3	
			2	VI	5.2	gcta2	18.4	
			2	VII	3.9	cctc3	14.2	
			4	VII	3.0	cagg9	9.6	
			Protein					
			1	III	6.1	cccc18	29.4	
			2	VI	3.5	acct7	15.8	
			2	A	3.9	cctc2	11.2	
			Maturity					
			1	II	6.3	ctgt3	15.8	
			2	II	3.8	cccc19	12.1	
			2	III	3.5	ccac8	10.4	
			4	VI	3.4	gggt4	9.8	
Resistance to Aphanomyces euteiches	Puget×90-2079	4 locations	Aph1	IVb	10–47*	–	20.4	Pilet-Nayel et al. 2002
			Aph2	V	8–32*	–	10.1	
			Aph3	Ia	11–14*	–	4.0	
			Aph4	Ib	6	–	3.2	
			Aph5	Ib	13	–	6.5	
			Aph6	VII	6	–	3.7	
			Aph7	B	7	–	4.2	

Table 7 (continued)

QTL	Population	Location	QTL identification			Marker		References
			Name	LG	Effect (%)	Name	LOD	
Resistance to Orobanche crenata	P-665×Messire	1 location, Cordoba, Spain	Ocp1	IX	9.6	STS-482/OPAC08$_{786}$	13.3	Valderrama et al. 2004
			Ocp2	I	11.4	OPB11$_{541}$/OPAE02$_{525}$	13.0	
Resistance to Ascochyta blight	3148-A88×Rovar	3 years in Medina, W Australia	Asc1.1	I	13–24*	c206	5.1	Timmerman-Vaughn et al. 2002
			Asc2.1	II	12–15*	SP2P5	5.1	
			Asc3.1	III	17	P10_711	5.7	
			Asc4.1	IV	9–13*	P628	2.9	
			Asc4.2	IV	10–13*	P9	4.0	
			Asc4.3	IV	8–12*	PI39	3.5	
			Asc5.1	V	21–24*	P3P8bM09	4.4	
			Asc7.1	VII	8	sAFP2-P2	2.6	
Green seed color	Primo×OSU442-15	1 location (F_2)	V	I	17	Lg-J/P482	2.6	McCallum et al. 1997
			U	III/IV	56	I05_530/K02_1700	4.7	
			Y	V	61	P108/r	13.5	
			V	V	27	P108/r	3.8	
			U	VII	16	P694	2.4	

* = range of values depending on experimental locations and/or phenotypic trait evaluated.

Green color hue and retention of green seed color is an important quality criterion of green dry peas. McCallum et al. (1997) studied the luminance (Y) and chrominance (U and V) components of color in an F_2 population developed from the cross Primo×OSU442-15. They detected four genomic regions, one each on LGs I, III/IV, V, and VII. The region on LG V contains two QTLs, one for luminance (Y) and the other for the V component of chrominance and each explain 61 and 27 % of the variation, respectively. The QTLs located on LG I (V component of chrominance), LG III/IV (U component of chrominance), and VII (U component of chrominance) explain 17, 56, and 16% of the variation, respectively.

Valderrama et al. (2004) reported two QTLs conferring partial resistance to *Orobanche crenata* in pea. They analyzed 115 F_2-derived F_3 families under field conditions for resistance to *O. crenata*. The two QTLs, *Ocp1* and *Ocp2*, were located on LG 1 and IX, respectively, of their map; however, alignment with the pea consensus map is not possible.

Agronomic characters that have been subjected to QTL analysis include seed size (Timmerman et al. 1996), grain yield, seed protein, maturity (Tar'an et al. 2004), plant height (Tar'an et al. 2003; Prioul et al. 2004), lodging (Tar'an et al. 2003), and flowering (Prioul et al. 2004). Timmerman et al. (1996) identified three QTLs contributing to seed size in the cross Primo×OSU445-15, one each on LG III/IV, IV and Va and each explained 26, 23, and 13% of the variation, respectively. In a second population derived from the cross, JI 1794×Slow, four regions were identified as contributing to seed size, one on LG I, two on LG III and one on LG VII. Only one genomic region contributing to seed size is common to the two populations; however, the region on LG VII in the JI 1794×Slow cross is not represented in the Primo×OSU445-15 cross.

Tar'an et al. (2004) located four QTLs each for days to maturity and yield and three QTLs for protein content in a population of 88 RILs derived from the cross Carneval×MP1401. The maturity QTLs were located on LG II (2QTL), III and VI and explained 16, 12, 10, and 10% of the variation, respectively. Prioul et al. (2004) reported three QTLs for flowering on LG II, III and VI explaining 38, 16, and 9% of the variation, respectively. Yield QTLs were identified on LGs II, VI and two on VII each explaining 12, 18, 14, and 10% of the variation, respectively. The seed-protein-content QTLs were located on LGs III, VI and an unlinked LG A each explaining 30, 16, and 11% of the variation, respectively. Plant height QTLs were identified by Tar'an et al. (2003) and Prioul et al. (2004), and Prioul et al. (2004) identified QTLs for plant height on LG II, III and VII while Tar'an et al. (2003) identified QTLs on LG III and two unlinked LGs, C and D. The QTLs on LG III accounted for 57 and 63% of the variation for plant height in each of the reports, respectively.

QTL analyses in pea have identified 72 QTLs for 11 traits. Many of the maps used in these analyses have made use of anchor markers to align LGs in the current study to those of previous studies. To date none of the QTLs identified and characterized in pea has been Mendelized. This is expected to be the subject of future work and will be necessary to identify specific markers that can be used reliably in MAS. Mendelization of QTLs in pea will be possible as more detailed maps are developed and fine mapping of individual traits is completed.

2.5 Marker-Assisted Breeding

A few markers have been converted to simple and easy-to-use PCR-based assays. Timmerman et al. (1994) converted a RAPD marker linked to *er-1*, a gene conferring resistance to powdery mildew caused by *Erysiphe pisi*, to sequence characterized amplified region (SCAR) marker ($PD10_{650}$). It is positioned 2 cM from the gene and serves as an excellent candidate for MAS. Rameau et al. (1998) converted RAPD markers located 16.7 cM from *Rms3* and 4.7 cM from *dne* to SCAR markers. Yu et al. (1995) converted a RAPD marker linked to *En*, a gene for resistance to pea enation mosaic virus, to a SCAR marker identified as $p256_{900}$ that was located 6 cM from *En*.

MAS has been successfully applied in selecting fresh pea germplasm with resistance to *Aphanomyces euteiches*, the causal organism of aphanomyces root rot (Rebecca McGee, personal communication). Few examples of MAS are available in pea from publicly funded programs despite the reported availability of useful molecular markers. Many traits for which markers are available are controlled by single genes and are effectively selected during field screening with relative ease. However, if markers based on a common methodology were available for a multitude of traits, it would be more economical and efficient to initiate MAS schemes.

Germplasm collections have been established in a number of institutes worldwide. The primary collections for pea are being curated at the John Innes Center (JIC), Norwich, UK, USDA-ARS Western Regional Plant Introduction Station, Pullman, WA, USA, and Victorian Institute of Dryland Agriculture, Horsham, Victoria, Australia. Molecular characterization of these collections would augment phenotypic characterization by identifying genetic diversity and reducing duplications. Dr. Mike Ambrose, curator of the JIC collection, summarized molecular data used to characterize the JIC collection for molecular diversity (Ambrose et al. 2004). Similar efforts with the intent of establishing a procedure for association mapping are anticipated for the US collection (Clare Coyne, personal communication). Detailed analysis of these collections for both phenotypic and molecular characteristics will improve collection use and allow more specialized and accurate core collections representing the overall diversity present in the collection.

2.6 Map-Based Cloning

Map-based cloning in plant species requires extensive knowledge of genome organization available through fine mapping in the region of interest and using those markers most closely linked to the gene of interest to search genomic libraries for candidate genes. Bacterial artificial libraries (BACs) are a convenient tool to catalog genomic DNA sequence in fragments averaging 100 kb in length. A BAC library for pea was created for PI 269818 and packaged in the binary vector V41 (pCLD04541) (Coyne et al. 2000). The PI269818 library has an average insert size of 110 kb and represents ca. 2.5 haploid genome equivalents (Coyne et al. 2004). Development of the BAC libraries in a binary vector facilitates direct transfer of selected sequences into plants via *A. tumefaciens*-mediated methods (Schroeder et al. 1993; McPhee et al. 2005).

To date few genes have been cloned from pea. Von Stackelberg et al. (2003) reported the initial stages of cloning the gene, *def*, that controls development of the abscission layer between the funiculus and hilum. It is suggested that this gene may have value regarding shatter resistance in pea. Stracke et al. (2004) used sequence information from *Lotus japonicus* and *Arabidopsis thaliana* in an attempt to clone the Sym19 gene from pea. Due to the wide distance (6 cM) between the flanking markers they used and unexpected breaches in collinearity within that region, they were unable to locate Sym19. Additional genes have been cloned from pea; however, sequence information is not currently available in the literature, potentially due to proprietary considerations. As additional information from related species and the model legume species becomes available, a rapid increase in the number of genes cloned from pea can be expected.

2.7 Advanced Works

Physical mapping in pea began with the assignment of individual LGs to the seven *Pisum* chromosomes beginning as early as the mid-1950s (Lamm and Miravalle 1959). Due to the early mistaken assignment of genes to LGs the correct assignment of LGs to chromosomes was not accomplished until the late 1990s (Hall et al. 1997; Fuchs et al. 1998; Ellis and Poyser 2002). It was still uncertain whether LG IV and VII should be assigned to chromosomes 4 and 7 or vice versa. Neumann et al. (2002) used three translocation lines from the John Innes Center and showed that LG IV corresponds to chromosome 4 and LG VII corresponds to chromosome 7. A further point of concern is the assignment of LG VI and LG I to chromosomes 1 and 2 (Table 8), respectively, due to the morphological similarity of these chromosomes.

Table 8. Assignment of individual linkage groups to seven known chromosomes of pea. Representative genes on each LG are also presented

Chromosome	Linkage group	Gene assignment
1	VI	*wlo-er1-gty-sbm1-pl*
2	I	*lb-d-f-af-i*
3	V	*bt-r-tl-gp-Pgdc*
4	IV	*fa-was-age-sym9*
5	III	*m-uni-st-np-le*
6	II	*Aatp-a-fum-wb-s*
7	VII	*Pgmc-fk-Rrn4-Skdh-amy*

Application of chromosome sorting technology via flow cytogenetics to gene mapping will likely allow greater definition of the LG assignment to individual chromosomes (Dolezel et al. 2004). Separation of individual chromosomes will serve as an invaluable tool to generate chromosome-specific genomic libraries and will allow chromosome-specific application of physical mapping techniques such as fluorescent in situ hybridization (Fuchs et al. 1998), primed in situ DNA labeling (PRINS), and cycling-PRINS (Kubalakova et al. 1997).

A few gene-specific sequences have been placed on the *Pisum* genetic map. Gilpin et al. (1997) mapped 29 ESTs and Timmerman-Vaughn et al. (2000) mapped eight RGA sequences on the *Pisum* genetic map. A coordinated effort is under way to combine ESTs from a number of plant tissues in pea and arrange them in a broadly applicable microarray; however, the sequences are not publicly available at the time of this publication. It is expected that this microarray will be valuable for a range of gene expression and identification studies.

2.8 Future Scope of Works

Comparative mapping in legumes has progressed rapidly with genomic developments in the model legumes, *Medicago truncatula* and *L. japonicus*. One of the earliest works of comparative mapping in pea was that of Weeden et al. (1992), where regions of similarity between *Lens* and *Pisum* were identified on five of the seven *Pisum* LGs. Significant collinearity between pea and *M. sativa* has been reported and demonstrated the conservation of gene order and distribution of homologous genes between the two related species (Kalo et al. 2004). Additionally, Choi et al. (2004) demonstrated significant genome conservation among the model legume species, *M. truncatula* and *L. japonicus*, with a broad group of legumes. Synteny among species offers the opportunity to transfer information on gene function gained in one species to another. However, as Choi et al. (2004) suggest, the utility of this transfer will be limited by the extent to which traits are controlled by similar genetic mechanisms and by the degree of gene-order conservation. The increasing body of knowledge surrounding genome sequence promises to allow greater understanding of the synteny between related crops and will provide insight into genome organization and functional genomics.

References

Ambrose MJ, Knox M, Vershinin A, Flavell AJ, Moulton V, Ellis THN (2004) Exploration of germplasm resources: progress and results from the application of molecular markers to the whole JIC Pisum collection. In: Proc 5th Eur Conf Grain Leg, Dijon, France, pp 157–160

Basten CJ, Weir BS, Zeng ZB (2001) QTL Cartographer, version 1.15. Department of Statistics, North Carolina State University, Raleigh, NC

Ben Ze'ev N, Zohary D (1973) Species relationships in the genus *Pisum* L. Isr J Bot 22:73–91

Bevan JR, Taylor JD, Crute IR, Hunter PJ, Vivian A (1995) Genetics of specific resistance in pea (*Pisum sativum*) cultivars to seven races of *Pseudomonas syringae* pv. *pisi*. Plant Pathol 44:98–108

Blixt S (1958) Cytology of Pisum II. The normal karyotype. Agric Hort Genet 16:221–237

Blixt S (1959) Cytology of Pisum III. Investigation of five interchange lines and coordination of linkage groups with chromosomes. Agric Hort Genet 17:47–75

Blixt S (1972) Mutation genetics of Pisum. Agric Hort Genet 30:1–293

Blixt S (1974) The pea. In: King RC (ed) Handbook of Genetics. Plenum, New York, pp 181–221

Burstin J, Deniot G, Potier J, Weinachter C, Aubert G, Baranger A (2001) Microsatellite polymorphism in *Pisum sativum*. Plant Breed 120:311–317

Caroli G, Blixt S (1953) The morphology of the chromosomes of *Pisum*. Agric Hort Genet 11:133–140

Choi H, Mun J, Kim D, Zhu H, Baek J, Mudge J, Roe B, Ellis N, Doyle J, Kiss GB, Young ND, Cook DR (2004) Estimating genome conservation between crop and model legume species. Proc Natl Acad Sci USA 101(43):15289–15294

Clulow SA, Matthews P, Lewis BG (1991) Genetical analysis of resistance to *Mycosphaerella pinodes* in pea seedlings. Euphytica 58:183–189

Coyne CJ, Inglis DA, Whitehead SJ, McClendon MT, Muehlbauer FJ (2000) Chromosomal location of *Fwf*, the fusarium wilt resistance gene in *Pisum sativum*. Pisum Genet 32:20–22

Coyne CJ, Meksem K, Lightfoot DA, Keller KE, Martin RR, McClendon MT, Inglis DA, Storlie EW, McPhee KE (2000) Construction of a bacterial artificial chromosome library for pea (*Pisum sativum* L.). Pisum Genet 32:23–26

Coyne CJ, Murray S, Timmerman-Vaughan GM (2004) Identification of RGAs from a pea BAC library using BAC pools. In: Fischer et al (eds) New Directions for a Diverse Planet. Proc 4th Int Crop Sci Congr, Brisbane, Queensland, Australia p 373. http://www.cropscience.org.au/icsc2004/

Dirlewanger E, Isaac PG, Ranade S, Belajouza M, Cousin R, deVienne D (1994) Restriction fragment length polymorphism analysis of loci associated with disease resistance genes and developmental traits in *Pisum sativum* L. Theor Appl Genet 88:17–27

Dolezel J, Kubalakova M, Bartos J, Macas J (2004) Flow cytogenetics and plant genome mapping. Chrom Res 12:77–91

Ellis THN, Poyser SJ (2002) An integrated and comparative view of pea genetic and cytogenetic maps. New Phytol 153:17–25

Ellis THN, Turner L, Hellens RP, Lee D, Harker CL, Enard C, Domoney C, Davies DR (1992) Linkage maps in pea. Genetics 130:649–663

FAOSTAT (2003) http://apps.fao.org/cgi-bin/nph-db.pl?subset=agriculture

Flavell AJ, Knox MR, Pearce SR, Ellis THN (1998) Retrotransposon-based insertion polymorphisms (RBIP) for high throughput marker analysis. Plant J 16(5):643–650

Folkeson D (1990 a) Assignment of linkage segments to chromosomes 3 and 5 in *Pisum sativum*. Hereditas 112:249–255

Folkeson D (1990 b) Assignment of linkage segments to the satellite chromosomes 4 and 7 in *Pisum sativum*. Hereditas 112:257–263

Fuchs J, Kuhne M, Schubert I (1998) Assignment of linkage groups to pea chromosomes after karyotyping and gene mapping by fluorescent *in situ* hybridization. Chromosoma 107:272–276

Gilpin BJ, McCallum JA, Frew TJ, Timmerman-Vaughn GM (1997) A linkage map of the pea (*Pisum sativum* L.) genome containing cloned sequences of known function and expressed sequence tags (ESTs). Theor Appl Genet 95:1289–1299

Goldenberg JB (1965) 'Afila', a new mutation in pea (*Pisum sativum* L.). Boletin Genetico I:27–28

Hall KJ, Parker JD, Ellis THN (1997 a) The relationship between genetic and cytogenetic maps of pea. I. Standard and translocation karyotypes. Genome 40:744–754

Hall KJ, Parker JS, Ellis THN, Turner L, Knox MR, Hofer JMI, Lu J, Ferrandiz C, Junter PJ, Taylor JD, Baird K (1997 b) The relationship between genetic and cytogenetic maps of pea. II. Physical maps of linkage mapping populations. Genome 40:755–769

Hellens RP, Ellis THN, Lee D, Turner L (1993) Repeated sequences as genetic markers in pooled tissue samples. Plant Mol Biol 22:153–157

Hunter PJ, Ellis N, Taylor JD (2001) Association of dominant loci for resistance to *Pseudomonas syringae* pv. *pisi* with linkage groups II, VI and VII of *Pisum sativum*. Theor Appl Genet 103:129–135

Janilla P, Sharma B (2004) RAPD and SCAR markers for powdery mildew resistance gene *er* in pea. Plant Breed 123(3):271–274

Kalo P, Seres A, Taylor SA, Jakab J, Kevei Z, Kereszt A, Endre G, Ellis THN, Kiss GB (2004) Comparative mapping between *Medicago sativa* and *Pisum sativum*. Mol Genet Genom 272:235–246

Kraft JM (1989) Registration of 86-638, 86-2197, 86-2231, and 86-2236 pea germplasms. Crop Sci 29(2):494–495

Kraft JM (1992) Registration of 90-2079, 90-2131, and 90-2332 pea germplasms. Crop Sci 32(4):1076

Kraft JM, Tuck JA (1986) Registration of 75-786, 84-1638, and 84-1930 pea germplasms. Crop Sci 26:1262–1263

Kubalakova M, Macas J, Dolezel J (1997) Mapping of repeated DNA sequences in plant chromosomes by PRINS and C-PRINS. Theor Appl Genet 94:758–763

Kvostova VV (1983) Genetics and breeding of peas [Translation by Amerind, New Delhi]. Nauka Publishers, Siberian Division, Novosibirsk, pp 293

Lamm R, Miravalle RJ (1959) A translocation tester set in Pisum. Hereditas 45:417–440

Lamprecht H (1948) The variation of linkage and the course of crossing over. Agric Hort Genet 6:10–49

Lamprecht H (1958) Genetics of Pisum. [Translated by Amstutz E] Ed: Lawyer L. California Packing, San Leandro, CA, pp 1–110

Lander ES, Green P, Abrahamson J, Barlow A, Daly MJ (1987) MAPMAKER: an interactive computer package for constructing primary genetic linkage maps of experimental and natural populations. Genomics 1:174–181

Laucou V, Haurogne K, Ellis N, Rameau C (1998) Genetic mapping in pea. 1. RAPD-based genetic linkage map of *Pisum sativum*. Theor Appl Genet 97:905–915

Lewitsky GA (1931) The morphology of the chromosomes. Bull Appl Bot Genet Plant Breed 27:103–173

Lincoln M, Daly M, Lander E (1992) Constructing genetic maps with Mapmaker/exp ver 3.0. Technical Report, 3rd edn, Whitehouse Institute, Cambridge, MA

Loridon K, McPhee K, Morin J, Dubreuil P, Pilet-Nayel ML, Aubert G, Rameau C, Baranger A, Coyne C, Lejeune I, Burstin J (2005) Microsatellite marker polymorphism and mapping in pea (*Pisum sativum* L.). Theor Appl Genet 111:1022–1031

Lu J, Knox MR, Ambrose MJ, Brown JKM, Ellis THN (1996) Comparative analysis of genetic diversity in pea assessed by RFLP- and PCR-based methods. Theor Appl Genet 93:1103–1111

Mahmoud SH, Gatehouse JA, Bourlter D (1984) Inheritance and mapping of isoenzymes in pea (*Pisum sativum* L.). Theor Appl Genet 68:559–566

Mendel G (1866) Experiments in hybridization [Translation by Royal Horticultural Society of London, Harvard University Press, Cambridge, MA, 1963]. Verh Naturforsch Ver. Brunn 4

McCallum J, Timmerman-Vaughn G, Frew T, Russell A (1997) Biochemical and genetic linkage analysis of green seed color in field pea. J Am Soc Hort Sci 122(2):218–225

McPhee KM, Gallasch S, Schroeder HE, Higgins TJV (2005) Gene technology in pea. In: Curtis I (ed) Transgenic Crops of the World: Essential Protocols. Kluwer, Dordrecht, pp 351–360

Muehlbauer FJ (1992) Use of introduced germplasm in cool-season food legume cultivar development. In: Shands

HL, Weisner LE (eds) Use of Plant Introductions in Cultivar Development, Part 2. Crop Science Society of America, Madison, WI, pp 51–76

Neumann P, Pozarkova D, Vrana J, Dolezel J, Macas J (2002) Chromosome sorting and PCR-based physical mapping in pea (*Pisum sativum* L.). Chrom Res 10:63–71

Pilet-Nayel ML, Muehlbauer FJ, McGee RJ, Kraft JM, Baranger A, Coyne CJ (2002) Quantitative trait loci for partial resistance to Aphanomyces root rot in pea. Theor Appl Genet 106:28–39

Prioul S, Frankewitz A, Deniot G, Morin G, Baranger A (2004) Mapping of quantitative trait loci for resistance to *Mycosphaerella pinodes* in pea (*Pisum sativum* L.), at the seedling and adult plant stage. Theor Appl Genet 108:1322–1334

Rameau C, Denoue D, Fraval F, Haurogne K, Josserand J, Laucou V, Batge S, Murfet IC (1998) Genetic mapping in pea. 2. Identification of RAPD and SCAR markers linked to genes affecting plant architecture. Theor Appl Genet 97:916–928

Schneider A, Walker SA, Sagan M, Duc G, Ellis THN, Downie JA (2002) Mapping of the nodulation loci *sym9* and *sym10* of pea (*Pisum sativum* L.). Theor Appl Genet 104:1312–1316

Schroeder HE, Schotz AH, Wardley-Richardson T, Spencer D, Higgins TJV (1993) Transformation and regeneration of two cultivars of pea (*Pisum sativum* L.). Plant Physiol 101:751–757

Stracke S, Sato S, Sandal N, Koyama M, Kaneko T, Tabata S, Parniske M (2004) Exploitation of collinear relationships between the genomes of *Lotus japonicus*, *Pisum sativum* and *Arabidopsis thaliana*, for positional cloning of a legume symbiosis gene. Theor Appl Genet 108:442–449

Tar'an B, Warkentin T, Somers DJ, Miranda D, Vandenberg A, Blade S, Woods S, Bing D, Xue A, DeKoeyer D, Penner G (2003) Quantitative trait loci for lodging resistance, plant height and partial resistance to mycosphaerella blight in field pea (*Pisum sativum* L.). Theor Appl Genet 107:1482–1491

Tar'an B, Warkentin T, Somers DJ, Miranda D, Vandenberg A, Blade S, Bing D (2004) Identification of quantitative trait loci for grain yield, seed protein concentration and maturity in field pea (*Pisum sativum* L.). Euphytica 136:297–306

Timmerman GM, Frew TJ, Miller AL, Weeden NF, Jermyn WA (1993) Linkage mapping of *sbm-1*, a gene conferring resistance to pea seed-borne mosaic virus, using molecular markers in *Pisum sativum*. Theor Appl Genet 85:609–615

Timmerman GM, Frew TJ, Weeden NF, Miller AL, Goulden DS (1994) Linkage analysis of *er-1*, a recessive *Pisum sativum* gene for resistance to powdery mildew fungus (*Erysiphe pisi* D.C.). Theor Appl Genet 88:1050–1055

Timmerman-Vaughn GM, McCallum JA, Frew TJ, Weeden NF, Russell AC (1996) Linkage mapping of quantitative trait loci controlling seed weight in pea (*Pisum sativum* L.). Theor Appl Genet 93:431–439

Timmerman-Vaughn GM, Frew TJ, Weeden NF (2000) Characterization and linkage mapping of R-gene analogous DNA sequences in pea (*Pisum sativum* L.). Theor Appl Genet 101:241–247

Timmerman-Vaughn GM, Frew TJ, Russell AC, Khan T, Butler R, Gilpin M, Murray S, Falloon K (2002) QTL mapping of partial resistance to field epidemics of ascochyta blight of pea. Crop Sci 42:2100–2111

Turner L, Ellis THN (1997) Repeated sequence markers in pea (*Pisum sativum*). New Phytol 137:35–41

Valderrama MR, Roman B, Aatovic Z, Rubiales D, Cubero JI, Torres AM (2004) Locating quantitative trait loci associated with *Orobanche crenata* resistance in pea. Weed Res 44:323–328

von Stackelberg M, Linemann S, Menke M, Riesselmann S, Jacobsen HJ (2003) Identification of AFLP and STS markers closely linked to the *def* locus in pea. Theor Appl Genet 106:1293–1299

Vos P, Hogers R, Bleeker M, Reijans M, Van de Lee T, Hornes M, Frijters A, Pot J, Peleman J, Kuiper M, Zabeau M (1995) AFLP: a new technique for DNA fingerprinting. Nucleic Acids Res 23:4407–4414

Weeden NF (1996) Genetic studies in pea and apple: the yin and yang of mapping in plants. Hort Sci 31(7):1111–1116

Weeden NF, Marx GA (1984) Chromosomal locations of twelve isozyme loci in *Pisum sativum*. J Hered 75:365–370

Weeden NF, Marx GA (1987) Further genetic analysis and linkage relationships of isozyme loci in pea. J Hered 78:153–159

Weeden NF, Muehlbauer FJ, Ladizinsky G (1992) Extensive conservation of linkage relationships between pea and lentil genetic maps. J Hered 83:123–129

Weeden NF, Swiecicki WK, Timmerman-Vaughn GM, Ellis THN, Ambrose M (1996) The current pea linkage map. Pisum Genet 28:1–4

Weeden NF, Timmerman-Vaughn GM, Ellis THN, Swiecicki WK, Rosov S, Berdnikov V (1998) A consensus map of the pea genome. Pisum Genet 30:1–4

Yu J, Gu WK, Provvidenti R, Weeden NF (1995) Identifying and mapping two DNA markers linked to the gene conferring resistance to pea enation mosaic virus. J Am Soc Hort Sci 120(5):730–733

Yu J, Gu WK, Weeden NF (1996) Development of an ASAP marker for resistance to bean yellow mosaic virus in *Pisum sativum*. Pisum Genet 28:31–32

Zohary D, Hopf M (1973) Domestication of pulses in the old world. Science 182:887–894

(Steele 1976). Interestingly, while West Africa appears to be the major center of diversity of cultivated forms of cowpea (Ng and Padulosi 1988) and was probably domesticated by farmers in this region (Ba et al. 2004), the center of diversity of wild *Vigna* species is southeastern Africa (Padulosi and Ng 1997). Some evidence that domestication occurred in northeastern Africa, based on studies of amplified fragment length polymorphism (AFLP) analysis, has also been presented (Coulibaly et al. 2002). The wild cowpea *Vigna unguiculata* ssp. *unguiculata* var. *spontanea* is the likely progenitor of cultivated cowpea (Pasquet 1999).

It is likely that the crop was first introduced to India during the Neolithic period, and therefore India appears to be a secondary center of genetic diversity (Pant et al. 1982). "Yardlong beans," a unique cultivar group (Sesquipedialis) of cowpea that produces very long pods widely consumed in Asia as a fresh green or "snap" bean, apparently evolved in Asia and is rare in African landrace germplasm. Cowpea has been cultivated in southern Europe at least since the 8th century BC and perhaps since prehistoric times (Tosti and Negri 2002). Cowpea was introduced to the West Indies in the 16th century by the Spanish and was taken to the USA about 1700 (Pursglove 1968). Presumably it was introduced into South America at about the same time.

Web sites for the International Institute of Tropical Agriculture (www.IITA.org) and for the United States Agency of International Development (USAID)-funded Bean/Cowpea Collaborative Research Support Program (Bean/Cowpea CRSP) (www.isp.msu.edu/CRSP) are excellent references for general information about cowpea and current cowpea research.

3.1.2 Morphological and Phenological Characteristics

Cowpea is an herbaceous warm-season annual that is similar in appearance to common bean except that leaves are generally darker green, shinier, and less pubescent. Cowpeas also are generally more robust in appearance than common beans with better developed root systems and thicker stems and branches. Plant growth habit can be erect, semierect, prostrate (trailing), or climbing depending mostly on genotype, although photoperiod and growing conditions can also affect plant stature. Most cowpea accessions have indeterminate stem and branch apicies. Early flowering cowpea genotypes can produce a crop of dry grain in 60 d, while longer season genotypes may require more than 150 d to mature depending on photoperiod. Flowers are borne on racemes on 15- to 40-mm peduncles that arise from the leaf axils. Two or three pods per peduncle are common, and often four or more pods are carried on a single peduncle if growing conditions are very favorable. The presence of these long peduncles is a distinguishing feature of cowpea, and this characteristic also facilitates hand harvesting.

Cultivated cowpea seed weighs between 8 and 32 mg and ranges from round to kidney shaped. Pods are cylindrical and may be curved or straight, with between 8 and 15 seeds per pod. The seed coat can be either smooth or wrinkled and of var-

Fig. 1. Diversity of seed types in cowpea. Shown in photograph is variation in seed shape, color, and texture observed in cowpea from around the world (picture courtesy of J.D. Ehlers)

ious colors including white, cream, green, buff, red, brown, and black (Fig. 1). Seed may also be speckled or patterned. Seeds of well-known cowpea types, such as "blackeye pea" and "pinkeye," are white with a round irregular-shaped black or red pigmented area encircling the hilum, giving the seed the appearance of an eye.

Emergence is epigeal (similar to common bean and lupin), where the cotyledons emerge from the ground during germination. This type of emergence makes cowpea more susceptible to seedling injury, since the plant does not regenerate buds below the cotyledonary node. The open display of flowers in and above the canopy and the presence of extrafloral nectaries contribute to the attraction of insects. Cowpea primarily is self-pollinating, but outcrossing rates as high as 5% have been recorded and care needs to be taken to avoid outcrossing during the production of breeder and foundation seed, or unacceptable levels of "off-types" will result.

Cowpea is a short day plant, and many cowpea accessions exhibit photoperiod sensitivity with respect to floral bud initiation and development, while others are day neutral (Ehlers and Hall 1996; Craufurd et al. 1997). For some genotypes, the degree of sensitivity to photoperiod (extent of delay in flowering) is modified by temperature (Wein and Summerfield 1980; Ehlers and Hall 1996). In West Africa, selection for differing degrees of photosensitivity or differences in juvenility has occurred in different climatic zones such that pod ripening coincides with the end of the rainy season in a given locale, regardless of planting date, which is often variable due to the variable onset of wet seasons (Steele and Mehra 1980). This attribute allows pods to escape damage from excessive moisture and pathogens. Photoperiod sensitivity, when appropriately deployed in a breeding program, can be valuable to ensure crop maturity after wet seasons or before drought or cold weather limits crop growth. However, it may constrain the direct usefulness of an otherwise desirable cultivar to a small area of adaptation or even to a specific season within this restricted area.

Cultivated cowpeas have been divided into five cultivar groups based mainly on pod and seed characteristics (Pursglove 1968; Pasquet 1999). Cultivar group Unguiculata is the largest and includes most medium- and large-seeded African grain and forage-type cowpeas. Cultivar group Melanophthalmus includes "blackeye pea"-type cowpea with large, somewhat elongated seeds with wrinkled seed coats and fragile pods (Pasquet 1998). Members of cultivar group Biflora (also known as "catjang") are common in India and characterized by their relatively small smooth seeds borne in short pods that are held erect until maturity. Cultivar group Textilis is a rather rare form of cowpea with very long peduncles that were used in Africa as a source of fiber. Cultivar group Sesquipedialis (known as "yardlong bean," "long bean," "Asparagus bean," or "snake bean") is widely grown in Asia for production of its very long (40 to 100 cm) green pods that are used as "snap" beans. Despite the striking differences in morphological characteristics among the cultivar groups, there are no practical barriers to hybridization or recombination between members of the different groups.

3.1.3
Production Systems and Economic Importance

Production of cultivar group Sesquipedialis (or yardlong) beans is widespread throughout Asia and is thought to be grown on about 300,000 ha. Dry grain production is the only commodity of cowpea formerly estimated on a worldwide basis. The United Nations Food and Agricultural Organization (FAO) estimates that nearly 4 million metric tons (mt) of dry cowpea grain is produced annually on about 10 million ha worldwide (www.faostat.fao.org/faostat). Cowpea grain production estimates by Singh et al. (2002) are slightly higher than FAO estimates, with worldwide production of 4.5 million (mt) on 12 to 14 million ha. About 70% of this production occurs in the drier Savanna and Sahelian zones of West and Central Africa, where the crop is usually grown as an intercrop with pearl millet [*Pennisetum glaucum* (L.) R.Br.] or sorghum [*Sorghum bicolor* (L.) Moench] and, less frequently, as a sole crop or intercropped with maize (*Zea mays* L.), cassava (*Manihot esculenta* Crantz), or cotton (*Gossypium* sp.) (Langyintuo et al. 2003). Other important production areas include lower elevation areas of eastern and southern Africa and in South America (particularly in northeastern Brazil and in Peru), parts of India, and the southeastern and southwestern regions of North America. Nigeria is the largest producer and consumer of cowpea grain, with about 5 million ha and over 2 million mt production annually, followed by Niger

(650,000 mt) and Brazil (490,000 mt) (Singh et al. 2002). Estimates of cowpea grain production in Latin America and East and southern Africa, regions of the world that produce significant quantities of common beans [*Phaseolus vulgaris* (L.)], may be underestimates because cowpea grain is not always distinguished from common bean grain during collection of production statistics. Trade in dry cowpea grain and cowpea hay are important to the economy of West Africa in particular, with substantial quantities of cowpea grain being traded at the local and regional level (Singh 2002; Langyintuo et al. 2003). The large urban centers of coastal West Africa are huge markets for cowpea produced further inland where climates are drier and favorable to production of high-quality grain. The United States produces about 80,000 mt, in several southern states (Alabama, Arkansas, Georgia, Louisiana, Missouri, Tennessee) and in Texas and California (Fery 2002).

A long-term drought in the Sahelian zone of West Africa has caused many farmers in this part of Africa to shift more of their production to cowpea because of its drought tolerance (Duivenbooden et al. 2002). As a result of this shift in production and the adoption of new varieties and improved production systems, worldwide cowpea production has gone from an annual average of about 1.2 million mt during the decade of the 1970s to ca. 3.6 million mt per annum (during the five-year period spanning 1998 to 2003) according to the FAO (http://faostat.fao.org/faostat). Rapidly growing populations with high per-capita cowpea consumption in the West and Central African regions have fueled demand for cowpea grain during this period, and the trend is expected to continue.

3.1.4 Nutritional Composition

The nutritional content of cowpea grain is important because it is eaten in quantity by millions of people who otherwise have diets lacking in protein, minerals, and vitamins. The nutritional profile of cowpea grain is similar to that of other pulses, with a relatively low fat content and a total protein content that is two to four times greater than cereal and tuber crops. Like other pulses, the protein in cowpea grain is rich in the amino acids lysine and tryptophan, compared to cereal grains. However, it is deficient in methionine and cystine when compared to animal proteins. In a study of 100 cowpea breeding lines in the IITA collection, seed protein content ranged from 23 to 32% of seed weight (Nielson et al. 1993). Similarly, protein content of 12 West African and US cultivars ranged from 22 to 29%, with most accessions having protein content values between 22 and 24% (Hall et al. 2003). These results suggest that sufficient genetic variation exists to develop new cowpea cultivars with protein content of at least 30%. Cowpea grain is also a rich source of minerals and vitamins (Hall et al. 2003) and it has one of the highest levels of any food of folic acid, a crucial B vitamin that helps prevent spinal tube defects in unborn children (http://www.cdc.gov/doc.do/id/0900f3ec8000d558).

Cowpea can be used at all stages of growth as a vegetable crop, and the leaves contain significant nutritional value (Ahenkora et al. 1998; Nielson et al. 1993). The tender green leaves are an important food source in Africa and are prepared as a pot herb, like spinach. Immature green pods are used in the same way as snap beans, often being mixed with cooked dry cowpeas or with other foods. Nearly mature "fresh-shelled" cowpea grains are boiled as a fresh vegetable or may be canned or frozen. Dry mature seeds are also suitable for boiling and canning. In many areas of the world, cowpea foliage is an important source of high-quality hay for livestock feed (Tarawali et al. 2002).

In developed countries, cowpea is expected to become increasingly important as consumers seek interesting and healthy "new" foods and rediscover "traditional" foods that are low in fat, high in fiber, and that have other health benefits. Fat contents of 100 advanced breeding lines from IITA showed a range in fat contents from 1.4 to 2.7% (Nielson et al. 1993), while fiber content is about 6% (Bressani 1985). Besides being low in fat and high in fiber, the protein in grain legumes like cowpea has been shown to reduce low-density lipoproteins that are implicated in heart disease (Phillips et al. 2003). In addition, because grain legume starch is digested more slowly than starch from cereals and tubers, their consumption produces fewer abrupt changes in blood glucose levels following consumption (Phillips et al. 2003). Innovative and appealing processed-food products using dry cowpea grain, such as cowpea-fortified baked goods, extruded snack foods, and weaning foods, have been developed (Phillips et al. 2003). Protein isolates from cowpea grains have good functional properties, including solubility emulsifying and foaming activities (Rangel et al. 2004), and could be a substitute for soy protein isolates for persons (especially infants) with soy protein allergies.

Varieties of cowpea with a "persistent-green" grain have been developed by breeding programs in the USA that are a versatile product for frozen vegetable applications (Ehlers et al. 2002 a). Persistent-green cowpea grains are green colored when dry but when soaked in water for several hours closely resemble fresh-shelled cowpea that can be used in frozen vegetable products to add color and variety. Because persistent-green cowpea grain can be harvested and stored dry until rehydration and freezing, it is a quite convenient and economical frozen vegetable compared to other frozen vegetable crops that require highly coordinated harvesting and processing operations and expensive long-term frozen storage.

3.1.5 Classical Genetics and Breeding

Significant long-term genetic improvement efforts of cowpea have taken place within national laboratories and universities in several West African countries, India, Brazil, and the USA, as well as at the International Institute of Tropical Agriculture (IITA), based in Ibadan, Nigeria. The accomplishments of some of these programs have been described recently (Ehlers et al. 2002 a; Singh et al. 2002; Hall et al. 2003).

Most cowpea breeders employ backcross, pedigree, or bulk breeding methods to handle segregating populations because cowpea is a self-pollinating species and varieties are pure lines. Grain yield and quality are primary breeding objectives of nearly all programs, but because losses to diseases and pests can be high, most programs are also concentrating on breeding for resistance to the major pests they face in their target environments. A comprehensive review of cowpea breeding that is still relevant was published in 1997 (Hall et al. 1997).

Sources of resistance to many viruses and fungal diseases have been identified, and screening techniques are well developed for many of these (Ehlers and Hall 1997). In general, good progress has been made using conventional techniques in breeding for resistance to the parasitic weeds *Striga gesneroides* (witchweed) and *Alectra vogelii*, root-knot nematodes, viruses, and fungal and bacterial diseases. Unfortunately, resistance to these pathogens and parasites is usually governed by single genes that are often only effective in a restricted region due to pathogen/parasite variability and may be overcome in a relatively short period of time. Marker-assisted selection (MAS) can be helpful in assembling more durable resistance by incorporating an array of resistance genes from other regions or defeated resistance genes, as discussed below.

Developing cultivars with sustainable resistance to insects is a key objective of breeding programs throughout the world for several reasons. Insect damage is the number one constraint for cowpea grain production in most cowpea-producing regions (Singh and van Emden 1979; Daoust et al. 1985). There is also concern that new and significantly more stringent restrictions on the use of some popular insecticides are forthcoming, and currently there is a lack of new alternative insect-control products registered for use on cowpea. The insecticides themselves, or the financial resources required to purchase them and the equipment required for proper application, are simply not available to the vast majority of farmers in Africa. In addition, there are concerns that the increased use of insecticides could cause major environmental and safety problems.

Breeding insect-resistant cowpeas would have a significant impact on food availability and nutritional status in many regions. Achieving this goal will not be easy, however, because of the number and diversity of pests that attack the crop and the nature of the pests. In many regions of the world, multiple pest resistance is needed to permit adequate grain production without the use of insecticides. This is because attacks by any one of the major pests can be devastating. For example, if cultivars were developed with a high level of resistance to flower thrips, capable of protecting their floral buds from damage, any resulting flowers and pods on these plants would likely be destroyed by pod bugs and pod borers. However, resistance to individual pests can reduce the number of sprays needed to obtain optimal yields and would generally increase yields without insect protection in regions where pest pressure is moderate, as in the case of the Sahel.

Screening methods have been developed for several major insect pests of cowpea (Ehlers and Hall 1997). However, despite the evaluation of hundreds to thousands of cowpea accessions, plants with high levels of resistance to most notable significant pests have not been identified. The notable exceptions are resistance to cowpea aphid (*Aphis craccivora*) and leaf hoppers (*Empoasca* sp.). Recurrent selection is being used to combine low to moderate levels of resistance to flower thrips, pod bugs, and Maruca pod borer identified in several genotypes (Singh et al. 2002). However, progress in this area is being hampered by the low heritability of the traits based on the field screening methods used. Identification of molecular markers for insect resistance could facilitate transfer and pyramiding of the resistance genes.

3.1.6 Germplasm Collections

Cowpea germplasm is maintained in collections around the world with varying levels of accessibility and documentation. The largest collections are held by the IITA with more than 14,000 accessions. The collection can be accessed via an electronic database maintained through the CGIAR-SINGER system (http://singer.cgiar.org). The United States Department of Agriculture (USDA) maintains a collection with ca. 8,000 accessions. Access to this collection is through the USDA Germplasm Resources Information Network or GRIN system (www.ars-grin.-gov). The University of California-Riverside has a collection with ca. 5000 accessions accessible on a Microsoft Access database. There is also a large collection of Mediterranean and African landraces (ca. 600 accessions) held at the Istituto di Genetica Vegetale at Bari, Italy (www.ba.cnr.it). Other centers maintaining seed of wild and cultivated cowpeas include the following: Agricultural University-Wageningen (Wageningen, The Netherlands), Botanical Research Institute (Pretoria, South Africa), Le Jardin Botanique National de Belgique (Meise, Belgium), International Plant Genetic Resources Institute (IPGRI) in Harare (Zimbabwe), Institut Français de la Recherché Scientifique pour le Développement en Coopération (ORSTOM; now IRD) in Montpellier (France), Empresa Brasileira de Pesquisa Agropecuária (EMBRAPA) in Goiana (Brazil), Zentralinstitut für Genetik und Kulturpflanzenforschung (GAT) in Gatersleben (Germany), and the National Bureau of Plant Genetic Resources in New Delhi (India).

In addition to the centers and facilities mentioned above, many national cowpea breeding programs in Africa (including programs in Botswana, Burkina Faso, Ghana, Kenya, Nigeria, and Senegal) also have substantial germplasm collections. The condition of some of these collections, which are important reserves of local diversity, could be improved with funding for germplasm maintenance and facility repair.

3.2 Molecular Phylogeny and Genome Organization

Cowpea (*Vigna unguiculata*) is one of several important cultivated species that constitute the genus *Vigna*. Other members include mungbean (*V. radiata*), adzuki bean (*V. angularis*), blackgram (*V. mugo*), and the bambara groundnut (*V. subterranea*). The genus was initially divided into several subgenera based upon morphological characteristics, extent of genetic hybridization/reproductive isolation, and geographic distribution of species (Marechal et al. 1978). The major groupings consist of the African subgenera *Vigna* and *Haydonia*, the Asian subgenus *Ceratotropis*, and the American subgenera *Sigmoidotropis* and *Lasiopron*. Under the scheme proposed by Marechal and his colleagues, cultivated cowpea was placed in the subgenus *Vigna*, whereas mungbean and blackgram were placed in the Asian subgenera.

The development and use of biochemical-based analytical techniques and molecular-marker technologies, such as analysis of restriction fragment length polymorphisms (RFLPs), random amplified polymorphic DNAs (RAPDs) (Williams et al. 1990), amplified fragment length polymorphisms (AFLPs) (Vos et al. 1995), minisatellites (Sonnante et al. 1994), and simple sequence repeats (SSRs) (Akkaya et al. 1992, 1995), have greatly facilitated the analysis of the structure of plant genomes and their evolution. This in turn has contributed significantly to our current understanding of cowpea genome organization. Using RFLP analysis, Fatokun et al. (1993 a) analyzed 18 *Vigna* species including five of the subgenus *Ceratotropis* in order to determine the taxonomic relationship between the subgenus *Ceratotropis* and other subgenera. These investigators showed that a high level of genetic variation exists within the genus, with a remarkably higher amount of variation associated with *Vigna* species from Africa relative to those from Asia. Their data supported the taxonomic separation of the Asian and Africa genera as proposed by Marechal et al. (1978) and underscored the previously held viewpoint that Africa is the likely center of diversity for *Vigna*. In general, the placement of species and subspecies based upon molecular taxonomic procedures by Fatokun et al. (1993a) substantiated prior classifications based on classical taxonomic criteria, such as morphological and reproductive traits.

Genetic variation in the subgenus *Ceratotropis* was subsequently reinvestigated by using RAPD analysis (Kaga et al. 1996a). Examining the extent of polymorphism in 23 accessions of five species within the subgenus *Ceratotropis*, these investigators identified ca. 404 amplified fragments capable of providing comparative information. Based on the degree of polymorphism at these informative loci, these investigators were able to separate the accessions into two main groups differing by ca. 70% at the molecular level. Within each of the main groups, the accessions could be further divided into five subgroups whose composition were in complete agreement with their taxonomic species classifications.

Sonnante et al. (1996) examined isozyme variation between *V. unguiculata* and other species in the subgenus *Vigna* and showed that *V. unguiculata* was more closely related to *V. vexillata*, a member of the subgenus *Plectotropis*, than to any other species belonging to section *Vigna*. This is not surprising since *V. vexillata* is thought to be the intermediate species between African and Asian *Vigna* species. Vaillancourt and Weeden (1996) reached a similar conclusion about the relatedness of these species. Based on an analysis of variation in chloroplast DNA structure (Vaillancourt and Weeden 1992) and isozyme polymorphisms (Vaillancourt et al. 1993), it was suggested that *V. vexillata* and *V. reticulata* were the closest relatives of *V. unguiculata*. While the close relationship between *V. unguiculata* and *V. vexillata* proposed by Vaillancourt and Weeden (1996) is consistent with previous observations (Marechal et al. 1978), *V. reticulata* was placed in a different cluster based upon RFLP analysis (Fatokun et al. 1993a).

Polymorphisms in 21 different enzyme systems were used by Pasquet (1999) to evaluate the relationship between 199 accessions of wild and cultivated cowpea differing in breeding system and growth characteristic (i.e., annual vs. perennial growth habit). Based on these allozyme data, perennial subspecies of cowpea (spp. *unguiculata* var. *unguiculata*) were shown to form a coherent group closely related to annual forms (ssp. *unguiculata* var. *spontanea*). Among the ten subspecies studied, *V. unguiculata* var. *spontanea* and ssp. *pubescens* were the closest taxa to cultivated cowpea. Most recently, Ajibade et al. (2000) used inter simple sequence repeat (ISSR) DNA polymorphism analysis to study the genetic relationships among 18 *Vigna* species. They showed that closely related species within each subgenus clustered together [e.g., *V. umbellata* and *V. angularis* (subgenus *Ceratotropis*), *V. adenantha* and *V. caracalla* (subgenus *Sigmoidotropis*), and *V. luteola* and *V. ambacensis* (subgenus *Vigna*)]. Cultivated cowpea grouped closely with the wild subspecies of *V. unguiculata*, and the entire species was separated from its most closely allied species *V. triphylla* and *V. reticulata*. ISSR polymorphism analysis split *Vigna* into groupings that differed in their composition from previous classifications. For example, the subgenus *Vigna* was split into three lineages, with *V. unguiculata/reticulata/friesorum* forming one group, *V. luteola/ambacensis* forming a second, and *V. subterranea* being far from the other two. *Ceratotropis* split into two sections, with three species (*V. radiata*, *V. mungo*, and *V. acontifolia*) in one section and two species (*V. angularis* and *V. umbellata*) in a second section. While such groupings had been suggested previously (Marechal et al. 1978; Fatokun et al. 1993a; Vaillancourt and Weeden 1996), it should be noted that ISSR analysis was not as effective at resolving genetic distance relationships at the subgeneric level as it was at resolving relationships at the species level and below. Therefore, the authors note that their conclusions regarding subgeneric classifications should be taken with some caution. Thus, there is still considerable need to develop appropriate strategies and molecular techniques to resolve exact taxonomic relationships among members of this important genus.

Repetitive DNA sequences have been shown to represent a substantial fraction of the nuclear genome of all higher plant species and to account for much of the variation in genomic DNA content observed among species (Flavell et al. 1994). Many of the repeat sequences found in plant genomes appear to have originated through the activity of transposable elements (transposons) that either move by first forming an RNA intermediate [i.e., retrotransposons (Boeke et al. 1985)] or by direct

DNA transposition intermediates [i.e., transposons (Federoff 1989)]. To gain insight into the genomic organization and evolution of species within *Vigna*, Galasso et al. (1997) examined the genomic organization and distribution of Ty1-*copia* type retrotransposons in seven different species and subspecies of *Vigna* and several related leguminous plants. Gel blot analysis of genomic DNA from *V. unguiculata*, *V. luteola*, *V. oblongifolia*, *V. ambacensis*, and *V. vexillata* probed with radioactively labeled probes to the reverse transcriptase gene amplified from *V. unguiculata* subsp. *unguiculata*, *V. unguiculata* subsp. *dekindtania*, *V. luteola*, and *V. vexillata* showed variable hybridization patterns and intensities generally correlating with their previously defined taxonomic position. Fluorescence in situ hybridization analysis of the distribution of the Ty1-*copia* type sequences showed that these elements represented a major fraction of the cowpea genome and were dispersed relatively uniformly over all of the chromosomes. Little or no hybridization was found associated with centromeric, subtelomeric, and nucleolar organizing regions of the chromosomes, indicating that these portions of the genome may not be suitable sites for transposition. Comparisons of retrotransposon structural similarity between *Vigna* and other genera of legumes generally supported the subdivision of the tribes Phaseoleae and Vicieae, with greater homology being seen between members of the Cicereae and Phaseoleae than *Cicer* species and those from the Vicieae (Galasso et al. 1997).

Ba et al. (2004) used RAPD analysis to characterize genetic variation in domesticated cowpea and its wild progenitor, as well as their relationships. They included 26 domesticated accessions representing the five cultivar groups and 30 wild/weedy accessions, including accessions from West, East, and southern Africa. A total of 28 primers generated 202 RAPD bands. One hundred and eight bands were polymorphic among the domesticated compared to 181 among wild/weedy cowpea accessions. Wild accessions were more diverse in East Africa, which is the likely area of origin of *V. unguiculata* var. *spontanea*. *V. unguiculata* var. *spontanea* is thought to have spread westward and southward, with a loss of variability that is counterbalanced in southern Africa by introgressions with local perennial subspecies. Although the variability of domesticated cowpea was the highest ever recorded, cultivar groups were poorly resolved, and several results obtained with isozyme data were not confirmed here. However, primitive cultivars were more diverse than evolved cultivars, suggesting two consecutive bottlenecks within domesticated cowpea evolution. As with isozymes and AFLP markers, the RAPD data confirmed the single domestication hypothesis, the gap between wild and domesticated cowpea, and the widespread introgression phenomena between wild and domesticated cowpea. Therefore, these RAPD markers, which could have indicated a narrow center of origin, demonstrated that there is a widely distributed cowpea crop-weed complex all over Africa, as do some isozyme (Pasquet 1999), cpDNA (Pasquet, unpubl. obs.), and AFLP (Coulibaly et al. 2002) markers. Taking into account that there appears to have been a single domestication event, the genetic similarity of some of these wild accessions to the domesticated group would be the result of postdomestication gene flow between wild and domesticated forms due to their sympatric distribution.

3.3 Genetic Maps

The first attempt to generate a comprehensive linkage map for cowpea was by Fatokun et al. (1993 b), who used polymorphisms detected by 87 random genomic DNA fragments, 5 cDNAs, and RAPDs to generate a map consisting of ten linkage groups (LGs) spanning 680 cM. Improvement on this initial map was made by Menéndez et al. (1997), who were able to develop a linkage map for *V. unguiculata* consisting of 181 loci falling into 12 LGs. The resolution of the map was to ca. 6.4 cM between loci. Similarly, Menancio-Hautea et al. (1993 a, b) used RFLP analysis to construct a genome map of mungbean (*V. radiata*). The map consisted of 172 markers placed into 11 linkage groups and provided 1570 cM coverage with an average distance of 9 cM between loci. It is worth noting that even at these early stages of genome comparison, significant colinearity was recognized to exist between the cowpea and mungbean genomes (Menancio-Hautea et al. 1993 b). A total of 132 markers (108 RAPDs, 19 RFLPs, and 5 morphological markers) have been mapped in azuki bean using an interspecific population generated from a cross of *V. angularis* × *V. nakashimae* (Kaga et al. 1996 b). Comparison of the linkage map of azuki bean with those of

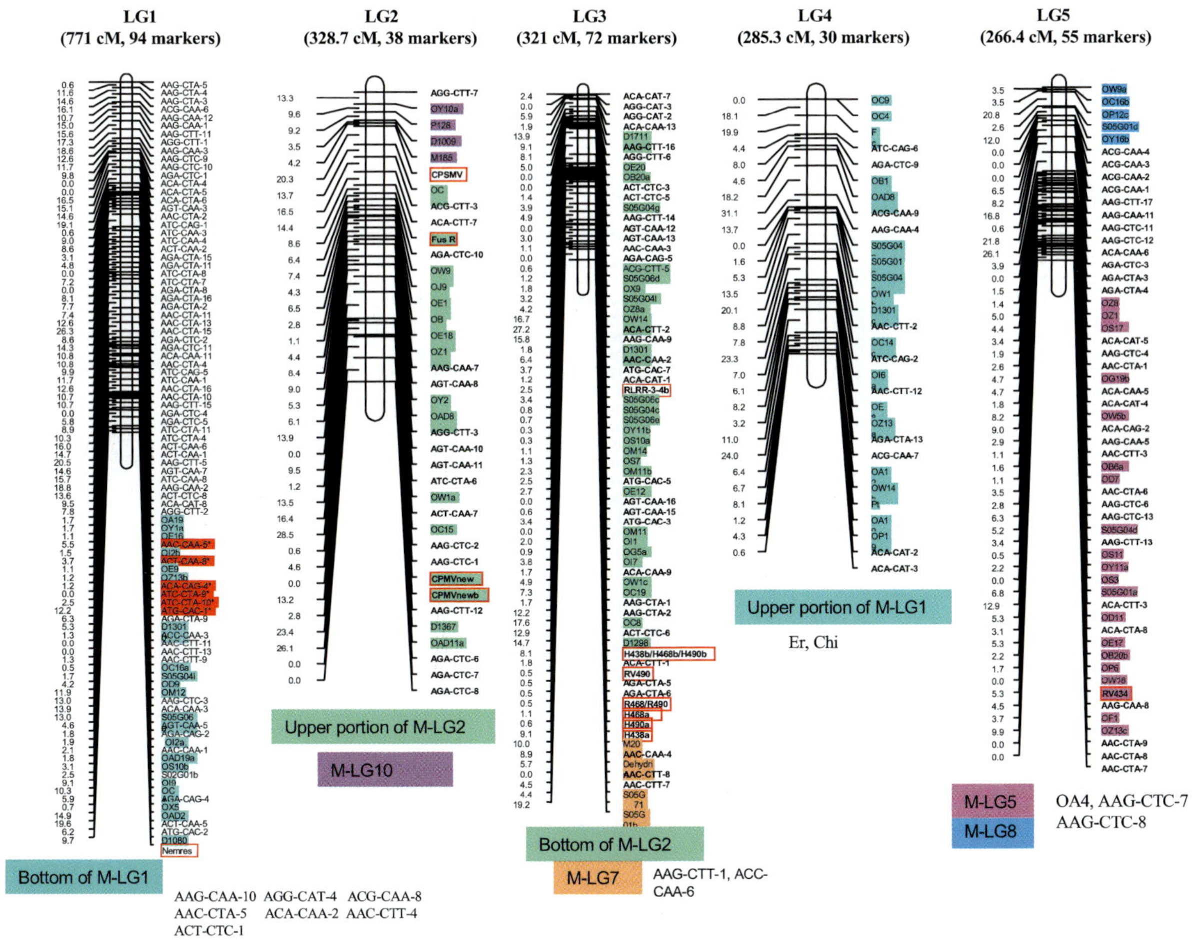

Fig. 2. Current genetic linkage map of cowpea. Shown are the 11 LGs comprising the genetic linkage map of cowpea as published by Ouédraogo et al. (2002 a). Above each LG is the length in centimorgans (cM) and number of markers comprising the LG. Distances (in cM) between adjacent markers are indicated to the *left*. Markers associated with LGs determined by Menéndez et al. (1997) are color coded in order to show their distribution on the current map. Markers linked to *Striga* resistance are given in *red* and marked by an asterisk. Loci for biological resistance/tolerance loci and resistance gene analogs (RGAs) are boxed in *red*. Markers that could not be placed with a LOD 3 score are listed under the LG they have the greatest affinity to. Unlinked markers are AAC-CTA-3, Parthcarp, AAC-CTT-10, ACA-CTA-7, ACG-CAA-10, AGG-CAT-1, R25, AAG-CTT-9

mungbean and cowpea using 20 RFLP markers indicated that, as might be expected, the three genomes share many linkage blocks in common.

Li et al. (1999) used DNA amplification fingerprinting (DAF) and AFLP analysis to identify additional molecular markers segregating in the F_8 recombinant inbred population derived from a cross between IT84S-2049 and 524B (Menéndez et al. 1997). These researchers screened 400 randomly generated DAF decamers and 128 AFLP primer combinations and were able to place 57 DAF and 90 AFLP markers to the existing cowpea genetic map. In addition, a map of the wild relative of cowpea *V. vexillata* has also been generated (Ogundiwin et al. 2000).

Building on the earlier version of the map developed by Menéndez et al. (1997), Ouédraogo et al. (2002 a) published what is the most current and complete map of *V. unguiculata* (http://pubs.nrc-cnrc.gc.ca/cgi-bin/rp/rp2_abst_f?gen_g01-102_45_ns_nf_gen1-02). This map was established in the recombinant inbred population IT84S-2049 × 524B developed by Tony Hall at the University of California-Riverside. IT84S-2049 is an advanced breeding line that was developed at IITA in Nigeria for multiple disease and pest resistance and has resistance to several races of blackeye cowpea mosaic virus (B1CMV) and to virulent root-knot nematodes in California (Menéndez et al. 1997). Line 524B is a

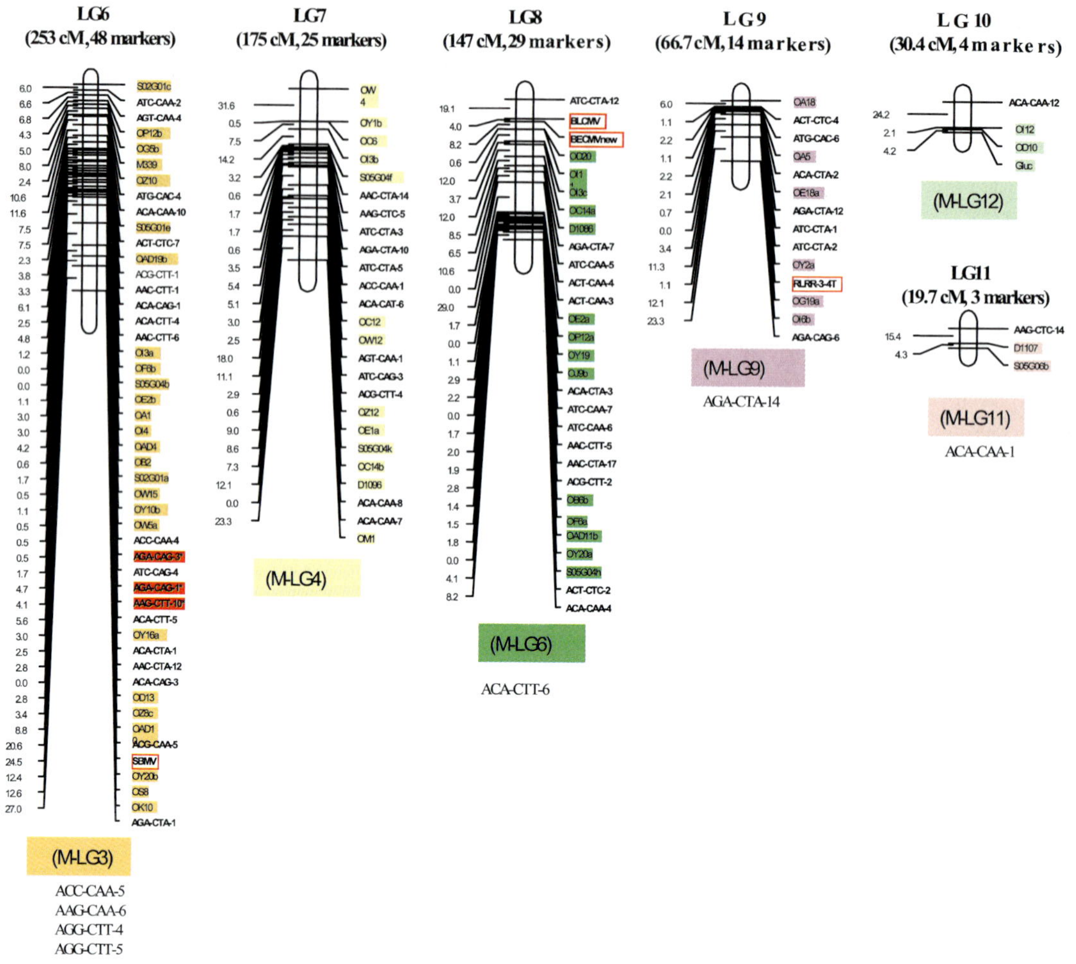

Fig. 2 (continued)

blackeye cowpea that shows resistance to Fusarium wilt and was derived from a cross between cultivars CB5 and CB3, which encompasses the genetic variability that was available in cowpea cultivars in California (Kelly et al. 2003).

The current map of cowpea consists of 11 LGs spanning a total of 2670 cM, with an average distance of ca. 6 cM between markers (Fig. 2). It includes 242 AFLP and 18 disease or pest-resistance-related markers (Ouédraogo et al. 2002 a), plus 133 RAPD, 39 RFLP, and 25 AFLP markers from the original map of Menéndez et al. (1997), for a total of 441 markers, of which 432 were assigned to a LG. Among these marker loci, genes for a number of biochemical and phenotypic traits have been located on this map. These include C, a general color factor, and P, for purple pod color, on LG4 [according to the numbering system of Ouédraogo et al. (2002 a), LGs on the bean and cowpea maps have been numbered independently; thus, LGs with the same number on the two maps probably refer to nonsyntenic groups], a 35-kDa dehydrin protein, implicated in chilling tolerance during emergence (LG2; Ismail et al. 1999), and markers for resistance to *Striga gesnerioides* races 1 and 3 (LG1 and LG6), cowpea mosaic virus (CPMV) and cowpea severe mosaic virus (CPSMV) (two distinct loci on LG2), B1CMV (LG8), southernbean mosaic virus (SBMV) (LG6), Fusarium wilt (LG3), and root-knot nematodes (gene *Rk*; NemR on LG1) (Ouédraogo et al. 2002 a). Candidate resistance genes (termed resistance gene analogs or RGAs) were also placed by RFLP analysis in various locations on the integrated cowpea map, including LG2, LG3, LG5, and LG9. Nevertheless, none of the RGA loci cosegregated with disease resistance phenotypes, suggesting that additional mapping for both RGAs and phenotypic disease resistance traits should be pursued in cowpea. Table 2 lists the various agronomic and disease resistance trait loci that have now been placed on the cowpea genetic map.

3.4 Transgenic Cowpea

Until recently cowpea remained one of the last major grain legume species for which an efficient genetic transformation/regeneration system had not been developed (Van Le et al. 2002; Avenido et al. 2004; Popelka et al. 2004), despite substantial efforts for more than ten years by several groups of researchers (Machuka 2002; Machuka et al. 2002). Ikea et al. (2003) reported the successful genetic transformation of cowpea using the particle-gun bombardment of shoot meristems. They were able to isolate several plants in the T3 generation that showed strong expression of the transgene "bar" that confers resistance to the herbicide Basta, but these studies were inconclusive. An efficient and stable cowpea transformation/regeneration system has been developed recently (Popelka et al. 2006), so that transgenic cowpea is now a reality.

Transgenic approaches should be undertaken to develop varieties of cowpeas with strong resistance to insect pests. Insect-resistant cowpeas would dramatically increase cowpea productivity in many developing countries and reduce costs, safety hazards, and environmental risks in virtually all cowpea-producing countries. Traditional plant breeding has made only limited progress in breeding for resistance to the major insect pests of cowpea and "new genes" are apparently needed to protect cowpea. The development and successful deployment of transgenic cultivars with genes conferring resistance to insects will be a major achievement.

The best current options for developing insect resistant cowpeas is to employ Bt technology against the Maruca pod borer (*Maruca testulalis*) and the alpha-amylase inhibitor gene from common bean that provides effective control of cowpea weevil (*Callosobruchus maculatus)* (T. J. Higgiens, personal communication, 2004). The soybean cysteine protease inhibitor soyacystatin N (scN) and alpha-amylase inhibitor (alphaAI) from wheat have synergistic effects against the cowpea weevil in artificial seed systems and are also potential genes that could be used to develop cowpea cultivars with resistance to this pest (Amirhusin et al. 2004).

Several different Cry1Ab, Cry1C, and CryIIA proteins that are produced by different forms of Bt genes are toxic to Maruca pod borer (L. E. N. Jackai, unpubl. data) and using these Bt genes in cowpea is considered a high priority for transformation (L. L. Murdock, personal communication, 2004). Genes producing plant lectins and plant proteinaceous inhibitors (PIs) of insect proteinases (serine, cysteine, aspartic, and metalloproteinases) are also considered potential candidates for gene transfer for resistance to Maruca pod borer (Machuka 2002).

3.5 Marker-Assisted Cowpea Breeding

Marker-assisted selection (MAS) is a tool to more efficiently assemble alleles of interest into an improved cultivar (Charcosset and Moreau 2004) and thereby increase the overall efficiency and effectiveness of crop improvement programs. Prior to applying MAS a realistic assessment of the cost-benefit ratio in comparison with phenotypic assays performed in the field, greenhouse, or laboratory needs to be conducted (Dekkers and Hospital 2002; Dreher et al. 2003). The expected economic return of MAS compared with phenotypic assessment decreases with the cost of genotyping (Moreau et al. 2000). In general, traits that are difficult or expensive to measure using phenotypic assays are good candidates for MAS. In some cases, MAS can allow smaller populations to be used, reduce the number of generations needed to reach a goal, or increase the accuracy of evaluations (Sharma et al. 2002). MAS offers the only practical method to combine multiple resistance genes into one cultivar when the genes mask the expression of one another, yet when together provide more durable resistance (Kelly et al. 2003). Other advantages of MAS are that a single technology can handle selection of diverse types of traits (e.g., pest resistance and grain quality parameters) and that cultivars developed through the use of MAS are not subjected to negative stereotyping as transgenic cultivars (Dubcovsky 2004). Also, selection of traits conferring resistance to quarantined pests can be conducted using MAS, eliminating the need for transfer of quarantined pests and assessment of resistance in expensive quarantine facilities.

MAS has yet to be implemented in cowpea, but some of the groundwork has been laid for its de-

Table 1. Races of *Striga gesnerioides* parasitic on cowpea in West Africa and differential responses of host cultivars and breeding lines [a]

Cowpea Cultivar	Race of *Striga gesnerioides* (Country of Origin) Race 1 (Burkina Faso, Mali)	Race 2 (Mali)	Race 3 (Niger, Nigeria)	Race 4 (Benin)	Race 5 (Cameroon, West Africa)
IT93K-693-2	R	R	R	R	R
B301	R	R	R	S	R
Suvita 2 (Gorom)	R	R	S	R	S
IT81D-994	R	R	S	R	S
IT82D-849	R	R	R	S	R
Tvu 14676	R	R	R	S	R
Tvx 3236	S	S	S	S	S
IT84S-2246	S	S	S	S	S
IT84S-2049	S	S	S	S	S

[a] Adapted from Lane et al. (1996, 1997)

Table 2. Agronomic, growth habit, and disease and pest resistance trait loci currently placed on the cowpea genetic map of Ouédraogo et al. (2002) and other traits mapped to probable nonanalogous linkage groups [1]

Trait	Locus designation	Linkage group/reference map
Pod pigmentation	P	LG1; (LG1-Menéndez et al. 1997)
Resistance to *Striga gesnerioides*-Race 1	*Rsg2-1*	LG1
Resistance to *Striga gesnerioides*-Race 3	*Rsg4-3, Rsg1-1*	LG1
Root-knot nematode (*Meloidogyne incognita*) resistance	*Rk*	LG1
Nodes to 1st Flower (D1301a)	NTF	LG2; (LG2-Menéndez et al. 1997)
Dehydrin protein	Dhy	LG2; (LG7-Menéndez et al. 1997)
Resistance to cowpea mosaic virus	CPMV	LG2
Resistance gene analog (pathogen unknown)	RGA-438	LG2
Resistance gene analog (pathogen unknown)	RGA-468	LG2
Resistance gene analog (pathogen unknown)	RGA-490	LG2
Resistance to *Fusarium oxysporum*	*Fus*R	LG3
Cowpea severe mosaic virus resistance	CPSMV (*ims*)	LG3
Cowpea mosaic virus resistance	CPMV	LG3
Resistance gene analog (pathogen unknown)	RLRR3-4B	LG3
General flower color factor	C	LG4; (LG1-Menéndez et al. 1997)
Seed weight (OB6a)	SW	LG5; (LG5-Menéndez et al. 1997)
Resistance gene analog (pathogen unknown)	RGA-434	LG5
Resistance to southern bean mosaic virus	SBMV (*sbc-1,2)*	LG6
Resistance to *Striga gesnerioides*-Race 1	*Rsg3-1, Rsg-994*	LG6
Resistance to blackeye cowpea mosaic virus	BlCMV	LG8
Resistance gene analogs (pathogen unknown)	RLRR3-4T	LG9
Traits mapped in other populations with probably nonanalogous linkage groups to map of Ouédraogo et al. 2002		
Resistance to cowpea aphid (*Aphid craccivora*)	*Rac1*	(LG1-Myers et al. 1996)
50% Flowering	50%FL	(LG7-Fatokun et al. 1993)
Seed weight	SW	(LG7-Fatokun et al. 1993)
Plant height	HT	(LG8-Fatokun et al. 1993)
Pod number per plant	PodN	(LG9-Fatokun et al. 1993)

[1] Adapted from genetic maps and data of Ouédraogo et al. (2002) and Menéndez et al. (1997) that used the same genetic population. There is insufficient marker data to integrate LGs of the maps of Fatokun et al. (1993) and data from Myers et al. (1996) with the map of Ouédraogo et al. (2002)

velopment (Kelly et al. 2003). A genetic map has been constructed (Ouédraogo et al. 2002 a) and loci controlling important pest and disease resistance genes and agronomic traits have been placed on the map (Section 3). In addition, markers closely linked to some resistance factors whose function has yet to be fully defined have been identified (Gowda et al. 2002). Many of these traits are controlled by single genes and therefore are potentially good candidates for MAS. Currently, no quantitative trait loci (QTLs) with linked markers have been identified for use in selecting for more complex traits such as grain yield.

At least five distinct races of the parasitic weed *Striga gesnerioides* have been identified within the cowpea-growing regions of West Africa (Lane et al. 1996, 1997) based on the differential response of various cowpea genotypes (cultivars and breeding lines) carrying specific resistance genes (Table 1). Similarly, "resistance-breaking" strains of the root-knot nematode *Meloidogyne incognita*, cowpea aphid (*Aphis craccivora*), cowpea weevil (*Callosobruchis maculatus*), and Fusarium wilt (*Fusarium oxysporum* f. sp. *tracheiphilum*) have been recognized in specific cowpea production areas. Markers for genes conferring resistance to the various strains of these pests would allow efficient development of varieties with resistance that is more broadly effective using MAS.

Currently, useful markers in cowpea for implementation of MAS are only available for some of the *Striga* resistance genes, and these are the first candidates for broad application in cowpea breeding programs (Fig. 3). Ouédraogo et al. (2001, 2002 b) found three AFLP markers linked to *Rsg2-1*, a gene that confers resistance to *Striga* Race 1

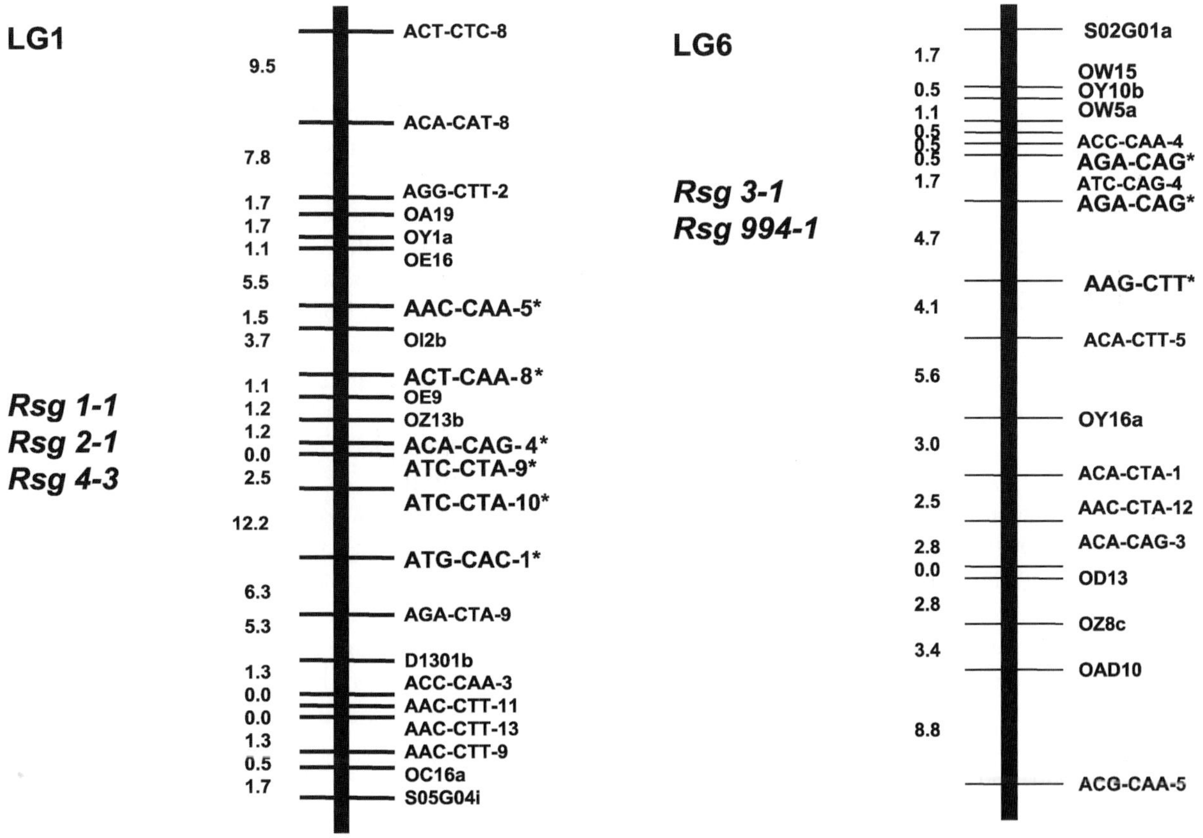

Fig. 3. Partial maps showing linkage of molecular markers to *S. gesnerioides* race-specific resistance genes in cowpea. Shown are the linkage of AFLP, RAPD, and other markers to *S. gesnerioides* race 1 (*Rsg2-1* and *Rsg1-1*) and race 3 (*Rsg4-3*) resistance genes on the partial map of LG1 of the cowpea genetic map (*left*) and the linkage of AFLP, RAPD, and other markers to *S. gesnerioides* race 1 (*Rsg3-1*) and race 3 (*Rsg994-1*) resistance genes on partial map of LG6 of the cowpea genetic map (*right*). AFLP markers linked to *Striga* resistance as reported in Ouédraogo et al. (2001) and Ouédraogo et al. (2002 b) are indicated by an asterisk. Map distances are shown in cM.

(SG1) present in Burkina Faso, and six AFLP markers linked to gene *Rsg4-3*, a gene that provides resistance to *Striga* Race 3 (SG3) from Nigeria. Two of the AFLP markers were associated with both *Rsg2-1* and *Rsg4-3*. Ouédraogo et al. (2002 a) were able to convert one of these markers to a SCAR (sequence-characterized amplified region) that has proven to be an effective and remarkably reliable marker for resistance to *Striga* SG1 and SG3 conferred by *Rsg2-1* and *Rsg4-3*. This SCAR marker, designated 61R (E-ACT/M-CAA), detects a single polymorphic band linked to SG1 and SG3 resistance in the resistant cultivars B301, IT82D-849, and Tvu 14676 and is being tested for use in breeding trials. Recently, two AFLP markers were identified that are closely linked to *Rsg1-1*, a gene that also confers resistance to SG3 in Nigeria (Boukar et al. 2004). One of the AFLP markers, designated E-ACT/M-CAC_{115} and determined to be 4.8 cM from *Rsg1-1*, was converted to a SCAR marker for ease of use in breeding programs (Boukar et al. 2004).

Chida et al. (2000) obtained three RAPD markers flanking a gene conferring resistance to cucumber mosaic cucumovirus (*Cry* gene) that could be useful in MAS. Linkage analyses of these molecular markers showed that genetic distances of the markers CRGA5, D13/E14-350, WA3-850, and OPE3-500 to the *Cry* locus were 0.7, 5.2, 11.5, and 24.5 cM, respectively.

Insect resistance is a good candidate for MAS in cowpea because assessments of host plant resistance to insects are often difficult to conduct in the field or greenhouse. Most insect resistance factors in cowpea do not provide immunity to the pest and often have low heritability under field conditions. Field screenings that rely on natural insect infestations are subject to natural fluctuations in pest pressure. When such variability is combined with incomplete resistance, field screens can lead to misclassification and selection of lines lacking the strongest resistance. For example, this has been the case with screening cowpea breeding lines and accessions for resistance to aphids, Lygus bug (*Lygus hesperus*), and pod-sucking bugs (such as *Nezara viridula, Clavigralla tomentosicollis, Riptortus dentipes*). In addition, colonies of insects may be difficult to rear without specialized facilities and trained entomologists to monitor the growth and uses. Such resources may not be available to cowpea breeding programs.

Resistance to the pod bug *Clavigralla tomentosicollis* has been identified in the wild cowpea (ssp. *dekindtiana*) germplasm line TVNu 151 (Koona et al. 2002). MAS could be used to introgress resistance factors from such wild cowpea into cultivated forms using a rapid backcrossing approach, based on simultaneous selection for the resistance genes (markers) and against markers associated with unwanted wild germplasm characteristics such as small seed size and seed shattering. Such an approach would require a substantial increase in the number of markers available in cowpea and the development of high-throughput markers such as SSR and SNP markers.

Implementation of MAS for resistance to root-knot nematodes (*Meloidogyne* spp.) in cowpea may be useful in some breeding programs. The genetic resistance to populations of these pests in the USA is well characterized (Roberts et al. 1996, 1997; Ehlers et al. 2002). At present, laboratory and field bioassays to assess resistance to root-knot nematodes in cowpea are effective and reasonably cost effective (Roberts et al. 1997; Ehlers et al. 2002 b). However, *Meloidogyne* populations are highly variable in response to resistance genes and resistance phenotyping is difficult for breeders to undertake without the close collaboration of nematologists for maintenance of cultures, preparation of inocula, and screening protocols. Current work to develop PCR-based markers tightly linked to the *Rk* locus that has multiple resistance specificities to *Meloidogyne* populations should lead to more effective breeding for nematode resistance in cowpea (Roberts et al. 1996, 1997; Ehlers et al. 2002).

The application of MAS for improvement of agronomic traits controlled by QTLs is much more difficult. Expression of many quantitative traits (such as yield) reflects the influence of many (often interacting) developmental processes over a substantial period of time such as a full growing season. As noted earlier there has been little progress toward the development of markers linked to QTLs useful in the selection of agronomic characteristics in cowpea. Progress has been faster in other related legumes (such as *Phaseolus*), and it is possible that some of this information may be leveraged since there is a significant degree of synteny between the bean and cowpea genomes (Kelly et al. 2003).

3.6 Future Prospects for Crop Improvement

One of the major goals of cowpea programs is to combine resistances to numerous pests and diseases and other desirable traits such as those governing maturity, photoperiod sensitivity, plant type, and seed quality. Parental lines with many desirable traits, such as resistance to cowpea weevil, cowpea aphid, and the parasitic weeds *Alectra vogeili* and *Striga gesnerioides*, along with resistances to bacterial blight, CABMV, and other pathogens, exist in different advanced breeding lines developed by cowpea breeding programs around the world. One of the biggest current challenges is to incorporate all of these desirable traits into individual cultivars with acceptable grain quality and adaptation to targeted farming systems and environments. MAS could be an important tool to facilitate this effort.

Cowpea remains to a large extent an underexploited crop where relatively large genetic gains can be made with only modest investments in both applied plant breeding and molecular genetics. Cowpea is grown mostly by poor farmers in developing countries and, as a consequence, has received relatively little attention from a research standpoint. Indeed, cowpea has been identified as an "orphan crop" that is recommended for increased public/donor support for biotechnology research (Naylor et al. 2004). A major challenge will be to apply the knowledge being gained from basic genomics research on "model species" such as Arabidopsis, rice (*Oryza sativa*), and *Medicago truncula*ta to cowpea.

Acknowledgements

The authors wish to thank Drs. Christopher J. Botanga, Bhavani S. Gowda, and Mohammad F. Ishiyaku for their comments. This work was supported in part by funds from the NSF (IBN-0322420) and a USAID University Linkage Grant with the International Institute of Tropical Agriculture (awarded to MPT), and grants from the Bean/Cowpea CRSP, USAID Grant no. DAN-G-SS-86-00008-00 to PAR. The opinions and recommendations are those of the authors and not necessarily those of USAID.

References

Ahenkora K, Adu-Dapaah HK, Agyemang A (1998) Selected nutritional components and sensory attributes of cowpea (*Vigna unguiculata* [L.] Walp.) leaves. Plant Foods Hum Nutr 52:221–229

Ajibade SR, Weeden NF, Chite SM (2000) Inter simple sequence repeat analysis of genetic relationships in the genus *Vigna*. Euphytica 111:47–55

Akkaya MS, Bhagwat AA, Cregan PB (1992) Length polymorphisms of simple sequence repeat DNA in soybean. Genetics 132:1131–1139

Akkaya MS, Shoemaker RC, Specht JE, Bhagwat AA, Cregan PB (1995) Integration of simple sequence repeat DNA markers into a soybean linkage map. Crop Sci 35:1439–1445

Amirhusin B, Shade RE, Koiwa H, Hasegawa PM, Bressan RA, Murdock LL, Zhu-Salzman K (2004) Soyacystatin N inhibits proteolysis of wheat alpha-amylase inhibitor and potentiates toxicity against cowpea weevil. J Econ Entomol 97:2095–2100

Avenido RA, Dimaculangan JG, Welgas JN, Del Rosario EE (2004) Plant regeneration via direct shoot organogenesis from cotyledons and cotyledonary node explants of pole sitao (*Vigna unguiculata* [L.] Walp. var *sesquipedalis* [L.] Koern.). Phillipine Agric Sci 87:457–462

Ba FS, Pasquet RE, Gepts P (2004) Genetic diversity in cowpea [*Vigna unguiculata* (L.) Walp.] as revealed by RAPD markers. Genet Resource Crop Evol 51:539–550

Boeke JD, Garfinkel DJ, Styles CA, Fink GR (1985) Ty elements transpose through an RNA intermediate. Cell 40:491–500

Boukar O, Kong L, Singh BB, Murdock L, Ohm HW (2004) AFLP and AFLP-derived SCAR markers associated with *Striga gesnerioides* resistance in cowpea. Crop Sci 44:1259–1264

Bressani R (1985) Nutritive value of cowpea. In: Singh SR, Rachie KO (eds) Cowpea Research, Production and Utilization. Wiley, New York, pp 353–359

Carsky RJ, Vanlauwe B, Lyasse O (2002) Cowpea rotation as a resource management technology for cereal-based systems in the savannas of West Africa. In: Fatokun CA, Tarawali SA, Singh BB, Kormawa PM, Tamo M (eds) Challenges and Opportunities for Enhancing Sustainable Cowpea Production. International Institute of Tropical Agriculture, Ibadan, Nigeria, pp 252–266

Charcosset A, Moreau L (2004) Use of molecular markers for the development of new cultivars and the evaluation of genetic diversity. Euphytica 137:81–94

Chida Y, Okazaki K, Karasawa A, Akashi K, Nakazawa-Nasu Y, Hase S, Takashi H, Ehara Y (2000) Isolation of molecular markers linked to the *Cry* locus conferring resistance to cucumber mosaic cucumovirus infection in cowpea. J Gen Plant Path 66:242–250

Ouédraogo JT, Gowda BS, Jean M, Close TJ, Ehlers JD, Hall AE, Gillespie AG, Roberts PA, Ismail AM, Bruening G, Gepts P, Timko MP, Belzile FJ (2002 a) An improved genetic linkage map for cowpea (*Vigna unguiculata* L.) combining AFLP, RFLP, RAPD, biochemical markers and biological resistance traits. Genome 45:175–188

Ouédraogo JT, Tignegre J-B, Timko MP, Belzile FJ (2002 b) AFLP markers linked to resistance against *Striga gesnerioides* race 1 in cowpea (*Vigna unguiculata*). Genome 45:787–793

Padulosi S, Ng NQ (1997) Origin, taxonomy, and morphology of *Vigna unguiculata* (L.) Walp. In: Singh BB, Mohan Raj DR, Dashiell KE, Jackai LEN (eds) Advances in Cowpea Research. Copublication of International Institute of Tropical Agriculture (IITA) and Japan International Research Center for Agricultural Sciences (JIRCAS). Sayce, Devon, UK, pp 1–12

Pant KC, Chandel KPS, Joshi BS (1982) Analysis of diversity in Indian cowpea genetic resources. SABRO J 14:103–111

Pasquet RS (1998) Morphological study of cultivated cowpea, *Vigna unguiculata* (L.) Walp. Importance of ovule number and definition of cv gr Melanophthalmus. Agronomie 18:61–70

Pasquet RS (1999) Genetic relationships among subspecies of *Vigna unguiculata* (L.) Walp. based on allozyme variation. Theor Appl Genet 98:1104–1119

Phillips RD, McWatters KH, Chinannan MS, Hung Y, Beuchat LR, Sefa-Dedeh S, Saki-Dawson E, Ngoddy P, Nnanyelugo D, Enwere J, Komey NS, Liu K, Mensa-Wilmot Y, Nnanna I, Okeke C, Prinyawiwatkul W, Saalia FK (2003) Utilization of cowpeas for human food. Field Crops Res 82:193–213

Popelka JC, Gollasch S, Moore A, Molvig L, Higgins TJV (2006) Genetic transformation of cowpea (*Vigna unguiculata* L.) and stable transmission of the transgenes to progeny. Plant Cell Rep 25:304–312

Popelka JC, Terryn N, Higgins TJV (2004) Gene technology for grain legumes: can it contribute to the food challenge in developing countries? Plant Science 167:195–206

Purseglove JW (1968) Tropical Crops – Dicotyledons. Longman, London, UK

Rangel A, Saraiva K, Schwengber P, Narciso MS, Domont GB, Ferreira ST, Pedrosa C (2004) Biological evaluation of a protein isolate from cowpea (*Vigna unguiculata*) seeds. Food Chem 87:491–499

Roberts PA, Matthews WC, Ehlers JD (1996) New resistance to virulent root-knot nematodes linked to the *Rk* locus of cowpea. Crop Sci 36:889–894

Roberts PA, Ehlers JD, Hall AE, Matthews WC (1997) Characterization of new resistance to root-knot nematodes in cowpea. In: Singh BB, Mohan Raj DR, Dashiell KE, Jackai LEN (eds) Advances in Cowpea Research. Copublication of International Institute of Tropical Agriculture (IITA) and Japan International Research Center for Agricultural Sciences (JIRCAS). Sayce, Devon, UK, pp 207–214

Sanden B (1993) Blackeye varietal and irrigation cutoff trial. In: University of California Dry Bean Research – 1993. Progr Rep California Dry Bean Advisory Board, Dinuba, CA, pp 120–121

Sanginga N, Dashiell KE, Diels J, Vanlauwe B, Lyasse O, Carsky RJ, Tarawali S, Asafo-Adjei B, Menkir A, Schulz S, Singh BB, Chikoye D, Keatinge D, Ortiz R (2003) Sustainable resource management coupled to resilient germplasm to provide new intensive cereal-grain-legume-livestock systems in the dry savanna. Agric Ecosyst Environ 100:305–314

Sharma HC, Crouch JH, Sharma KK, Seetharama N, Hash CT (2002) Applications of biotechnology for crop improvement: prospects and constraints. Plant Sci 163:381–395

Singh BB (2002) Recent genetic studies in cowpea. In: Fatokun CA, Tarawali SA, Singh BB, Kormawa PM, Tamo M (eds) Challenges and Opportunities for Enhancing Sustainable Cowpea Production. International Institute of Tropical Agriculture, Ibadan, Nigeria, pp 3–13

Singh SR, van Emden HF (1979) Insect pests of grain legumes. Annu Rev Entomol 24:255–278

Singh BB, Tarawali SA (1997) Cowpea and its improvement: key to sustainable mixed crop/livestock farming systems in West Africa. In: Renard C (ed) Crop Residues in Sustainable Mixed Crop/Livestock Farming Systems. CAB in Association with ICRISAT and ILRI, Wallingford, UK, pp 79–100

Singh BB, Ehlers JD, Sharma B, Freire Filho FR (2002) Recent progress in cowpea breeding. In: Fatokun CA, Tarawali SA, Singh BB, Kormawa PM, Tamo M (eds) Challenges and Opportunities for Enhancing Sustainable Cowpea Production. International Institute of Tropical Agriculture, Ibadan, Nigeria, pp 22–40

Sonnante G, Stockton T, Nodari RO, Becerra Velásquez VI, Gepts P (1994) Evolution of genetic diversity during the domestication of common bean (*Phaseolus vulgaris* L.). Theor Appl Genet 89:629–635

Sonnante G, Piergiovanni AR, Ng QN, Perrino P (1996) Relationships of *Vigna unguiculata* (L.) Walp., *V. vexillata* (L.) A. Rich., and species of section *Vigna* based on isozyme variation. Genet Resource Crop Evol 43:157–165

Steele WM (1976) Cowpea, *Vigna unguiculata* (Leguminosae-Papillionatae). In: Simmonds NW (ed) Evolution of Crop Plants. Longman, London, pp 183–185

Steele WM, Mehra KL (1980) Structure, evolution and adaptation to farming systems and environments in *Vigna*. In: Summerfield RJ, Bunting AH (eds) Advances in Legume Science. Royal Botanic Gardens, Kew, UK, pp 393–404

Tarawali SA, Singh BB, Peters M, Blade SF (1997) Cowpea haulms as fodder. In: Singh BB, Mohan Raj DR, Dashiell KE, Jackai LEN (eds) Advances in Cowpea Research. Copublication of International Institute of Tropical Agriculture (IITA) and Japan International Research Center for

Agricultural Sciences (JIRCAS). Sayce, Devon, UK, pp 313–325

Tarawali SA, Singh BB, Gupta SC, Tabo R, Harris F, Nokoe S, Fernández-Rivera S, Bationo A, Manyong VM, Makinde K, Odion EC (2002) Cowpea as a key factor for a new approach to integrated crop-livestock systems research in the dry savannas of West Africa. In: Fatokun CA, Tarawali SA, Singh BB, Kormawa PM, Tamo M (eds) Challenges and Opportunities for Enhancing Sustainable Cowpea Production. International Institute of Tropical Agriculture, Ibadan, Nigeria, pp 233–251

Tosti N, Negri V (2002) Efficiency of three PCR-based markers in assessing genetic variation among cowpea (*Vigna unguiculata* ssp. *unguiculata*) landraces. Genome 45:656–660

Vaillancourt RE, Weeden NF (1992) Chloroplast DNA polymorphism suggests a Nigerian center of domestication for the cowpea, *Vigna unguiculata* (Leguminosae). Am J Bot 79:1194–1199

Vaillancourt RE, Weeden NF (1996) *Vigna unguiculata* and its position within the genus *Vigna*. In: Pickersgill B, Lock JM (eds) Advances in Legume Systematics 8: Legumes of Economic Importance. Royal Botanic Gardens, Kew, UK, pp 89–93

Vaillancourt RE, Weeden NF, Barnard JD (1993) Isozyme diversity in the cowpea species complex. Crop Sci 33:606–613

Van Le B, de Carvalho MHC, Zully-Fodil Y, Thi ATP, Van KTT (2002) Direct whole plant regeneration of cowpea [*Vigna unguiculata* (L.) Walp] from cotyledonary node thin layer explants. J Plant Physiol 159:1255–1258

Vos P, Hogers R, Bleeker M, Reijans M, van de Lee T, Hornes M, Fritjers A, Pot J, Peleman J, Kuiper M, Zabeau M (1995) AFLP: a new technique for DNA fingerprinting. Nucleic Acids Res 23:4407–4414

Wein HC, Summerfield RJ (1980) Adaptation of cowpeas in West Africa: Effects of photoperiod and temperature responses in cultivars of diverse origin. In: Summerfield RJ, Bunting AH (eds) Advances in Legume Science. Royal Botanic Gardens, Kew, UK, pp 405–417

Williams JGK, Kubelik AR, Livak JJ, Rafalski JA, Tingey SV (1990) DNA polymorphisms amplified by arbitrary primers are useful as genetic markers. Nucleic Acids Res 18:6531–6535

4 Mungbean

C. J. Lambrides, I. D. Godwin

School of Land and Food Sciences, Plant Improvement Group, The University of Queensland, St Lucia, QLD 4072, Australia
e-mail: chris.lambrides@uq.edu.au

4.1 Introduction

Mungbean, *Vigna radiata* (L.) Wilczek, is an important grain legume crop in Asian agriculture, particularly in India, Southeast and East Asia, where it complements cereal-based diets with a large proportion of digestible protein through use as a pulse. In many parts of the developing and developed world, mungbean is also used in sprouted form as a salad vegetable. While substantial yield improvements have been made, mungbean yields are still low which has restricted its wider use as an alternative pulse crop in Asian farming systems. In addition to abiotic limitations associated with drought and heat stress and soil adaptation, mungbean has a number of specific productivity constraints, caused by foliar diseases and insect pests on-farm and postharvest. There are major product quality issues that have also slowed the speed of yield advance with conventional breeding programs.

The development of molecular plant improvement technologies has provided mungbean researchers with methods to make greater yield and quality improvements. This review will provide background information on mungbean growth, cultivation and breeding and give a detailed discussion of the molecular research of mungbean carried out to date.

4.1.1 Morphological Description and Growth Characteristics

Mungbean is a short-duration (70 to 110 d), warm-season grain legume adapted to tropical and subtropical conditions (Lawn 1979 a,b, 1983). The common name 'mungbean' has several synonyms (Table 1). Despite a photoperiod response to short day length, mungbean can be grown over a range of latitudes provided temperatures exceed 15 °C and production areas are frost free (Imrie 1996). Mungbean crops are short-stature, less than 1.25 m, depending on the variety and growing conditions. Plants are generally branched and habit can vary from erect to suberect in the cultivated types to

Table 1. Common names, wild progenitors and center of domestication for Asiatic *Vignas* belonging to the subgenus *Ceratotropis* (modified from Lawn 1995)

Species	Common names	Wild progenitor	Center of domestication
V. radiata (L.) Wilczek	Mungbean, greengram, moong, goldengram, Oregon pea, Chickasono pea, Chickasaw pea, chiroko	Var. *sublobata*	India
V. mungo (L.) Hepper	Urdbean, blackgram, urid, mash, mungo bean	Var. *silvestris*	India
V. angularis (Willd.)	Adzuki bean, red bean	Var. *niponensis*	NE Asia
V. umbellata (Thunb.)	Rice bean, red bean	Var. *gracilis*	SE Asia
V. aconitifolia (Jacq.)	Moth bean, mat bean	–	South Asia
V. trilobata (L.)	Pillipesara bean, jungli bean	–	South Asia
V. glabrescens (Marechal et al.)		*V. radiata* × *V. umbellata*	SE Asia

Genome Mapping and Molecular Breeding in Plants, Volume 3
Pulses, Sugar and Tuber Crops
C. Kole (Ed.)

prostrate in wild progenitors. Leaves are trifoliate and roots bear nodules that fix atmospheric nitrogen via a symbiotic relationship with the bacterium *Rhizobium*. Flowers are yellow and have typical legume 'butterfly' floral morphology with a large standard petal, two wing petals and two fused petals that form the keel, ten anthers and a single style. Seeds are smaller (< 8.0 g/100) than those of many other grain legumes. The seed appearance can vary greatly depending on the color of the testa and presence or absence of a texture layer. The texture layer is a secretion from the epidermis of the seed testa (Watt et al. 1977) and gives the seed a dull or buff appearance when present. Cultivated types are generally green or golden and can be shiny or dull depending on the presence of a texture layer.

4.1.2 Center of Origin, Botanical Origin, Domestication and Dissemination and Gene Pools

The evolution and taxonomy of grain legumes including the genus *Vigna* have been well documented by Smartt (1990). *Vigna* falls within the tribe *Phaseoleae* and family *Fabaceae*. The extensive studies of Verdcourt (1970) and Maréchal et al. (1978) (as cited by Smartt 1990) are largely responsible for the radical reorganisation and considerable amplification of the genus *Vigna*. The genus now includes the former Asian *Phaseolus* species and the entire genus *Voandzeia*. *V. radiata* was previously known as *Phaseolus aureus* Roxb. While some confusion still exists, it is considered the revision has brought some "order out of chaos" with molecular evidence lending strong support to the current groupings (Fatokun et al. 1993).

There are now three distinct groups of cultivated *Vigna* spp., two monospecific subgenera of African origin and one Asiatic group. The African species include the cowpea *V. unguiculata* (L.) Walp. and *V. subterranea* (L.) Verdc., both assigned to the subgenus *Vigna*. The Asiatic *Vigna*s (formerly Asian *Phaseolus* species) include 11 species assigned to the subgenus *Ceratotropis* (Table 1), including the cultivated species, *V. angularis* adzuki bean, *V. umbellata* rice bean, *V. aconitifolia* moth bean, *V. mungo* urdbean and *V. radiata* mungbean.

V. radiata has been further subdivided into three subgroups: subspecies *radiata* consisting of greengrams (including the cultivated mungbean) and golden grams, subspecies *sublobata* and subspecies *glabra* (Verdcourt 1970). Table 2 gives the complete taxonomic hierarchy for mungbean.

Table 2. Taxonomic hierarchy for mungbean *Vigna radiata* L.

Kingdom	Plantae – plant kingdom
Subkingdom	Tracheobionta – vascular plants
Division	*Magnoliophyta* – flowering plants
Class	*Magnoliopsida* – dicotyledons
Subclass	*Rosidae*
Order	*Fabales*
Family	*Fabaceae*
Tribe	*Phaseoleae*
Genus	*Vigna*
Subgenus	*Ceratotropis*
Species	*Radiata*
Subspecies	*Radiata*

Subspecies *sublobata* has been confused taxonomically with *V. trinervia* (Heyne ex Wight & Arn.), but recent molecular studies showed these two taxa to be distinct (Saravanakumar et al. 2004). In a phylogenetic study with molecular markers Fatokun et al. (1993) showed that the diversity assayed from the African *Vigna* spp. was far greater than that from the Asiatic *Vigna* spp., supporting the widely held view that Africa is the center of diversity for *Vigna* spp. and that the Asiatic *Vigna* spp. evolved more recently.

Germplasm resources for crop plants have been conveniently divided into three groupings: primary, secondary and tertiary (Harlan and de Wet 1971). The primary gene pool (GP1) includes the cultigen itself and any wild form that will readily hybridize with it and is considered conspecific. GP1 has been further divided to consider the distinction between domesticated (GP1A) and wild components (GP1B). The secondary gene pool (GP2) includes all species that will permit gene flow via interspecific hybridization. The tertiary gene pool (GP3) includes those taxa amongst which gene flow is possible but not by normal introgressive processes. All Asiatic *Vigna* spp. have both a GP1A (domesticated) and GP1B (wild) component, and some success with interspecific hybridization has led to the conclusion that secondary and tertiary groups also exist for the Asiatic *Vigna* spp. (Smartt 1984).

It is generally accepted that ssp. *sublobata* (GP1B) is the wild form of the cultigen *V. radiata* ssp. *radiata* (GP1A) (Purseglove 1968; Verdcourt

1970). The close relationship between these two subspecies has been verified by molecular marker comparisons (Fatokun et al. 1993). *V. radiata* is also considered to have both secondary and tertiary gene pools (Smartt 1990); GP2 includes *V. mungo* and GP3 includes *V. umbellata* and *V. angularis*. Mungbean was most likely domesticated in the Indian subcontinent (Smartt 1984), with archaeological evidence suggesting use in these regions for over 3500 years (Vishnu-Mittre 1974). Early in the domestication process mungbean cultivation spread to other parts of Asia and into north Africa. Cultivated mungbean developed through domestication and selection from *V. radiata* ssp. *radiata*, which is widely distributed throughout southern and eastern Asia, Africa and Austronesia.

4.1.3 Uses and Nutritional Information

Mungbean is a pulse or food legume crop used primarily as dried seed and occasionally as forage or green pods and seeds for vegetables (Lawn 1995). On a dry-weight basis mungbean contains 25 to 28% protein, 1.0 to 1.5% fat, 3.5 to 4.5% fiber, 4.5 to 5.5% ash and 60 to 65% carbohydrate. The nutrient composition of the edible portion of mungbean is given in Table 3. The seed protein is rich in lysine but low in sulphur amino acids methionine and cystine. The seeds are also rich in ascorbic acid (vitamin A), potassium, iron, phosphorus and calcium but low in sodium. Generally, mungbean provides an excellent complement for cereal-based diets, particularly in Asia where it used in various ways (Lawn and Ahn 1985). Dried seeds may be eaten whole or split, cooked, fermented or milled and ground into flour to make products like dahl, soups, porridge, confections, curries and alcoholic beverages. In western cultures the beans are popular for sprouting with major use as a fresh salad vegetable.

A range of anti-nutritional factors are present in mungbean (Akapunam 1996), although none of these factors cause significant obstacles to food preparation. Seeds contain trypsin inhibitor and haemaglutinins that can be inactivated by cooking. Oligosaccharides including raffinose, stachyose and verbascose are associated with flatulence in human diets that include raw or poorly processed pulses. While present in mungbean, these oligosaccharides are sol-

Table 3. Nutritional composition of raw, mature mungbean seeds

Nutrient	Units	Value per 100 g edible portion	Number of data points	Standard error
Proximates				
Water	g	9.05	60	0.439
Energy	kcal	347	0	0
Energy	kj	1453	0	0
Protein	g	23.86	82	0.371
Total lipid (fat)	g	1.15	65	0.144
Ash	g	3.32	58	0.099
Carbohydrate, by difference	g	62.62	0	0
Fiber, total dietary	g	16.3	0	0
Sugars, total	g	6.60	0	0
Minerals				
Calcium, Ca	mg	132	49	13.66
Iron, Fe	mg	6.74	55	0.872
Magnesium, Mg	mg	189	29	11.294
Phosphorus, P	mg	367	48	7.089
Potassium, K	mg	1246	18	42.146
Sodium, Na	mg	15	12	2.113
Zinc, Zn	mg	2.68	26	0.117
Copper, Cu	mg	0.941	21	0.044
Manganese, Mn	mg	1.035	19	0.076
Selenium, Se	mcg	8.2	0	0
Vitamins				
Vitamin C, total ascorbic acid	mg	4.8	5	0
Thiamin	mg	0.621	19	0.076
Riboflavin	mg	0.233	25	0.024
Niacin	mg	2.251	18	0.193
Pantothenic acid	mg	1.910	8	0.046
Vitamin B-6	mg	0.382	18	0.012
Folate, total	mcg	625	8	42.927
Folic acid	mcg	0	0	0
Folate, food	mcg	625	8	42.927
Folate, DFE	mcg_DFE	625	0	0
Vitamin B-12	mcg	0.00	0	0
Vitamin A, IU	IU	114	0	0
Vitamin A, RAE	mcg_RAE	6	0	0
Retinol	mcg	0	0	0
Vitamin E (α-tocopherol)	mg	0.51	0	0
Vitamin K (phylloquinone)	mcg	9.0	0	0
Phytosterols	mg	23	0	0
Amino acids				
Tryptophan	g	0.260	61	0

Table 3 (continued)

Nutrient	Units	Value per 100 g edible portion	Number of data points	Standard error
Threonine	g	0.782	51	0
Isoleucine	g	1.008	46	0
Leucine	g	1.847	46	0
Lysine	g	1.664	383	0
Methionine	g	0.286	423	0
Cystine	g	0.210	30	0
Phenylalanine	g	1.443	46	0
Tyrosine	g	0.714	26	0
Valine	g	1.237	46	0
Arginine	g	1.672	44	0
Histidine	g	0.695	44	0
Alanine	g	1.050	21	0
Aspartic acid	g	2.756	21	0
Glutamic acid	g	4.264	21	0
Glycine	g	0.954	21	0
Proline	g	1.095	19	0
Serine	g	1.176	21	0
Other				
Carotene, beta	mcg	68	0	0

Source: USDA National Nutrient Database for Standard Reference, Release 17 (2004) http://www.nal.usda.gov/fnic/foodcomp/cgi-bin/list_nut_edit.pl

uble in water and can be eliminated either by adequate presoaking, germination or fermentation.

4.1.4 Economic Importance

Statistics on world mungbean production are difficult to obtain. For example, FAOSTAT includes mungbean under the category drybeans, which can include species from the genus *Phaseolus*. Almost 90% of mungbean production on a world scale is produced in Asia, with India, the world's largest producer (Table 4), accounting for more than 50% of world production (Vijayalakshmi et al. 2003). Most of India's production is traded and consumed locally. In 2000, India produced about 2,887,000 ha of mungbean at an average yield of 381 kg/ha, and annual production has been constant over the 20-year period 1980-2000 (Table 4). Generally, mungbean yields in South Asia are low at about 0.4 t/ha. Thailand is the world's largest exporter of mungbean (Srinives 1991). In recent years mungbean production has increased dramatically in Myanmar, with an average annual increase of 22% in the period 1980–2000 (Table 4). Myanmar now has an area of 650,000 ha, which ranks it third behind India and China (Table 4).

Table 4. World production of mungbean in 2000 and its growth rate 1980–2000 (Weinburger 2003)

Country	Production ×1000 ha	Yield kg/ha	Annual average production growth rate % 1980–2000
South Asia			
Bangladesh	55	654	8.0
India	2887	381	-0.1
Pakistan	203	468	5.8
Sri Lanka	33	800	7.6
Southeast Asia			
Myanmar	650	674	22.0
Indonesia	324	895	1.5
Thailand	289	782	-1.8
Phillipines	39	755	-1.3
East Asia			
China	772	1154	1.7
Other			
Australia	40*		
USA	50**		

* Estimate

** Cupka and Edward 1988

4.1.5 Karyotype, Genome Size, DNA Content, Genome Organization

Vigna species including mungbean, belonging to the subgenus *Ceratotropis*, have chromosome complements typical of the tribe *Phaseolae* with 2n = 2x = 22 with the exception of the polypoid *V. glabrescens* 2n = 4x = 44 (Smart 1990). Mungbean chromosomes are small and difficult to study, and as a result published reports of the mungbean karyotype vary considerably. A comprehensive review of karyotype studies in mungbean by Poehlman (1991) showed that the haploid, total chromosome length reported varied from 12 to 37 μ. These studies also reported different ratios of chromosomes with median or submedian centromeres and satellites. Morphologically, although chromosomes of *V. radiata* and *V. mungo* are similar, *radiata* has longer total chromosome length and a distinctive giemsa C-banding pattern (Lavania and Lavania 1982).

The nuclear DNA content of mungbean and a range of economic plant species has been determined using flow cytometric analysis (Arumuganathan and Earle 1991). The genome size of mungbean was estimated to be 1.2 pg/2C or 579 Mb, which is about 4 times larger than the model species *Arabidopsis thaliana*, 0.3 pg/2C or 145 Mb, and about 30 times smaller than bread wheat (*Triticum aestivum*) 33.09 pg/2C or 15,996 Mb, but similar in size to other *Vigna* and *Phaseolus* species.

Angiosperm genomes vary enormously in nuclear DNA content. Generally, most of the variation in plant genome size is caused by differences in the amounts of repetitive DNA (Bennetzen 2002), particularly the class of mobile DNAs also known as long terminal repeat (LTR) retrotransposon. It is therefore likely that the small genome size of mungbean is associated with a lower content of LTR retrotransposons. This conclusion is in part supported by studies of DNA reassociation kinetics.

Reassociation kinetics used to characterize the linear arrangement of single-copy and repetitive sequences in the mungbean genome showed that the nuclear DNA content was estimated to contain 65% single-copy sequences, a considerably greater amount compared to genomes such as wheat and pea (Murray et al. 1979). Forty-six percent of the single-copy DNA of mungbean was interspersed within repetitive sequences at long periods, greater than 6.7 kb. The repetitive sequence families covered a range of about 50 to several thousand copies per haploid genome. The complete DNA sequence of the model plant species *Arabidopsis* has given researchers an extensive insight into the organisation of plant genomes. Despite the small size of the *Arabidopsis* genome (140 Mb), considerable regions were shown to be duplicated. Not unexpectedly, the RFLP study of Menancio-Hautea et al. (1993) also suggests duplication as an important mechanism in the evolution of the mungbean genome.

Duplicate markers were found on more than one linkage group (LG), with evidence of tandem duplication and other linkage arrangements of duplicated loci. The study of Menancio-Hautea et al. (1993) concluded that the evidence of duplication was consistent with mungbean being a secondary polyploid. They also suggested that insertion/deletion was also an important mechanism in the transition of cultivated mungbean from its wild progenitor ssp. *sublobata*. The RFLP evidence for insertion/deletion was threefold: (i) the co-segregation of loci with null phenotypes suggested large blocks of deletion, (ii) six-base cutter restriction enzymes *Eco*RI, *Eco*RV and *Hind*III produced larger restriction fragments and greater polymorphism than *Dra*I (4 base cutter) and (iii) 7% of the probes detected RFLPs with all restriction enzymes used implicating insertion/deletion rather than base substitution.

4.1.6 Cytoplasmic Genomes

The organisation of the mungbean chloroplast (cp) genome has been investigated by mapping the cleavage sites for a range of restriction enzymes (Palmer and Thompson 1981). The cp genome in mungbean is 150 kb, containing a 23-kb inverted repeat separated by single-copy regions of 21 and 83 kb. The chloroplast arrangement of mungbean differs from broad bean *Vicia faba*, which has lost the inverted repeat.

Although coding regions of mitochondrial (mt) genes are highly conserved among higher plants, gene order is not. Usually, a subpopulation of mt genomes exists within a species due to homologous recombination events between segments of repetitive DNA, resulting in a rearrangement of coding sequences (Palmer 1992). In this regard the mungbean mt-genome is no exception.

The effect of cytoplasmic genomes on the agronomic performance of mungbean is yet to be determined and should be considered in future studies given the finding that wild species cytoplasms can confer lower transpiration efficiency and an increased chance of drought susceptibility in sunflower (*Helianthus annuus* L.) (Lambrides et al. 2004 a).

4.1.7 Breeding Priorities

Several disadvantages of mungbean crops have prevented the widespread distribution of production in tropical and subtropical environments (Lawn 1995). These include low seed yield potential, photoperiod sensitivity, asynchronous reproductive development and susceptibility to weather damage (Williams et al. 1995 a–c). Mungbean is also susceptible to insect pests such as bruchids *Callosobruchus* spp. and bean fly (*Ophiomyia phaseolin*) and diseases such as mungbean yellow mosaic virus (MYMV) and cercospora leaf spot (*Cercospora canescens*). Because mungbean is used primarily as a food, extensive research is being done on seed quality traits (Imrie 1983, 1992; Imrie et al. 1984, 1985, 1988; Lambrides 1996; Humphry et al. 2005) such as size, shape, color, lustre, sprouting quality, hard-seededness, protein quantity and quality with emphasis on improving the methionine (sulphur amino acid) content of seeds.

Today, many of these breeding objectives have not been realised. In the early 1990s research was initiated in the lab of Dr Nevin Young, University of Minnesota, USA, to develop molecular expertise to give mungbean researchers additional tools to address the more challenging breeding objectives. Since this time there has been steady progress in molecular mungbean improvement with research efforts spawned in other parts of the world including Australia, Japan, Thailand and India.

4.2 Construction of Genetic Maps

4.2.1 Mungbean Maps

Six genetic linkage maps of mungbean based on RFLP (restriction fragment length polymorphism, Botstein et al. 1980) and RAPD (randomly amplified polymorphic DNA, Williams et al. 1990) markers have been published (Menancio-Hautea et al. 1993; Boutin et al. 1995; Lambrides et al. 2000; Humphry et al. 2002); their characteristics are provided in Table 5. These studies used mapping populations derived from two intersubspecific crosses, either VC3890A×TC1966 or Berken×ACC41. VC3890A and Berken are cultivated mungbean types from *V. radiata* ssp. *radiata*, and TC1966 and ACC41 are accessions of the wild progenitor *V. radiata* ssp. *sublobata*. The map of Humphry et al. (2002), constructed using a Berken×ACC41 population (made by CJL), is the most comprehensive map of mungbean published to date and is presented in Fig. 2. This and other mungbean mapping projects have numbered linkage groups according to the first published map of mungbean (Menancio-Hautea et al. 1993) constructed at the University of Minnesota.

Heterologous DNA probes (genomic and cDNA) developed from other legume species and genera, such as cowpea (*V. unguiculata*), soybean (*Glycine max*), bean (*Phaseolus vulgaris*) and lab lab (*Lablab purpureus*) in addition to mungbean were used as RFLPs in these projects. A basic strategy in these projects was to use probes made from *Pst*I libraries. *Pst*I is a methylation-sensitive restriction enzyme that is useful for selecting genomic fragments that hybridize with expressed genes or low copy sequences.

A large range in total map length of 680 cM to 1570 cM and average marker interval of 3.0 cM to 9.0 cM were observed. The large variation in map size can most likely be attributed to the level of genetic dissimilarity between the parents used to make the mapping populations, with less recombination expected between more distantly related genotypes. Using this logic, the genetic distance between Berken and ACC41 was likely to be greater than between VC3890A and TC1966, although other effects on recombination rates such as those of the environment cannot be excluded. This interpretation is consistent with the observation that the Australian forms of *sublobata* are more morphologically distinct from cultivated mungbean than are Asian forms (Lawn and Cottrell 1988; Rebetzke 1994). It is also consistent with other studies (James et al. 1999) and perhaps reflects long-term geographical isolation of the Australian forms from the presumed centers of origin of mungbean. The genetic dissimilarity between wild and cultivated

Table 5. Genetic linkage maps in genus *Vigna*

Year of publication	Species	Type of cross	Pedigree	Marker type	Population size/ Generation	Number of loci mapped	Map length cM	Number of linkage groups	Average marker interval cM	Reference
Mungbean										
1993	*V. radiata*	Inter-subspecific	VC3890A (ssp. *radiata*) ×TC1966 (ssp. *sublobata*)	RFLP	58 F2	172	1570	14	9.0	Menancio-Hautea et al. (1993)
1995	*V. radiata*	Inter-subspecific	VC3890A (ssp. *radiata*) ×TC1966 (ssp. *sublobata*)	RFLP	58 F2	102	1020	13	10.0	Boutin et al. (1995)
1995	*V. radiata*	Inter-subspecific	VC3890A (ssp. *radiata*) ×TC1966 (ssp. *sublobata*)	RFLP	58 F2	133	1240	12	9.3	Boutin et al. (1995)
2000	*V. radiata*	Inter-subspecific	Berken (ssp. *radiata*) ×ACC41 (ssp. *sublobata*)	RFLP, RAPD	67 F2	110	758	12	6.8	Lambrides et al. (2000)
2000	*V. radiata*	Inter-subspecific	Berken (ssp. *radiata*) ×ACC41 (ssp. *sublobata*)	RAPD	67 RIL	115	692	12	6.0	Lambrides et al. (2000)
2002	*V. radiata*	Inter-subspecific	Berken (ssp. *radiata*) ×ACC41 (ssp. *sublobata*)	RFLP	80 RIL	255	738	13	3.0	Humphry et al. (2002)
Cowpea										
1993	*V. unguiculata*	Inter-subspecific	IT2246-2×TVn1963 (ssp. *dekindtiana*)	RFLP, RAPD, cDNA	58 F2	89	680	10	7.6	Fatokun et al. (1993)
1997	*V. unguiculata*	Intraspecific	IT84S-2049×524B	RFLP, RAPD, AFLP	94 F8	181	972	12	6.4	Menendez et al. (1997)
2002	*V. unguiculata*	Intraspecific	IT84S-2049×524B	RFLP, RAPD, AFLP	94 F9	440	2670	11	6.4	Ouedraogo et al. (2002)
Other *Vigna* species										
1996	*V. angularis* ×*V. nakashimae*	Interspecific	cv. Erimoshouzu (*V. angularis* ×*V. nakashimae*	RAPD, RFLP	80 F2	132	1250	14	10.6	Kaga et al. (1996a)
2000	*V. angularis* ×*V. umbellata*	Interspecific	cv. Erimoshouzu (*V. angularis*) ×cv. Kagoshima (*V. umbellata*)	RFLP, RAPD	86 F2	189	1702	14	9.7	Kaga et al. (2000)

Fig. 1. Specimens of cv Berken *V. radiata* ssp. *radiata* (*left*) and the wild progenitor ACC41 *V. radiata* ssp. *sublobata*

forms of *V. radiata* was noted elsewhere by Kaga et al. (1996b). They studied intra- and inter specific variation in the subgenus *Ceratotropis* as revealed by RAPD analysis and found that the largest intraspecific variation was found within *V. radiata* in which wild forms of *sublobata* were very different from one another and from cultivated forms. Genetic distance between parents is an important issue in mapping studies as it can determine the levels of segregation distortion.

For Mendelian segregation of traits, it is assumed that meiosis during mega- and micro-gametogenesis in the F_1 proceeds normally and that there is neither preferential transmission of gametes (gametic selection) nor postzygotic selection. Often, meiotic abnormalities that cause gametic and/or zygotic selection are associated with hybrid breakdown. Hybrid breakdown is the loss of fertility that is common to wild×cultivated hybridizations. The manifestation of hybrid breakdown is segregation distortion.

Segregation distortion has been a common observation in studies of mungbean, particularly in crosses between wild and cultivated forms (Menancio-Hautea et al. 1993; Lambrides et al. 2000, 2004b; Humphry et al. 2002). Mapping studies with Berken×ACC41 (Lambrides et al. 2000, 2004b; Humphry et al. 2002) showed that 24 to 30% of mapped markers segregated with distortion. Markers on LGs 1 and 3 segregated with severe distortion in favor of alleles inherited from the cultivated parent. Conversely, distortion in favor of alleles inherited from the wild parent occurred on LG 2. In many cases, markers segregating with distortion occurred in linked blocks. The segregation distortion noted on LGs 1 and 3 was related to heavy selection against alleles homozygous for the wild parent, which was possibly related to a lethal condition at the gametic level. Gametic selection has been well documented in wheat (Miller 1983), where preferential transmission of chromosomes occurred with wild by cultivated crosses. Wild type (alien) chromosomes that failed to be transmitted were referred to as 'cuckoo' chromosomes. Segregation distortion and the 'cuckoo' effect have been reported in *Lablab purpureus* (Konduri et al. 2000).

The level of segregation distortion noted with Berken×ACC41 was greater than in the study of VC3890A×TC1966, where only 12% of markers segregated with distortion (Menancio-Hautea et al. 1992). There did not appear to be common regions of distortion in the maps derived from the different mapping populations (Lambrides et al. 2000).

While these first-generation maps have provided mungbean researchers an excellent base for further study, some LGs have few markers and none of the maps produced to date have the same number of LGs as the haploid chromosome number (n=11) of mungbean. Consequently, the numbering of mungbean LGs has not been finalized. None of the mungbean maps can be considered 'saturated', and the percent genome coverage has not been re-

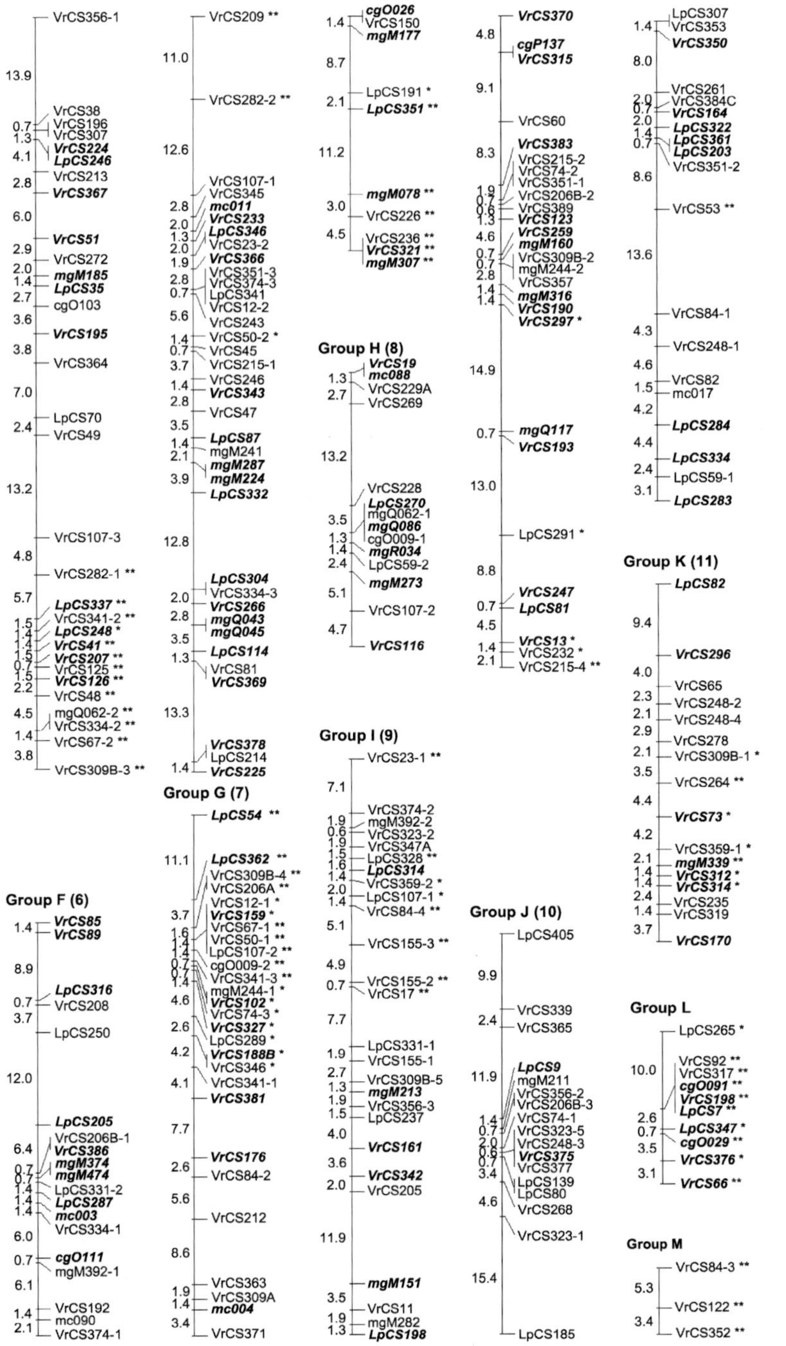

Fig. 2. A mungbean [*Vigna radiata* (L.) Wilczek] linkage map constructed using RFLP markers (Humphry et al. 2002). LGs are ordered primarily based on their similarity to a previously published mungbean map (numbers in *brackets* indicate previous ordering; see text). The long *vertical bars* indicate LGs and small *horizontal bars* indicate locus positions. Genetic distances (cM) are located to the *left* of the LGs and locus names are listed to the *right*. Marker names containing hyphens followed by a number indicate different non-allelic loci detected by single probes and those followed by the letters A or B indicate different amplification products from the same clone. Markers showing skewed segregation away from the predicted 1:1 ratio are indicated by an *asterisk* (*) if deviating at a significance level of 0.05 and a *double asterisk* (**) if deviating at a 0.01 significance level. Markers originating from clones that detected single loci are shown in *bold italics*

ported. Clearly, additional markers will need to be placed on the maps to address these deficiencies. Interestingly, more than twice the number of markers was added to the Berken×ACC41 map of Humphry et al. (2002) compared to the Berken× ACC41 maps of Lambrides et al. (2000), yet there was little difference in total map length, 738 cM compared to 692 and 758 cM (Table 5).

4.2.2 Other *Vigna* Maps

A large range in total map length, 680 to 2670 cM, was also observed among genetic linkage maps of cowpea *V. unguiculata* (Table 5; Fatokun et al. 1992a; Menendez et al. 1997; Ouédraogo et al. 2002). Maps developed from interspecific crosses, azuki bean *V. angularis*×*V. nakashimae* (Kaga et al. 1996a) and azuki bean×rice bean (*V. umbellata*) have produced map lengths of 1250 and 1702 cM, respectively. All of these maps failed to condense to the basic chromosome number n = 11, with the exception of the cowpea map constructed by Ouédraogo et al. (2002), which now has 440 markers placed across 11 LGs.

4.2.3 Microsatellites and Single Nucleotide Polymorphisms (SNPs)

All higher eukaryote genomes including plants consist of repetitive DNA of different kinds. A subgroup of repetitive sequences containing iterations of 1 to 5 bp motifs (core sequences), e.g. $(AT)_n$, are referred to as simple sequence repeats (SSRs) (Jacob et al. 1991) or microsatellites (Litt and Luty 1989). Polymerase chain reaction (PCR, Mullis and Faloona 1987) primers designed around the conserved flanking regions allow for the amplification of the SSR locus. The frequency of SSR loci in plant genomes has been estimated to be one in every 6 to 7 kb (Cardle et al. 2000) and usually have a high level of allelic diversity. With a genome size of 579 Mb the number of SSR markers in the mungbean genome would be in the order of 0.8 to 1.0×10^3. This may be an overestimate as the mungbean genome has been reported to contain about 65% single-copy sequences (Murray et al. 1979). Methods for the isolation of SSR loci in *Vigna* have been developed (Kumar et al. 2002a,b; Miyagi et al. 2004; Wang et al. 2004). Prasad and Jawali (1998) estimated that the average distance between two SSR loci in the mungbean genome was 150 kb.

The studies of Kumar et al. (2002a) used degenerate primers to PCR-amplify SSR sequences in the mungbean genome. A set of random PCR fragments was cloned into TOPO TA and sequenced. Primers were then designed to amplify either di- (TG, CT, TC, GA, AG) or tetra- (AACC, GGTT, CCAA, TGGT, CCAA) nucleotide repeat sequences. Seven polymorphic SSR loci detecting dinucleotide repeats amplified between two and five alleles per locus, while the observed heterozygosity ranged from 0 to 0.9048 when scored across eight *Vigna* species. Seven polymorphic SSR loci detecting tetranucleotide repeats amplified between two and six alleles per locus while the observed heterozygosity ranged from 0 to 0.561 when scored across the same eight *Vigna* species.

Wang et al. (2004) used an SSR enrichment method based on oligo-primed second-strand synthesis to develop SSR markers in azuki bean (*V. angularis*). Using this methodology 49 primer pairs were made to detect dinucleotide (AG) SSR loci. The average number of alleles in complex, wild and town populations of azuki bean was 3.0 to 3.4, 1.1 to 1.4 and 4.0 respectively. The genome size of azuki bean is 539 Mb; therefore, the number of (AG)n and (AC)n motif loci per haploid genome was estimated to be 3500 and 2100 respectively. It is likely the SSR markers developed in this study would be an extremely useful resource for mapping studies in mungbean.

Single base pair differences between individuals of a population are referred to as SNPs. SNP markers are ubiquitous and span the entire genome. In human populations it has been estimated that any two individuals have one SNP every 1000 to 2000 bps. Generally, there are an enormous number of potential SNP markers for any given genome. SNPs are highly desirable in genomes that have low levels of polymorphism using conventional marker systems, e.g. wheat and sorghum. SNP markers are biallelic (A/T or G/C) and therefore are highly amenable to automation and high-throughput genotyping. There have been no published reports of the development of SNP markers in mungbean, but they should be considered by research groups who envisage long-term plant improvement programs.

Because mungbean is such an important crop in mainly developing countries, the cost of gearing up in these technologies may not be warranted at this time. In the meantime, cheaper and less resource-rich technologies such as RAPD should not be overlooked. Studies by Kaga and Ishimoto (1998) and Lambrides et al. (1998) showed that linked RAPD markers are easily and efficiently converted to more robust markers such as RFLP and SCAR, probably because of the high complement of single-copy DNA that exists in the mungbean genome.

4.2.4 Comparative Mapping

The use of heterologous RFLP probes has facilitated a number of comparative genome studies with various species of the genus *Vigna* (Menancio-Hautea et al. 1993; Boutin et al. 1995; Kaga et al. 1996a, 2000; Humphry et al. 2002). In some instances, there is greater homology between mungbean and species in different genera compared to mungbean and related species from the subgenus *Ceratotropis*.

Chromosomal rearrangements could explain many of the differences between the mungbean and cowpea genomes (Menancio-Hautea et al. 1993) and the mungbean and azuki bean genomes (Kaga et al. 2000). These studies show that conserved blocks of genes appear on several LGs, although no entire LG was conserved between mungbean and cowpea and mungbean and azuki bean. The genetic complement of mungbean and cowpea was similar at the nucleotide level, although copy number changed and the linear arrangement of conserved linkage blocks also changed (Menancio-Hautea et al. 1993). Interestingly, this study showed that there were greater differences between the mungbean and cowpea genomes compared to sorghum and maize, which are in different genera.

Prior to the taxonomic reorganisation of *Phaseolus* and *Vigna* mungbean had been placed in the genus *Phaseolus*. Not surprisingly, with respect to marker order and conserved LGs, mungbean shows greater homology to common bean than to cowpea (Boutin et al. 1995) and azuki bean (Kaga et al. 1996a, 2000). All mungbean LGs consisted of one, two or three LGs of common bean with the average conserved linkage block of 36. 2 cM and the longest conserved linkage block of 103.5 cM occurring on LG 8 of mungbean and LG K of common bean. A balanced translocation between mungbean (LGs 2 and 6) and common bean (LGs F and H) was also detected. Mungbean LGs 8, 9, 10 (Menancio-Hautea et al. 1993) were all composed of markers from L, G, K of common bean, suggesting that these mungbean LGs may represent segments of one LG. The studies of Kaga et al. (2000) and Lambrides et al. (2000) also provided evidence that mungbean LGs 8 and 9 are segments of one LG.

Although Menancio-Hautea et al. (1993) found that 88% (125/142) of soybean genomic probes hybridized to mungbean, there was substantial genome rearrangement between these two genomes (Boutin et al. 1995). Only short and scattered linkage blocks were conserved between mungbean and soybean. For example, markers from up to 16 different LGs of soybean were found on LG 1 of mungbean. The average conserved linkage block between the mungbean and soybean genomes was about a third the size of conserved linkage blocks between mungbean and common bean.

There is surprising homology between mungbean and a more distantly related member of *Fabacae* lab lab (*Lablab purpureus*) (Humphry et al. 2002). Large conserved linkage blocks were observed between the two genomes, and linkage order was retained in the majority of cases, although evidence was presented to suggest the genomes differed by at least one inversion and other complicated chromosomal rearrangements. Different copy numbers detected in each of the genomes suggested that they had also accumulated a large number of deletions/duplications after they diverged.

4.3 Gene Mapping

4.3.1 Bruchid Resistance

Seed feeding weevil-like insects belonging to the genus *Callosobruchus* are known as bruchids and are a common pest of stored pulses. Bruchid infestations are common in mungbean crops throughout Austral-Asia, causing a downgrading of seed quality and in extreme cases causing complete loss of income (Talekar 1988). Adult bruchids lay eggs on the outside of the seed or pod after which the larvae hatch, chew through the seed coat and begin to feed on the cotyledonary tissue. The pupal stage occurs inside the seed and the adults emerge from small 'windows' cut in the seed coat by the larvae.

on LG 8 was associated with the powdery mildew response, where, in contrast to the previously identified markers, the allele from the wild parent TC1966 contributed to resistance. Markers sgK472, mgM208 and mgQ39 together explained 58% of the phenotypic variation for powdery mildew resistance.

Chaitieng et al. (2002) studied powdery mildew resistance in VC1210A, a line resistant to field levels of powdery mildew in northeast Thailand. Parents, F_1 and 96 F_2s of the cross VC1210A×TC1966 were screened for resistance in the field (Nakhon Ratchasima, Thailand) to a natural infestation of powdery mildew. A continuous distribution of F_2 phenotypes was observed. Quantitative trait locus (QTL) analysis with a framework set of RFLP probes failed to identify any significant resistance factors above the threshold of logarithm of odds (LOD) = 2.0, although one peak named *PMR2* near common bean probe Bng 065 located on LG 2 of the mungbean map (Menancio-Hautea et al. 1993) was detected at a lower threshold LOD = 1.79. PMR2 accounted for only 10% of the phenotypic variation for powdery mildew resistance.

However, when a bulked segregate analysis, AFLP (amplified fragment length polymorphism, Vos et al. 1995) approach was used, four polymorphic bands were identified. These bands were cloned into vector pGEM-T Easy and used as RFLP probes. Of the four bands cloned, two yielded a total of five RFLP probes that were linked to powdery mildew resistance and to each other, forming a LG of 9.6 cM. These markers were unlinked to any of the framework markers used earlier in the study. QTL analysis of the 9.6-cM LG identified an additional resistance factor, *PMR1*, at LOD = 20.2 which accounted for 63% of the phenotypic variation for powdery mildew resistance. The average disease resistance score of F_2 individuals heterozygous at *PMR1* was slightly less than that of the homozygous resistant class, suggesting partial dominance of *PMR1* in line VC1210A. It is not known which LG of the mungbean linkage map the *PMR1* locus falls in.

The study of Humphry et al. (2003) also identified a major locus conditioning powdery mildew resistance in line ATF3640. 147 F_7 and F_8 RILs from the cross ATF3640×Berken were screened with glasshouse strains of powdery mildew. Disease scores were taken at two developmental stages, 42 and 62 d after sowing, and in two different seasons. Using this population a framework set of 51 RFLP markers was placed on ten LGs covering 350 cM and used for QTL analysis. A single QTL that identified between 82 and 86% of the variation was identified on LG K (LG 11). This location differs from the location of three QTLs for powdery mildew resistance identified by Young et al. (1993), suggesting that resistance genes are distributed across the mungbean genome.

While none of the resistant lines used in these three studies showed complete immunity to powdery mildew, it should be possible to develop mungbean material with higher levels of powdery mildew resistance by pyramiding resistant loci from different sources. For example, a cross between VC3890A and ATF3640 should enable progeny to be selected with three powdery-mildew-resistant loci (two from VC3890A and one from ATF3640). Such a breeding approach will be greatly facilitated by marker-assisted selection (MAS).

4.3.3 MYMV Mungbean Yellow Mosaic Virus

MYMV is a gemini virus (Fauquet et al. 2000) transmitted by whitefly (*Bemisia tabaci*). Gemini viruses are a group of plant viruses that belong to the family Geminiviridae and have circular single-stranded genomes. MYMV belongs to the genera *Begomovirus*, which is characterized by a bipartite genome A and B (Fauquet et al. 2000). MYMV is the most serious disease of mungbean in Asia, particularly south Asia, but is not present in some mungbean producing countries such as Australia. The most reliable source of resistance to MYMV has been NM92, a line developed from the mutagenesis program using gamma rays at NIAB, Pakistan (Ali et al. 1997). In Pakistan about 50% of the mungbean acreage is sown to NM92, with a yield advantage of more than 50% over local cultivars. The resistance in NM92 was controlled by a pair of recessive alleles at a single locus (Khattak et al. 2000). Studies to identify markers linked to MYMV resistance in mungbean have been initiated.

Lambrides et al. (1999) crossed NM92 with Australian commercial cultivars Berken and Emerald to develop two recombinant inbred populations of about 90 individuals each. F_3 and F_6 seeds of each population were sent to Pakistan and screened in the field using natural epidemics of MYMV. F_3 and F_6 progeny rows were scored as resistant (<40%

leaflet infection) and susceptible (>40 % leaflet infection). Progeny rows that appeared to be segregating in the F_6 were also noted. Lines were selected from the Berken×NM92 recombinant inbred (RI) population for bulked segregate analysis (BSA) (Michelmore et al. 1991) using RAPD markers. Ten F_6 lines resistant to MYMV as both F_3 and F_6 progeny rows were selected for the resistant pool and ten lines susceptible to MYMV as both F_3 and F_6 progeny rows were selected for the susceptible pool. DNA isolated from each pool was screened with about 200 RAPD primers. Only 1/200 primers provided a marker linked to MYMV resistance. Primer OPAJ20 (Operon Technologies) provided one marker linked to resistance at 8.3 cM. This marker was also linked to resistance in the Emerald×NM92 RI population at about 10 cM. To date, the MYMV resistance locus in NM92 has not been placed on the mungbean genetic linkage map. A possible clue to its location has been provided by a study in soybean that tagged the locus *Rsv3* that confers resistance to potyvirus, soybean mosaic virus (Jeong et al. 2002). In this study a mungbean genomic clone Mng247 (=pM247, Menancio-Hautea et al. 1993) used as an RFLP marker was closely linked to a virus resistance locus. Mng247 was sequenced and corresponds to a leucine rich repeat (LRR)-type sequence which is typically associated with plant defence systems. Mng247 was placed on LG 7 of the mungbean linkage map (Menancio-Hautea et al. 1993).

4.3.4 Seed Weight

Strong consumer demand for large seed types has seen a six-fold increase in the seed weight of modern mungbean varieties compared to the wild progenitor ssp. *sublobata* (Fatokun et al. 1992b; Lambrides and Imrie 2000). Large seed size has therefore become a key trait in mungbean breeding programs, particularly when accessions of ssp. *sublobata* are used.

Seed weight in mungbean is a highly heritable, quantitative trait (see review by Fery 1980). The distribution of phenotypes in *radiata*×*sublobata* crosses is typically continuous with small seededness partially dominant. Mapping studies have identified 4 to 11 independent QTLs (Table 6) associated with seed weight (Fatokun et al. 1992b; Humphry et al. 2005). The field study of Fatokun et al. (1992b) using 58 F_3 families identified seed weight QTLs on LGs 1, 2, 3 and 4, accounting for 49% of the phenotypic variation. Evidence was also presented for orthologous genes for seed weight with a QTL of large effect identified on homologous genomic regions of LG 2 in both mungbean and cowpea. This genomic region was later shown to be associated with seed weight in pea, *Pisum sativum* L. (Timmerman-Vaughan et al. 1996).

Humphry et al. (2005) used 227 RILs to map 11 QTLs conditioning seed weight from field and glasshouse trials, with 7 loci common to both data sets. Of the 7 QTLs common to both data sets, one QTL was located on each of LGs 1, 9, 10, 11 and E and two QTLs were located on LG 2. The 11 QTLs collectively accounted for over 80% of the phenotypic variation. None of the loci identified in this study appeared to co-localise with any of the QTLs identified in the study of Fatokun et al. (1992b), although several QTLs did map to equivalent linkage groups. The lack of coincidence between these two studies may have reflected the different genetic material under study and/or the small population size (58 individuals) and, therefore, the lack of mapping resolution in the study of Fatokun et al. (1992b).

4.3.5 Hard-Seededness

Weather damage to mungbean seeds after physiological maturity can result in a decline in seed yield and quality, thus reducing the value of a crop (Mohammed-Lassim et al. 1984; Imrie 1992). Wetting and drying cycles related to sporadic rainfall events can cause seeds to crack and become more readily infected by seed borne fungi and bacteria. The soft-seeded varieties of mungbean are especially vulnerable to this type of damage, whereas hard-seeded mungbean varieties are largely impervious to moisture uptake and weathering. Therefore, hard-seededness has been identified as a key trait in mungbean breeding programs where the aim is to develop weather-tolerant varieties (Williams 1989; Imrie et al. 1991).

Previous studies with mungbean (Kumari and Dahiya 1981; Mohammed-Lassim et al. 1984; Williams 1989) and blackgram, *V. mungo* L. (Tomer and Kumari 1991) indicated that after physiological maturity, combinations of high temperature and humidity and periods of wetting and drying can affect levels of hard-seededness. For the field grown study, plants during pod-fill were exposed to high temperature (>30 °C), high relative humidity (>50%) and cycles of wetting and drying mediated by heavy dews and rainfall events. This may explain why the overall level of hard-seededness was lower for seeds harvested from the field trial compared to those harvested from the glasshouse trial and why there was only a moderate correlation between the two trials. A similar disparity between glasshouse and field trials for hard-seededness was reported by Imrie et al. (1988).

Humphry et al. (2005) used 227 RILs to map four QTLs for hard-seededness from field data and a single QTL from glasshouse data. The single QTL from the glasshouse data (hsA) co-localised with one of the QTLs from the field data. This locus on LG 1 (=LG K) explained up to 23% of the variation in the field trial and 11% of the variation in the glasshouse trial. Lambrides (1996) also mapped a hard-seededness locus near hsA on LG 1 in an F_2 population of Berken×ACC41. These results support the previous observations that very few genes appear to control hard-seededness in mungbean (Lawn et al. 1988; Williams 1989; Lambrides 1996). The inheritance pattern of hard-seededness in this population is similar to that observed in soybean, where one major QTL was identified which explained 30% of the variation with other minor QTLs contributing to the continuous distribution of phenotypes (Keim et al. 1990).

4.3.6 Seed Appearance Traits

Several traits other than seed weight and hard-seededness are associated with seed quality in mungbean. Processors demand large shiny seed with green testa, largely for aesthetic reasons and sprouts which are free from anthocyanin pigmentation. Some mungbean genotypes possess a ridged texture layer on the seed surface consisting mainly of carbohydrate and secreted from the epidermis of the testa during grain fill (Watt et al. 1977). Sprouters have found this trait undesirable as fewer microbial contaminants are found with shiny coated seeds (O'Brien et al. 1992). Generally, the appearance of the seed is determined by the presence or absence of the texture layer, the pigmentation of the texture layer and the color of the testa. These traits have been mapped in the Berken×ACC41 population using 67 F_2 and 67 RIL individuals (Table 6) (Lambrides 1996; Lambrides et al. 2000).

Presence of Texture Layer. When individuals of the population Berken (shiny seed)×ACC41 (dull seed) were scored for the amount of texture layer using a scoring system of 0 (shiny seed, no texture layer) to 5 (deep texture layer), a continuous range of phenotypes was observed (Lambrides 1996). QTL analysis detected three regions on LGs 1-3, 2 and 8-9, collectively accounting for 71.5% of the phenotypic variation for the texture layer score. The QTL near pO9b (LGs 8 and 9) was detected at LOD=5.94, where the allele from ACC41 showed dominant gene action for deep texture layer. Unexpectedly, the gene action at a QTL detected near marker pM78 (LGs 1-3) was suggestive of overdominance. This mode of gene action had not been reported previously for a seed coat trait in mungbean. Overdominance may explain the unexpected appearance of dull seeded segregates from hybrids between shiny seed coat parents (Imrie personal communication). Pseudo-overdominance (Crow 1952) has been proposed as an alternative explanation to overdominant gene action. For pseudo-overdominance, dominant and recessive alleles occur in coupling phase at closely linked loci. In the case of the texture layer in mungbean, high-resolution mapping would be required to distinguish between a case of pseudo-overdominance and overdominance.

Testa Color. The green speckled black testa color of ACC41 was dominant over green testa color and controlled by a single locus that mapped to LG 2 (Table 6) (Lambrides et al. 2000). RFLP marker pA235 was linked to the testa color locus at 2.4 cM (LOD=15.0), and this region of the mungbean genome was shown to segregate with distortion (Lambrides et al. 2004b). An analysis of 207 RILs showed that the segregation distortion occurred after each generation of self-pollination from F_1 thru to F_7, resulting in F_7 phenotypic frequencies of 151:56 instead of the expected 103.5:103.5.

Pigmentation. The pigmentation of the texture layer of ACC41 was dominant over no pigmentation and controlled by a single locus which also mapped to LG 2, but it was not linked to the testa color locus. RFLP marker pA204 was linked at 6.4 cM, LOD=8.0.

4.4 Tissue Culture and Transgenic Approaches

Like many grain legume or pulse crops, mungbean is regarded as a recalcitrant species for regeneration from tissue culture. As a result, it is not a crop which lends itself easily to high-frequency transformation. With the exception of soybean, where genetically modified organism (GMO) approaches to herbicide resistance have been very successful, there are no commercially grown grain legume crops. Methods have been published for the transformation of many of pulse crop species, including mungbean (Jaiwal et al. 2001), but as yet most have not been able to attain an acceptable frequency of transformation for reliable use.

Transformation methods require a system for the introduction of transgene(s) into one or a few cells, and then the ability to select for the growth, proliferation and regeneration of whole fertile plants which contain and express the transgene(s). The basic requirement for most species is therefore a tissue culture and regeneration protocol from suitable cells or tissues, plus a means of delivering the transgene DNA such that it is incorporated into the host nuclear genome.

The first report of stable transformation of mungbean was from scientists at the Bose Institute in India, who reported the development of transgenic shoots with reporter gene expression (Pal et al. 1991). Genes for β-glucuronidase expression and kanamycin antibiotic resistance were introduced using *Agrobacterium tumefaciens* as a vector, and these genes were expressed in regenerating shoots. The authors were unable to produce whole plants using their methodology, and for the next decade, those workers who attempted to produce transgenic mungbean struggled to regenerate many genotypes from various tissue culture explants (Tivarekar and Eapen 2001).

In recent times, the most effective means of producing high-frequency regeneration in mungbean has been through the use of either immature or mature cotyledons. Mature cotyledon tissues from seed have obvious advantages as convenient sources of explant, with repeatable regeneration frequencies achieved for three different genotypes (Amutha et al. 2003). However, whether these types of tissues are readily amenable to transformation is yet to be determined.

An international collaboration between scientists from India and Switzerland published the first report of whole transgenic mungbean plants expressing β-glucuronidase under the control of a constitutive promoter (Jaiwal et al. 2001). It would appear that the methodology is not without its problems, and the authors presented data for only two transgenic T_0 plants. There is as yet no follow-up publication reporting on the stability of transgene expression among progenies of the plants regenerated.

4.5 Future Scope of Research

While the mapping studies outlined above have provided some useful information on the genomic location of traits in mungbean, some of the information has been obtained from experiments that used small population sizes (< 70 individuals). The limitations of QTL analysis in small populations have been recognised (Beavis 1994). Essentially, errors can arise due to statistical sampling error associated with small populations (< 100 individuals). Future studies, particularly those involving quantitative traits, should aim to use large segregating populations (> 200) grown over a range of environments to improve estimates of heritability. Analysis of large populations will allow linkage order and map distance to be estimated with greater precision and importantly add significantly to the statistical power of QTL analysis. Clearly, the size of such studies will be limited by resources available to mungbean researchers.

Generally the marker approach enables the inheritance of simple and complex traits to be determined with far greater precision than traditional methods. Furthermore, this approach allows for the greater manipulation of target traits. 'Landmark' PCR-based markers linked to traits of interest can be traced in any population under any environment. Consequently, all future experiments enhance the base of knowledge for any given trait. Knowledge of inheritance can help determine the most efficient breeding strategy, and with MAS several desirable traits can be expediently added to elite cultivars free from the effects of environmental influences. MAS in mungbean breeding programs will be an appropriate and relevant strategy as the mapping studies described above have shown that a large number of independent loci (Table 6) are associated with the traits of interest and that many of the desirable traits are recessively inherited.

Much of the marker work in mungbean to date has been based on studies involving populations derived from ssp. *radiata*×ssp. *sublobata* crosses. These populations have been useful because they offer high levels of DNA polymorphism and segregation for important seed traits, e.g. bruchid resistance, seed weight and hard-seededness. Accessions of ssp. *sublobata* also have the potential to broaden the germplasm base for traits that confer tolerance to environmental stresses (Lawn et al. 1988). Tolerance to environmental stresses in *sublobata* has arisen as a result of natural selection and the accumulation of favorable alleles through persistent, long-term exposure of wild populations to their environments. Collections of *sublobata* made from degraded saline soils and limestone-derived soils have shown greater survival in saline conditions and tolerance of alkalinity, respectively, than cultivated varieties under controlled conditions.

While *sublobata* will continue to provide mungbean researchers with a rich source of traits, attention should be paid to which *sublobata* accessions are used. As noted above, a large range of variation exists among forms of *V. radiata* including ssp. *sublobata*. Low recombination rates and segregation distortion are two genetic phenomena that can occur when distant crosses are made, both of which were observed in studies of Berken (ssp. *radiata*) ×ACC41 (ssp. *sublobata*). Low levels of recombination may adversely affect the efficiency of a backcrossing program because of linkage drag. Ideally, recombination events around target loci are required without the inclusion of additional and possibly deleterious segments of chromosome from donor parents (e.g. the wild parent ACC 41). To overcome the effects of linkage drag, additional backcrosses or large segregating populations are required to be confident of selecting segregates with the desired combination of traits. Low levels of recombination would also inhibit map-based cloning strategies where it is desirable to have key recombination events occurring near and flanking target loci. Observed recombination rates in populations of Berken×ACC41 (Lambrides et al. 2000; Humphry et al. 2002) were about 50% less than those observed in VC3890A×TC1966 populations (Menancio-Hautea et al. 1993). For this reason populations derived from Berken×ACC41 would be less efficient than VC3890A×TC1966 for map-based cloning experiments.

The presence of segregation distortion in mungbean populations between cultivated and wild parents (such as Berken×ACC 41) may impact on a mungbean improvement program. The impact will be positive if the favorable trait from a wild parent is located in a region or regions of the genome where segregation distortion favors wild parent alleles. However, if the converse is true, the introgression of a favorable trait from a wild parent may be difficult because of the systematic elimination of wild parent alleles after each round of meiosis during the breeding process.

As indicated above, mungbean researchers have yet to make extensive use of high-throughput marker systems such as SSR and SNP. Stable, reproducible and routine transformation systems in *Vigna* have not been developed and will restrict the mungbean improvement options that require transgenic (GMO) approaches. Because mungbean is not a large crop on a world scale, investment in these molecular technologies should be shared between institutions on a national and international level. Wherever possible the large synteny between mungbean and *Phaseolus vulgaris* should be exploited. In addition, the comparative genomics approach for estimating genome conservation between crop legumes soybean, bean, pea and mungbean with model legume species *Medicago truncatulata* and *Lotus japonicus* (Choi et al. 2004) will no doubt provide mungbean researchers new and inexpensive avenues for developing perfect markers for useful traits.

References

Agrios GN (1988) Plant Pathology, 3rd edn. Academic, San Diego

Akapunam M (1996) Mung bean (*Vigna radiata* (L.) Wilczek). In: Nwololo E, Smartt J (eds) Food and Feed from Legumes and Oilseeds. Chapman and Hall, London

Ali M, Malik IA, Sabir HM, Ahmad B (1997) The mungbean green revolution in Pakistan. Asian Vegetable Research and Development Center, Shanhua, Taiwan. AVRDC Technical Bulletin No. 24. Publication No. 97-459

Amutha S, Ganapathi A, Muruganathan M (2003) *In vitro* organogenesis and plant formation in *Vigna radiata* (L.) Wilczek. Plant Cell Tiss Org Cult 72:203–207

Arumuganathan K, Earle ED (1991) Nuclear DNA content of some important plant species. Plant Mol Biol Rep 9:208–218

Beavis WD (1994) The power and deceit of QTL experiments: lessons from comparative QTL studies. In: Proc

49th Annual Corn and Sorghum Industry Research Conference ASTA, Washington, DC, pp 250–266

Bennetzen JL (2002) Mechanisms and rates of genome expansion and contraction in flowering plants. Genetica 115:29–36

Botstein D, White RL, Skolnik M, Davis RW (1980) Construction of a genetic linkage map in man using restriction fragment length polymorphisms. Am J Hum Genet 32:314–331

Boutin SR, Young ND, Olson TC, Yu Z-H, Shoemaker RC, Vallejos CE (1995) Genome conservation among three legume genera detected with DNA markers. Genome 38:928–937

Cardle L, Ramsay L, Milbourne D, Macaulay M, Marshall D, Waugh R (2000) Computational and experimental characterization of physically clustered simple sequence repeats in plants. Genetics 156:847–854

Chaitieng B, Kaga A, Han OK, Wang XW, Wongkaew S, Laosuwan P, Tomooka N, Vaughan DA (2002) Mapping a new source of resistance to powdery mildew in mungbean. Plant Breed 121:521–525

Choi H-K, Mun J-H, Kim D-J, Zhu H, Baek J-M, Mudge J, Roe B, Ellis N, Doyle J, Kiss GB, Young ND, Cook DR (2004) Estimating genome conservation between crop and model legume species. Proc Natl Acad Sci USA 101:15289–15294

Cregan PB, Mudge J, Fickus EW, Marek LF, Danesh D, Denny R, Shoemaker RC, Matthews BF, Jarvik T, Young ND (1999) Targeted isolation of simple sequence repeat markers through the use of bacterial artificial chromosomes. Theor Appl Genet 98:919–928

Crow JF (1952) Dominance and overdominance. In: Cowen JW (ed) Heterosis. Iowa State College Press, Ames, IA, pp 282–297

Cupka TB, Edward LH (1988) Production and breeding of mungbean in USA. In: Proc 2nd Int Symp on Mungbean, 16–20 Nov 1987, Bangkok, Thailand, pp 675–680

Daglish GJ, Erbacher JM, Eelkema M (1993) Efficacy of protectants against *Callosobruchus phaseolin* (Gyll.) and *C. maculatus* (F.) (*Coleoptera:Bruchidae*) in mungbeans. J Stored Prod Res 29:345–349

Fatokun CA, Danesh D, Menancio-Hautea DI, Young ND (1992a) A linkage map of cowpea [*Vigna unguiculata* (L.) Walp.] based on DNA markers (2N=22). In: O'Brien SJ (ed) Genome Maps. Cold Spring Harbor Laboratory, Cold Spring Harbor, New York, pp 6.256–6.258

Fatokun CA, Menancio-Hautea DI, Danesh D, Young ND (1992b) Evidence for orthologous seed weight genes in cowpea and mung bean based on RFLP mapping. Genetics 132:841–846

Fatokun CA, Danesh D, Young ND, Stewart EL (1993) Molecular taxonomic relationships in the genus *Vigna* based on RFLP analysis. Theor Appl Genet 86:97–104

Fauquet CM, Maxwell DP, Gronenborn B, Stanley J (2000) Revised proposal for naming gemini viruses. Arch Virol 145:1743–1761

Fery RL (1980) Genetics of *Vigna*. Hort Rev 2:311–394

Fujii K, Ishimoto M, Kitamura K (1989) Patterns of resistance to bean weevils (Bruchidae) in *Vigna radiata* – *sublobata* complex inform the breeding of new resistant varieties. Appl Entomol Zool 24:126–132

Harlan JR, de Wet JMJ (1971) Toward a rational classification of cultivated plants. Taxonomy 20:509–517

Humphry ME, Konduri V, Lambrides CJ, Magner T, McIntyre CL, Aitken EAB, Liu CJ (2002) Development of a mungbean (*Vigna radiata*) RFLP linkage map and its comparison with lablab (*Lablab purpureus*) reveals a high level of co-linearity between the two genomes. Theor Appl Genet 105:160–166

Humphry ME, Magner T, McIntyre CL, Aitken EAB, Liu CJ (2003) Identification of a major locus conferring resistance to powdery mildew (*Erysiphe polygoni* DC) in mungbean (*Vigna radiata* L. Wilczek) by QTL analysis. Genome 46:738–744

Humphry ME, Lambrides CJ, Liu CJ (2005) Relationships between loci conditioning hard-seededness and seed size in mungbean (*Vigna radiata* (L.) Wilczek) assessed by QTL analysis. Plant Breed 124:292–298

Imrie BC (1983) Response to selection for weathering resistance in mungbean. In: Driscoll CJ (ed) Proc 8th Australian Plant Breed Conf, University of Adelaide, Adelaide, Australia, pp 298–299

Imrie BC (1992) Reduction in hardseededness in mungbean by short duration high temperature treatment. Aust J Exp Agric 32:483–486

Imrie BC (1996) Mung bean. In: Hyde K (ed) The New Rural Industries. A Handbook for Farmers and Investors. Rural Industries Research and Development Corporation, Canberra, Australia, pp 355–360

Imrie BC, Lawn RJ, Williams RW (1984) Plant breeding for seed production and quality. In: Australian Seeds Research Conference, Queensland Agricultural College, Lawes, Australia, pp 165–180

Imrie BC, Lawn RJ, Ahmed ZU (1985) Breeding for weather resistance in mungbean. In: Yates JJ (ed) Crop and Pasture Production: Science and Practice. Working papers of the 3rd Australian Agronomy Conference, Australian Society of Agronomy, Hobart, Australia, pp 326

Imrie BC, Lawn RJ, Williams RW (1988) Breeding for resistance to weather damage in mungbean. In: Shanmugasundaram S, McLean BT (eds) Mungbean. Proc 2nd Int Symp on Mungbean, AVRDC, Taiwan, pp 131–135

Imrie BC, Lawn RJ, Yeates S (1991) Mungbean breeding and cultivar development in Australia. In: Imrie BC, Lawn RJ (eds) Mungbean: The Australian Experience. Proc 1st Australian Mungbean Workshop, CSIRO Division of Tropical Crops and Pastures, Brisbane, Australia, pp 14–20

Ishimoto M, Teraishi M, Kaga A, Aoyama K, Ijima N, Sugawara F, Kawasaki S (2000) Biochemical and genetic basis of the insect resistance in mungbean (*Vigna radiata* (L.) Wilczek). In: Proc 6th Int Congress of Plant Mol Biol, Quebec, Australia, S16–25

Jacob HJ, Lindpaintner K, Lincoln SE, Kusumi K, Bunker RK, Yi-Pei Mao, Ganten D, Dzau VJ, Lander ES (1991) Genetic mapping of a gene causing hypertension in the stroke-prone spontaneously hypertensive rat. Cell 67:213–224

Jaiwal PK, Kumari R, Ignacimuthu S, Potrykus I, Sautter C (2001) Agrobacterium-mediated genetic transformation of mungbean – a recalcitrant grain legume. Plant Sci 161:239–247

James AT, Lawn RJ, Williams RW, Lambrides CJ (1999) Cross fertility of Australian accessions of wild mungbean (*Vigna radiata* ssp. *sublobata*) with green gram (*V. radiata* ssp. *radiata*). Aust J Bot 47:601–610

Jeong SC, Kristipati S, Hayes AJ, Maughan PJ, Noffsinger SL, Gunduz I, Buss GR, Saghai Maroof MA (2002) Genetic and sequence analysis of markers tightly linked to the *soybean mosaic virus* resistance gene, Rsv3. Crop Sci 42:265–270

Kaga A, Ohnishi M, Ishii T, Kamijima O (1996 a) A genetic linkage map of azuki bean constructed with molecular and morphological markers using an interspecific population (*Vigna angularis*×*V. nakashimae*). Theor Appl Genet 93:658–663

Kaga A, Tomooka N, Egawa Y, Hosake K, Kamijima O (1996 b) Species relationships in the subgenus *Ceratotropis* (genus *Vigna*) as revealed by RAPD analysis. Euphytica 88:17–24

Kaga A, Ishimoto M (1998) Genetic localisation of a bruchid resistance gene and its relationship to insecticidal cyclopeptide alkaloids, the vignatic acids, in mungbean (*Vigna radiata* L. Wilczek). Mol Gen Genet 258:378–384

Kaga A, Ishii T, Tsukimoto K, Tokoro E, Kamijima O (2000) Comparative molecular mapping in *Ceratotropis* species using an interspecific cross between azuki bean (*Vigna angularis*) and rice bean (*V. umbellata*). Theor Appl Genet 100:207–213

Keim P, Diers BW, Shoemaker RC (1990) Genetic analysis of soybean hard seededness with molecular markers. Theor Appl Genet 79:465–469

Khattak GSS, Haq MA, Ashraf M, Elahi T (2000) Genetic of *mungbean yellow mosaic virus* (MYMV) in mungbean (*Vigna radiata* L. Wilczek). J Genet Breed 54:237–243

Kitamura K, Ishimoto M, Sawa M (1988) Inheritance of resistance to infestation with azuki bean weevil in *Vigna sublobata* and successful incorporation to *V. radiata*. Jpn J Breed 38:459–464

Konduri V, Godwin ID, Liu CJ (2000) Genetic mapping of the Lablab purpureus genome suggests the presence of 'cuckoo' gene(s) in this species. Theor Appl Genet 100:866–871

Kumari P, Dahiya BS (1981) Preliminary observations on production of hard seeds in green gram (*Vigna radiata* L.). In: Anonymous (ed) 1st National Workshop on Seed Technology, HAU, Hisar

Kumar SV, Tan SG, Quah SC, Yusoff K (2002 a) Isolation of microsatellite markers in mungbean, *Vigna radiata*. Mol Ecol Notes 2:96–98

Kumar SV, Tan SG, Quah SC, Yusoff K (2002 b) Isolation and characterisation of seven tetranucleotide microsatellite loci in mungbean, *Vigna radiata*. Mol Ecol Notes 2:293–295

Lambrides CJ (1996) Breeding for improved seed quality traits in mungbean (*Vigna radiata* L. Wilczek) using DNA markers. PhD dissertation, University of Queensland, Brisbane, Australia

Lambrides CJ, Imrie BC (2000) Susceptibility of mungbean varieties to the bruchid species *Callosobruchus maculatus* (F.), *C. phaseoli* (Gyll.), *C. chinensis* (L.), and *Acanthoscelides obtectus* (Say.) (Coleoptera: Chrysomelidae). Aust J Agric Res 51:85–89

Lambrides CJ, Ealing PM, Imrie BC (1998) Marker-assisted selection for bruchid resistance in mungbean. In: Proc 4th Asia Pacific Conf on Agri Biotechnol. Agricultural Biotechnology: Laboratory, Field and Market, 13–16 July, Darwin, Australia, p 280

Lambrides CJ, Diatloff AL, Liu CJ, Imrie BC (1999) Molecular marker studies in mungbean *Vigna radiata*. In: Proc 11th Australasian Plant Breeding Conf, 19–23 April, Adelaide, Australia

Lambrides CJ, Lawn RJ, Godwin ID, Manners J, Imrie BC (2000) Two genetic linkage maps of mungbean using RFLP and RAPD markers. Aust J Agri Res 51:415–425

Lambrides CJ, Chapman SC, Shorter R (2004 a) Genetic variation for carbon isotope discrimination in sunflower: association with transpiration efficiency and evidence for cytoplasmic inheritance. Crop Sci 44:1642–1653

Lambrides CJ, Godwin ID, Lawn RJ, Imrie BC (2004 b) Segregation distortion for seed testa colour in mungbean (*Vigna radiata* L. Wilczek). J Hered 95:532–535

Lavania UC, Lavania S (1982) Chromosome banding patterns in some Indian pulses. Ann Bot 49:235–239

Lawn RJ (1979 a) Agronomic studies on *Vigna* spp. in south-eastern Queensland I. Phenological response of cultivars to sowing date. Aust J Agric Res 30:855–870

Lawn RJ (1979 b) Agronomic studies on *Vigna* spp. in south-eastern Queensland II. Vegetative and reproductive response of cultivars to sowing date. Aust J Agric Res 30:871–882

Lawn RJ (1983) Agronomic studies on *Vigna* spp in south-eastern Queensland. III. Response to sowing arrangement. Aust J Agric Res 34:505–515

Lawn RJ (1995) The Asiatic *Vigna* species. *Vigna* spp. (*V. radiata*, *V. mungo*, *V. angularis*, *V. umbellata* and *V. acontifolia*) (Leguminosae-Papilionoideae). In: Smartt J, Simmonds NW (eds) Evolution of Crop Plants, 2nd edn. Longman, London, pp 321–326

Lawn RJ, Ahn CS (1985) Mungbean (*Vigna radiata* (L.) Wilczek/*Vigna mungo* (L.) Hepper). In: Summerfield RJ, Roberts EH (eds) Grain Legume Crops. Longman, London, pp 584–623

Lawn RJ, Cottrell A (1988) Wild mungbean and its relatives in Australia. Biologist 35(5):267–273

Lawn RJ, Williams RW, Imrie BC (1988) Potential of wild germplasm as a source of tolerance to environmental stresses in mungbean. In: Shanmugasundaram S, McLean BT (eds) Proc 2nd International Symposium on Mungbean. AVRDC, Taiwan, pp 136–145

Litt M, Luty JA (1989) A hypervariable microsatellite revealed by *in vitro* amplification of a dinucleotide repeat within the cardiac muscle action gene. Am J Hum Genet 44:391–401

Maréchal R, Mascherpa JM, Stanier F (1978) Etude taxonomique d'un groupe complexe d'espècies des genres Phaseolus et Vigna (Papilionaceae) sur la base de données morphologiques et polliniques traitées par l'analyse informatique. Boissiera 28:1–273

Menancio-Hautea D, Kumar L, Danesh D, Young ND (1992) A genome map for mungbean [*Vigna radiata* (L.) Wilczek] based on DNA genetic markers (2N=2X=22). In: O'Brien SJ (ed) Genome maps. Cold Spring Harbor Laboratory, Cold Spring Harbor, NY, pp 6.259–6.261

Menancio-Hautea D, Fatokun CA, Kumar L, Danesh D, Young ND (1993) Comparative genome analysis of mungbean (*Vigna radiata* L. Wilczek) and cowpea (*V. unguiculata* L. Walpers) using RFLP mapping data. Theor Appl Genet 86:797–810

Menendez CM, Hall AE, Gepts P (1997) A genetic linkage map of cowpea (*Vigna unguiculata*) developed from a cross between two inbred, domesticated lines. Theor Appl Genet 95:1210–1217

Michelmore RW, Paran I, Kesseli RV (1991) Identification of markers linked to disease-resistance genes by bulked segregant analysis: a rapid method to detect markers in specific genomic regions by using segregating populations. Proc Natl Acad Sci USA 88:9828–9832

Miller TE (1983) Preferential transmission of alien chromosomes in wheat. Kew Chromosome Conference II, London, pp 173–182

Miura K, Ishimoto M, Yamanaka N, Miyazaki S, Hiramatsu M, Nakajima Y, Hirano K, Kitamura K, Miyazaki S (1996) Effects of bruchid-resistant mungbean meal on growth and blood-biochemical values in mice. Jircas J 3:23–31

Miyagi M, Humphry M, Ma ZY, Lambrides CJ, Bateson M, Liu CJ (2004) Construction of bacterial artificial chromosome libraries and their application in developing PCR-based markers closely linked to a major locus conditioning bruchid resistance in mungbean (*Vigna radiata* L. Wilczek). Theor Appl Genet 110:151–156

Mohammad-Lassim MB, Chin HF, Abdullah WD (1984) The effects of weathering on mungbean (*Vigna radiata* (L.) Wilczek) seed quality. Pertanika 7:77–81

Mullis KB, Faloona FA (1987) Specific synthesis of DNA *in vitro* via a polymerase-catalysed chain reaction. Meth Enzy 155:335–350

Murray MG, Palmer JD, Cuellar RE, Thompson WF (1979) Deoxyribonucleic acid sequence organisation in the mung bean genome. Biochemistry 18:5259–5266

O'Brien GP, Webb RI, Uwins PJR, Desmarchelier PM, Imrie BC (1992) Suitability of the environmental scanning electron microscope for studies of bacteria on mungbean seeds. J Comp-Ass Microsc 4:225–229

Ouedraogo JT, Gowda BS, Jean M, Close TJ, Ehlers JD, Hall AE, Gillespie AG, Roberts PA, Ismail AM, Bruening G, Gepts P, Timko MP, Belzile FJ (2002) An improved genetic linkage map for cowpea (*Vigna unguiculata* L.) combining AFLP, RFLP, RAPD, biochemical markers, and biological resistance traits. Genome 45:175–188

Pal M, Ghosh U, Chandra M, Pal A, Biswas BB (1991) Transformation and regeneration of mung bean (*Vigna radiata*). Ind J Biochem Biophys 28:449–455

Palmer JD, Thompson WF (1981) Rearrangements in the chloroplast genome of mung bean and pea. Proc Natl Acad Sci USA 78:5533–5537

Palmer JD (1992) Comparison of chloroplast and mitochondrial genome evolution in plants. In: Herrman RG (ed) Plant Gene Research: Cell Organelles. Springer, Berlin Heidelberg New York, pp 99–133

Poehlman JM (1991) The Mungbean. Oxford and IBH, New Delhi

Prasad S, Jawali N (1998) Frequency of di-, tri-, and tetranucleotide repeats in mungbean. In: DNA Technologies: Forensic and Other Applications. 23–24 Feb 1998, Hyderabad, India

Purseglove JW (1968) Tropical Crops. Dicotyledons I. Longmans, Harlow, UK, p 331

Rebetzke G (1994) Attributes of potential adaptive and agronomic significance in the wild mungbean (*Vigna radiata* (L.) Wilczek ssp. *sublobata* (Roxb.) Verdc.). Masters Dissertation, Univ of Queensland, Brisbane, Australia

Saravanakumar P, Kaga A, Tomooka N, Vaughan DA (2004) AFLP and RAPD analysis of intra- and interspecific variation in some *Vigna* subgenus *Ceratotropis* (*Leguminosae*) species. Aust J Bot 52:417–424

Shizuya H, Birren B, Kim U-J, Mancino V, Slepak T, Tachiiri Y, Simon M (1992) Cloning and stable maintenance of 300-kilobase-pair fragments of human DNA in Escherichia coli using an F-factor-based vector. Proc Natl Acad Sci USA 89:8794–8797

Smartt J (1984) Gene pools in grain legumes. Econ Bot 38:24–35

Smartt J (1990) Grain Legumes. Evolution and Genetic Resources. Cambridge University Press, Cambridge, p 371

Soria JA, Quebral FC (1973) Occurrence and development of powdery mildew on mungbean. Philippine Agric 57:158–177

Sugawara F, Ishimoto M, Le-Van N, Koshino H, Uzawa J, Yoshida S, Kitamura K (1996) Insecticidal peptide from mungbean: a resistant factor against infestation with azuki bean weevil. J Agric Food Chem 44:3360–3364

Srinives P (1991) Breeding for resistance to diseases and pests of mungbean in Thailand. In: Green SK, Kim D (eds) Proc Int Workshop on Mungbean Yellow Mosaic Virus, 2–3 July 1991, Bangkok, Thailand, pp 59–67

Talekar NS (1988) Biology, damage and control of bruchid pests of mungbean. In: Shanmugasundaram S, McLean BT (eds) Mungbean. Proc 2nd Int Symp on Mungbean, AVRDC, Taiwan, pp 329–342

Teraishi M, Kaga A, Aoyama K, Sugawara F, Kawasaki S, Ishimoto M (2000) Genomic information of the bruchid resistance locus, Br, in mungbean (*Vigna radiata* (L.) Wilkzek). In: Proc 6th Int Congr of Plant Mol Biol, Quebec, S16–26

Timmerman-Vaughan GM, McCallum JA, Frew TJ, Weeden NF, Russell AC (1996) Linkage mapping of quantitative trait loci controlling seed weight in pea (*Pisum sativum* L.). Theor Appl Genet 93:431–439

Tivarekar S, Eapen S (2001) High frequency plant regeneration form immature cotyledons of mungbean. Plant Cell Tiss Org Cult 45:165–168

Tomer RPS, Kumari P (1991) Hard seed studies in black gram (Vigna mungo L.). Seed science and technology 19:51–56

Verdcourt B (1970) Studies in the Leguminosae-Papilionoideae for the flora of tropical East Africa. IV. Kew Bull 24:507–569

Vijayalakshmi P, Amirthaveni S, Devadas RP, Weinberger K, Tsou SCS, Shanmugasundaram S (2003) Enhanced bioavailability of iron from mungbeans and its effects on health of school children. AVRDC Technical Bulletin No. 30, Shanhua, Taiwan

Vishnu-Mittre A (1974) Palaeobotanical evidence in India. In: Hutchison J (ed) Evolutionary Studies in World Crops. Cambridge University Press, Cambridge, UK, pp 3–30

Vos P, Hogers R, Bleeker M, Reijans M, van de Lee T, Hornes M, Frijters A, Pot J, Peleman J, Kuiper M, Zabeau M (1995) AFLP: a new technique for DNA fingerprinting. Nucleic Acids Res 23:4407–4414

Wang XW, Kaga A, Tomooka N, Vaughan DA (2004) The development of SSR markers by a new method in plants and their application to gene flow studies in azuki bean [*Vigna angularis* (Willd.) Ohwi & Ohashi]. Theor Appl Genet 109:352–360

Watt EE, Poehlman JM, Cumbie BG (1977) Origin and composition of a texture layer on seeds of mung bean. Crop Sci 17:121–125

Weinberger K (2003) Impact analysis on mungbean research in south and southeast Asia. AVRDC Processing No. 99.9117.5, Shanhua, Taiwan

Williams RW (1989) A study of the causes of, and selection for resistance to weather damage in mungbean (*Vigna radiata* (L.) Wilczek; *V. mungo* (L.) Hepper). PhD dissertation, University of Queensland, Brisbane, Australia

Williams JGK, Kubelik AR, Livak KJ, Rafalski JA, Tingey SV (1990) DNA polymorphisms amplified by arbitrary primers are useful as genetic markers. Nucleic Acids Res 18:6531–6535

Williams RW, Lawn RJ, Imrie BC, Byth DE (1995 a) Studies on water damage in mungbean. I. Effect of weathering on seed quality and viability. Aust J Agric Res 46:887–899

Williams RW, Lawn RJ, Imrie BC, Byth DE (1995 b) Studies on weather damage in mungbean. II. Electrical conductivity of seed leachate as an assay of level of damage. Aust J Agric Res 46:901–907

Williams RW, Lawn RJ, Imrie BC, Byth DE (1995 c) Studies on weather damage in mungbean. III. Development of a system for measuring genotypic variation in resistance to weathering. Aust J Agric Res 46:909–920

Young ND, Danesh D, Menancio-Hautea D, Kumar L (1993) Mapping oligogenic resistance to powdery mildew in mungbean with RFLPs. Theor Appl Genet 87:243–249

Young ND, Kumar L, Menancio-Hautea D, Danesh D, Talekar NS, Shanmugasundaram S, Kim D-H (1992) RFLP mapping of a major bruchid resistance gene in mungbean (*Vigna radiata*, L. Wilczek). Theor Appl Genet 84:839–844

5 Lentil

R. Ford[1], Rubeena[2], R.J. Redden[3], M. Materne[3], P.W.J. Taylor[1]

[1] BioMarka, Faculty of Land and Food Resources, The University of Melbourne, Victoria 3010, Australia, *e-mail*: rebeccaf@unimelb.edu.au

[2] Department of Crop and Soil Sciences, Washington State University, 291 Johnson Hall, PO Box 646420, Pullman, WA 99164, USA

[3] Australian Temperature Field Crops Collection, Department of Primary Industries, Horshan, Victoria 3401, Australia

5.1 Introduction

5.1.1 Origin, Domestication and Taxonomy

Lentil, *Lens culinaris* ssp. *culinaris* Medic., was domesticated with wheat, barley and other pulses in the Fertile Crescent of the Near East and spread through southern Europe, the Middle East, North Africa and across the Indo-Gangetic plain before 1000 BC (Cubero 1981). The current distribution of lentil ranges over subtropical and temperate cropping regions of Europe, Asia, northern Africa and Ethiopia, the Indian subcontinent, North and South America, southern Africa and Australia, grown either as a spring crop or a winter crop (Blain 1988).

The earliest evidence of lentil cropping is in association with wheat and barley at Mureybit in Syria 8500–7500 BC, Hacilar and Cayonu in Turkey 7500-6500 BC, and other sites from western Iran to Palestine before 7000 BC (Cubero 1981). In many of these sites, wild lentil is rare and it is likely that the seed remains are of cultivated lentil, especially in association with cultivated wheat. The migration of lentil to Europe and Asia matched the initial spread of cereal and other legume crops (Hancock 2004).

Domestication of lentil led to the evolution of two major seed groups: macrosperma (6 to 9 mm diameter) in the Mediterranean Europe and Africa plus Asia minor and microsperma (2 to 6 mm diameter) in western Asia and in northern Africa (Cubero 1981). Macrosperma and microsperma lentils are useful functional groups within cultivated lentil as they relate to processing and consumer end uses that are different for each type (Barulina 1930; Erskine 1996).

According to the United States Department of Agriculture (USDA) and the International Legume Database and Information Service (ILDIS), the genus *Lens* comprises the cultivated *Lens culinaris* ssp. *culinaris* and the wild relative subspecies and species; *L.c.* ssp. *orientalis*, *L.c.* ssp. *odemensis*, *L.c.* ssp. *tomentosus*, *L.c.* ssp. *nigricans*, *L. ervoides*, *L. montbretii*, and *L. lamottei*. The USDA also classifies *L. nigricans* as a separate species and places all the above species in the family Fabaceae, but with the family Papilionaceae as an alternate. Historically the *Lens* genus has also been placed in the genus *Ervum*, and some species have been placed in genus *Vicia* as botanically there is a continuum between the *Vicia* and *Lens* genera (Cubero 1981).

The *Lens* genus is primarily self-pollinated, with a chromosome number of $2n = 14$. Wide intraspecific variations in chromosomal rearrangements occur within the wild species, which affects the success of both intraspecific and interspecific crossing (Ladizinsky 1979). Hancock (2004) listed three groups: Group 1 comprises *L.c.* ssp. *culinaris* and *L.c.* ssp. *orientalis* with overlapping ranges, similar morphology and a high percentage of shared molecular markers. These species are intercrossable and also crossable with *L.c.* ssp. *odemensis*, but the interspecific hybrids are only partially fertile (Ladizinsky et al. 1984). Group 2 contains the intercrossable species *L. nigricans*, *L. ervoides* and *L. lamottei*. Hybrids of *L. culinaris*, *L. orientalis* and *L. odemensis* with *L. nigricans* have irregular meiosis and do not produce viable seed (Ladizinsky et al. 1984, 1985). However, hybrids of *L. culinaris* with *L. ervoides* can produce viable seed but only with the use of embryo rescue culture (Ladizinsky et al. 1985). Group 3 contains *L. tomentosus* as a single species group. Between these groups no viable crosses have been achieved.

Genome Mapping and Molecular Breeding in Plants, Volume 3
Pulses, Sugar and Tuber Crops
C. Kole (Ed.)

The probable progenitor species for domesticated lentil is *Lens c.* ssp. *orientalis*, which has a center of origin in the Middle East Fertile Crescent (Zohary 1972). Although many endemic and rare forms of lentil occur in the Hindu-Kush region, the geographic distribution of *L. c.* ssp. *orientalis* is centered in southern Turkey – Lebanon. *L. c.* ssp. *orientalis* is able to form interspecific hybrids with the domesticated subspecies, which undergo normal meiosis resulting in seven bivalents (Ladizinsky 1979). Morphologically, *L. c.* ssp. *orientalis* has the appearance of miniaturized *L. c.* ssp. *culinaris*; however, some accessions differ from the cultivated subspecies by one or two chromosomal inversions and were found to be cross incompatible (Ladizinsky 1979).

5.1.2 Botany and Ecology

Domesticated lentil is an annual bushy herb, has very branched with slender, soft-haired angular stems, 25 to 75 cm tall, has compound pinnate leaves often ending in a tendril, one to four flowers per raceme with colors of white to blue and purple, one- to two-seeded oblong flattened pods, and biconvex rounded seed 4 to 8 mm in diameter and 2 to 3 mm thick (Duke 1981). The plant may exhibit considerable variation in the growth habit: low and bushy, suberect or erect. Flowering and branching are indeterminate. The 100-seed weight ranges from 2 to 8 g, with the seed coat varying in color from green, green-brown, to light red with black speckles (Duke 1981). Cotyledons may be red/orange, yellow or green.

Seed protein content varies from 22 to 35%, with relatively high levels of lysine, leucine and sulphur-amino acids (Muehlbauer and Tullu 1997). Anti-nutritional factors include trypsin inhibitors, haemoglutinins and oligosaccharides. Williams et al. (1994) suggested that seed nutritional quality might be improved by increasing both protein content and amino acid balance and by reducing the levels of anti-nutritional factors.

Lentil is produced as a winter crop in the semi-arid tropics, mild temperate and Mediterranean regions but is spring grown in climates with very cold winters. Lentil grows best on neutral to alkaline well-drained clay to sandy loam soils. Lentil is a quantitative long-day plant with a flowering response pattern that is independently controlled by photoperiod and temperature. A wide genetic variation for these traits was detected within a world collection (Erskine et al. 1990).

5.1.3 Genetic Resources

The largest and most representative collection of lentil landraces is maintained by the International Center for Agricultural Research in the Dry Areas (ICARDA), Syria (Erskine et al. 1989), currently at 3351 accessions. Large collections are also held in the USA (2807), Russia (2857), Australia (2847) and Turkey (1435) (European Genetic Resources Database 2004: www.ecpgr.cgiar.org).

The first systematic analysis of genetic diversity in lentil was performed by Barulina (1930) with the Vavilov collection, defining the macrosperma and microsperma types as subspecies and describing regional groups (*grex*) according to geographic distribution of qualitative traits, chiefly seed characteristics, pubescence and size of vegetative organs, phenology, height and pod shape. Landraces held at the ICARDA were later characterized and four major regional groups were identified through analysis of variability in quantitative and qualitative morphological traits (Erskine and Witcombe 1984; Erskine et al. 1989). These were the Levantine group (Egypt, Jordan, Lebanon, Syria), northern group (Greece, Iran, Turkey, USSR, Chile), Indian subcontinent group and Ethiopian group. The most discriminating traits were in time to maturity, lowest pod height and 100-seed weight, but seed color was also important. The ecological environment was a major evolutionary force in the spread of cultivated lentil (Erskine et al. 1989). For example, the importance of flowering response for adaptation in south Asia caused a bottleneck in the introduction of lentil and subsequently low genetic variability among landraces into this region (Erskine 1997; Erskine et al. 1998). Extensive variation within landraces and between landraces within a region was found for both morphological traits and allelic variation of isozymes, indicating a complex organisation of lentil populations with co-adapted multilocus allelic combinations mediated by a very low frequency of outcrossing (Erskine and Choudhary 1986; Erskine and Muehlbauer 1991).

5.1.4 Economic Importance and Current Breeding Objectives

World production of lentil was predicted to reach 3.3 million metric tons in 2004 (Agriculture Canada Market Analysis). Lentil is currently grown in the Indian subcontinent, Middle East, northern Africa, southern Europe, North and South America, Canada, Australia and New Zealand. India is the leading producer of lentil, producing about one third of the total world production, predominantly for its own domestic usage. With the exception of Chile (Barulina 1930), lentil production in the Americas is a relatively recent event (Muehlbauer and McPhee 2002). However, Canada has since become a major world producer and the largest exporter of lentil. There has also been a considerable increase in the area (125,000 ha in 2002) under lentil in Australia since 1991 (293 ha) (Carter and Materne 1997).

Water Availability and Waterlogging. Lentil yields are very dependent on available soil moisture during the growing season (Erskine and Saxena 1993; Erskine et al. 1994 a). Although mechanisms such as good early vigour and early flowering and maturity (Erskine et al. 1994 a), higher osmotic adjustment (Leport et al. 1998), deeper rooting (Buddenhagen and Richards 1988; Turner et al. 2001) or tolerance to subsoils constraints (Materne et al. 2002) have been advocated as ways of increasing legume productivity under moisture limiting conditions, they are unproven over seasons. Selection is typically based on grain yield under variable rainfed conditions to increase water use efficiency (Erskine and Saxena 1993). Furthermore, drought may occur during plant establishment, intermittently during the vegetative growth period, terminally or progressively during the season depending on the environment (Rahman and Mallick 1988; Erskine et al. 1994 a).

Excess water and waterlogging during winter and late spring to summer can reduce lentil yields in Mediterranean and subtropical environments respectively. The sensitivity of lentil to waterlogging and anaerobic conditions is thought to account for the poor response of the crop to irrigation, although responsive genotypes have been identified with larger parenchyma in their roots (Erskine and Saxena 1993; Erskine et al. 1994 a; Bejiga and Anbessa 1995).

Diseases. A wide range of pathogens infect lentil crops worldwide (Brouwer et al. 2000). The diseases rust, caused by *Uromyces viciae-fabae* (Pers.), vascular wilt, caused by *Fusarium oxysporum* f.sp. lentis Vasd. and Srin. and ascochyta blight, caused by *Ascochyta lentis* Vassilievsky, are the key fungal diseases of lentil worldwide (Erskine and Saxena 1993; Erskine et al. 1994 c). Other important fungal diseases are powdery mildew, caused by *Erysiphe polygoni* D.C. and downy mildew, caused by *Peronospora lentis* Gauman (Khare et al. 1993). Anthracnose, caused by *Colletotrichum truncatum* (Schwein.) Andrus and Moore, and stemphylium blight, caused by *Stemphylium botrysum* Wallr., are locally of major economic importance in Canada and Bangladesh respectively. *Botrytis cinerea* is regarded as a major problem in Pakistan (Brouwer et al. 2000) and *Botrytis fabae* is the major Botrytis pathogen on lentil in Australia (Materne et al. 2002). Lentil also has viral pathogens such as pea seed-borne mosaic virus (PSbMV), bean leaf roll virus (BLRV) and pea enation mosaic virus (PEMV) as well as several bacteria, nematodes, insects and parasitic weeds (Erskine et al. 1994 a). Tolerance to many viruses has been identified at ICARDA and in Australia (ICARDA 1994; Latham and Jones 2001; Latham et al. 2001).

Research, including breeding, has been initiated for many diseases in lentil. The highest level of understanding of these diseases in terms of the host-pathogen relationships and genetic improvement can be demonstrated using ascochyta blight as an example. Many sources of foliar resistance to ascochyta blight have been recorded, but resistance to seed infection is more limited (Andrahennadi et al. 1996; Nasir and Bretag 1997 a). Pathogen variability also exists (Ahmed et al. 1996; Ahmed and Morrall 1996) and of most significance was the identification of an isolate that was virulent on accession ILL5588 (Nasir and Bretag 1997 b), a major source of resistance used in breeding programs around the world. Importantly, the accession ILL7537 was shown to be highly resistant to this isolate (Nasir and Bretag 1997 b; Nguyen et al. 2001) and, with ILL5588 and Indianhead, is a key source of resistance genes for breeding. Breeding is well advanced for ascochyta blight, and resistant cultivars have been released in many countries. Individually these genes can be selected for in well-designed field screening nurseries; however, differentiating genotypes with a combination of these genes is more difficult. Resistance to anthracnose was recently found in accession PI320937 and is being selected for in elite genotypes along with resistance to ascochyta blight (Tullu et al. 2003).

Temperature Stress. Hot or dry weather during flowering and pod fill causes severe constraints on the productivity of lentil crops in many regions of the world, including the Mediterranean (Erskine 1985 b). In colder areas such as in the USA and Turkey, large yield increases can be achieved by sowing lentil in winter rather than spring (Muehlbauer and McPhee 2002). However, problems associated with a lack of winter hardiness, increased incidence of diseases and weed control issues must be addressed before winter sowing becomes widespread. Kahraman et al. (2004) identified the most winter-hardy accessions in the USA and Turkey as WA8649041, WA8649090, ILL1878 and ILL669. Inheritance was shown to be quantitative, indicating that several genes of varying effect may be required for survival. Frost at flowering is a major limitation to lentil production in Australia, but genetic variation has not been identified.

Flowering. The timing of flowering is a particularly important event as it determines the duration of the vegetative phase, which establishes the potential of the crop and determines the climatic conditions to which the crop will be exposed during reproductive growth. The optimal flowering response differs between regions and is simple to measure within a target region (Erskine et al. 1994 b). Presently, the understanding of the genetic control of flowering is limited (Sarker et al. 1999).

Nutrient Imbalances. Boron toxicity is increasingly being recognised as a problem in the arid areas of west Asia and Australia, where lentil is widely grown (Yau and Erskine 2000). Tolerance to high soil boron was identified in lentil (Yau and Erskine 2000; Hobson et al. 2003), and breeding was initiated in Australia using soil-based screening methods. Conversely, boron deficiency has been identified as a limitation to production in Nepal. However, genetic variability was identified and field-based selection is currently under way (Srivastava et al. 1999). There is increasing evidence that the same genetic mechanisms are likely to control tolerance to both boron deficiency and toxicity, predominantly boron exclusion (Yau and Erskine 2000; Dannel et al. 2002).

Lentil is sensitive to zinc and iron deficiency and poor nitrogen fixation due to factors that affect the host, rhizobia and the symbiosis between the two. Interactions between rhizobia and host have been identified with the potential for future selection of genotypes with improved nitrogen fixation (Slattery, personal communication, 2004).

The major lentil growing areas of the world are regions with a high frequency of saline or sodic soils (Saxena 1993). Variation in tolerance to a range of salts was initially identified in the USA (Jana and Slinkard 1979), and NaCl-tolerant accessions have been identified by others (Rai et al. 1985; Ashraf and Waheed 1990). The inheritance of salt tolerance has been investigated and recessive genes were implicated (Ashraf and Waheed 1993).

Harvestability. Hand harvesting is considered a major constraint to lentil production in the Middle East and northern Africa (Erskine and Goodrich 1988; Erskine et al. 1991). The development of cultivars for mechanised harvesting is a prime research goal in these countries and in countries where lentil is machine harvested, such as Australia and North America (Erskine and Goodrich 1988; Erskine et al. 1991; Materne et al. 2002). Plant breeding can assist harvest mechanisation through the development of taller, non-lodging cultivars that retain their pods and seeds at maturity and mature uniformly. Genetic variability exists for height (Erskine and Witcombe 1984; Muehlbauer et al. 1995), lodging resistance (Muehlbauer et al. 1995), pod dehiscence (Erskine 1985 a) and uniformity of maturity.

Quality. Lentil quality is largely defined by visual characteristics such as size, shape and color and contamination of the sample. The preferred appearance is most often that which mimics the local product. Lentil seeds are susceptible to mechanical damage due to their thin shape and sharp edges. Therefore, the development of rounder seeded cultivars offers potential for reduced mechanical damage, while these types are also the preferred quality in many markets.

5.1.5 Limitations of Classical Breeding Approaches and the Need for Molecular Breeding

Although lentils are an old crop species, genetic improvement other than farmer selection has only occurred relatively recently, led by establishment of an international breeding program at ICARDA during the late 1970s. In the last 20 years limitations to lentil production have been defined in many countries, genetic variability sought for economically important traits and breeding initiated to overcome these limitations. Lentil is also a relatively small crop compared to crops such as wheat, rice, maize and soybean, and consequently funding for research is more limited and often directed at applied research to address the large number of limitations to production. Also, much of the crop is grown in economically poor countries where research funding and expertise in novel molecular breeding approaches such as biotechnology and genomics have often been limited. This lack of funding and expertise, the short period of breeding history and the large number of limitations to be addressed have meant that molecular research for lentil has been much less extensive than for many other crop species. However, even with these restrictions, some substantial advances have been made towards understanding the lentil genome and the development and application of molecular markers in breeding.

Much of the focus has been on developing molecular markers for resistance to a few major diseases, especially where research was complementary between countries, for example ascochyta blight. Molecular markers have also been invaluable in understanding the inheritance of resistance and in identifying novel genes (Nguyen et al. 2001). In the near term, potential exists to develop markers for many traits where genetic variability exists, for example boron and salinity tolerance, and resistance to botrytis grey mould, anthracnose, rust, Fusarium and stemphylium diseases. Where classical breeding attempts have been limited, markers will become invaluable if genetic variability for difficult-to-measure traits, such as frost and drought tolerance, is identified. The devastating effect that the introduction of ascochyta blight had on chickpea production in Australia and North America highlighted the potential effect of exotic diseases on lentil cultivation. Molecular markers offer potential to improve pre-emptive breeding strategies and overcome the difficulties associated with screening in distant localities.

The implementation of markers for routine use in lentil breeding programs is currently very limited, often because the traits investigated can be phenotyped relatively cost effectively. The key to both research and implementation of markers for lentil lies in the integration of the markers within the breeding program to ensure that cost-effective utilisation of the technology is achieved.

5.2 Genetic Markers and Lentil Genome Mapping

5.2.1 Morphological and Biochemical Markers

Morphological and biological markers have been used by lentil researchers and breeders as useful tools for the purposes of diversity analysis, taxonomy and trait selection (Barulina 1930; Ladizinsky and Sarkar 1982; Erskine and Choudhary 1986; Erskine et al. 1989; Ferguson and Robertson 1999). Cotyledon color has been used to estimate the percentage of natural outcrossing (Wilson and Law 1972). Other useful morphological markers include number of days to flowering (*Sn*), seed coat pattern (*Scp*) and pubescent peduncle (*Pep*). These were found to be linked together and mapped in linkage group (LG) 5 of the lentil genome, whereas tendrilled leaf (*Tnl*) was linked with colored stem (*Gs*) in LG 1 (Sarker et al. 1999). Linkage was also reported between spreading-erect growth habit, brown-green stem and brown green leaf (Emami and Sharma 1999).

Similar to morphological markers, biochemical markers have also demonstrated allelic variation of gene expression products, and isozyme markers have been used for lentil genome mapping and trait association (Vaillancourt and Slinkard 1993; Tahir and Muehlbauer 1994; Rodriguez et al. 1997). However, accuracy of linkage with these markers may be limited due to the relatively small number of morphological and isozyme loci which may be assessed and potential restriction of their expression to a specific development stage or tissue type.

represent a far better solution for reproducible and multiple environment lentil trait mapping, where, after recombination, each locus is fixed and self-pollination can create an infinite amount of genetically identical individuals in a relatively short period of time. However, the production of a lentil DH population is dependent on amenability to another culture, and in general grain legumes are more recalcitrant to in vitro manipulation than many other species (reviewed by Christou 1997).

Many *Lens* genome linkage maps have been produced over the past 20 years (Table 1). Historically, *Lens* maps comprise a small number of markers, covering a relatively small portion of the lentil genome. The first map of *Lens* was constructed using morphological and isozyme markers (Zamir and Ladizinsky 1984; Tadmor et al. 1987) and the first map with DNA-based markers (RFLP) was developed by Havey and Muehlbauer (1989). PCR-based markers subsequently revolutionized the construction of linkage maps, and the first extensive linkage map of *Lens* comprised 177 markers (RAPD, AFLP, RFLP and morphological markers) and was developed from an RIL population created from an intersubspecific cross (Eujayl et al. 1998a). The first intraspecific linkage map of lentil was constructed with 114 RAPD, ISSR and RGA markers (Rubeena et al. 2003, Fig. 1). Recently, two more extensive molecular linkage maps have been reported, one using an intraspecific population (Kahraman et al. 2004) and the other based on an intersubspecific population (Durán et al. 2004). The lentil linkage map, produced by Durán et al. (2004), contained 62 RAPD, 29 ISSR, 65 AFLP, 4 morphological and 1 SSR markers and spanned a distance of 2172 cM within 10 LGs. This was the first time a lentil-specific SSR marker was mapped. The map of Kahraman et al. (2004) covered 1192

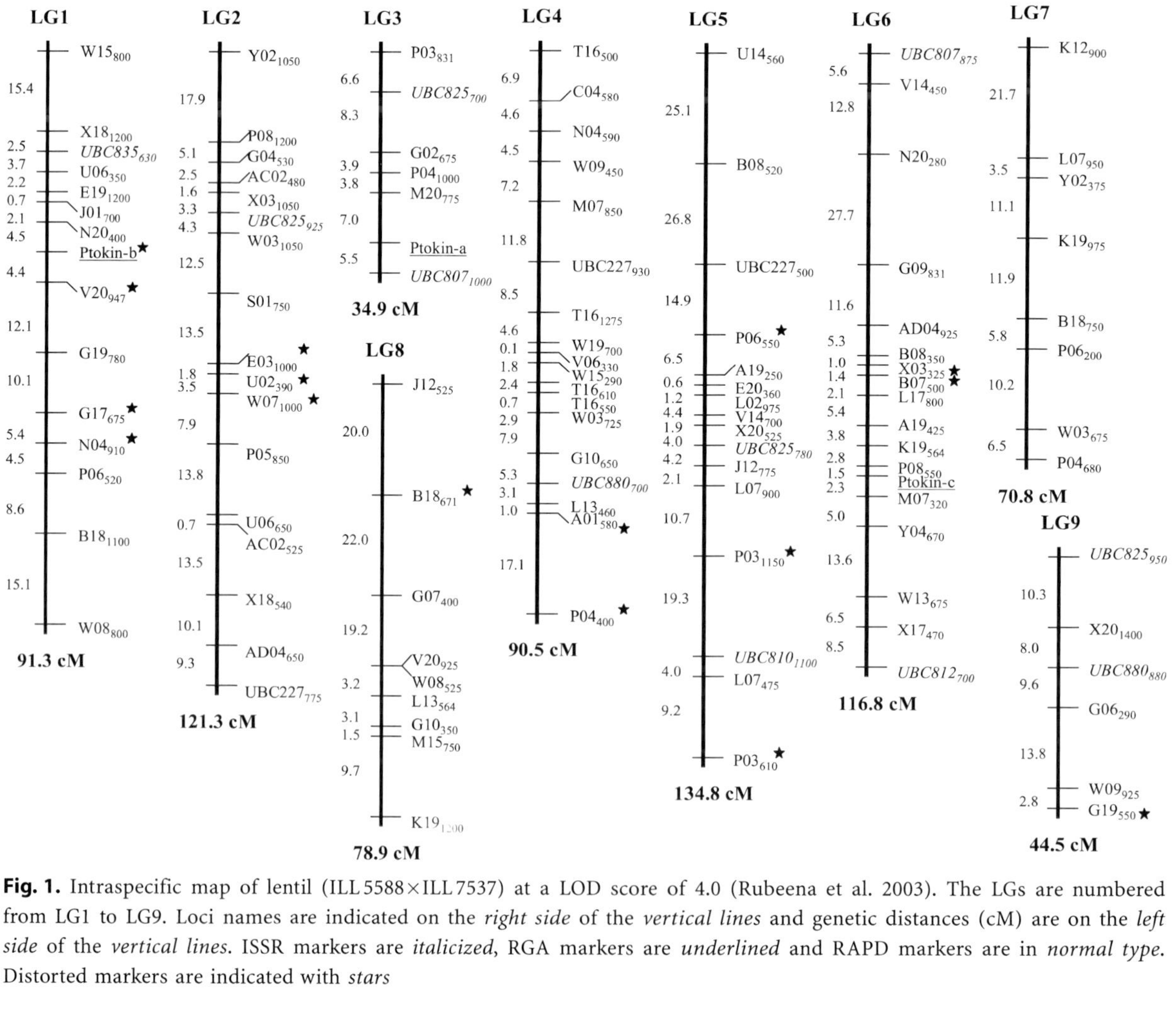

Fig. 1. Intraspecific map of lentil (ILL 5588×ILL 7537) at a LOD score of 4.0 (Rubeena et al. 2003). The LGs are numbered from LG1 to LG9. Loci names are indicated on the *right side* of the *vertical lines* and genetic distances (cM) are on the *left side* of the *vertical lines*. ISSR markers are *italicized*, RGA markers are underlined and RAPD markers are in *normal type*. Distorted markers are indicated with *stars*

cM within 9 LGs and comprised a total of 130 arbitrarily produced (RAPD, ISSR and AFLP) markers. Most recently a comprehensive intersubspecific *Lens* map was developed by enriching the previous map of Eujayl et al. (1998 a) with 39 new lentil-specific SSR and 50 new AFLP markers (Hamwieh et al. 2005). The map comprised a total of 283 markers spanning 751 cM within 14 LGs (8 with more than 3 markers).

To date, all *Lens* and lentil genome maps developed have more LGs than the species haploid chromosome number (n=7). The amount of the genome mapped varies from 751 to 2172 cM with an average marker density of 2.7 to 15.87 cM (Table 1). The expected full genome length is as yet unknown. However, given the close phylogeny among the species, perhaps the expected length would be close to that of field pea, which is 700 to 800 cM as determined by cytological studies (Hall et al. 1997 a, b).

Other important characteristics of the current maps include the clustering of markers at various regions and the inclusion of distorted markers. Clustering may be indicative of centromeric and telomeric regions, which experience up to ten-fold less recombination than other areas of the genome (Tanksley et al. 1992). This was also observed in *Pisum* and *Cicer* maps (Laucou et al. 1998 and Winter et al. 2000 respectively). Segregation distortion is the consequence of unequal inheritance of parts of chromosomes, which may affect the ordering of markers within a LG (Lorieux et al. 1995). Factors that contribute to marker distortion include recessive alleles, structural rearrangements or differences in DNA content, abortion of male and female gametes and the selective fertilization of a particular gametic genotype (Tadmor et al. 1987; Barzen et al. 1995; Berry et al. 1995; Quillet et al. 1995; Jenczewski et al. 1997; Xu et al. 1997).

Until recently, a major limitation to *Lens* mapping has been the unavailability of locus-specific PCR-based and co-dominant markers such as expressed sequence tag (EST), cleaved amplified polymorphic sequences (CAPS), single nucleotide polymorphism (SNP) or SSR microsatellite markers, which are more robust and informative than arbitrary DNA markers. The lack of such markers has largely hampered the ability to compare various published linkage maps. However, the development of SSR markers for lentil was recently achieved by researchers at the ICARDA (Hamwieh et al. 2004, 2005). At present, one SSR marker has been mapped by Durán et al. (2004), and a further 39 SSR loci have been mapped by Hamwieh et al. (2005).

5.2.4 Towards a *Lens* Consensus Map

The existing maps have not been well linked to each other due to the lack of common markers. However, morphological markers and the recently developed lentil SSR markers (Hamwieh et al. 2005) should prove useful in assigning common LGs. Of the seven morphological markers already mapped, cotyledon color (orange vs. yellow; *Yc*), presence or absence of anthocyanin in the stem (*Gs*), seed coat pattern or spotting (*Scp*), pod dehiscence-indehiscence (*Pi*), ground color (brown vs. tan) of the seed (*Ggc*), erect or prostrate growth habit (*Gh*) and presence or absence of anthocyanin in the pod (*Pdp*), four have been placed on multiple maps (*Yc*, *Gs*, *Scp* and *Pi*) (Table 2).

Other markers that may be useful for consensus mapping include the repetitive DNA sequences that have been localised by fluorescent in situ hybridization (FISH) (Patil et al. 1995) and other gene-specific markers such as expressed sequence tag (EST) markers. Also, gene-specific markers transferable from related model legume crop species such as *Medicago truncatula* and *Lotus japonicus* and converted to SNP or cleaved amplified polymorphism (CAP) type markers. Such markers will also be useful for comparative mapping across species, to assign genetic LGs to specific *Lens* chromosomes and for integrating information from both physical and genetic maps (Galasso et al. 2001).

Table 2. Morphological markers mapped on different LGs in *Lens* genome may be useful as anchor markers

Marker locus	Linkage group in different studies		
	Tahir et al. 1993	Eujayl et al. 1998	Durán et al. 2004
Scp	V	III	I
Yc	II	–	II
Gs	I	–	IV
Pi	IV	II	–

5.3 Marker-Assisted Trait Mapping and Selection

5.3.1 Trait Mapping

Many simply inherited traits have been placed upon *Lens* genome maps. By knowing the map position of a gene, one can diagnose the presence of the gene using flanking DNA markers without waiting for the gene effect to be present in the phenotype (Paterson et al. 1991).

Bulked segregant analysis (BSA), first described by Michelmore et al. (1991), is one method used for linking molecular markers to phenotypic traits controlled by single major genes. This method relies on the availability of two bulked DNA samples collected from individuals that segregate for an extreme phenotype within a single population. One bulk contains the DNA of the trait being targeted, while the other contains DNA from individuals lacking the trait. DNA polymorphisms between the bulks are therefore likely to be linked to genes that govern the trait. In lentil, this method has been used to identify markers that are tightly linked to genes for resistance to Fusarium vascular wilt and ascochyta blight (Eujayl et al. 1998 b; Ford et al. 1999; Chowdhury et al. 2001).

Eujayl et al. (1998 b) used an RIL mapping population to identify linked molecular markers to the single dominant gene conditioning Fusarium vascular wilt resistance (*Fw*). They subsequently identified a RAPD marker ($OPS16_{750}$) that was 9.1 cM from the radiation-frost tolerance locus (*Frt*) (Eujayl et al. 1999). However, most probably due to insufficient genome map coverage, the *Frt* locus and the linked RAPD marker were unable to be placed on the existing linkage map developed by Eujayl et al. (1998 a).

Ford et al. (1999) identified RAPD markers, RV01 and RB18, approximately 6 and 14 cM, respectively away from and flanking the foliar ascochyta blight resistance locus *Ral*1 (AbR_1) in ILL5588. These were subsequently converted to locus-specific sequence characterized amplified region (SCAR) markers and screened for applicability across parental lines in the Australian breeding program. Although the linkage was not maintained across all parental genotypes, great potential exists for the targeted use of these markers in breeding and the pyramiding of resistance genes in ILL5588-derived genetic backgrounds. Subsequently, two RAPD markers, $UBC227_{1290}$ and $OPD\text{-}10_{870}$, were identified that flanked and were linked in repulsion phase to the resistance gene *ral*2 in the cultivar Indianhead at 12 and 16 cM respectively (Chowdhury et al. 2001). Most recently, molecular markers were developed linked to the complementary dominant resistance genes in ILL7537 (Rubeena, unpubl.). The resistance sources within these genotypes were shown to be novel using pathogenicity tests (Nguyen et al. 2001). Thus the potential exists to use markers to pyramid ascochyta blight resistance genes to develop durably resistant varieties.

Two RAPD markers, $OPE06_{1250}$ and $UBC704_{700}$, linked at 6.4 cM (in repulsion) and 10.5 cM (in coupling) respectively, were recently identified for selecting the anthracnose resistance locus *LCt-2* in accession PI 320937 (Tullu et al. 2003). Three AFLP markers were also identified linked in repulsion phase to *LCt-2*; however, the closest of these was 21.5 cM. Most recently, an SSR marker and an AFLP marker were identified flanking the *Fw* locus in ILL5588 at 8.0 and 3.5 cM respectively (Hamwieh et al. 2005).

5.3.2 Quantitative Trait Loci Mapping

When a trait is governed by multiple and quantitative trait loci (QTL) and/or co-dominantly inherited genes, a more holistic genome mapping approach may be undertaken to identify relative genome loci location, interaction and subsequent molecular markers for accurate trait selection.

A few QTL studies have been reported thus far for lentil. The first one employed a genetic linkage map developed from an intersubspecific population (*L. c.* ssp. *culinaris*×*L. c.* ssp. *orientalis*). A total of 22 QTLs were placed upon the map including five for height of the first ramification, three for plant height, five for flowering, seven for pod dehiscence, one for shoot number and one for F_3 seed diameter (Durán et al. 2002). QTLs governing winter hardiness were recently mapped using an F_6-derived RIL population of 106 individuals of a cross between WA8649090 and Precoz (Kahraman et al. 2004). For this, a framework map was produced with 9 LGs comprising a total of 130 markers and spanning 1192 cM. Quantitative survival and injury data were

Table 3. Putative QTL for winter hardiness in lentil (Kahraman et al. 2004)

Location	Linkage group	QTL position (cM)[a]	LOD	R^2 (%)
Haymana 1997–98	3	28	2.3	10.9
	4	118	7.3	28.8
	6	80	3.2	17.7
Total				**33.4**
Pullman 1998–99	4	110	2.5	11.5
Haymana 1999–2000	1	38	2.3	9.5
	1	146	2.2	10.1
	4	132	2.0	9.5
Total				**22.9**
Combined	4	116	6.9	28.8
	6	80	3.1	14.2
Total				**31.5**

[a] QTL position from bottom of LG

collected at two locations in 1997 (Pullman, WA, USA and Haymana, Turkey) and three locations in 1998 and 1999 (Pullman, WA, USA, Haymana and Sivas, Turkey). Five independent QTLs were detected to account for survival with a LOD score >2. These were located on LGs 1, 3, 4 and 6 with two QTLs on LG 1 (Table 3). One QTL on LG 4 was common among locations, although the effect and position differed. The maximum of the winter survival phenotype variation that these accounted for was 33.4%. One ISSR marker, ubc808-12, was identified that may be useful for MAS of winter survival. A further four QTLs were reported to influence winter injury in the USA location and together accounted for 42.7% of the trait variation.

Preliminary QTL analysis of the ascochyta blight resistance in ILL7537 was conducted using a population comprising 153 F_2 individuals [ILL 7537 (R)×ILL 6002 (S)] and a linkage map comprising 72 markers spanning 412.5 cM anchored to a pre-existing map (Rubeena et al. 2003). The disease reaction was scored using a 1–9 scale on each of the F_2 individuals at 14, 21 and 28 d after inoculation, and three QTL peaks (two on LGI and one on LGII) were observed using composite interval mapping (CIM) (Fig. 2). Two QTLs (QTL-1 and QTL-2) were observed on LG I in close proximity; since these were

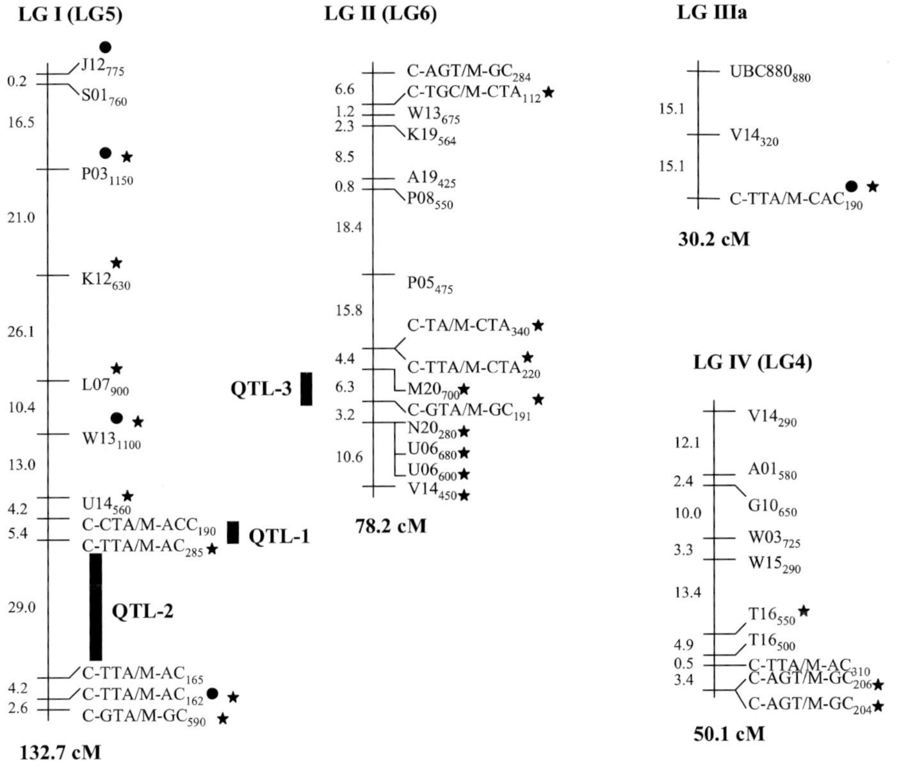

Fig. 2. Four LGs with significant markers shown by *stars* and distorted markers by *dots*. QTL regions are shown as *filled vertical bars* and named QTL-1, QTL-2 and QTL-3. All three QTL were observed by CIM, whereas only QTL-2 and QTL-3 were detected by MIM

Table 4. Putative QTL for ascochyta blight resistance identified in the F_2 population (ILL 7537/ILL 6002) by composite interval mapping (CIM)

Test parameter	QTL	Linkage group	Interval length (cM)[a]	Flanking markers	QTL position (cM)[b]	LR[c]	Additive effect	$R^{2\,d}$ (%)
14 DAI	QTL-1	LG I	5.4	C-CTA/M-ACC_{190} – C-TTA/M-AC_{285}	2.0	12.78	–0.48	7.82
	QTL-2	LG I	29.0	C-TTA/M-AC_{285} – C-TTA/M-AC_{165}	14.0	13.83	–1.03	26.80
	QTL-3	LG II	6.3	$M20_{700}$ – C-GTA/M-GC_{191}	0.0	10.49	–0.59	6.19
Total								**40.81**
21 DAI	QTL-1	LG I	5.4	C-CTA/M-ACC_{190} – C-TTA/M-AC_{285}	4.0	20.35	–0.65	11.02
	QTL-2	LG I	29.0	C-TTA/M-AC_{285} – C-TTA/M-AC_{165}	12.0	26.36	–0.95	33.62
	QTL-3	LG II	6.3	$M20_{700}$ – C-GTA/M-GC_{191}	0.0	17.38	–0.72	9.25
Total								**53.89**
28 DAI	QTL-1	LG I	5.4	C-CTA/M-ACC_{190} – C-TTA/M-AC_{285}	2.0	31.19	–0.89	16.41
	QTL-2	LG I	29.0	C-TTA/M-AC_{285} – C-TTA/M-AC_{165}	8.0	34.70	–1.06	30.70
	QTL-3	LG II	6.3	$M20_{700}$ – C-GTA/M-GC_{191}	0.0	20.80	–0.73	10.25
Total								**57.36**
MDS	QTL-1	LG I	5.4	C-CTA/M-ACC_{190} – C-TTA/M-AC_{285}	2.0	24.05	–0.74	13.76
	QTL-2	LG I	29.0	C-TTA/M-AC_{285} – C-TTA/M-AC_{165}	12.0	29.12	–0.98	33.80
	QTL-3	LG II	6.3	$M20_{700}$ – C-GTA/M-GC_{191}	0.0	18.73	–0.70	10.10
Total								**57.66**

[a] Interval between the two flanking markers (cM)
[b] QTL position from left flanking marker (cM)
[c] Peak value of maximum likelihood ratio (LR) test statistic observed for QTL
[d] Proportion of phenotypic variance explained by QTL

> 10 cM apart, they were considered to be separate QTLs (Table 4). They accounted for ca. 47%, whereas QTL-3 on LG II accounted for ca. 10% of the variance of the trait. The position of the QTL changed slightly over the different scoring periods after inoculation. The AFLP marker C-TTA/M-AC_{285} was found to be 3.4 cM away from QTL-1 and 12 cM away from QTL-2. The RAPD marker $M20_{700}$ was located at the same position as QTL-3. When multiple interval mapping (MIM) was performed, only two significant QTLs (QTL-2 and QTL-3) were identified. These two QTLs may potentially be the major effects of the two co-dominant resistant genes previously identified to govern resistance in ILL 7537 (Nguyen et al. 2001). However, the QTLs identified must be validated in different genetic backgrounds and populations before incorporation into breeding programs.

Recently, resistance gene analogues belonging to the nucleotide binding site gene families were isolated from the lentil genotype ILL5588 (Yaish et al. 2004). Mapping of RGA, together with the ascochyta blight resistance trait, may be useful to validate the location of genes that are functional in the resistance mechanism, a step towards map-based cloning of the active resistance genes.

5.3.3 Marker-Assisted Selection and Trait Pyramiding

Marker-assisted selection (MAS) is the ability to select for and breed for a desirable trait with a marker, or suit of markers, from within a plant genotype without the need to express the associated phenotype. Therefore, MAS offers great opportunity for improved efficiency and effectiveness in the selection of plant genotypes with a desired combination of traits. This approach relies upon the establishment of a tight linkage between a molecular marker and the chromosomal location of the gene(s) governing the trait to be selected in a particular environment. Once this has been achieved, selection can be conducted in the laboratory and does not require the expression of the associated phenotype. For example, using MAS, disease resistance can be evaluated in the absence of the disease and in the early stage of plant development.

Markers used for MAS are also termed sequence tagged sites (STSs). These are mapped loci for which all or part of the corresponding DNA sequences has been determined. This information can be used to design PCR primers for amplification of all or part of the original sequence. They are more robust and reproducible than the arbitrary sequences they are designed from, such as RAPD markers, as they are developed from the known sequences and produce an amplicon from longer primers. Differences in the lengths of amplified fragments serve as genetic markers for the locus. If no length polymorphism is detected, the amplified fragments can be cleaved with restriction enzymes to observe subsequent length differences. This technique is often referred to as cleaved amplified polymorphic sequences or CAPS (Jarvis et al. 1994).

The use of converted locus-specific PCR markers is also referred to as a specific polymorphic locus amplification test (SPLAT), as well as sequence characterized amplified region (SCAR) markers and allele-specific associated primer (ASAP) markers. SPLAT markers are designed from sequencing the insert of a polymorphic RFLP marker (Gale and Witcombe 1992), whereas SCAR and ASAP markers are developed from sequencing specific RAPD markers (Paran and Michelmore 1993; Gu et al. 1995; Ford et al. 1999). The conversion of more technically demanding RFLP markers into PCR-based markers (e.g. SPLAT) may provide a more rapid, cost-effective and efficient tool in lentil breeding.

Nguyen et al. (2001) first converted an arbitrarily produced lentil sequence to a SCAR marker (SCARW19) for selecting resistance to ascochyta blight in lentil accession ILL5588. Tar'an et al. (2003) converted the $RB18_{680}$ RAPD marker, formerly also shown to be linked to the *AbR1* gene (Ford et al. 1999), into a robust SCAR marker. They subsequently used SCAR markers linked to the *AbR1* gene and the *ral2* gene (Chowdhury et al. 2001), together with a marker linked in repulsion to a gene for anthracnose resistance (*LCt2*), to pyramid the traits in an RIL population. Using the linked markers, 11 of 156 RILs were shown to retain all three resistance genes. Of these, 82%, which contained the markers linked to *AbR1* and *ral2*, were resistant to a highly virulent *A. lentis* isolate. Furthermore, 85% of the lines that did not contain the marker linked to the *LCt2* gene were resistant to the virulent 95B36 isolate of *C. truncatum*. This is the first evidence of validating the use of molecular markers for marker-assisted trait selection in lentil. Pyramiding of multiple resistance genes to foliar fungal pathogens should provide a broader and more durable resistance, as similarly shown in rice against bacterial blight (Singh et al. 2001).

5.4 Future Scope of Works

To date, no gene-specific, SSR or arbitrarily produced molecular marker has been identified in extremely close proximity (<1 cM) to any mapped lentil quality, disease resistance or stress tolerance gene locus. Hence, in order to increase the accuracy of MAS, the identification of tightly linked markers is an ongoing goal for researchers targeting many traits.

Of the specific traits for which molecular markers could provide considerable benefit to current breeding practices, water usage and drought tolerance are paramount. Furthermore, breeding for a flowering response that gives broad adaptation is currently a goal of the lentil breeding program in Australia, and markers that are proven stable across multiple environments have great potential in improving grain yield across variable locations and years in water-limited regions of the world.

To densely map the genomic areas surrounding the genes governing traits of interest, highly robust, unilocus, co-dominant and transferable markers are required. In particular, those that may be transferred among multiple different genetic backgrounds, and hence applicable across a broad range of breeding programs, would be most useful. Markers such as SSRs and ESTs are only now being developed for lentil by several groups. Alternatively, sequences from the model species *Medicago trancatula* are showing great promise for their transferability to *Lens* (R. Oliver, personal communication). Such sequences are particularly useful since many have already been assigned function and associated with desirable traits. Furthermore, the gene space of the *M. trunculata* genome will be fully sequenced using a bacterial artificial chromosome (BAC) library approach by the end of 2006 (Young et al. 2004), enabling the elucidation of syntenic relationships among the grain legumes.

Another requirement necessary for the future of fine genome mapping in lentil is the availability of large fixed mapping populations, such as single seed descent RILs and DHs. Such populations would allow for the identification and validation of trait-associated markers across different environments and at different plant growth stages. For map-based cloning of genes shown to be associated with desirable traits, a BAC library of lentil is required with good genome coverage (>5). Furthermore, the ability to tile the lentil BAC clones along the *M. truncatula* sequenced genome will also negate the necessity to sequence the entire lentil genome. Naturally the choice of genotypes for mapping populations and BAC library construction will need to be carefully considered in order to produce tools that are compatible with as many breeding priorities and programs throughout the world as possible.

Once genes have been isolated from the lentil genome they will need to be validated for function using a differential or transgenic approach. Already differential array chips exist that were constructed from expressed sequences within related legume species such as soybean (Vodkin et al. 2004). Through gene conservation and intergeneric synteny, these chips may be useful for determining the orthologous genes up-regulated in lentil in response to similar stimuli, such as exposure to a fungal pathogen. However, correct gene identification is still very much reliant on the sequences already present within existing databases. Agrobacterium-mediated and particle bombardment gene transfer protocols have been developed for lentil (Sarker et al. 2003 and Gulati et al. 2002 respectively), and ultimately these will serve as a means to prove gene function via gene silencing and gene expression in alternate genetic backgrounds.

References

Ahmed S, Morrall RAA (1996) Field reactions of lentil lines and cultivars to isolates of *Ascochyta fabae* f. sp. *lentis*. Can J Plant Pathol 18:362–369

Ahmed S, Morrall RAA, Sheard JW (1996) Virulence of *Ascochyta fabae* f.sp. *lentis* on lentil. Can J Plant Pathol 18:354–361

Andrahennadi CP, Slinkard AE, Vandenberg A (1996) Ascochyta resistance in lentil. LENS Newslett 23:5–7

Arumuganathan K, Earle ED (1991) Nuclear DNA content of some important plant species. Plant Mol Biol Rep 9:208–218

Ashraf M, Waheed A (1990) Screening of local/exotic accessions of lentil (*Lens culinaris* Medic.) for salt tolerance at two growth stages. Plant Soil 128:167–176

Ashraf M, Waheed A (1993) Responses of some local/exotic accessions of lentil (*Lens culinaris* Medic.) to salt stress. J Agron Crop Sci 170:103–112

Barulina H (1930) Lentil of the U.S.S.R. and other countries. Bull Appl Bot, Genet Plant Breed, Leningrad Suppl 40:1–319

Barzen E, Mechelke W, Ritter E, Schulte-Kappert E, Salamini F (1995) An extended map of the sugar beet genome containing AFLP and RFLP loci. Theor Appl Genet 90:189–193

Bejiga G, Anbessa Y (1995) Waterlogging tolerance in lentil. LENS Newslett 22:8–10

Berry ST, Leon AJ, Hanfrey CC, Challis P, Burkholz A, Barnes SJ, Rufener GK, Lee M, Caligari PDS (1995) Molecular marker analysis of *Helianthus annuus* L. 2. Construction of an RFLP linkage map for cultivated sunflower. Theor Appl Genet 91:195–199

Blain HL (1988) Future trends in supply and demand of pea and lentil. In: Summerfield RJ (ed) World Crops: Cool Season Food Legumes. ISBN 90-247-3641-2:501–511

Brouwer JB, Sharma B, Malik BA, Hill GD (2000) Region 6: Asia-Pacific: Meeting the challenge. In: Knight R (ed) Linking Research and Marketing Opportunities for Pulses in the 21st Century. Proc 3rd Int Food Legumes Res Conf on Current Plant Science and Biotechnology in Agriculture. Kluwer, Dordrecht, vol 34, pp 115–129

Buddenhagen IW, Richards RA (1988) Breeding cool season food legumes for improved performance in stress environments. In: Summerfield RJ (ed) World Crops: Cool Season Food Legumes. Kluwer, Spokane, WA, pp 81–95

Carter JM, Materne MA (1997) Lentil growers guide: a guide for the production of lentils. Department of Natural Resources and Environment, Victoria, Australia

Chowdhury MA, Andrahennadi CP, Slinkard AE, Vandenberg A (2001) RAPD and SCAR markers for resistance to ascochyta blight in lentil. Euphytica 118:331–337

Christou P (1997) Biotechnology applied to grain legumes. Field Crops Res 53:83–97

Collard BCY, Pang ECK, Ades PK, Taylor PWJ (2003) Preliminary investigation of QTLs associated with seedling resistance to ascochyta blight from *Cicer echinospermum*, a wild relative of chickpea. Theor Appl Genet 107:719–729

Cubero JI (1981) Origin, taxonomy and domestication. In: Webb C, Hawtin G (eds) Lentils. CAB, Slough, UK, pp 15–38

Dannel F, Pfeffer H, Romheld V (2002) Uptake on boron in higher plants – uptake, primary translocation and compartmentation. Plant Biol 4:193–204

Duke JA (1981) Handbook of Legumes of World Economic Importance. Plenum, New York, pp 110–113, 317–338

Durán Y, Fratini R, Morales S, Fernández M, García P, Pérez de la Vega M (2002) A genetic map of lentil. In: 1st Int Conf Legume Genom Genet: Translation to Crop Improv, 2–6 June 2002, Minneapolis-St. Paul, MN

Durán Y, Fratini R, García P, Pérez de la Vega M (2004) An intersubspecific genetic map of *Lens*. Theor Appl Genet 108:1265–1273

Emami MK, Sharma B (1999) Linkage between three morphological markers in lentil. Plant Breed 118:579–581

Erskine W (1985a) Selection for pod retention and pod dehiscence in lentils. Euphytica 34:105–112

Erskine W (1985b) Perspectives in lentil breeding. In: Saxena MC, Varma S (eds) Faba Beans, Kabuli Chickpeas and Lentils in the 1980's. ICARDA Aleppo, Syria, pp 91–100

Erskine W (1996) Seed-size effects on lentil (*Lens culinaris*) yield potential and adaptation to temperature and rainfall in West Asia. J Agric Sci 126:335–341

Erskine W (1997) Lessons for breeders from land races of lentil. Euphytica 93:107–112

Erskine W, Choudhary MA (1986) Variation between and within lentil landraces from Yemen Arab Republic. Euphytica 35:695–700

Erskine W, Goodrich WJ (1988) Lodging in lentil and its relationship with other characters. Can J Plant Sci 68:929–934

Erskine W, Muehlbauer FJ (1991) Allozyme and morphological variability, outcrossing rate and core collection formation in lentil germplasm. Theor Appl Genet 83:119–125

Erskine W, Saxena MC (1993) Problems and prospects of stress resistance breeding in lentil. In: Singh KB, Saxena MC (eds) Breeding for Stress Tolerance in Cool-Season Food Legumes. ICARDA/Wiley, Chichester, UK, pp 51–62

Erskine W, Witcombe JR (1984) Lentil Germplasm Catalog. ICARDA, Aleppo, Syria

Erskine W, Adham Y, Holly L (1989) Geographic distribution of variation in quantitative traits in a world lentil collection. Euphytica 43:97–103

Erskine W, Ellis RH, Summerfield RJ, Roberts EH, Hussain A (1990) Characterisation of responses to temperature and photoperiod for time to flowering in a world lentil collection. Theor Appl Genet 80:193–199

Erskine W, Diekmann J, Jegatheeswaran P, Salkini A, Saxena MC, Ghanaim A, Ashkar FEL (1991) Evaluation of lentil harvest systems for different sowing methods and cultivars in Syria. J Agric Sci 117:333–338

Erskine W, Tufail M, Russell A, Tyagi MC, Rahman MM, Saxena MC (1994a) Current and future strategies in breeding lentil for resistance to biotic and abiotic stresses. Euphytica 73:127–135

Erskine W, Hussain A, Tahir M, Bahksh A, Ellis RH, Summerfield RJ, Roberts EH (1994b) Field evaluation of a model of photothermal flowering responses in a world lentil collection. Theor Appl Genet 88:423–428

Erskine W, Tufail M, Russell A, Tyagi MC, Rahman MM, Saxena MC, Muehlbauer FJ (1994c) Expanding the production and use of cool season food legumes. In: Kaiser WJ (ed) Proc 2nd Int Food Legume Res Conf on Pea, Lentil, Faba Bean, Chickpea, and Grasspea, Cairo, Egypt, pp 559–571

Erskine W, Chandra S, Chaudhury M, Malik IA, Sarker A, Sharma B, Tufail M, Tyagi MC (1998) A bottleneck in lentil: widening its genetic base in South Asia. Euphytica 101:207–211

Eujayl I, Baum M, Erskine W, Pehu E, Muehlbauer FJ (1997) The use of RAPD markers for lentil genetic mapping and the evaluation of distorted F_2 segregation. Euphytica 96:405–412

Eujayl I, Baum M, Powell W, Erskine W, Pehu E (1998a) A genetic linkage map of lentil (*Lens* sp.) based on RAPD and AFLP markers using recombinant inbred lines. Theor Appl Genet 97:83–89

Eujayl I, Erskine W, Bayaa B, Baum M, Pehu E (1998b) Fusarium vascular wilt in lentil: inheritance and identification of DNA markers for resistance. Plant Breed 117:497–499

Eujayl I, Erskine W, Baum M, Pehu E (1999) Inheritance and linkage analysis of frost injury in lentil. Crop Sci 39:639–642

Ferguson ME, Robertson LD (1999) Morphological and phenological variation in the wild relatives of lentil. Genet Resource Crop Evol 46:3–12

Ferguson ME, Maxted N, van Slageren M, Robertson LD (2000) A re-assessment of the taxonomy of *Lens* Mill. (*Leguminosae*, *Papilionoideae*, *Vicieae*). Bot J Linn Soc 133:41–59

Feuillet C, Schachermayr G, Keller B (1997) Molecular cloning of a new receptor-like kinase gene encoded at the *Lr10* disease resistance locus of wheat. Plant J 11:45–52

Ford R, Pang ECK, Taylor PWJ (1997) Diversity analysis and species identification in *Lens* using PCR generated markers. Euphytica 96:247–255

Ford R, Pang ECK, Taylor PWJ (1999) Genetics of resistance to ascochyta blight (*Ascochyta lentis*) of lentil and the identification of closely linked RAPD markers. Theor Appl Genet 98:93–98

Ford R, Garnier-Géré P, Nasir M, Taylor PWJ (2000) The structure of *Ascochyta lentis* in Australia revealed with RAPD markers. Aust Plant Pathol 29:36–45

Galasso I, Schmidt T, Pignone D (2001) Identification of *Lens culinaris* ssp. *culinaris* chromosomes by physical mapping of repetitive DNA sequences. Chrom Res 9:199–209

Gale MD, Witcombe JR (1992) DNA markers and marker-mediated applications in plant breeding, with particular reference to pearl millet breeding. In: Moss JP (ed) Biotechnology and Crop Improvement in Asia. ICRISAT, Patancheru, AP, India, pp 323–332

Gu WK, Weeden NF, Yu J, Wallace DH (1995) Large-scale, cost-effective screening of PCR products in marker-assisted selection applications. Theor Appl Genet 91:465–470

Gulati A, Schryer P, McHughen A (2002) Production of fertile transgenic lentil (*Lens culinaris* Medik.) plants using particle bombardment. In Vitro Cell Dev Biol Plant 38:316–324

Hall KJ, Parker JS, Ellis THN (1997a) The relationship between genetic and cytogenetic maps of pea. I. Standard and translocation karyotypes. Genome 40:744–754

Hall KJ, Parker JS, Ellis THN, Turner L, Knox MR, Hofer JMI, Lu J, Ferrandiz C, Hunter PJ, Taylor JD, Baird K (1997b) The relationship between genetic and cytogenetic maps of pea. II. Physical maps of linkage mapping populations. Genome 40:755–769

Hamwieh A, Choumane W, Udapa SM, Dreyer F, Jung C, Baum M (2004) Development of microsatellite markers for the genus *Lens*. In: Plant and Animal Genome XII Conf, San Diego, p 135

Hamwieh A, Udapa SM, Choumane W, Sarker A, Dreyer F, Jung C, Baum M (2005) A genetic linkage map of lentil based on microsatellite and AFLP markers and localization of *Fusarium* vascular wilt resistance. Theor Appl Genet 110: 669–677

Hancock JF (2004) Plant Evolution and the Origin of Crop Species, 2nd edn. CABI, Wallingford, UK

Havey MJ, Muehlbauer FJ (1989) Linkages between restriction fragment length, isozyme, and morphological markers in lentil. Theor Appl Genet 77:395–401

Hobson K, Armstrong R, Connor D, Nicolas M, Materne M (2003) Genetic variation in tolerance to high concentrations of soil boron exists in lentil germplasm. In: Solutions for a Better Environment. Proc 11th Aust Agron Conf, Victoria, Australia

ICARDA (1994) Legume viruses. In: Germplasm Program: Legumes. Annual report for 1994, ICARDA, Syria, pp 285–294

Jana MK, Slinkard AE (1979) Screening for salt tolerance in lentils. LENS Newslett 6:25–27

Jarvis P, Lister C, Szabó V, Dean C (1994) Integration of CAPS markers into the RFLP map generated using recombinant inbred lines of *Arabidopsis thaliana*. Plant Mol Biol 24:685–687

Jenczewski E, Gherardi M, Bonnin I, Prosperi JM, Olivieri I, Huguet T (1997) Insight on segregation distortions in two intraspecific crosses between annual species of *Medicago* (Leguminosae). Theor Appl Genet 94:682–691

Jones N, Ougham H, Thomas H (1997) Markers and mapping: we are all geneticists now. New Phytopathol 137:165–177

Kahraman A, Kusmenoglu I, Aydin N, Aydogan A, Erskine W, Muehlbauer FJ (2004) QTL mapping of winter hardiness genes in lentil. Crop Sci 44:13–22

Kanazin V, Marek LF, Shoemaker RC (1996) Resistance gene analogs are conserved and clustered in soybean. Proc Natl Acad Sci USA 93:11746–11750

Khare MN, Bayaa B, Beniwal SPS (1993) Selection methods for disease resistance in lentil. In: Singh KB, Saxena MC (eds) Breeding for Stress Tolerance in Cool-Season Food Legumes. ICARDA/Wiley, Chichester, UK, pp 107–121

Ladizinsky G (1979) The origin of lentil and its wild gene pool. Euphytica 28:179–187

Ladizinsky G, Sakar D (1982) Morphological and cytogenetical characterisation of *Vicia montbrettii* (synonym: *Lens montbrettii* [Fisch. And Mey.] Davis and Plitmann). Bot J Linn Soc 85:209–212

Ladizinsky G, Braun D, Goshen D, Muehlbauer FJ (1984) The biological species of the genus *Lens* L. Bot Gaz 145:253–261

Ladizinsky G, Cohen D, Muehlbauer FJ (1985) Hybridization in the genus *Lens* by means of embryo culture. Theor Appl Genet 70:97–101

Latham LJ, Jones RAC (2001) Alfalfa mosaic and pea seedborne mosaic viruses in cool season crop, annual pasture, and forage legumes: susceptibility, sensitivity and seed transmission. Aust J Agric Res 52:771–790

Latham LJ, Jones RAC, McKirdy SJ (2001) Cucumber mosaic cucumovirus infection of cool season crop, annual pasture, and forage legumes: susceptibility, sensitivity, and seed transmission. Aust J Agric Res 52:683–697

Laucou V, Haurogné K, Ellis N, Rameau C (1998) Genetic mapping in pea. 1. RAPD-based genetic linkage map of *Pisum sativum*. Theor Appl Genet 97:905–915

Leister D, Ballvora A, Salamini F, Gebhardt C (1996) A PCR-based approach for isolating pathogen resistance genes from potato with potential for wide application in plants. Nat Genet 14:421–429

Leport L, Turner NC, French RJ, Tennant D, Thomson BD, Siddique KHM (1998) Water relations, gas exchange and growth of cool-season grain legumes in a Mediterranean-type environment. Eur J Agric 9:295–303

Lorieux M, Goffinet B, Perrier X, Gonzalez de Leon D, Lanaud C (1995) Maximum-likelihood models for mapping genetic markers showing segregation distortion. 1. Backcross populations. Theor Appl Genet 90:73–80

Materne M, McMurray L, Nitschke S, Regan K, Heuke L, Dean G, Carpenter D (2002) The future of Australian lentil production. In: Brouwer JB (ed) Proc Lentil Focus 2002 Meeting, Horsham, Victoria, Australia

Michelmore RW, Paran I, Kesseli RV (1991) Identification of markers linked to disease-resistance genes by bulked segregant analysis: a rapid method to detect markers in specific genomic regions by using segregating populations. Proc Natl Acad Sci USA 88:9828–9832

Morgan TH (1911) Random segregation versus coupling in Mendelian inheritance. Science 34:384

Muehlbauer FJ, McPhee KE (2002) Future of North American lentil production. In: Brouwer JB (ed) Proc Lentil Focus 2002 Meeting, Horsham, Victoria, Australia

Muehlbauer FJ, Tullu A (1997) *Lens culinaris* Medik. In: NewCROP FactSHEET, www.hort.purdue.edu/newcrop/cropfactsheet/lentil.html p 8\CEnote{ This link didn't work. Please send valid link.}

Muehlbauer FJ, Weeden NF, Hoffman DL (1989) Inheritance and linkage relationships of morphological and isozyme loci in lentil (*Lens* Miller). J Hered 80:298–303

Muehlbauer FJ, Kaiser WJ, Clement S L, Summerfield RJ (1995) Production and breeding of lentil. Adv Agron 54:283–332

Nasir M, Bretag TW (1997a) Prevalence of *Ascochyta fabae* f.sp. *lentis* on lentil seed from Victoria, Australia. Aust Plant Pathol 26:117–120

Nasir M, Bretag TW (1997b) Pathogenic variability in Australian isolates of *Ascochyta lentis*. Aust Plant Pathol 26:217–220

Nguyen TT, Taylor PWJ, Brouwer JB, Pang ECK, Ford R (2001) A novel source of resistance in lentil (*Lens culinaris* ssp. *culinaris*) to ascochyta blight caused by *Ascochyta lentis*. Aust Plant Pathol 30:211–215

Paran I, Michelmore RW (1993) Development of reliable PCR-based markers linked to downy mildew resistance genes in lettuce. Theor Appl Genet 85:985–993

Paterson AH, Tanksley SD, Sorrells ME (1991) DNA markers in plant improvement. Adv Agron 46:39–90

Patil PB, Vrinten PL, Scoles GJ, Slinkard AE (1995) Variation in the ribosomal RNA units of the genera *Lens* and *Cicer*. Euphytica 83:33–42

Quillet MC, Madjidian N, Griveau Y, Serieys H, Tersac M, Lorieux M, Bervillé A (1995) Mapping genetic factors controlling pollen viability in an interspecific cross in *Helianthus* sect. *Helianthus*. Theor Appl Genet 91:1195–1202

Rahman MM, Mallick RN (1988) Factors which limit cool season food legume productivity in Bangladesh. In: Summerfield RJ (ed) World Crops: Cool-Season Food Legumes. Kluwer, Spokane, WA, pp 230–234

Rai R, Nasar SKT, Singh SJ, Prasad V (1985) Interactions between Rhizobium strains and lentil (*Lens culinaris* Linn.) genotypes under salt stress. J Agric Sci 104:199–205

Rodriguez MM, Paredes OM, Becerra VL (1997) Isozyme diversity of Chilean lentil germplasm (*Lens culinaris* Medik). In: Int Food Legume Res Conf III, Adelaide, Australia, p 135

Rubeena, Ford R, Taylor PWJ (2003) Construction of an intraspecific linkage map of lentil (*Lens culinaris* ssp. *culinaris*). Theor Appl Genet 107:910–916

Sarker A, Erskine W, Sharma B, Tyagi MC (1999) Inheritance and linkage relationships of days to flower and morphological loci in lentil (*Lens culinaris* Medikus subsp. *culinaris*). J Hered 90:270–275

Sarker RH, Biswas A, Mustafa BM, Mahbub S, Hoque MI (2003) *Agrobacterium*-mediated transformation of lentil (*Lens culinaris* Medik.). Plant Tiss Cult 13:1–12

Saxena MC (1993) The challenge of developing biotic and abiotic stress resistance in cool-season food legumes. In: Singh KB, Saxena MC (eds) Breeding for Stress Tolerance in Cool-Season Food Legumes. ICARDA/Wiley, Chichester, UK, pp 3–14

Sharma SK, Knox MR, Ellis THN (1996) AFLP analysis of the diversity and phylogeny of *Lens* and its comparison with RAPD analysis. Theor Appl Genet 93:751–758

Singh S, Sidhu JS, Huang N, Vikal Y, Li Z, Brar DS, Dhaliwal HS, Khush GS (2001) Pyramiding three bacterial blight resistance genes (*xa5*, *xa13* and *Xa21*) using marker-assisted selection into *Indica* rice cultivar PR106. Theor Appl Genet 102:1011–1015

Srivastava SP, Joshi M, Johansen C, Rego TJ (1999) Boron deficiency of lentil in Nepal. LENS Newslett 26:22–24

Tadmor Y, Zamir D, Ladizinsky G (1987) Genetic mapping of an ancient translocation in the genus *Lens*. Theor Appl Genet 73:883–892

Tahir M, Muehlbauer FJ (1994) Gene mapping in lentil with recombinant inbred lines. J Hered 85:306–310

Tahir M, Simon CJ, Muehlbauer FJ (1993) Gene map of lentil: a review. LENS Newslett 20:3–10

Tanksley SD, Ganal MW, Prince JP, de Vicente MC, Bonierbale MW, Broun P, Fulton TM, Giovannoni JJ, Grandillo S, Martin GB, Messeguer R, Miller JC, Miller L, Paterson AH, Pineda O, Röder MS, Wing RA, Wu W, Young ND (1992) High density molecular linkage maps of the tomato and potato genomes. Genetics 132:1141–1160

Tar'an B, Buchwaldt L, Tullu A, Banniza S, Warkentin TD, Vandenberg A (2003) Using molecular markers to pyramid genes for resistance to ascochyta blight and anthracnose in lentil (*Lens culinaris* Medik). Euphytica 134:223–230

Tullu A, Buchwaldt T, Warkentin T, Tar'an B, Vandenberg A (2003) Genetics of resistance to anthracnose and identification of AFLP and RAPD markers linked to the resistance gene in PI 320937 germplasm of lentil (*Lens culinaris* Medikus). Theor Appl Genet 106:428–434

Turner NC, Wright GC, Siddique KHM (2001) Adaptation of grain legumes (Pulses) to water-limited environments. Adv Agron 71:193–231

Vaillancourt RE, Slinkard AE (1993) Linkage of morphological and isozyme loci in lentil, *Lens culinaris* L. Can J Plant Sci 73:917–926

Vodkin LO, Khanna A, Shealy R, Clough SJ, Gonzalez DO, Zabala G, Thibaud-Nissen F, Sidarous M, Srömvik MV, Shoop E, Schmidt C, Retzel E, Erpelding J, Shoemaker R, Rodriguez-Huete AM, Polacco JC, Coryell V, Keim P, Gong G, Liu L, Pardinas J, Schweitzer P (2004) Microarrays for global expression constructed with a low redundancy set of 27,500 sequenced cDNAs representing an array of developmental stages and physiological conditions of the soybean plant. http://www.biomedcentral.-com/1471-2164/5/73

Weeden NF, Muehlbauer FJ, Ladizinsky G (1992) Extensive conservation of linkage relationships between pea and lentil genetic maps. J Hered 83:123–129

Wilson VE, Law AG (1972) Natural crossing in *Lens esculenta* Moench. J Am Soc Hort Sci 97:142–143

Williams PC, Bhatty RS, Deshpande SS, Hussein SA, Savage GP (1994) Improving the nutritional quality of cool season food legumes. In: Muehlbauer F, Kaiser WJ (eds) Ex-chickpea and southeastern Turkey as the center of origin for the crop. Van der Maesen (1987) also recognized the southeastern part of Turkey adjoining Syria as the possible center of origin of chickpea based on the presence of the closely related annual species, *C. reticulatum* and *C. echinospermum*.

Investigations into the genetic relatedness among *Cicer* species were initially done by assessing similarities in plant morphology (Ladizinsky and Adler 1976 b; Robertson et al. 1997), karyotype (Ocampo et al. 1992), crossability (Ladizinsky and Adler 1976 b; Ahmad et al. 1988; Singh and Ocampo 1993) and isozymes (Kazan and Muehlbauer 1991; Ahmad et al. 1992; Labdi et al. 1996; Tayyar and Waines 1996). Results from karyomorphological examination (Ocampo et al. 1992), interspecific hybridization (Singh and Ocampo 1993, 1997), isozyme pattern analysis (Labdi et al. 1996) and chromatin characterization (Galasso et al. 1996) between chickpea and the wild relatives indicated that *C. arietinum* evolved from *C. reticulatum*. More recently, analyses using DNA-based molecular markers to directly evaluate genetic variation (Serret et al. 1997; Ahmad 1999; Choumane et al. 2000; Iruela et al. 2002; Sudupak et al. 2002; Nguyen et al. 2004) confirmed that *C. reticulatum* was likely the progenitor of cultivated chickpea.

Using AFLP markers, Nguyen et al. (2004) demonstrated that all assessed *Cicer* species fitted into three main groups: Group I (the primary and secondary crossability groups documented by Ladizinsky et al. 1976 b) clusters *C. arietinum* between *C. echinospermum* and *C. reticulatum*. Group II (the annual tertiary group) includes *C. judaicum*, *C. pinnatifidium* and *C. bijugum*. Group III (mostly perennial tertiary) includes *C. anatolicum*, *C. oxyodon*, *C. canariensis*, *C. flexuosum*, *C. macracanthum*, *C. microphyllum*, *C. multijugum*, *C. nuristanicum* and *C. songaricum* as well as two annual species *C. yamashitae* and *C. cuneatum*. Using C-banding and in situ hybridization techniques, Galasso et al. (1996) suggested that both *C. arietinum* and *C. echinospermum* were derived from *C. reticulatum* via loss of one pair of satellite chromosomes. Subsequent accumulation of chromosomal rearrangements in *C. echinospermum* gave rise to a hybrid sterility barrier, further differentiating this species from the other two *Cicer* species.

6.1.2 Traditional and Current Breeding Objectives

Despite the economic value of chickpea and the relatively long period of traditional breeding, the productivity of the crop has not greatly improved in recent years. In Australia, the annual yield substantially dropped during the past decade and has only now recovered to the 1995 level (FAOSTAT Database 2004). This was most probably due to disease and the effects of poor environmental adaptation. Therefore, chickpea breeding programs are focused on improving the genetic potential of the crop by producing genetic combinations that offer protection against key abiotic and biotic stresses. In particular, the crop is highly susceptible to the fungal diseases fusarium wilt and ascochyta blight.

Fusarium wilt, caused by *Fusarium oxysporum* Schlecht. emend Snyd. & Hans. f.sp. *ciceri* (Padwick) Snyd. & Hans., is prevalent in many chickpea-growing regions of the world including South and North America, Asia, Africa and the Mediterranean. The disease is capable of causing total crop loss and is difficult to effectively manage via crop rotation or foliage fumigation since the pathogen is soil borne (Benko-Iseppon et al. 2003; Sharma et al. 2004).

Ascochyta blight, caused by *Ascochyta rabiei* (Pass.) Lab, is one of the most damaging diseases of chickpea worldwide. Under favorable conditions of high humidity and moderate temperature (20 to 22 °C), the disease can cause total yield loss (Singh and Reddy 1990). The use of resistant cultivars is considered to be the most practical and cost-effective method of control (Jimenez-Diaz et al. 1993). However, resistance is thought to be inherited in a complex and quantitative manner (Flandez-Galvez et al. 2003 b). Although moderate resistance has been detected in existing accessions within germplasm collections (Singh and Reddy 1994; Tekeoglu et al. 2000; Flandez-Galvez et al. 2003 a), the probable recessive nature of the resistance genes has most probably hindered the development of highly resistant and superior cultivars.

As an initial step towards broadening the genetic diversity of chickpea and achieving better yielding varieties, there has been a surge in the comprehensive understanding of the amount and pattern of natural genetic variation that exists within and between the available cultivated chickpea germplasm collections. Researchers have concluded that the cultivars and wilder accessions of *C. arietinum* are lacking in necessary diversity that may include desirable traits needed for effective genetic improvement of the crop (Robertson et al. 1997; Collard et al. 2003). Nguyen et al. (2004) detected 102 polymorphic AFLP loci among accessions of *C. pinnatifidium*, with an overall gene diversity of h = 0.126, whereas only 20 polymorphic AFLP loci were detected among accessions of *C. arietinum* (h = 0.036). By looking to crosses with wild relative species members, the genetic bases of breeding programs may potentially be broadened and desirable traits integrated (Singh and Ocampo 1997).

Singh et al. (1994) reported that many *Cicer* species contained sources of resistance to many pathogens. In particular, accessions of the wild annual species *C. pinnatifidium, C. judaicum, C. bijugum* and *C. echinospermum* have been shown to contain useful levels of resistance to *Ascochyta rabiei* (Singh et al. 1981; Haware et al. 1992; Stamigna et al. 1998; Collard et al. 2003; Nguyen et al. 2005).

Recently, 11 of 56 accessions of *Cicer* species (Berger et al. 2003) assessed were found to contain moderate to high levels of resistance to an aggressive Australian isolate of *A. rabiei*. Interestingly all of the resistant *C. pinnatifidum* accessions (ATC 48694, ATC 46901, ATC 46960), the resistant *C. echinospermum* accession (ATC 42328) and one resistant *C. judaicum* accession (ATC 46864) originated from Turkey. The other resistant *C. judaicum* accessions originated from Syria (ATC 46868, ATC 46934, ATC 46935), Israel (ATC 46892, ATC 46899) and Jordan (ATC 46893) (Nguyen et al. 2005).

To be useful for potential desirable gene transfer, two accessions from different species need to be crossable. Among the annual relatives, only two wild species, *C. reticulatum* and *C. echinospermum*, have been crossed with *C. arietinum* to produce fertile interspecific hybrids. Plants from *C. arietinum* × *C. reticulatum* crosses were completely fertile (Singh and Ocampo 1993, 1997); however, plants from *C. arietinum* × *C. echinospermum* crosses varied from complete sterility to complete fertility (Singh and Ocampo 1993). Confirming their close phylogenetic origin, Tayyar et al. (1994) observed similar C-banded karyotypes of these wild species with that of cultivated chickpea. A preliminary comparative mapping between interspecific *C. arietinum* × *C. reticulatum* (Winter et al. 2000) and intraspecific *C. arietinum* genomes likewise revealed the speciation of chickpea from *C. reticulatum* via inversion of DNA sequences and minor chromosomal translocation (Flandez-Galvez et al. 2003b).

In order to make optimal use of any desirable traits detected within novel germplasm, the genomic location of the governing genes should be identified. The use of molecular markers closely linked to genes of agronomic importance, such as disease resistance and drought tolerance, may be used to screen large numbers of new genetic combinations within breeding programs. Marker-assisted selection will lead to a more effective selection and pyramiding of such desirable genes within agronomically superior genotype backgrounds. The first step towards this is the production of an informative genome map.

6.2 Construction of Genetic Maps

6.2.1 A Brief History of *Cicer* Mapping Efforts

There has been considerable progress on genetic mapping in chickpea in recent years. However, due to the extremely low level of genetic polymorphism detected within the cultivated gene pool (Ahmad et al. 1992; Udupa et al. 1993; Labdi et al. 1996), mapping has historically focused on using interspecific *Cicer* populations. These have included populations derived from crosses between *C. arietinum* and the wild relatives *C. reticulatum* (Kazan et al. 1993; Simon and Muehlbauer 1997; Winter et al. 1999, 2000; Santra et al. 2000; Banerjee et al. 2001; Tekeoglu et al. 2002) and *C. echinospermum* (Collard et al. 2003) (Table 1).

However, a genetic map constructed from an interspecific cross may not represent the true recombination distance (cM) map order of the cultivated genome. Due to the uneven recombination of homoeologous chromosomes during meiosis, DNA markers for linkage analysis may have a high degree of segregation distortion in the mapping population. This results in biased estimations of the linkage marker distances (Lorieux et al. 1995; Liu et al. 1996). Irregularities during meiosis in interspecific F_1 plants have been reported in *Cicer*, like the lacking of one chromosome during Anaphase I in 6% of pollen mother cells (Pundir and Mengesha 1995) and the occasional observation of two univa-

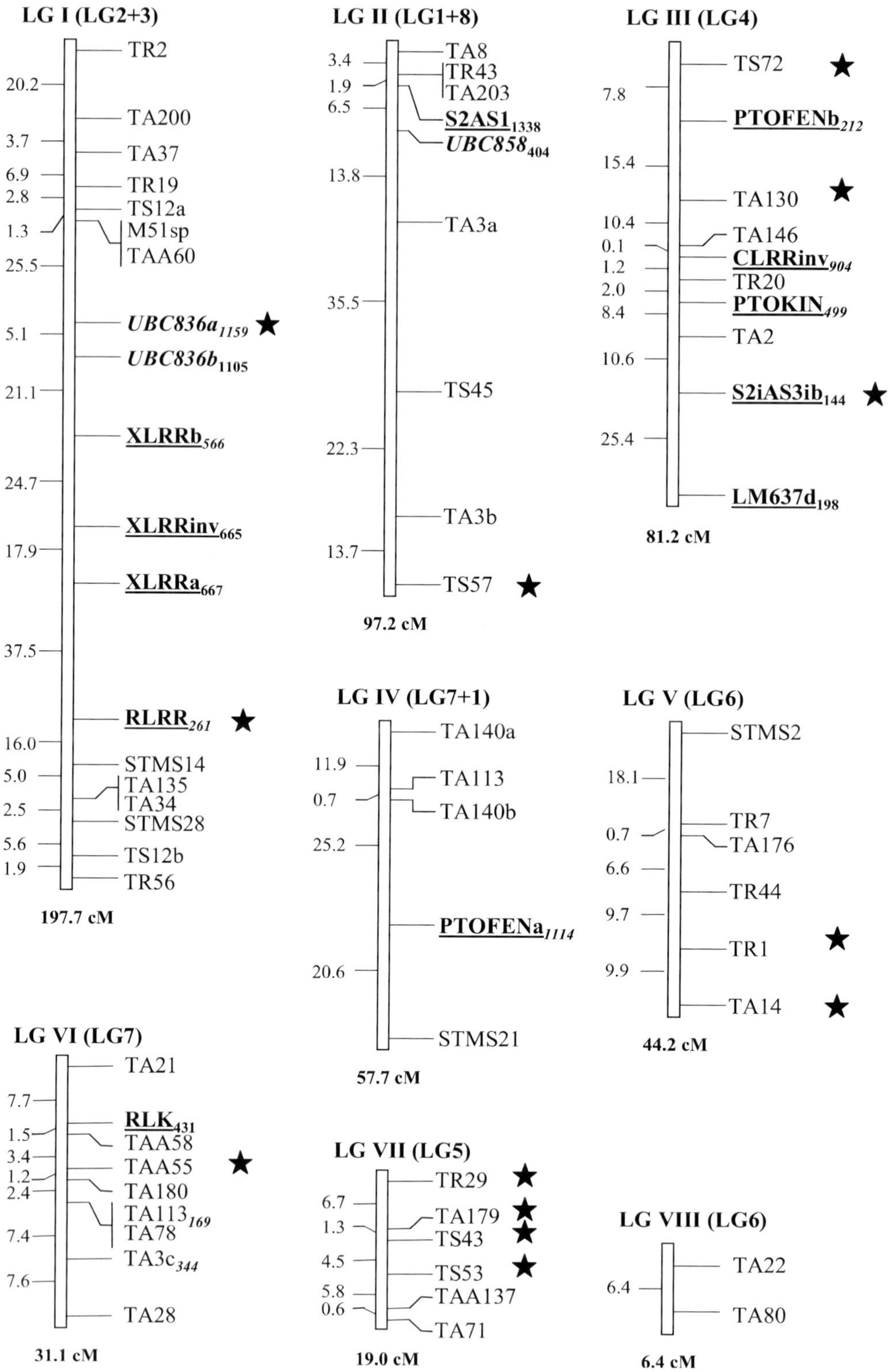

Fig. 1. Intraspecific map of chickpea genome (Lasseter×ICC12004) based on 51 chickpea-STMS, 12 RGA (*bold*) and 3 ISSR markers (*italics*). Marker distance was set in cM by the Kosambi function (Lander et al. 1987, with 1 cM ~1.4 Mbp. *Stars*: Loci that showed segregation distortion. The LGs are numbered from LG I to LG VIII according to their total cM lengths, with the LG number of the interspecific (*C. arietinum*×*C. reticulatum*) map by Winter et al. (2000) noted in parentheses. After Flandez-Galvez et al. (2003 a)

ker or cluster may indicate the location of ascochyta blight resistance genes in the region. However, there was no sufficient data to conclude that the loci mapped using the RGA by Flandez-Galvez et al. (2003 a) were the same as those mapped by Santra et al. (2000).

Preliminary alignment of the intraspecific linkage map of chickpea with the *C. arietinum*×*C. reticulatum* interspecific linkage map (Winter et al. 2000) revealed high linkage conservation in at least three LGs and colinearity in one LG (Flandez-Galvez et al. 2003 a). In particular, markers on LGs III and V of the Flandez-Galvez et al. (2003 a) map and LGs 4 and 6 of the Cho et al. (2002) map were syntenic to LG 4 and LG 6 of the Winter et al. (2000) map respectively. As well LGs I, II and VII of the Flandez-Galvez et al. (2003 a) map showed synteny to LGs 2+3, 1 and 5 of the Winter et al. (2000) map.

6.3 Gene Mapping and QTL Detection for Disease Resistance

Molecular markers associated with quantitative trait loci (QTL) for resistance to ascochyta blight and fusarium wilt have been located on *Cicer* sp. interspecific linkage maps (Santra et al. 2000; Winter et al. 2000; Banerjee et al. 2001; Tekeoglu et al. 2002; Collard et al. 2003) and chickpea intraspecific maps (Flandez-Galvez et al. 2003 a; Udupa and Baum 2003; Sharma et al. 2004). The estimated location of these QTL for ascochyta blight resistance (ABR) and fusarium wilt resistance (foc) were assigned to the LGs of the chickpea intraspecific map of Flandez-Galvez et al. (2003 a) by reciprocal comparisons of STMS markers on the interspecific maps (Santra et al. 2000; Winter et al. 2000; Tekeoglu et al. 2002) and intraspecific map (Udupa and Baum 2003).

To identify the QTL which conditioned ascochyta blight resistance in a pure *C. arietinum* genetic background, Flandez-Galvez et al. (2003 b) used a segregating population that was either infected with a single-spore *A. rabiei* isolate in a controlled environment or naturally infected in the field. The genetic effects of single-locus QTL and QTL in combination were determined relative to their map positions in the chickpea genome.

By single-locus composite interval mapping (CIM) analysis, six QTL for ABR were located on the chickpea map (Table 2, Fig. 2). However, multiple interval mapping (MIM) resolved only two QTL that had simultaneous as well as epistatic interaction effects for resistance, for each disease scoring system and blight condition. Although different QTL peaks were detected on LG III using either initial disease score (IDS) or area under the disease progress curve (AUDPC) disease scoring systems, the region was

Table 2. Putative QTL for ABR identified by composite interval mapping (CIM) at critical LR threshold of 10.8 (α=0.05). After Flandez-Galvez et al. (2003 b)

Test parameter[a]	Linkage group (QTL no.)	Interval length (cM)[b]	Flanking markers	QTL position (cM)[c]	LR[d]	R^2[e] (%)
Growth room						
IDS	II (3)	22.3	TS45-TA3b	14.0	15.61	14.2
	III (6)	0.1	TA146-CLRRinv$_{904}$	0.04	35.18	36.3
AUDPC	II (2)	35.5	TA3a-TS45	16.0	11.70	48.9
	III (5)	10.4	TA130-TA146	6.0	39.46	47.8
Field trial						
IDS	I (1)	5.6	STMS28-TS12b	4.0	13.31	8.0
	III (5)	10.4	TA130-TA146	2.0	51.85	37.3
AUDPC	I (1)	5.6	STMS28-TS12b	4.0	12.11	6.5
	III (4)	15.4	PTOFENb$_{212}$-TA130	10.0	67.09	50.2

[a] IDS = initial disease score, AUDPC = area under the disease progress curve
[b] Interval between the two flanking markers (cM)
[c] QTL position from the left flanking marker (cM)
[d] Peak value of the maximum likelihood ratio (LR) test statistic observed for the QTL in question
[e] Proportion of phenotypic variance explained by the QTL

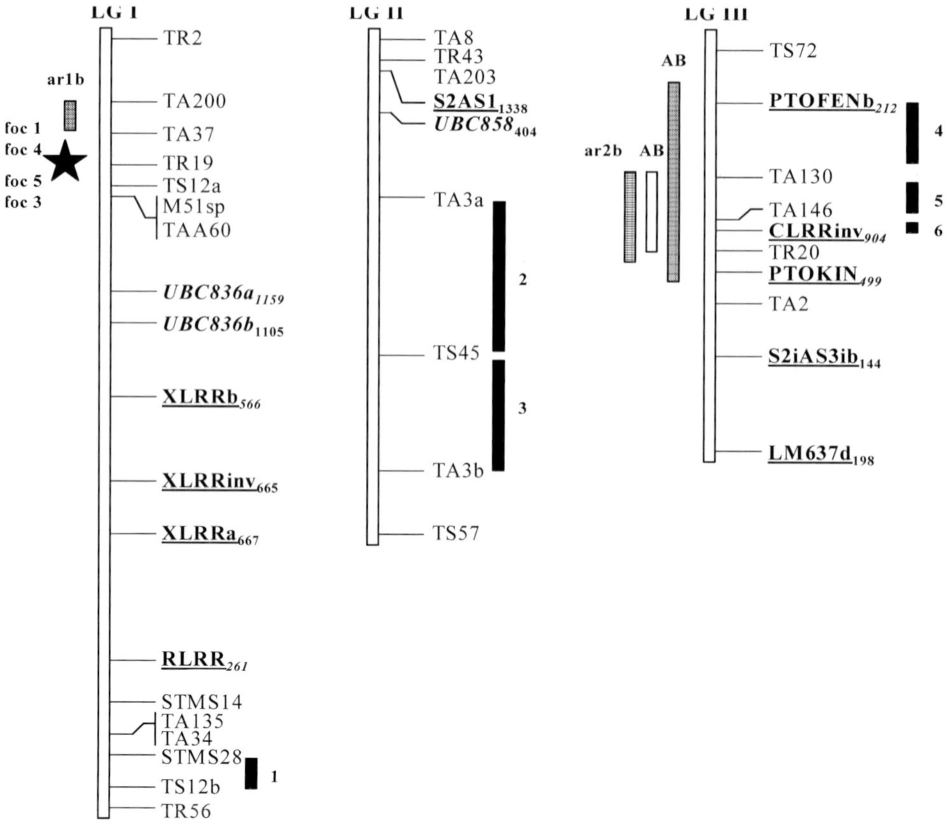

Fig. 2. Location of putative QTL for ascochyta blight resistance (1 to 6, Flandez-Galvez et al. 2003 b) in the intraspecific map of the chickpea genome (Lasseter×ICC12004; Flandez-Galvez et al. 2003 a). The estimated location of QTL for ascochyta blight resistance and fusarium wilt resistance was assigned to LGs (*left*) by reciprocal comparisons of STMS markers on interspecific maps of Santra et al. (2000) (AB *shaded box*), Collard et al. (2003) (AB *open box*) and Winter et al. (2000) (foc 4 and 5 *star*); and the intraspecific chickpea maps of Cho et al. (2004) (ar1b *horizontal shaded box*), Udupa and Baum (2003) (ar2b *horizontal shaded box*) and Sharma et al. (2004) (foc 3 *star*)

consistently mapped for adjacent QTL during both controlled environment inoculations in a growth room and the field trial with natural ascochyta blight infection. The shift of QTL peaks could be due to the specificity of the two disease scoring systems or to the limitation of the linkage map used as a framework in the analysis of the QTL. IDS measured the initial disease reaction of the chickpea plant to ascochyta blight, while AUDPC measured the total disease severity over time.

Although LG III seemed to be well covered with markers, the whole linkage map was still far from marker saturation to precisely locate the ABR-QTL. LG III could be conservatively declared as a major QTL region for ABR, whereas LGs II and I contained QTL regions more specific for either a controlled environment or field resistance. QTL 2 and QTL 3 on LG II were detected only in the controlled environment trial, while QTL 1 on LG I was detected only in the field trial.

All QTL for ascochyta blight resistance verified by MIM were in regions similar to those mapped by Santra et al. (2000) in an interspecific *C. arietinum*×*C. reticulatum* population. QTL 1 was mapped on the LG containing the marker UBC836 – the QTL 2-ISSR marker previously detected in the interspecific population. QTL 2 and 3 were also mapped adjacent to UBC858, the QTL3-ISSR marker detected by Santra et al. (2000). Although no marker was detected between the two regions that was common among studies, the major QTL region in the Flandez-Galvez et al. (2003 b) study (QTL 4, 5 and 6) could also be associated with QTL 2 of Santra et al. (2000) based on the alignment of a few STMS-anchor markers between the *C. reticulatum* interspecific linkage maps (Tekeoglu et al. 2002).

In another intraspecific population of chickpea (Udupa and Baum 2003) and an interspecific cross involving *C. echinospermum* as the resistance donor (Collard et al. 2003), the same major QTL region was also strongly associated with ABR in glasshouse inoculation trials. In this region, Udupa and Baum (2003) also mapped one of two recessive major loci (ar2b) which conferred resistance to pathotype II of *A. rabiei*. The other QTL (ar2a) was located on LG I near the resistance for pathotype I. Pathotype II is the same pathotype that occurred in Australia, and a virulent isolate of this pathotype was used by Flandez-Galvez et al. (2003b) to map six QTL for ascochyta blight resistance, including QTL 4, 5 and 6 on LG III. Studies on resistance loci to pathotype I located QTL (ar1 – Udupa and Baum 2003; and ar1a, ar1b – Cho et al. 2004) on LG I near the site of foc 4 and 5 resistant loci.

Two QTL for seedling resistance to ascochyta blight were detected by interval mapping on LG III of an interspecific population established from a cross between *C. arietinum* and a resistant *C. echinospermum* accession (Collard et al. 2003). Five markers were also associated with stem resistance with four of these being linked to seedling resistance. QTL 1 for seedling resistance was flanked by an STMS marker (TR20) which flanked the QTL 6 peak in the chickpea map (Flandez-Galvez et al. 2003b) (Fig. 2). As well, QTL 2 from Santra et al. (2000) was determined to be in the same location as QTL 1 in LG III.

In summary, LG III appears to be a major site for ascochyta blight resistance loci. There appears to be a clustering of resistance genes that have originated from different resistant sources and expressed from plants challenged with different isolates of the pathogen. Fine mapping around this site and a functional genomics approach are needed to determine the similarity or the uniqueness of these QTL in this region and the effect that these loci have on resistance.

The mapping of QTL adjacent to RGA markers could provide a preliminary understanding of the most likely biological significance of these statistical estimates and evidence of the location of candidate ABR genes in the region. The mapping of QTL 6 at ca. 0.06 cM from CLRRinv (Flandez-Galvez et al. 2003b) may suggest that this region of the genome contained many resistance genes which also contained a *cf9*-like gene. Further nucleotide sequencing of this area and functional analysis of putative resistance genes are required to confirm the presence of major resistance genes conferring resistance to ascochyta blight. RGA marker CLRRinv was designed from the LRR regions of the *Cf9* gene in tomato against *Cladosporium fulvum* (Chen et al. 1998). The *Cf9* gene was shown to be involved both in first-defence pathogen recognition and in interaction with a *Pto* gene product to activate a protein kinase cascade of defense responses (Hammond-Kosack and Jones 1997). Indeed, QTL 4 was mapped adjacent to QTL 5 and 6 and was linked to the RGA marker PTOFEN. The PTOFEN marker was designed from the protein kinase domain of the *Pto* gene in tomato which conferred resistance to *Pseudomonas syringae* pv. *tomato* (Chen et al. 1998) via both signal transduction and binding/deactivating capacities to *AvrPto*-proteins (Scofield et al. 1996; Tang et al. 1996). Furthermore, the strong association of the other RGA markers that flanked these major QTL by one-way ANOVA might explain the consistent association of this region for ABR across different genetic backgrounds, disease scoring systems and possibly different *A. rabiei* pathotypes.

Cho and Muehlbauer (2002) also identified defence-related sequences in chickpea populations segregating for ABR and fusarium resistance that were homologous to a LRR receptor-like kinase. Pfaff and Khal (2003) mapped 47 gene-specific markers derived from sequences of plant defence response (DR) proteins in the interspecific cross of Winter et al. (2000). Although the markers were distributed over all LGs, none was linked to fusarium wilt resistance, foc4/foc5. However, the catalase marker mapped between STMS markers TR20 and TA2 flanked the major locus for ascochyta blight resistance on LG 3 (Fig. 2).

Flandez-Galvez et al. (2002b) determined that resistance of chickpea to ascochyta blight was conferred by an epistatic interaction between the additive gene action of the major QTL and the dominance gene action of the other (minor) QTL. In all QTL, the resistance-enhancing alleles were found to be derived from the susceptible parent, Lasseter, validating the previously reported recessive inheritance of ABR in chickpea (Singh and Reddy 1983; Tewari and Pandey 1986; Dey and Singh 1993; Tekeoglu et al. 2000). Results from QTL analysis revealed that at least three recessive and complementary major genes, and four minor genes specific for either glasshouse or field resistance, were controlling the ABR in chickpea. This is in agreement with the findings of Tekeoglu et al. (2000) that in intra- and interspecific recombinant inbred lines (RIL), ABR is conferred by at least three recessive and complementary major genes with several modifiers. Udupa and Baum (2003) reported that ascochyta blight resistance to pathotype I was controlled by one major gene, and resistance to pathotype II was controlled by two recessive genes with complementary gene action.

Fusarium wilt caused by *Fusarium oxysporum* f. sp. *ciceris* is known to contain eight races; however, resistance is known for some of these to be conferred by more than one gene in specific genotypes. The resistance genes (foc) for fusarium wilt races foc 1, 3, 4 and 5 have been shown to be clustered into two groups on LG I: foc 1, 4, and foc 3, 5 (Fig. 2; Winter et al. 2000; Sharma et al. 2004). The clustering of fusarium wilt resistance genes (foc gene clusters) and ascochyta blight resistance genes (ar1, ar2a, ar1a, ar1b) against two different pathotypes of *A. rabiei* on this LG I indicates the presence of another major site for pathogen defence in chickpea.

6.4 Future Scope of Works

The exploitation of genome mapping in chickpea to improve breeding strategies is in its infancy. Saturated linkage maps have yet to be developed, and only a few traits have been putatively mapped. A lot more work is needed to fine map LGs that contain functional genes and identify flanking markers to these loci that can be used in marker-assisted selection or for map-based gene cloning. The use of transcriptome mapping and functional array technologies will enhance the identification of functional genes.

The precise chromosomal localisation of fusarium blight resistance genes and the quantitative ascochyta blight resistance genes will be developed through the production of dense linkage maps of chickpea. The complexity of the resistance genes in chickpea is associated with the complex pathotype or race structure of the pathogens.

For ascochyta blight and fusarium resistance, the development of tightly linked molecular markers to major resistance genes will lead to pyramiding of resistance genes through the breeding of elite lines containing multiple resistance loci. Selected lines containing multiple resistance genes, as well as other desirable physiological and agronomic traits, will be assessed for resistance level and sustainability to multiple isolates and different pathotypes of *A. rabiei* under varying environmental conditions. The developed lines will be used by breeders to incorporate high levels of resistance into commercial cultivars through conventional breeding methods.

References

Ahmad F (1999) Random amplified polymorphic DNA (RAPD) analysis reveals genetic relationships among the annual *Cicer* species. Theor Appl Genet 98:657–663

Ahmad F, Slinkard AE, Scoles GJ (1988) Investigations into the barrier(s) to interspecific hybridization between *Cicer arietinum* L. and eight other annual *Cicer* species. Plant Breed 100:193–198

Ahmad F, Gaur PM, Slinkard AE (1992) Isozyme polymorphism and phylogenetic interpretations in the genus *Cicer* L. Theor Appl Genet 83:620–627

Arumuganathan K, Earle ED (1991) Nuclear DNA content of some important plant species. Plant Mol Biol Rep 9:208–218

Banerjee H, Pai RA, Moss JP, Sharma RP (2001) Use of random amplified polymorphic DNA markers for mapping the chickpea genome. Biol Plant 44:195–202

Benko-Iseppon AM, Winter P, Hettel B, Staginnus C, Muehlbauer FJ, Kahl G (2003) Molecular markers closely linked to Fusarium resistance genes in chickpea show significant alignments to pathogenesis-related genes located on *Arabidopsis* chromosomes 1 and 5. Theor Appl Genet 107:379–386

Bent AF (1996) Plant disease resistance genes: function meets structure. Plant Cell 8:1757–1771

Berger J, Abbo S, Turner NC (2003) Ecogeography of annual wild *Cicer* species: the poor state of the world collection. Crop Sci 43:1076–1090

Carter JM (1999) Chickpea growers guide: a guide for the production of chickpeas. Department of Natural Resources and Environment, Agriculture, Victoria, Australia, p 20

Chen XM, Line RF, Leung H (1998) Genome scanning for resistance-gene analogs in rice, barley and wheat by high-resolution electrophoresis. Theor Appl Genet 97:345–355

Cho S, Muehlbauer FJ (2002) 'P479: Cloning and mapping of fungal resistance genes in chickpea (*Cicer arietinum*). In: Plant, Animal & Microbe Genomes X Conf, San Diego, p 197

Cho S, Kumar J, Shultz JL, Anupama K, Tefera F, Muehlbauer FJ (2002) Mapping genes for double podding and other morphological traits in chickpea. Euphytica 128:285–292

Cho S, Chen W, Muehlbauer FJ (2004) Pathotype-specific genetic factors in chickpea (*Cicer arietinum* L.) for quantitative resistance to ascochyta blight. Theor Appl Genet 109:733–739

Choumane W, Winter P, Weigand F, Kahl G (2000) Conservation and variability of sequence-tagged microsatellite sites (STMSs) from chickpea (*Cicer arietinum* L.) within the genus *Cicer*. Theor Appl Genet 101:269–278

Collard BCY, Pang ECK, Ades PK, Taylor PWJ (2003) Preliminary investigation of QTLs for seedling resistance to ascochyta blight from *Cicer echinospermum*, a wild relative of chickpea. Theor Appl Genet 107:719–729

De Candolle A (1883) Origine des Plantes Cultivees. G. Bailliere, Paris, pp 208–260

Dey SK, Singh G (1993) Resistance to Ascochyta blight in chickpea - genetic basis. Euphytica 68:147–153

Ellis JG, Lawrence GJ, Finnegan EJ, Anderson PA (1995) Contrasting complexity of two rust resistance loci in flax. Proc Natl Acad Sci USA 92:4185–4188

FAOSTAT Database (2004) http://apps.fao.org/

Faris JD, Li WL, Liu DJ, Chen PD, Gill BS (1999) Candidate gene analysis of quantitative disease resistance in wheat. Theor Appl Genet 98:219–225

Feuillet C, Schachermayr G, Keller B (1997) Molecular cloning of a new receptor-like kinase gene encoded at the *Lr10* disease resistance locus of wheat. Plant J 11:45–52

Flandez-Galvez H, Ades PK, Ford R, Pang ECK, Taylor PWJ (2003 a) QTL analysis for ascochyta blight resistance in an intraspecific population of chickpea (*Cicer arietinum* L.). Theor Appl Genet 107:1257–1265

Flandez-Galvez H, Ford R, Pang ECK, Taylor PWJ (2003 b) An intraspecific linkage map of the chickpea (*Cicer arietinum* L.) genome based on sequence tagged microsatellite site and resistance gene analog markers. Theor Appl Genet 106:1447–1456

Galasso I, Frediani M, Maggiani M, Cremonini R, Pignone D (1996) Chromatin characterization by banding techniques, *in situ* hybridization, and nuclear DNA content in *Cicer* L. (Leguminosae). Genome 39:258–265

Galvez HF, Ford R, Pang ECK, Brouwer JB, Taylor PWJ (2002) Mapping and QTL analysis of ascochyta blight (*Ascochyta rabiei*) resistance genes in chickpea. In: Plant, Animal & Microbe Genomes X Conf, San Diego, p 197

Hammond-Kosack KE, Jones JDG (1997) Plant disease resistance genes. Annu Rev Plant Physiol Plant Mol Biol 48:575–607

Haware MP, Narayana Rao J, Pundir RPS (1992) Evaluation of wild *Cicer* species for resistance to four chickpea diseases. Int Chickpea Newslett 27:16–18

Hüttel B, Winter P, Weising K, Choumane W, Weigand F, Kahl G (1999) Sequence-tagged microsatellite site markers for chickpea (*Cicer arietinum* L.). Genome 42:210–217

Hüttel B, Santra D, Muehlbauer F, Kahl G (2002) Resistance gene analogues of chickpea (*Cicer arietinum* L.): isolation, genetic mapping and association with a *Fusarium* resistance gene cluster. Theor Appl Genet 105:479–490

Iruela M, Rubio J, Cubero JI, Gil J, Millan T (2002) Phylogenetic analysis in the genus *Cicer* and cultivated chickpea using RAPD and ISSR markers. Theor Appl Genet 104:643–651

Jimenez-Diaz RM, Crino P, Halila MH, Mosconi C, Trapero-Casas AT (1993) Screening for resistance to fusarium wilt and ascochyta blight in chickpea. In: Singh KB, Saxena MC (eds) Breeding for Stress Tolerance in Cool-season Food Legumes. Wiley, Chichester, UK, pp 77–95

Kanazin V, Marek LF, Shoemaker RC (1996) Resistance gene analogs are conserved and clustered in soybean. Proc Natl Acad Sci USA 93:11746–11750

Kazan K, Muehlbauer FJ (1991) Allozyme variation and phylogeny in annual species of *Cicer* (Leguminosae). Plant Syst Evol 175:11–21

Kazan K, Muehlbauer FJ, Weeden NF, Ladizinsky G (1993) Inheritance and linkage relationships of morphological and isozyme loci in chickpea (*Cicer arietinum* L.). Theor Appl Genet 86:417–426

Kearns B (1991) Growing grain legumes in 1991 - Chickpeas. Department of Natural Resources and Environment, Agriculture, Victoria, Australia, p 12

Kupicha F (1977) The delimitation of the tribe *Vicieae* and the relationships of *Cicer* L. Bot J Linn Soc 74:131–162

Labdi M, Robertson LD, Singh KB, Charrier A (1996) Genetic diversity and phylogenetic relationships among the annual *Cicer* species as revealed by isozyme polymorphism. Euphytica 88:181–188

Ladizinsky G, Adler A (1976 a) The origin of chickpea (*Cicer arietinum* L.). Euphytica 25:211–217

Ladizinsky G, Adler A (1976 b) Genetic relationships among the annual species of *Cicer* L. Theor Appl Genet 48:197–203

Lander ES, Green P, Abrahamson J, Barlow A, Daly J, Lincoln SE, Newsburg L (1987) MAPMAKER: an interactive computer package for constructing primary genetic linkage maps of experimental populations. Genomics 1:174–181

Leister D, Ballvora A, Salamini F, Gebhardt C (1996) A PCR-based approach for isolating pathogen resistance genes from potato with potential for wide application in plants. Nat Genet 14:421–429

Liu CJ, Devos KM, Witcombe JR, Pittaway TS, Gale MD (1996) The effect of genome and sex on recombination rates in *Pennisetum* species. Theor Appl Genet 93:902–908

Lorieux M, Goffinet B, Perrier X, Gonzalez de Leon D, Lanaud C (1995) Maximum-likelihood models for mapping genetic markers showing segregation distortion. 1. Backcross populations. Theor Appl Genet 90:73–80

Mahadevappa M, DeScenzo RA, Wise RP (1994) Recombination of alleles conferring specific resistance to powdery mildew at the *Mla* locus in barley. Genome 37:460–468

Maisonneuve B, Bellec Y, Anderson P, Michelmore RW (1994) Rapid mapping of two genes for resistance to downy mildew from *Lactuca serriola* to existing clusters of resistance genes. Theor Appl Genet 89:96–104

Mansfield DC, Brown AF, Green DK, Carothers AD, Morris SW, Evans HJ, Wright AF (1994) Automation of genetic linkage analysis using fluorescent microsatellite markers. Genomics 24:225–233

Muehlbauer FJ (1993) Use of wild species as a source of resistance in cool-season food legume crops In: Singh KB, Saxena MC (eds) Breeding for Stress Tolerance in Cool-Season Food Legumes. Wiley, Chichester, UK, pp 359–372

Muehlbauer FJ, Tullu A (1997) *Cicer arietinum* L., NewCROP FactSHEET. Washington State University, USDA-ARS, Seattle, WA, p 6

Nguyen TT, Taylor PWJ, Redden RJ, Ford R (2004) Genetic relationships in the *Cicer* genus revealed with AFLP analysis. Plant Breed 123:173–179

Nguyen TT, Taylor PWJ, Redden RJ, Ford R (2005) Resistance to *Ascochyta rabiei* in a wild *Cicer* germplasm collection. Aust J Exp Agri 45:1291–1296

Ocampo B, Venora G, Errico A, Singh KB, Saccardo F (1992) Karyotype analysis in genus *Cicer*. J Genet Breed 46:229–240

Oplinger ES, Hardman LL, Oelke EA, Kaminski AR, Schulte EE, Doll JD (1997) Chickpea (garbanzo bean). Alternative Field Crops Manual: http://www.hort.purdue.edu/newcrop/afcm/chickpea.html, p 8

Pfaff T, Kahl G (2003) Mapping of gene-specific markers on the genetic map of chickpea (*Cicer arietinum* L.). Mol Genet Genom 269:243–251

Pryor T (1987) Origin and structure of disease resistance genes in plants. Trends Genet 3:157–161

Pundir RPS, Mengesha MH (1995) Cross compatibility between chickpea and its wild relative, *Cicer echinospermum* Davis. Euphytica 83:241–245

Rajesh PN, Tekeoglu M, Gupta VS, Ranjekar PK, Muehlbauer FJ (2002) Molecular mapping and characterization of an RGA locus RGAPtokin1-2_{171} in chickpea. Euphytica 128:427–433

Rayapati PJ, Gregory JW, Lee M, Wise RP (1994) A linkage map of diploid *Avena* based on RFLP loci and a locus conferring resistance to nine isolates of *Puccinia coronata* var. '*avenae*'. Theor Appl Genet 89:831–837

Robertson LD, Ocampo B, Singh KB (1997) Morphological variation in wild annual *Cicer* species in comparison to the cultigen. Euphytica 95:309–319

Santra DK, Tekeoglu M, Ratnaparkhe M, Kaiser WJ, Muehlbauer FJ (2000) Identification and mapping of QTLs conferring resistance to ascochyta blight in chickpea. Crop Sci 40:1606–1612

Scofield SR, Tobias CM, Rathjen JP, Chang JH, Lavelle, Michelmore RW, Staskawicz B J (1996) Molecular basis of gene-for-gene specificity in bacterial speck disease of tomato. Science 274:2063–2065

Serret MD, Udupa SM, Weigand F (1997) Assessment of genetic diversity of cultivated chickpea using microsatellite-derived RFLP markers: implications for origin. Plant Breed 116:573–578

Sharma KD, Winter P, Kahl G, Muehlbauer FJ (2004) Molecular mapping of *Fusarium oxysporum* f. sp. *ciceris* race 3 resistance gene in chickpea. Theor Appl Genet 108:1243–1246

Simon CJ, Muehlbauer FJ (1997) Construction of a chickpea linkage map and its comparison with maps of pea and lentil. J Hered 88:115–119

Singh KB, Ocampo B (1993) Interspecific hybridization in annual *Cicer* species. J Genet Breed 47:199–204

Singh KB, Ocampo B (1997) Exploitation of wild *Cicer* species for yield improvement in chickpea. Theor Appl Genet 95:418–423

Singh KB, Reddy MV (1983) Inheritance of resistance to ascochyta blight in chickpea. Crop Sci 23:9–10

Singh KB, Reddy MV (1990) Pattern of resistance and susceptibility to races of *Ascochyta rabiei* among germplasm accessions and breeding lines of chickpea. Plant Dis 74:127–129

Singh KB, Reddy MV (1994) Registration of eight ascochyta blight-resistant, early-maturing, large-seeded chickpea germplasms. Crop Sci 34:1416–1417

Singh KB, Hawtin GC, Nene YL, Reddy MV (1981) Resistance in chickpeas to *Ascochyta rabiei*. Plant Dis 65:586–587

Singh KB, Malhotra RS, Halila H, Knights EJ, Verma MM (1994) Current status and future strategy in breeding chickpea for resistance to biotic and abiotic stresses. Euphytica 73:137–149

Stamigna C, Mancinelli R, Crino P, Infantino A, Porta-Puglia A, Saccardo F (1998) Multiple resistance to diseases in wild relatives of chickpea (*Cicer arietinum* L.). In: Proc 3rd Eur Conf Grain Legumes, 14–19 Nov 1998, Valladolid, Spain

Sudapak MA, Bennetzen JL, Hulbert SH (1993) Unequal exchange and meiotic instability of disease-resistance genes in the *Rp1* region of maize. Genetics 133:119–125

Sudupak MA, Akkaya MS, Kence A (2002) Analysis of genetic relationships among perennial and annual *Cicer* species growing in Turkey using RAPD markers. Theor Appl Genet 105:1220–1228

Tang X, Frederick RD, Zhou J, Halterman DA, Jia Y, Martin GB (1996) Initiation of plant disease resistance by physical interaction of *avrPto* and *Pto* kinase. Science 274:2060–2063

Tayyar RI, Waines JG (1996) Genetic relationships among annual species of *Cicer* (Fabaceae) using isozyme variation. Theor Appl Genet 92:245–254

Tayyar RI, Lukaszewsky AJ, Waines JG (1994) Chromosome banding patterns on the annual species of *Cicer*. Genome 37:656–663

Tekeoglu M, Santra DK, Kaiser WJ, Muehlbauer FJ (2000) Ascochyta blight resistance inheritance in three chickpea recombinant inbred line populations. Crop Sci 40:1251–1256

Tekeoglu M, Rajesh PN, Muehlbauer FJ (2002) Integration of sequence tagged microsatellite sites to the chickpea genetic map. Theor Appl Genet 105:847–854

Tewari SK, Pandey MP (1986) Genetics of resistance to ascochyta blight in chickpea (*Cicer arietinum* L.). Euphytica 35:211–215

Udupa SM, Baum M (2003) Genetic dissection of pathotype-specific resistance to ascochyta blight disease in chickpea (*Cicer arietinum* L.) using microsatellite markers. Theor Appl Genet 106:1196–1202

Udupa SM, Sharma A, Sharma RP, Pai RA (1993) Narrow genetic variability in *Cicer arietinum* L. as revealed by RFLP analysis. J Plant Biochem Biotechnol 2:83–86

van der Maesen LJG (1972) *Cicer* L., a Monograph of the genus with special reference to chickpea (*Cicer arietinum* L.), its ecology and cultivation. Mededlingen landbouwhogeschool, Wageningen, The Netherlands, pp 72–100

van der Maesen LJG (1987) Origin, history and taxonomy of chickpea. In: Saxena MC, Singh KB (eds) The Chickpea. CAB Intl, UK, pp 11–34

Vavilov NI (1926) Studies on the origin of cultivated plants. Bull Appl Bot Plant Breed (St. Petersburg) 16:139–248

Winter P, Benko-Iseppon AM, Hüttel B, Ratnaparkhe M, Tullu A, Sonnante G, Pfaff T, Tekeoglu M, Santra D, Sant VJ, Rajesh PN, Kahl G, Muehlbauer FJ (2000) A linkage map of the chickpea (*Cicer arietinum* L.) genome based on recombinant inbred lines from a *C. arietinum*×*C. reticulatum* cross: localization of resistance genes for *fusarium* wilt races 4 and 5. Theor Appl Genet 101:1155–1163

Winter P, Pfaff T, Udupa SM, Hüttel B, Sharma PC, Sahi S, Arreguin-Espinoza R, Weigand F, Muehlbauer FJ, Kahl G (1999) Characterization and mapping of sequence-tagged microsatellite sites in the chickpea (*Cicer arietinum* L.) genome. Mol Gen Genet 262:90–101

Yu YG, Buss GR, Maroof MAS (1996) Isolation of superfamily of candidate disease-resistance genes in soybean based on a conserved nucleotide-binding site. Proc Natl Acad Sci USA 93:11751–11756

7 Genome Mapping and Molecular Breeding in *Lathyrus*

Beata Skiba[1], Allison M. Gurung[2], Edwin C.K. Pang[3]

[1] School of Medical Sciences, RMIT University, Bundoora, Victoria 3083, Australia
[2] Faculty of Land and Food Resources, The University of Melbourne, Parkville, Victoria 3010, Australia, *e-mail*: a.gurung@unimelb.edu.au
[3] School of Applied Sciences, RMIT University, Bundoora, Victoria 3083, Australia

7.1 Introduction

7.1.1 History of Crop

The genus *Lathyrus* L. is a member of the Vicieae tribe (family Fabaceae), the other members of which are *Pisum*, *Lens* and *Vicia*. *Lathyrus* consists of ca. 160 species (Allkin et al. 1986). The most comprehensive infrageneric classification of *Lathyrus* by Kupicha (1983) separated the species into 13 sections based on morphological traits (Orobus, Lathyrostylis, Lathyrus, Orobon, Pratensis, Aphaca, Clymenum, Orobastrum, Viciopsis, Linearicarpus, Nissolia, Neurolobus and Notolathyrus). Recent molecular-based studies have supported the classification system of Kupicha (1983) for some species (Croft et al. 1999); however, studies of the chloroplast DNA of 42 *Lathyrus* species (Asmussen and Liston 1998) and AFLP analysis of 18 species (Badr et al. 2002) suggest that reclassification of some species to different sections may be required.

The most economically important and widely cultivated *Lathyrus* species is *L. sativus*. *L. sativus* is grown as a forage crop and is also cultivated for human consumption and as a stock feed. Other important *Lathyrus* species are *L. cicera* (dwarf chickling), *L. tingitanus*, *L. ochrus*, *L. hirsutus*, and *L. sylvestris* as grain and/or forage crops and *L. odoratus* (sweet pea) and *L. latifolius*, which are ornamentals. The main center of diversity of *Lathyrus* species is the eastern Mediterranean, with smaller centers of diversity in North and South America (Kupicha 1983). *L. sativus* is thought to have originated in southwest and central Asia; however, its natural distribution has been obscured by widespread cultivation, even in its presumed center of origin (Smartt 1984). Domestication of *L. sativus* is evidenced by remains of *L. sativus* seed from the Balkans, dated at 8000 BC (Kislev 1989), and from Jarmo (Iraq), dated at 6000 BC (Helbaek 1965). *L. sativus* is now widely distributed throughout the world. However, despite the widespread distribution of *L. sativus* and the length of time that it has been cultivated, it has not progressed as a pulse crop to the same extent as other Mediterranean pulses (field pea, lentil and chickpea). Smartt (1984) hypothesized that the lack of progress of *L. sativus* as a pulse crop might have been due to its other and perhaps more important use as a forage crop; thus the countervailing selection pressures for both purposes may have cancelled each other out.

7.1.2 Botanical Descriptions

L. sativus is known by many common names, including grasspea, chickling pea, khesari (India, Bangladesh), matri (Pakistan), sabberi and guaya (Ethiopia) and san lee dow (China) (Campbell 1997). It is an herbaceous annual which may grow to 1.5 m tall (Smartt 1984). It is much-branched with a straggling or climbing habit (Campbell 1997). The morphology of the plant is reviewed in detail by Campbell (1997). In summary, the stems are angular and winged, leaves are pinnate with a terminal tendril, flowers are solitary and range in color from blue to red to white, the pods are oblong and contain three to five seeds and the seeds are angled and wedge-shaped, ranging in color from white to brownish-grey and may be spotted or mottled (Fig. 1).

L. sativus is predominantly self-pollinated, although outcrossing does occur. The average outcrossing rate observed by Chowdhury and Slinkard

Genome Mapping and Molecular Breeding in Plants, Volume 3
Pulses, Sugar and Tuber Crops
C. Kole (Ed.)

Fig. 1. *Lathyrus sativus* plant at flowering, young pods and variation in seeds

(1997) was about 2.2%; however, the rate of outcrossing has been found to vary significantly between *L. sativus* genotypes, ranging from 9.8 to 27.8% (Rahman et al. 1995). Bees are presumed to be the predominant pollinator of *L. sativus* (Rahman et al. 1995; Chowdhury and Slinkard 1997).

All known annual species and most perennial species in *Lathyrus* are diploid with $2n=14$ chromosomes. There are a few polyploids among the perennials, including *L. palustris*, a hexaploid with $4x=42$, and *L. venosus*, a tetraploid with 28 chromosomes (Narayan and Durrant 1983). Large increases in chromosome size have occurred during the evolution of diploid *Lathyrus* species; Narayan and Durrant (1983) measured a four-fold difference in DNA between 25 *Lathyrus* species (16.78 pg of 2C nuclear DNA in *L. sativus*). However, despite differences in chromosome size, chromosome shape and karyotype arrangement between complements are similar among *Lathyrus* species (Narayan and Durrant 1983).

L. sativus is very tolerant of drought conditions and is not affected by excessive rainfall (Campbell et al. 1994). It has a penetrating root system and can be grown on a wide range of soil types (Ali et al. 2001) and is suited to a range of environments from temperate to subtropical regions. In subtropical regions of south Asia it is broadcast sown into a standing rice crop before harvest and grows through the winter months. In Mediterranean regions it is sown after the first autumn rains (Campbell 1997).

7.1.3 Economic Importance

L. sativus is an important pulse crop of economic significance in Bangladesh, India, Pakistan, Nepal and Ethiopia (Campbell et al. 1994; Yadav and Mehta 1995). It is also cultivated in central, south and eastern Europe, west Asia and north Africa (Campbell 1997). Additionally, it is being investigated as a potential crop for other parts of the world, including Australia (Siddique et al. 1996; Hanbury et al. 1999). The crop is commonly grown for its grain, but it is also used for fodder or green manure (Campbell 1997). As *L. sativus* is a legume crop it fixes atmospheric nitrogen; therefore the crop itself is important in the farming system as a source of nitrogen for later crops.

For human consumption the plant is utilized in a number of ways, including boiling the split seeds to produce a soup-like dish, using the ground seed to produce bread, roasting the seed and cooking the young shoots, green pods and seeds as a fresh vegetable (Campbell 1997). There are few studies on the nutritional composition of *L. sativus* seeds. In a recent investigation of 117 lines, the protein and ash contents ranged from 23.0 to 31.1% and 2.1 to 4.0% respectively (Granati et al. 2003). These values are similar to the few previously published reports reviewed by Campbell (1997).

Statistics on *L. sativus* area, production and yield are not readily accessible; however, the information that is available indicates that India is the largest producer of *L. sativus* grain. Between 1980/

81 and 1994/95 the area sown to *L. sativus* in India ranged from 0.9 to 1.4 million ha, with an average yield of 0.4 to 0.6 t/ha (Gautam et al. 1997). Bangladesh and Ethiopia are the second and third largest producers of *L. sativus* grain, with 0.24 m ha (0.7 t/ha) in Bangladesh (as cited in Campbell 1997) and 0.14 m ha (0.7 t/ha) in Ethiopia (as cited in Tadesse and Bekele 2003).

7.1.4 Breeding Objectives

The major breeding objective for *L. sativus* improvement is reducing concentrations of the neurotoxin β-N-oxalyl-L-α,β-diaminopropanoic acid (ODAP) in the seed. ODAP is a non-protein amino acid associated with a neurodegenerative disease called neurolathyrism. The disease is not lethal but causes a degeneration of upper motor neurons, which is manifested as irreversible spastic paraparesis of the lower limbs (Spencer et al. 1986). A recent epidemic of neurolathyrism occurred in Ethiopia after the drought of 1995/96, which wiped out most of the food crops except *L. sativus* (Getahun et al. 1999). Overconsumption of *L. sativus* was found to be the primary casual factor for neurolathyrism incidence (Getahun et al. 1999).

Other major breeding objectives for *L. sativus* improvement are: increasing grain yield through the incorporation of yield components such as double pods per node and increased seeds per pod and increasing biomass yield (for forage varieties) (Campbell 1997).

L. sativus lines with low levels of ODAP have been produced through classical breeding programs. There is a large natural range of variation in ODAP concentrations in seed, ranging from ca. 0.22 to 7.20 g/kg (Campbell 1997). Selection for low ODAP concentration by Dr Clayton Campbell at Agriculture Canada resulted in a line with 0.03% ODAP (Campbell 1987). ICARDA has also had a breeding program for low ODAP since 1989/90, which involves germplasm evaluation, introduction of variation through somaclonal variation and hybridization (to introduce the low ODAP trait into locally adapted germplasm) (Ali et al. 2001).

Identification of molecular markers linked to alleles for the low ODAP trait would hasten the transfer of this trait into locally adapted germplasm. ODAP concentration is assessed on seed samples; thus markers that distinguished between plants with alleles conditioning low and high ODAP concentration would allow assessment at the seedling stage, saving space and allowing more germplasm to be screened faster.

Screening of *L. sativus* lines by Hanbury et al. (1999) indicated that genotype is the most important determinant of ODAP concentration and that environment has less influence, with genotype-environment interactions having no effect on seed ODAP concentrations, which is positive news for breeding for low ODAP concentration and for the use of molecular markers to assist selection. ODAP concentration was found to be quantitatively inherited in crosses involving four low ODAP lines and one high ODAP line (Tiwari and Campbell 1996). The experiments also indicated that there were common genes for ODAP concentration among the low lines and that there was a maternal effect on ODAP concentration in some crosses. Quantitative inheritance of ODAP concentration suggests that molecular mapping would be a very useful tool for breeding low ODAP lines.

The biochemical pathway(s) for ODAP production are not well characterized. Further research on identification of the enzyme(s) responsible for ODAP production would complement efforts to map the gene(s) responsible for controlling ODAP concentration. This could eventually lead to the elimination of ODAP production in *L. sativus*.

An additional use of molecular markers would be as diagnostic probes for identifying the presence of *L. sativus* DNA. This is particularly important as the split seed and flour of *L. sativus* is reported to be used to adulterate higher-priced alternatives on the market (Campbell 1997). Thus, *L. sativus*-specific markers would allow detection of *L. sativus* in mixtures with other pulses.

7.1.5 Germplasm Screening

Morphological characters and characters associated with phenology, yield and quality of *L. sativus* accessions from a wide range of geographical origins are highly variable (Jackson and Yunus 1984; Hanbury et al. 1999), which is promising for selection and breeding of improved *L. sativus* varieties. Hanbury et al. (1999) assessed 407 *L. sativus* lines and found that there was good potential for selection for seed weight, seedling vigor, yield, seed ODAP concentration and phenological traits. The *L. sati-*

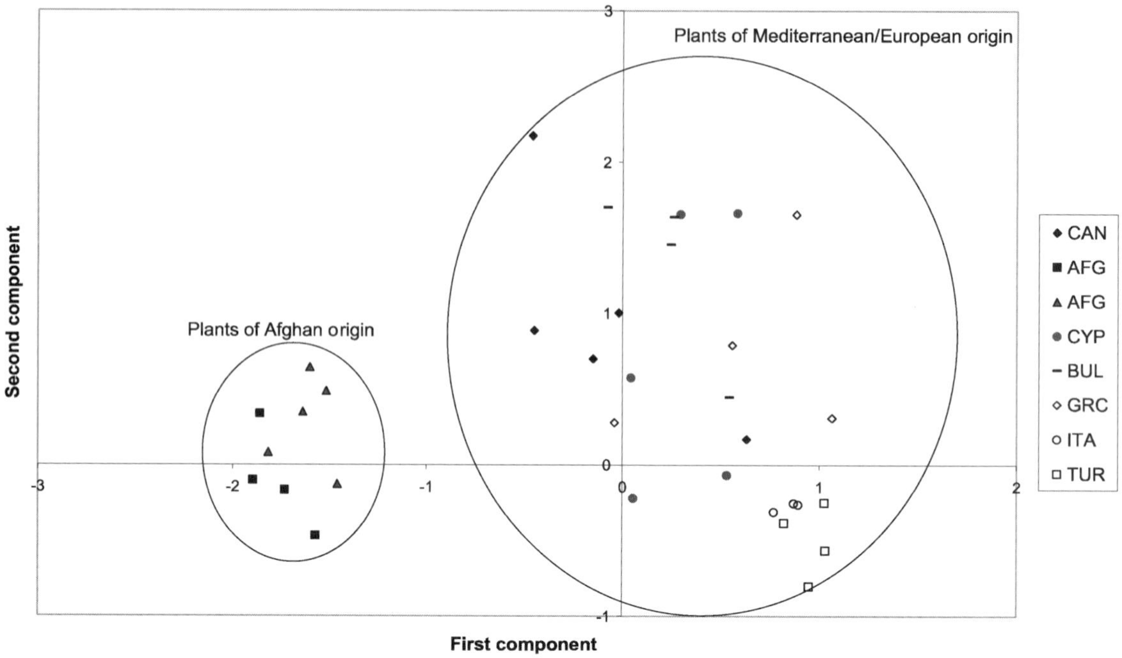

Fig. 2. Principal component analysis of *L. sativus* genotypes, based on RAPD data presented in Croft et al. (1999)

vus lines could be divided into two geographical origins: Indian subcontinent and Mediterranean/ European, based on the agronomic and quality traits (Hanbury et al. 1999). Jackson and Yunus (1984) also observed a distinct separation of accessions into two main groups, based primarily on the size of vegetative parts, flower color and seed color. One group consisted of blue flowered types from south-west and south Asia and Ethiopia while the other group comprised the white and white and blue flowered types, with a more westerly distribution around the Mediterranean basin.

Germplasm screening has also been undertaken using isozymes and molecular markers. Isozyme analysis by Yunus et al. (1991) showed polymorphisms between accessions and also showed no correlation between isozyme polymorphisms and morphology or geographical origin. Molecular (random amplified polymorphic DNA, RAPD) analysis of variation within and between *L. sativus* accessions by Croft et al. (1999) showed that genetic variation exists within and between accessions, with generally a higher level of variation between accessions than within accessions. Principal component analysis of this data supports the observation of Hanbury et al. (1999) that *L. sativus* lines from the Indian subcontinent (south and south-western Asia) are distinct to Mediterranean/European lines (Fig. 2).

7.2 Construction of Genetic Maps

One of the main objectives of plant geneticists is the development of genetic linkage maps for important agricultural crop species, and in recent times these have become valuable tools for plant breeders. Relative to other pulse species, limited published information exists reporting the generation of linkage maps for any *Lathyrus* species. The first reports on linkage in *Lathyrus* examined linkage of morphological traits in *L. odoratus* (sweet pea) (Punnett 1923, 1932; Bhat 1948) and *L. sativus* (grasspea) (Sen and Ghosh 1962). Punnett (1923, 1932) reported the arrangement of 17 morphological characters into five linkage groups (LGs) and two unassociated characters, a result that illustrated that the haploid number of chromosomes in *L. odoratus* is seven. Bhat (1948) later presented an improved genetic map of Punnett's B chromosome in *L. odoratus*, based on Punnett's data for only three characters, using Fisher's scoring method to estimate linkage. More recently, the inheritance and linkage of isozymes in four F_2 populations of *L. sativus* has been investigated (Chowdhury and Slinkard 2000). However, it has only been in the last 4 to 5 years that attempts have been made to use mo-

Table 1. Molecular linkage maps generated for *Lathyrus sativus*

Population	Marker type	Marker coverage (cM)	Average marker interval (cM)	No. linkage groups	Reference
F_2 (100 individuals)	71 RAPDs 3 isozymes 1 morphological	898	17.2	14	Chowdhury and Slinkard (1999)
BC_1 (92 individuals)	47 RAPDS 7 STMSs 13 STS/CAPS	803.1	15.8	9	Skiba et al. (2004)

lecular (DNA) markers to construct linkage maps of the *L. sativus* genome, and only two molecular maps have been published (Table 1).

Chowdhury and Slinkard (1999) constructed the first genetic linkage map of *L. sativus* using 1 morphological, 71 RAPD and 3 isozyme markers. An F_2 mapping population was generated by crossing a white-flowered parent (PI 42689.1.1.3) with a blue-flowered parent (PI 283564c.3.2). Map distances were calculated using the Haldane function, and the resultant map consisted of 14 LGs covering 898 cM, with an average distance between markers of 17.2 cM.

The second molecular linkage map generated for *L. sativus* was based on a backcross population derived from a cross between one plant resistant to ascochyta blight, of accession ATC 80878, and one susceptible plant of accession ATC 80407 (Skiba et al. 2004). The resistant accession ATC 80878 was used as the recurrent parent. This linkage map incorporating 64 molecular markers, including 47 RAPD, 7 STMS and 13 STS/CAPS markers, comprised 9 LGs covering 803.1 Kosambi cM, with an average spacing between markers of 15.8 cM. Unfortunately, due to a lack of common markers, the two maps by Chowdhury and Slinkard (1999) and Skiba et al. (2004) could not be aligned.

These two maps represent the first generation of molecular linkage maps for *L. sativus.* Future mapping studies should focus on further saturating the current *L. sativus* maps with markers to create a high-density linkage map, whereby the number of LGs equals the haploid chromosome number of *L. sativus.* Potential anchor markers from these two maps should also be tested in order to attempt comparative mapping. A key feature of the map generated by Skiba et al. (2004) is that it shows the location of 13 STS/CAPS markers, which were generated by primers from EST sequences obtained from a *L. sativus* cDNA library (Skiba et al. 2003). These STS/CAPS markers were located on all but two LGs, and since they were designed from ESTs derived from coding DNA, which generally have a high degree of sequence conservation, these STS/CAPS markers may potentially be transferable among species, to serve as anchors, thereby facilitating comparative mapping.

7.3 Gene Mapping

As stated earlier, limited literature exists reporting the use of molecular techniques to generate linkage maps for the identification of genes or regions in the *Lathyrus* genome linked to desirable traits. The first and currently only report of the identification of a molecular marker linked to a trait in *Lathyrus* was made by Hanada and Hirai (2003). Using bulked segregant analysis, Hanada and Hirai (2003) identified a RAPD marker linked to the gene responsible for the tendril trait of *L. odoratus.* The distance between the marker and the gene for tendrils in the F_2 population tested was demonstrated to be 7.7 cM. Breeding *L. sativus* varieties with zero or low ODAP content is of primary importance to *Lathyrus* breeders. Developing a molecular marker linked to this trait may promote the application of marker-assisted selection (MAS) in *Lathyrus* breeding. The use of molecular markers may accelerate the generation of new varieties by three plant generations by allowing the selection of offspring that contain the desired combination of genetic characters (Tanksley et al. 1989).

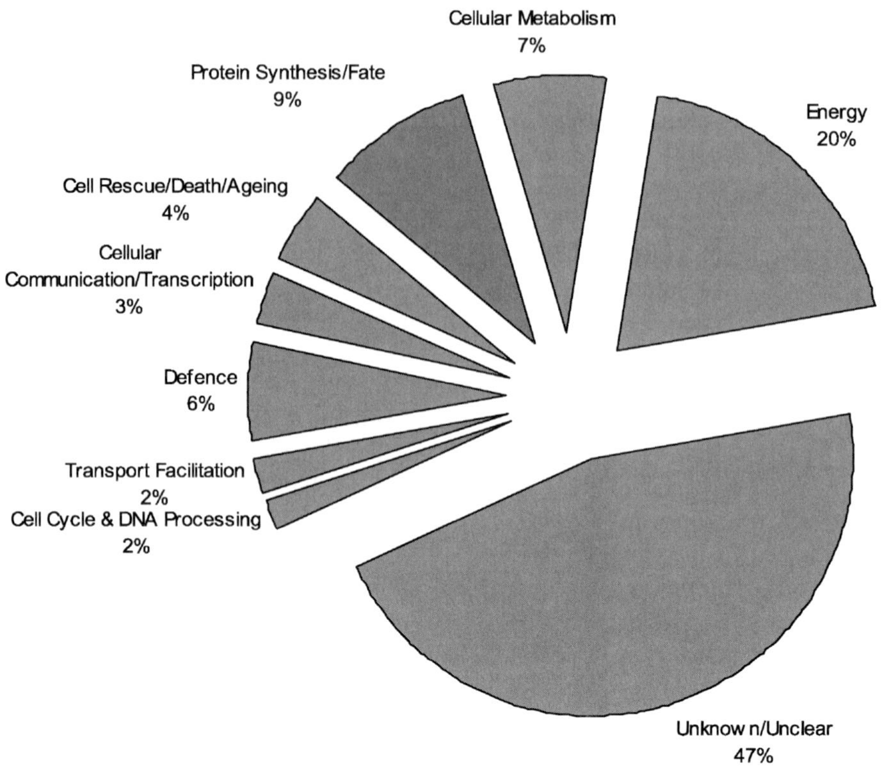

Fig. 4. Functional classification of 818 *Lathyrus sativus* ESTs

ESTs encompasses the entire spectrum of functional categories (Fig. 4), with a slight over-representation of defence and cell rescue/death/aging ESTs. This library will therefore be a particularly useful resource in future studies on biotic and abiotic stress factors in *Lathyrus*. A large percentage of ESTs (47%) could not be assigned any function, and may represent pseudogenes, untranslated regions or novel genes. ESTs of interest from this library may be spotted onto microarrays and be used for a variety of expression studies, for instance for responses to drought and salinity. Defence-related ESTs from this library have also been converted into cleaved amplified polymorphic sequence (CAPS) markers (Skiba et al. 2003) for linkage mapping purposes. As these markers are transferable between closely related species, they may be useful as anchor markers in comparative genomic studies within *Lathyrus* and between other pulses. Further, a more comprehensive STS/CAPS map of *L. sativus* will allow a comparison of the genome organization of stress-related genes of this species with that of model plant species such as *Arabidopsis thaliana* and *Medicago truncatulata*. However, despite the usefulness of this library for studying biotic/abiotic stresses, it may not be adequate for the study of other important traits in *L. sativus* breeding, e.g. seed size/quality and ODAP content. For such traits, it may be necessary to construct specific EST libraries from seed and ovary tissues. Microarrays constructed from such libraries may then be used to examine e.g. the effects of different environmental conditions on the expression of genes in the ODAP biosynthetic pathway or the expression of genes upregulated during seed filling.

Despite the economic importance of *Lathyrus*, in particular *L. sativus*, molecular approaches to the improvement of this crop have largely been lacking. Progress has been made in other areas, particularly in germplasm evaluation and the reduction of seed ODAP content. The achievement of the latter is significant, as high seed ODAP levels has been the main impediment for the adoption of *L. sativus* as a safe and viable crop in many countries. As this main hurdle is removed, new challenges and opportunities may be found in breeding new varieties of *L. sativus* for cultivation in their traditional regions, but perhaps also for countries such as Australia and Canada, where the crop is yet to gain wide acceptance. In this future arena, genetic engineering, MAS and genomics may provide the means for accelerating variety development. More effort must therefore be expended to develop these techniques as routine tools for *Lathyrus* breeding.

References

Ali M, El Moneim A, Van Dorrestein B, Baum M, Ryan J, Bejiga G (2001) Role of ICARDA in improving the nutritional quality and yield potential of grasspea (*Lathyrus sativus* L.) for subsistence farmers in dry areas. Lathyrus Lathyrism Newslett 2:55–58

Allkin R, Goyder DJ, Bisby FA, White RJ (1986) Names and synonyms of species and subspecies in the Vicieae. Vicieae Database Project 7:1–75

Asmussen CB, Liston A (1998) Chloroplast DNA characters, phylogeny, and classification of *Lathyrus* (Fabaceae). Am J Bot 85:387–401

Badr A, El Shazly H, El Rabey H, Watson LE (2002) Systematic relationships in *Lathyrus* sect. *Lathyrus* (Fabaceae) based on amplified fragment length polymorphism (AFLP) data. Can J Bot 80:962–969

Bhat HR (1948) An improved genetical map of Punnett's 'B' chromosome in the sweet pea (*Lathyrus odoratus* L.). Biochem J 83:225–345

Campbell CG (1987) Registration of low neurotoxin content *Lathyrus* germplasm LS8246. Crop Sci 27:821

Campbell CG (1997) Grasspea *Lathyrus sativus* L. Promoting the conservation and use of underutilized and neglected crops. 18. Institute on Plant Genetics and Crop Plant Research, Gatersleben/International Plant Genetic Resources Institute, Rome

Campbell CG, Mehra RB, Agrawal SK, Chen YZ, Abd El Moneim AM, Khawaja HIT, Yadov CR, Tay JU, Araya WA (1994) Current status and future strategy in breeding grasspea (*Lathyrus sativus*). Euphytica 73:167–175

Chowdhury MA, Slinkard AE (1997) Natural outcrossing in the grasspea. J Hered 88:154–156

Chowdhury MA, Slinkard AE (1999) Linkage of random amplified polymorphic DNA, isozyme and morphological markers in grasspea (*Lathyrus sativus*). J Agric Sci 133:389–395

Chowdhury MA, Slinkard AE (2000) Genetics of isozymes in grasspea. J Hered 91:142–145

Croft AM, Pang ECK, Taylor PWJ (1999) Molecular analysis of *Lathyrus sativus* L. (grasspea) and related *Lathyrus* species. Euphytica 107:167–176

Gautam PL, Singh IP, Karihaloo JL (1997) Need for a crop network on *Lathyrus* genetic resources for conservation and use. In: Mathur PN, Ramanatha Rao V, Arora RK (eds) Lathyrus Genetic Resources Network. Proc IPGRI-ICARDA-ICAR Regional Working Group Meeting, 8–10 December 1997, New Delhi, India, pp 15–21

Getahun H, Mekonnen A, TekleHaimanot R, Lambein F (1999) Epidemic of neurolathyrism in Ethiopia. Lancet 354:306–307

Granati E, Bisignano V, Chiaretti D, Crin P, Polignano GB (2003) Characterization of Italian and exotic *Lathyrus* germplasm for quality traits. Genet Resource Crop Evol 50:273–280

Grant JE, Cooper PA (2003) Genetic transformation in pea. Appl Genet Leguminosae Biotechnol 10B:23–34

Gurung AM, Pang ECK, Taylor PWJ (2002) Examination of *Pisum* and *Lathyrus* species as sources of ascochyta blight resistance for field pea (*Pisum sativum*). Aust Plant Pathol 31:41–45

Haley CS, Knott SA (1992) A simple regression model for mapping quantitative trait loci in line crosses using flanking markers. Heredity 69:315–324

Hanada H, Hirai M (2000) Classification of sweet pea (*Lathyrus odoratus* L.) and everlasting pea (*Lathyrus latifolium* L.) by random amplified polyporphic DNA (RAPD) analysis. J Jpn Soc Hort Sci 69:758–763

Hanada H, Hirai M (2003) Development of genetic marker linked to the tendril trait of sweet pea (*Lathyrus odoratus* L.). Breed Sci 53:7–13

Hanbury CD, Siddique KHM, Galwey NW, Cocks PS (1999) Genotype-environment interaction for seed yield and ODAP concentration of *Lathyrus sativus* L. and *L. cicera* L. in Mediterranean-type environments. Euphytica 110:45–60

Helbaek H (1965) Early Hassunan vegetable remains at es-Sawwan near Samarra. Sumer 20:45–48

Henry RJ (1997) Practical applications of plant molecular biology. Chapman and Hall, London, pp 99–132

Jackson MT, Yunus AG (1984) Variation in the grass pea (*Lathyrus sativus* L.) and wild species. Euphytica 33:549–559

Kislev ME (1989) Origins of the cultivation of *Lathyrus sativus* and *L. cicera* (Fabaceae). Econ Bot 43:262–270

Kupicha FK (1983) The infrageneric structure of *Lathyrus*. Notes RBG Edinb 41:209–244

Manly KF, Cudmore Jr RH, Meer JM (2001) Map Manager QTX, cross-platform software for genetic mapping. Mammal Gen 12:930–932

Muehlbauer FJ, Tullu A (1997) NewCROP FactSHEET – *Lathyrus sativus* L. Internet publication. http://www.hort.purdue.edu/newcrop/CropFactSheets/grass-pea.html

Mewes H, Frishman D, Guldener U, Mannhaupt G, Mayer K, Mokrejs M, Morgenstern B, Munste M, Rudd S, Weil B (2002) MIPS: a database for genomes and protein sequences. Nucleic Acids Res 30:31–34

Narayan RKJ, Durrant A (1983) DNA distribution in chromosomes of *Lathyrus* species. Genetica 61:47–53

Popelka JC, Terryn N, Higgins TJV (2004) Gene technology for grain legumes: can it contribute to the food challenge in developing countries. Plant Sci 167(2):195–206

Punnett RC (1923) Linkage in the sweet pea (*Lathyrus odoratus*). J Genet 13:101–123

Punnett RC (1932) Further studies of linkage in the sweet pea. J Genet 26:97–114

Rahman MM, Kumar J, Rahman MA, Afzal MA (1995) Natural outcrossing in *Lathyrus sativus* L. Indian J Genet 55:204–207

Robertson LD, Singh KB, Erskine W, Abd el Moneim AH (1996) Useful genetic diversity in germplasm collections of food and forage legumes from West Asia and North Africa. Genet Resource Crop Evol 43:447–460

Sen NK, Ghosh AK (1962) Genetics of grass pea (*Lathyrus sativus* L.). J Genet 58:51–64

Siddique KHM, Loss SP, Herwig SP, Wilson JM (1996) Growth, yield and neurotoxin (ODAP) concentration of three *Lathyrus* species in Mediterranean-type environments of Western Australia. Aust J Exp Agric 36:209–218

Skiba B, Ford R, Pang ECK (2003) Amplification and detection of polymorphic sequence-tagged sites in *Lathyrus sativus*. Plant Mol Biol Rep 21:391–404

Skiba B, Ford R, Pang ECK (2004) Construction of a linkage map based on a *Lathyrus sativus* backcross population and preliminary investigation of QTLs associated with resistance to ascochyta blight. Theor Appl Genet 109:1726–1735

Skiba B, Ford R, Pang ECK (2005) Construction of a cDNA library of *Lathyrus sativus* inoculated with *Mycosphaerella pinodes* and the expression of potential defence-related expressed sequence tags (ESTs). Physiol Mol Plant Pathol 66:55–67

Smartt J (1984) Evolution of grain legumes. I. Mediterranean pulses. Expl Agric 20:275–296

Spencer PS, Roy DN, Ludolph A, Hugon J, Schaumberg HH (1986) Lathyrism: evidence for role of the neuroexcitatory amino acid BOAA. Lancet ii:1066–1067

Tadesse W, Bekele E (2003) Variation and association of morphological and biochemical characters in grass pea (Lathyrus sativus L.). Euphytica 130:315–324

Tanksley SD, Young ND, Paterson AH, Bonierbale MW (1989) RFLP mapping in plant breeding: new tools for old science. Biotechnology 7:257–264

Tanksley SD, Ganal M, Martin GB (1995) Chromosome landing – a paradigm for map-based cloning in plants with large genomes. Trends Genet 11:63–68

Tiwari KR, Campbell CG (1996) Inheritance of neurotoxin (ODAP) content, flower and seed coat colour in grass pea (*Lathyrus sativus* L.). Euphytica 91:195–203

Weimer JL (1947) Resistance of *Lathyrus* spp. and *Pisum* spp. to *Ascochyta pinodella* and *Mycosphaerella pinodes.* J Agric Res 75:181–190

Winter P, Kahl G (1995) Molecular marker technology for plant improvement. World J Microbiol Biotechnol 11:438–448

Yadav VK, Mehta SL (1995) *Lathyrus sativus*: a future pulse crop free of neurotoxin. Current Sci 68:288–292

Yunus AG, Jackson MT, Catty JP (1991) Phenotypic polymorphisms of six enzymes in the grasspea (*Lathyrus sativus* L.). Euphytica 55:33–42

8 Pigeonpea

Milind B. Ratnaparkhe[1], Vidya S. Gupta[2]

[1] Virginia Polytechnique Institute and State University, Blacksburg, VA 24061, USA, *e-mail*: ratnaparkhem@missouri.edu
[2] National Chemical Laboratory, Pune 411008, India

8.1 Introduction

Pigeonpea (*Cajanus cajan* L.) Millspaugh (2n = 22) is a major grain legume crop of the semiarid tropics. It is grown extensively in India and other developing countries of Asia, Africa, and Latin America. Pigeonpea belongs to the genus *Cajanus* of the subtribe *Cajaninae*, tribe Phaseoleae of the subfamily Papilionoideae under the family Leguminosae. It is the only cultivated food crop of the *Cajaninae* subtribe. In terms of global grain legume production, it ranks sixth following *Phaseolus* bean, peas, chickpeas, broad beans, and lentils. The highest production of pigeonpea is recorded by India, where it is the most widely grown grain legume second only to chickpea. Pigeonpea is grown on about 5.25 million ha, yielding 3 million tons, and contributes to about 5% of total world production of pulses. It is grown for food, fodder, fuel wood, hedges, windbreaks, soil conservation, green manuring, and roofing. Pigeonpea is cultivated as an annual or semiperennial legume crop, usually in mixed cropping systems. The crop shows a great variation in maturity types (90 to 300 d) (Saxena and Sharma 1990). Traditional cultivars or landraces are medium to long duration and are harvested 6 to 12 months after sowing. The perennial habit of short-duration pigeonpea enables the production of multiple harvests in tropical areas. Short-duration pigeonpea takes 100 to 140 d to mature, which now makes it possible to grow a sole crop of pigeonpea. More recently, short- and extra-short-duration genotypes have been developed that mature in 90 d. The shorter-duration genotypes generally have less sensitivity to photoperiod than long-duration types.

Pigeonepea is a perennial erect bush, 0.5 to 4 m tall with strong stem. It can be found growing on an otherwise arid landscape due to its ability to tolerate drought and high temperatures. Its deep root system allows for optimum moisture and nutrient utilization and breaks the plow pans, thus improving soil structure (Nene and Sheila 1990). It has a fast growing, strong, woody taproot with well-developed lateral roots in the superficial layers of the soil. The extensive root system allows plant to grow and produce grain in very arid conditions when no other crop can survive. This crop is ideal for intercropping or mixed cropping because of its slow initial growth, allowing companion crop, usually a coarse grain cereal, to grow unhindered. In Asia, pigeonpea is frequently intercropped with sorghum or millet, while in Africa maize is the predominant intercrop. In addition to cereals, pigeonpea is also intercropped with other legumes and with root and fiber crops. Pigeonpea grows mainly by utilizing the residual soil moisture left after the harvest of companion crop. Local landraces and cultivars of 180 to 280 d duration are often grown in this manner to exploit residual moisture in soil when it is not feasible to raise another crop in South Asia. Elsewhere, they are generally grown as perennial hedge crop. Their photoperiod sensitivity allows them to be grown as a winter crop in mild winter environments. Pigeonpea is mostly self-pollinating, but a range of 3 to 95% outcrossing has been reported, probably depending on environment and populations of pollinating insects. When pure lines need to be maintained, it may be necessary to cover plants with muslin bags to exclude insects.

Pigeonpea is part of many farming systems throughout the tropics and subtropics. Although India produces around 80% of the total crop, 2.6 million tons from nearly 3.5 million ha, it is also grown less intensively, for instance, in home gardens, elsewhere in Asia, throughout Africa, and in Latin America. Pigeonpea is best known as a human food. In India, decorticated, split dried peas are an important protein source. Pigeonpea is an

Genome Mapping and Molecular Breeding in Plants, Volume 3
Pulses, Sugar and Tuber Crops
C. Kole (Ed.)

excellent fodder species. Pigeonpea can be grown in a wide range of soil textures, from sandy soils to heavy clays. It grows best at a soil pH of 5.0 to 7.0 but tolerates a wider range. It does well in low-fertility soils, making it a favorite among subsistence farmers. As with most legumes, it does not tolerate waterlogged or flooded conditions for too long. In general, pigeonpea is tolerant to drought and high temperature as compared to most other pulse crops. It thrives under annual rainfall of between 24 and 40 inches (600 to 1000 mm). It is generally grown where the temperatures are in the range of 64 to 85 °F (18 to 30°C), but under moist soil conditions it can withstand temperatures of 95 °F (35 °C) or above. Once established, it is one of the most drought-tolerant legumes, and it can be grown in rainfed conditions or with minimal irrigation. Pigeonpea is also known for its ability to access insoluble phosphates in soils low in P, increasing the availability of soluble P for the following cash crops in the rotation. Pigeonpea assimilates more nitrogen per unit of plant biomass than most other legumes and can nodulate in most soils. Yields of pigeonpea vary considerably among locations, cultivars, seasons, and cropping systems. Pigeonpea seed contains 20 to 30% proteins, is rich in essential amino acids, carbohydrates, minerals, and has high amounts of vitamins A and C.

8.1.1 Classification, Nomenclature, and Origin

Pigeonpea originated in India and later spread to Africa. The revised genus *Cajanus* comprises 32 species, with 18 species. 18 species distributed in Asia, 15 in Australia, and one in West Africa. Of these, *C. cajan* is the only domesticated species. Eleven related genera including *Rhynchosia, Eriosema, Dunbaria, Flemingia*, and *Paracalyx* have been described in the subtribe *Cajaninae* (van der Maesen 1990). *Cajanus cajanifolius* is the most probable progenitor of pigeonpea. Krishna and Reddy (1982) studied the esterase isozyme pattern of seven *Cajanus* species, showing a closer homology between pigeonpea and *C. cajanifolius* than when compared with other species. The genomic relationships between *C. cajan* and its wild relatives have been discussed on the basis of karyotypes, crossability relationships, and meiotic behavior (Pundir and Singh 1985a–c, 1986; Dundas 1990), isozyme analysis (Krishna and Reddy 1982), seed protein profiles (Ladizinsky and Hemel 1980; Jha and Ohri 1996), trypsin and chymotrypsin inhibitor patterns (Kollipara et al. 1994), random amplified polymorphic DNA (RAPD) analysis (Ratnaparkhe et al. 1995), and restriction fragment length polymorphism (RFLP) studies (Nadimpalli et al. 1993; Parani et al. 2000; Sivaramakrishna et al. 2002). Six of the In-

Table 1. Gene pools of pigeonpea.

Primary gene pool	Secondary gene pool	Tertiary gene pool
Pigeonpea cultivar collection	*C. acutifolius* (F.v.Muell) van der Maesen	*C. geonsis* Dalz.
	C. albicans (W.&A.) van der Maesen	*C. heynei* (W.&A.) van der Maesen
	C. cajanifolius (Haines) van der Maesen	*C. mollis* (Benth.) van der Maesen
	C. lanceolatus (W.V. Fitzg) van der Maesen	*C. platycarpus* (Benth) van der Maesen
	C. latisepalus (Reynolds & Pedley) van der Maesen	*C. rugosus* (W.&A.) van der Maesen
	C. lineatus (W.&A.) van der Maesen	*C. volubilis* (Blanco) Blanco
	C. reticulatus (Drylander) F. v. Muell var. *grandifolius* (F. v. Muell) van der Maesen var. *reticulatus*	*C. kerstingii* Harms
	C. scarabaeoides (L.) Thouars var. *Penduncu-latus* (Reynolds & Pedley) van der Maesen var. *scarabaeoides*	*Rhyncosia* spp.
	C. sericeus (Benth. Ex Bak.) van der Maesen	*Dunbaria* spp.
	C. trinervius (DC) van der Maesen	*Erisema* spp.
		Flemingia spp.

Source: van der Maesen (1990)

dian wild species, *C. cajanifolius*, *C. scarabaeoides*, *C. trinervius*, *C. albicans*, *C. lineatus*, and *C. sericeus*, were successfully crossed with pigeonpea. Interspecific hybrids have also been obtained with the Australian species *C. acutifolius*, *C. confertiflorus*, *C. lanceolatus*, *C. latisepalus*, and *C. reticulatus*. However, the hybrids had higher levels of meiotic abnormalities than reported in hybrids between pigeonpea and Indian species (Dundas 1990). Recently, *C. platicarpus* has been successfully crossed with pigeonpea using embryo rescue (Mallikarjuna and Moss 1995). Table 1 shows the wild relatives of pigeonpea and their classification in secondary and tertiary gene pool.

The gene bank of the International Crop Research Institute For Semi-Arid Tropics (ICRISAT), Patancheru, India, currently holds 213 accessions representing 20 species, besides about 13,300 accessions of cultivated pigeonpeas. The wild gene pool assembled at ICRISAT also includes other related genera *Rhynchosia* (35 species, 303 accessions), *Flemingia* (8 species, 18 accessions), *Eriosema* (4 species, 7 accessions), and *Dunbaria* (2 species, 12 accessions). On the basis of the success of hybridization, the species in the secondary gene pool are interfertile with pigeonpea. Those species that do not readily cross with pigeonpea are placed in the tertiary gene pool.

8.1.2 Yield Constraints and Breeding Objectives in Pigeonpea

Diseases. More than 50 pigeonpea diseases caused by a number of casual agents have been documented. These include fungi, bacteria, viruses, nematode, and mycoplasma like organisms. Fusarium wilt, phytophthora blight, and sterility mosaic are identified as the most important diseases of pigeonpea. Fusarium wilt is caused by *Fusarium udum* Butler and is the most important soilborne disease of pigeonpea. Phytophthora blight is caused by the phycomycetous fungus *Phytophthora drechsleri* Tucker f. sp. cajani. The disease seems to be relatively more serious in short-duration pigeonpea than in medium- and long-duration pigeonpea. Sterility mosaic (SMD) is the most damaging disease of pigeonpea. It is caused by pigeonpea sterility mosaic virus (PPSMV) transmitted by the eriophyid mite, *Aceria cajani* (Kumar et al. 2005). SMD is responsible for yield losses worth over US$ 300 million per annum and is endemic in all the pigeonpea-growing areas of South Asia. SMD-affected plants show characteristic mosaic symptoms in leaves with reduced or no flowering (Jones et al. 2004). Other pigeonpea diseases include Cercospora leaf spot, witches' broom, collar rot, dry root rot, phoma stem canker, alternaria leaf spot, powdery mildew, rust, bacterial leaf spot and stem canker, yellow mosaic, cyst nematode, and root-knot nematode. Though considerable progress has been made on some of the important diseases, much remains to be done.

Insect Pests of Pigeonpea. Insect pests are a major constraint to pigeonpea production. Pigeonpea plants and seeds attract over 200 species of insects that damage their roots, shoots, flowers, and seeds. Most of these insect species are sporadic in their distribution and therefore may not all be regarded as pests. Most of the economically important insect pests attack pigeonpea at the reproductive phase and in storage, when they damage flower buds, flowers, pods, and seeds. Broadly speaking, the key insect pests of pigeonpea at the reproductive phase are grouped into three categories: (1) the flower- and pod-feeding Lepidoptera larvae (mainly *Helicoverpa armigera* Hübner, *Maruca vitrata*, *Etiella zinkenella*), (2) the pod-sucking Hemiptera (mainly *Clavigralla* spp.), and (3) the seed-feeding Diptera (*Melanagromyza* sp.) and Hymenoptera.

The most important yield constraint on pigeonpea is from the lepidopteran pest *Heliocoverpa armigera*. It is commonly found throughout the tropics and subtropics in Asia. In addition to its wide distribution and host range, high levels of insecticide resistance make this species one of the most difficult pests to manage. Annual pigeonpea losses due to *H. armigera* have been estimated at US$ 317 million worldwide. The pigeonpea pod fly, *Melanagromyza obtuse* Malloch, is another major pest of pigeonpea that appears to be restricted to Asia. Losses due to pod fly damage have been estimated at US$ 256 million annually. In addition to pod borer and pod fly, the legume pod borer or spotted caterpillar, *M. vitrata* (Geyer), is a serious pest of grain legumes in the tropics and subtropics because of its extensive host range and destructiveness. The other important pests of pigeonpea are plume moth [*Exelastis atomosa* (Walsingham)], blister beetles (*Mylabris* spp.), pod-sucking bugs [*Clavigralla* spp., *Nezara Viridula* (L.)], weevils (*Myllocerus* spp.,

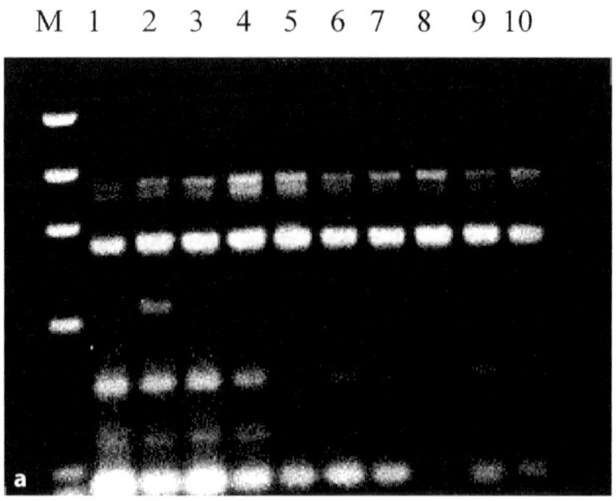

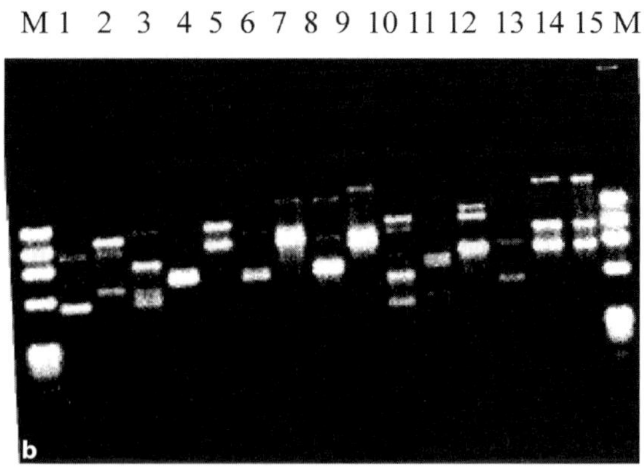

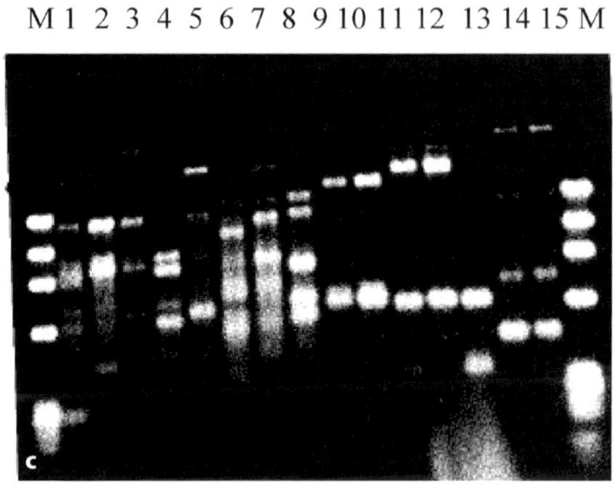

Fig. 1. a Agarose gel electrophoretic pattern of PCR amplification products obtained with different pigeonpea accessions using Primer OPA1. *M* represents the phi X 174 marker. Lanes 1–10 indicate pigeonpea cultivars PT-14, T21, PT-31, DPPA 85-12, ICPL 75-125, TAT-14, Bahar, MTH-12, ICPL-87, and C-11. **b, c** Agarose gel electrophoretic pattern of PCR amplification products obtained with different wild species of pigeonpea using Primer OPA11 and OPA15. *M* represents the phi X 174 marker. Lanes 1: *C. acutifolius*, 2: *C. albicans*, 3: *C. goensis*, 4: *C. grandifolius*, 5: *C. lineatus*, 6: *C. reticulates*, 7: *C. sericeus*, 8: *C. volubilis*, 9: *D. ferruginea*, 10: *R. rothii*, 11: *R. bracteata*, 12: *R. melacophylla*, 13: *F. stricta*, 14: *C. cajan*-1, 15: *C. cajan*-2

8.3.1 First-Generation DNA Markers

Nadimpalli et al. (1993) provided the first report of the use of molecular markers (RFLP) to study the phylogenetic relationships between the wild relatives of pigeonpea. Twenty-four accessions representing 12 species of four genera (*Cajanus, Dunbaria, Eriosema* and *Rhynchosia*) were examined and the RFLP markers generated sufficient polymorphism within the wild species. The overall grouping pattern of species provides partial support for the sectional classification of the genus as proposed by van der Maesen (1986). Parsimony and distance matrix analysis was used for establishing genetic relationships among species. Both the analysis generated trees with similar, but not identical topologies. Following parsimony analysis species *C. cajanifolius* and *C. cajan* grouped together, as do species belonging to other sections (*C. albicans, C. scarabaeoides, C. lineatus, C. sericeus*). Parsimony analysis supports previous reports by van der Maesen (1986, 1990) and De (1974) that *C. cajanifolius* is one of the progenitors of *C. cajan*. Results from distance matrix analysis, however, differed with those from parsimony analysis where *C. cajanifolius* clusters with *C. scarabaeoides* and not with *C. cajan*. Similarly *C. albicans* clusters with *C. sericeus* and *C. lineatus*, not with *C. scarabaeoides*, *C. cajanifolius* and *C. cajan*. In this study the grouping of *C. volubilis* with species of *Dunbaria* and *Rhynchosia* corroborates the view that the genomes of these species are genetically similar. The data also indicate that *C. cajan* is a monophyletic group with a single line of descent and that *C. cajanifolius* and *C. cajan* diverged more rapidly than *C. cajan* and *C. scarabaeoides*. In general RFLP data supported studies utilizing seed storage profiles, and to lesser extent, crossability relationships and cytology. These findings emphasize the importance of characterizing the genomes of related species to assist in effective utilization of wild germplasm in pigeonpea improvement.

Later, Ratnaparkhe et al. (1995) studied the utility of RAPD markers for studying genetic relationships among pigeonpea cultivars and wild species.

Thirteen wild species representing the genus *Cajanus, Rhynchosia, Flemingia* and *Dunbaria* were genotyped with RAPDs. Figures 1 a–c are representative gel pictures of RAPD studies in pigeonpea and its wild species. Figure 1 a shows electrophoretic pattern of PCR amplification products obtained with different pigeonpea accessions using primer OPA1. Pigeonpea cultivar specific bands were observed using different RAPD primers. Figures 1 b and c are electrophoretic patterns of PCR amplification products obtained with different wild species of pigeonpea using primer OPA11 and OPA15, respectively. Extensive genetic variability was observed in different wild relatives of pigeonpea. The study indicated that RAPD could be useful for genetic fingerprinting of pigeonpea accessions and to study the phylogenetic relationship between pigeonpea and its wild relatives. The similarity matrix between wild species ranges from 0.22 to 0.85. The results based on similarity matrix for different wild species indicates that *C. albicans* clusters with *C. lineatus* and *C. sericeus. Cajanus grandifolius, C. reticulates* and *C. acutifolius* are less closely related to *C. cajan* as compared to *C. albicans, C. sericeus* and *C. lineatus. Rhynchosia* species grouped together, suggesting that all these species have been derived from a common ancestor. Figure 2 shows the genetic relationships among selected wild species of pigeonpea using RAPD markers.

RAPD markers were recently used for the identification of two pigeonpea cytoplasmic male sterile lines derived from crosses between the wild (*Cajanus scarabaeoides* and *C. sericeus*) and the cultivated species of *Cajanus cajan* (Souframanien et al. 2003). The male sterile (A) line and its maintainer (B) line could be easily differentiated with certain random primers. Authors reported that sufficient polymorphism exists to allow the distinction between the CMS (A) line, maintainer line, putative R line and other genotypes.

Parani et al. (2000) studied phylogenetic relationship between pigeonpea and wild relatives using ribosomal DNA. Eight species of pigeonpea were probed with ribosomal gene from wheat and rDNA intergenic spacer region of *Vicia faba*. The dendrogram revealed a close relationship between *C. cajan* and *C. scarabaeoides*. The Australian species *C. scarabaeoides* and *C. retuculatus* were closely related to *C. cajan* and *C. platycarpus* while *C. goensis* and *C. lineatus* were distantly related to the cultivated species.

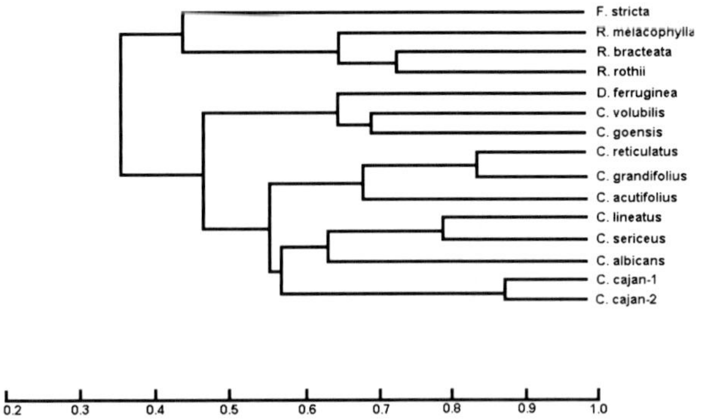

Fig. 2. Genetic relationships among selected wild species of pigeonpea using RAPD markers (Ratnaparkhe et al. 1995)

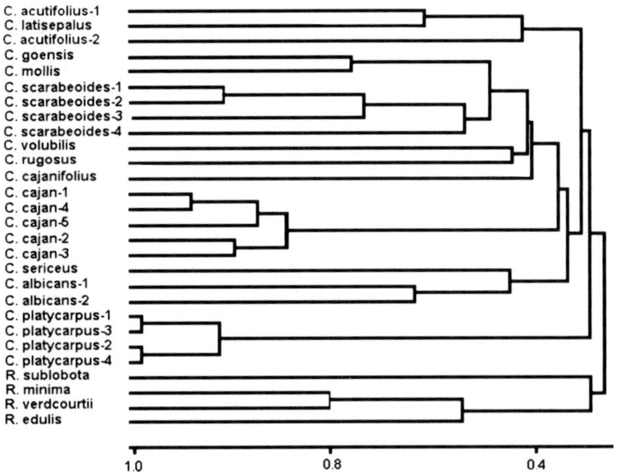

Fig. 3. Genetic relationship among wild relatives of pigeonpea using mtDNA RFLP studies (Sivaramakrishnan et al. 2002)

The RFLP study of mitochondrial DNA has proved to be an excellent method to characterize closely related species (Grabau et al. 1992; Deu et al. 1995). In pigeonpea, Sivaramakrishnan et al. (1997) successfully used maize mtDNA probes to characterize the various CMS systems.

Finally, Sivaramakrishnan et al. (2002) studied the diversity in 28 accessions representing 12 species of the genus *Cajanus*. The six sections of the *Cajanus* could easily be distinguished based on their characteristic hybridization patterns. The study suggests that RFLP of mtDNA could be used to distinguish *Cajanus* belonging to the secondary and tertiary gene pool. Cluster analysis revealed the clear separation of the two genera *Cajanus* and *Rhyncosia* from each other. The four species of *Rhyncosia* formed one major group. In *Cajanus*, the four accessions of *C. platycarpus* belonging to section *Rhyncosoides* formed a separate group. *Cajanus rugosus* was together with *C. volubilis*. *Cajanus cajanifolius* clustered with section *Cantharospermum* in a separate subgroup. The close relationship of *C. albicans*, *C. sericeus*, *C. cajanifolius* and *C. cajan* was similar to those observed by others using different molecular methods. Figure 3 shows the genetic relationship among wild relatives of pigeonpea using mtDNA RFLP studies.

8.3.2 Retrotransposons and Microsatellite Markers: The Second Generation of Molecular Markers

Retrotransposons constitute a large fraction of many plant genomes. Retrotransposons are now being utilized as molecular tools in DNA fingerprinting, genetic linkage mapping, phylogenetic studies, and molecular breeding. Most retrotransposon insertions are irreversible leading thus to relatively fixed changes, which is a desired attribute for analyzing phylogenetic relationships. Retrotransposons are known to cause gene inactivation by inserting near or within genes or by causing alterations in transcript processing and/or stability. The presence of retroelements in high copy number and in heterogeneous populations; their dispersed distribution throughout the genome; and their insertion into new genomic sites without the loss of parental copies are some of the properties of retrotransposons which make them particularly suitable candidates for generating molecular markers in a variety of crop plants. Lall et al. (2002) reported isolation and characterization of the first complete retrotransposon from pigeonpea, called *Panzee*. *Panzee* is a typical retrotransposon, with two long terminal repeats (LTRs) flanking its internal coding region. Genomic Southern hybridization experiments using probes derived from three different regions of the element show that *Panzee* or *Panzee*-related elements are present in high copy numbers in the pigeonpea genome.

Microsatellite or SSRs have recently become important genetic markers in plant genome research. SSRs are particularly attractive for distinguishing between cultivars because the level of polymorphism detected at SSRs loci is higher than any other molecular marker assay. They are also mostly codominant, specific molecular markers. They are easier to use than restriction fragment length polymorphism (RFLPs) owing to the smaller amount of DNA required, higher polymorphism and the ability to automate assays and they are more reproducible from one laboratory to another than the RAPD markers. SSR markers can easily be exchanged between researchers because each locus is defined by the primer sequences. However, isolation of useful simple sequence repeat (SSR) loci can be a time-consuming and expensive process. In pigeonpea, the isolation of SSR was first reported by Burns et al. (2001). In this study, $(CA)_{15}$ or $(CT)_{15}$ probe was used to identify colonies containing SSR motifs that were then sequenced. The screening for polymorphic SSRs was conducted on a set of 12 diverse pigeonpea accessions using 20 primers pairs derived from the obtained sequences. A set of 10 SSR markers has been successfully developed, which exhibits polymorphism across a range of pigeonpea varieties. The development of SSR loci in pigeonpea is believed to facilitate the rapid assessment of gene flow between populations, allowing the rates of genetic erosion to be monitored, and the development of risk assessment strategies that are required before any transgenic pigeonpea can be grown in the field. It remains to be checked if some of these primers can also be used to amplify SSR loci in close relatives of *C. cajan*.

8.4 Pigeonpea Transformation

The introduction of specific genes into pigeonpea to improve pest and disease resistance and also to improve nutritional quality could be achieved by genetic engineering. Although a number of reports are available on genetic transformation of legumes, there are few reports of pigeonpea transformation.

Initial studies on regeneration in pigeonpea have been reported from callus cultures (Kumar et al. 1983), direct differentiation from leaf (Eapen and George 1993), cotyledon-node explants (Shiva Prakash et al. 1994), and from different seedling explants (Eapen et al. 1998; Geetha et al. 1998; Mohan and Krishnamurthy 1998). George and Eapen (1994) reported organogenesis and embryogenesis from diverse explants of pigeonpea. Somatic embryogenesis has been reported in pigeonpea by Patel et al. (1994), Mallikarjuna et al. (1996), Srinivasu et al. (1998), Anbazhagan and Ganapathi (1999), and Mohan and Krishnamurthy (2002). Multiple shoot production was achieved from cotyledon-node explants (Mehta and Mohan Ram 1980; Kumar et al. 1984; Shiva Prakash et al. 1994; Naidu et al. 1995; Sudarsana Rao et al. 2001) and from epicotyls explants (Kumar et al. 1984; Naidu et al. 1995).

Reports describing genetic transformation of pigeonpea are very recent. Geetha et al. (1999) reported *Agrobacterium*-mediated genetic transformation of pigeonpea and development of transgenic plants via direct organogenesis. Lawrence and Koundal (2001) and Mohan and Krishnamurthy (2003) reported the genetic transformation of pigeonpea and the analysis of regenerated plants. Dayal et al. (2003) developed an efficient protocol for shoot regeneration and genetic transformation of pigeonpea using leaf explant. Sivamani et al. (2001) described the development of a novel gene transfer system for pigeonpea cell lines using particle bombardment. In this study, molecular analysis of transformed pigeonpea callus lines engineered with the heterologous oat arginine decarboxylase activity (*adc*) cDNA demonstrated the successful introduction of transgenes into the pigeonpea genome. Kumar et al. (2004) reported the genetic transformation of pigeonpea with the rice chitinase gene. Prasad et al. (2004) reported the expression of the biologically active hemagglutin-neuraminidase protein of *Peste des petits ruminants* virus in transgenic pigeonpea. The study shows that pigeonpea is a good experimental system in view of the biological activity of hemagglutin-neuraminidase and that pigeonpea satisfies the required conditions as source of edible vaccine. Biotechnological approaches such as gene transfer for enhanced disease and pest resistance offer opportunities for rapid improvement of pigeonpea.

8.5 Pigeonpea Research at ICRISAT

ICRISAT has been engaged in research for pigeonpea improvement for the last 25 years. During this period, concerted team efforts have metamorphosed the crop from a traditional medium- to long-duration, bushy, low-yielding crop to pest- and disease-resistant, early-maturing, photoperiod-insensitive, and relatively high-yielding crop. The two major achievements of ICRISAT were the identification of various sources of genetic male sterility and their use for the development of the hybrid cultivar of pigeonpea. Many efforts are devoted to the identification of a CMS system to improve large-scale seed production of hybrid cultivars. Male sterile lines have been developed and are used in hybridization programs. A major germplasm collection has been developed. This collection is characterized by a network of collaborators for different agronomic traits. Transgenic pigeonpea resistant to the destructive insect pest legume pod borer (*H. armigera*) has been developed. New molecular and biotechnological approaches are being used for pigeonpea improvement.

8.6 Future Outlook: Comparative Mapping and Genomics in Legumes

The grain legume crops are second to cereals in agricultural importance based on area harvested and total production. In 2004, more than 300 million metric tons of grain legumes were produced on 190 million ha. Traditionally, the legume family has been divided into three subfamilies: Caesalpinieae, Mimosoideae, and Papilionoideae. Most cultivated grain legumes are found within the Papilionoideae, which include the so-called tropical or phaseloid legumes and the temperate or galegoid legumes, with the genera *Melilotus, Trifolium, Medicago, Pisum, Vicia, Lotus, Cicer,* and *Lens* (Young et al. 2003). The phaseoloid/millettioid clade includes several legumes that are better adapted to tropical climates (warm-season legumes), such as common bean, cowpea (*Vigna unguiculata* L. Walp.), pigeonpea (*Cajanus cajan* L. Millsp.), and soybean (Doyle and Luckow 2003). Phylogenetic relationships within the legume family (Wojciechowski et al. 2004) are reflected in relatively high similarity or synteny at the genome level among the cool-season legumes, including *Medicago* sp. and pea (Kalo et al. 2004), or between the warm-season legumes common bean and soybean (Lee et al. 2001), but limited synteny is present among other legumes (for example, between cool-season and warm-season legumes) (Choi et al. 2004; Zhu et al. 2005).

Within the legume family, there are four well-studied model systems, *Medicago truncatula*, *Lotus japonicus*, *Glycine max*, and *Pisum sativum*, the first two being currently the targets of large-scale genome sequencing projects. While these cover only a tiny portion of the legume diversity, they are providing valuable tools for the genetic characterization and comparative mapping of some of the more traditional crops. Genomics and bioinformatics approaches allow the scientists to investigate sequence collinearity between the main crops and these model species. On a similar line, useful information can also be obtained through the construction of comparative consensus maps integrating the information of anchor markers and the results of different mapping populations within the same species, and in different species. The molecular genetic map of the *Medicago* genome, together with the knowledge of agronomically important genes, should pave the way for comparative legume genomics (Thoquet et al. 2002). A large set of markers from highly conserved *M. truncatula* gene regions is being created and used to establish a worldwide framework for comparative genomic studies in legumes. Gutierrez et al. (2005) investigated the potential for cross-species amplification of 209 expressed sequence tag (EST)-based and 33 bacterial artificial chromosome (BAC)-based microsatellites from *M. truncatula* in the three most important legume pulses, pea, faba bean, and chickpea, that might facilitate future comparative mapping. The results revealed significant transferability of *M. truncatula* microsatellites to the three pulses (40% in faba bean, 36.3% in chickpea, and 37.6% in pea). The studies from *M. truncatula*, soybean, and other legumes may provide important information for the development of the pigeonpea genetic map and in genome analysis. The molecular maps for legume crops of the semiarid tropics (chickpea, groundnut, and pigeonpea) that are not very dense as compared to the other related legume crops (i.e., soybean) could be improved by the development of new markers derived from the sequences of the model species such as *Medicago*. For legumes, such as pigeonpea, in which the availability of genomic sequences and DNA markers are limited, venturing into such approaches may provide better alternatives than to wait for the genomic information. The establishment of a comprehensive genetic linkage map will be a major tool in the improvement of selection efficiency of present breeding programs. It would also help to understand pigeonpea genome and its relationships to the wild species genome.

Acknowledgements
We thank Dr A.-F. Adam Blondon (URGV, INRA-CNRS, Evry, France) for critical review of the manuscript.

References

Anbazhagan VR, Ganapathi A (1999) Somatic embryogenesis in cell suspension culture of pigeonpea (*Cajanus cajan*). Plant Cell Tiss Org Cult 56:179–184

Ariyanayagam RP, Nageshwar A, Zaveri PP (1993) Cytoplasmic-genic male sterility in interspecific matings of *Cajanus* crop. Crop Sci 35:981–985

Burns MJ, Edwards KJ, Newbury HJ, Ford-Lloyd BV, Baggott CD (2001) Development of simple sequence repeat (SSR) markers for the assessment of gene flow and genetic diversity in pigeonpea (*Cajanus cajan*). Mol Ecol Notes 1(4):283–285

Choi HK, Mun JH, Kim DJ, Zhu H, Baek JM, Mudge J, Roe B, Ellis N, Doyle J, Kiss GB, et al (2004) Estimating genome conservation between crop and model legume species. Proc Natl Acad Sci USA 101:15289–15294

Dabak MM, Ranade SA, Dhar MS, Gupta VS, Ranjekar PK (1988) Molecular characterization of pigeonpea genome. Ind J Biochem Biophys 25(3):230–236

Dayal S, Lavanya M, Devi P, Sharma KK (2003) An efficient protocol for shoot regeneration and genetic transformation of pigeonpea (*Cajanus cajan* (L.) Millsp) using leaf explants. Plant Cell Rep 21:1072–1079

De DN (1974) Pigeonpea. In: Hutchinson J (ed) Evolutionary Studies in World Crops: Diversity and Change in the Indian Subcontinent. Cambridge University Press, London, pp 79–87

Deu M, Hamon P, Chantereau J, Dufer P, D'hont A, Lanaud C (1995) Mitochondrial DNA diversity in wild and cultivated sorghum. Genome 38:635–645

Doyle JJ, Luckow MA (2003) The rest of the iceberg: Legume diversity and evolution in a phylogenetic context. Plant Physiol 131:900–910

Dundas IS (1990) Pigeonpea, cytology and cytogenetics – perspective and prospects. In: Nene YL, Hall SD, Sheila VK (eds) The Pigeonpea. CAB, Wallingford, UK, pp 117–136

Eapen S, George L (1993) Plant regeneration from leaf discs of peanut and pigeonpea: Influence of benzyladenine, indoleacetic acid-amino acid conjugate. Plant Cell Tiss Org Cult 35:223–227

Eapen S, Tivarekar S, George L (1998) Thidiazuron-induced shoot regeneration in pigeonpea (*Cajanus cajan* L.). Plant Cell Tiss Org Cult 53:217–220

Geetha N, Venkatachalam P, Prakash V, Sita GL (1998) High frequency induction of multiple shoots and plant regeneration from seedling explant of pigeonpea (*Cajanus cajan* L.). Curr Sci 75:1036–1041

Geetha N, Venkatachalam P, Prakash V, Sita GL (1999) Agrobacterium mediated genetic transformation of pigeonpea (*Cajanus cajan* L.) and development of transgenic plants via direct organogenesis. Plant Biotechnol 16:213–218

George L, Eapen S (1994) Organogenesis and embryogenesis from diverse explants in pigeonpea (*Cajanus cajan* L.). Plant Cell Rep 13:417–420

Grabau EA, Davis WH, Phelps ND, Gengenbach BG (1992) Classification of soybean cultivars based on mitochondrial restriction fragment length polymorphisms. Crop Sci 32:271–274

Greilhuber J, Obermayer R (1998) Genome size variation in *Cajanus cajan* (Fabaceae): a reconsideration. Plant Syst Evol 212:135–141

Gutierrez MV, Vaz Patto MC, Huguet T, Cubero JI, Moreno MT, Torres AM (2005) Cross-species amplification of *Medicago truncatula* microsatellites across three major pulse crops. Theor Appl Genet 110:1210–1217

Jha SS, Ohri D (1996) Phylogenetic relationships of *Cajanus cajan* (L.) Millsp. (Pigeonpea) and its wild relatives. Genet Resour Crop Evol 43:275–281

Jones AT, Kumar PL, Saxena KB, Kulkarni NK, Muniyappa V, Waliyar F (2004) Sterility mosaic disease – the green plague of pigeonpea: advances in understanding the etiology, transmission and control of a major virus disease. Plant Dis 88:436–445

Kalo P, Seres A, Taylor SA, Jakab J, Kevei Z, Kereszt A, Endre G, Ellis THN, Kiss GB (2004) Comparative mapping between *Medicago sativa* and *Pisum sativum*. Mol Genet Genom 272:235–246

Kannaiyan J, Nene YL, Raju TN, Sheila VK (1981) Screening for resistance to phytophthora blight of pigeonpea. Plant Dis 65:61–62

Kollipara KP, Singh L, Hymowitz T (1994) Genetic variation of trypsin and chymotrypsin inhibitors in pigeonpea, *Cajanus cajan* (L.) Millsp, and its wild relatives. Theor Appl Genet 88:986–993

Krishna TG, Reddy LJ (1982) Species affinities between *Cajanus cajan* and some *Atylosia* species based on esterase isozyme. Euphytica 31:709–713

Kumar PL, Latha TKS, Kulkarni NK, Raghavendra N, Saxena KB, Waliyar F, Rangaswamy KT, Muniyappa V, Doriswamy S, Jones AT (2005) Broad-based resistance to pigeonpea sterility mosaic disease in wild relatives of pigeonpea (*Cajanus*: *Phaseoleae*). Ann Appl Biol 146 (3):371–379

Kumar AS, Reddy TP, Reddy GM (1983) Plantlet regeneration from different callus cultures of pigeonpea (*Cajanus cajan* L.). Plant Sci Lett 32:271–278

Kumar AS, Reddy TP, Reddy GM (1984) Multiple shoots from cultured explants of pigeonpea and *Atylosia* species. SABRAO J 16:101–105

Kumar SM, Kumar BK, Sharma KK, Devi P (2004) Genetic transformation of pigeonpea with rice chitinase gene. Plant Breed 123:485–489

Ladizinsky G, Hamel A (1980) Seed protein profiles of pigeonpea (*Cajanus cajan*) and some *Atylosia* species. Euphytica 29:313–317

Lall IP, Maneesha, Upadhyaya KC (2002) Panzee, a copia-like retrotransposon from the grain legume, pigeonpea (*Cajanus cajan* L.). Mol Genet Genom 267(3):271–280

Lawrence PK, Koundal KR (2001) *Agrobacterium tumefaciens* mediated transformation of pigeonpea (*Cajanus cajan* (L.) Millsp.) and molecular analyses of regenerated plants. Curr Sci 80:1428–1432

Lee JM, Grant D, Vallejos CE, Shoemaker RC (2001) Genome organization in dicots. II. Arabidopsis as a "bridging species" to resolve genome evolution events among legumes. Theor Appl Genet 103:765–773

Mallikarjuna N, Reena MJT, Sastri DC, Moss JP (1996) Somatic embryogenesis in pigeonpea *Cajanus cajan* L. Ind J Exp Biol 34:282–284

Mallikarjuna N, Moss JP (1995) Production of hybrids between *Cajanus platycarpus* and *Cajanus cajan*. Euphytica 83(1):43–46

Mallikarjuna N, Saxena KB (2002) Production of hybrids between *Cajanus acutifolius* and *C. cajan*. Euphytica 124: 107–110

Mallikarjuna N, Saxena KB (2005) A new cytoplasmic nuclear male-sterility system derived from cultivated pigeonpea cytoplasm. Euphytica 142(1/2):143–148

Mehta U, Mohan Ram HY (1980) Regeneration of plantlets from the cotyledons of *Cajanus cajan*. Ind J Exp Biol 18:800–802

Mohan ML, Krishnamurthy KV (1998) Plant regeneration in pigeonpea (*Cajanus cajan* (L.) Millsp.) by organogenesis. Plant Cell Rep 17:705–710

Mohan ML, Krishnamurthy KV (2002) Somatic embryogenesis and plant regeneration in pigeonpea (*Cajanus cajan* L. Millsp.). Biol Plant 45:19–25

Mohan ML, Krishnamurthy KV (2003) Plant regeneration from decapitated mature embryo axis and Agrobacterium mediated genetic transformation of pigeonpea. Biol Plant 46(5):519–527

Nadimpalli RG, Jarret JL, Pathak SC, Kochert G (1993) Phylogenetic relationships of pigeonpea (*Cajanus cajan*) based on nuclear restriction fragment length polymorphism. Genome 36:216–223

Naidu RB, Kulkarni DD, Krishnamurthy KV (1995) Genotype dependent morphogenetic potentiality of various explant of a food legume, the pigeonpea (*Cajanus cajan* L.). In Vitro Cell Dev Biol 31:26–30

Naithani ST (1941) Cytological studies on Indian pulses. Part I: The somatic chromosomes and the pro chromosomes of Cajanus. Proc Natl Acad Sci India 11:67–73

Nene YL, Sheila VK (1990) Pigeonpea: geography and importance. In: Nene YL, Hall SD, Sheila VK (eds) The Pigeonpea. CAB, Wallingford, UK, pp 1–15

Ohri D, Jha SS, Kumar S (1994) Variability in nuclear DNA content within pigeonpea. Plant Syst Evol 189:211–216

9 Quinoa (*Chenopodium quinoa*)

Peter J. Maughan[1], Alejandro Bonifacio[2], Craig E. Coleman[1], Eric N. Jellen[1], Mikel R. Stevens[1], Daniel J. Fairbanks[1]

[1] Brigham Young University, Provo, UT 84602, USA, *e-mail*: Jeff_Maughan@byu.edu
[2] Fundación PROINPA, Edificio Marconi, Piso 2, La Paz, Bolivia

9.1 Brief History of Crop

Quinoa (*Chenopodium quinoa* Willd.) is a domesticated staple food in Andean South America (Fig. 1). It is principally a grain crop, harvested and consumed in a manner similar to that for cereal grains, although its leaves are also used as a potherb. Quinoa was domesticated by ancient Andean civilizations in the region surrounding the Bolivian and Peruvian Altiplano (high plain). Gandarillas (1968) examined genetic diversity in quinoa landraces and found that the greatest diversity is native to an area between Cuzco, Peru and Potosí, Bolivia, with the largest number of landraces located in the area of the Altiplano surrounding Lake Titicaca in Bolivia and Peru. According to Gandarillas (1974) and the National Research Council (1989), there is a consensus that quinoa's center of origin is in the Andean Altiplano and that the area of ancient cultivation extends from Andean Altiplano to regions of Bolivia, Peru, Ecuador, Northern

Fig. 1. Quinoa (*Chenopodium quinoa*) cv. Real, growing under traditional cultivation practices (groups of plants spaced 1×1 m) near the Salar de Uyuni, Bolivia at ca. 3656 m above sea level

Genome Mapping and Molecular Breeding in Plants, Volume 3
Pulses, Sugar and Tuber Crops
C. Kole (Ed.)

Chile, and Colombia. The oldest archeological remains of domesticated quinoa date to 5000 BC (Tapia 1979).

Anciently, quinoa was known by a number of names in local languages. Pulgar-Vidal (1954) mentioned that people of the Chibcha (Bogota) culture called quinoa "suba" or "supha," the Tiahuancotas (Bolivia) called it "jupha," and the inhabitants of the Atacama desert (currently in Chile) knew it by the name "dahue." León (1964) wrote that the names "quinua" and "quinoa" were used in Bolivia, Peru, Ecuador, Argentina, and Chile.

Although it was and is the most widely cultivated grain chenopod, quinoa is not the only domesticated species of *Chenopodium*. Cañihua (*C. pallidicaule*) was domesticated in the same region as quinoa but is not as productive, nor is it as highly domesticated as quinoa. Because of its extreme frost tolerance, cañihua is adapted to high-altitude environments greater than 4000 m above sea level and is still cultivated in these areas (Galwey 1995). Huazontle (*C. berlandieri* subsp. *nuttalliae*) was domesticated in Mexico, where it served anciently as a seed crop and potherb but is currently cultivated principally for its leaves and immature inflorescence. A domesticated form of *C. album* is cultivated in the Himalayas as a seed grain and potherb (Partap and Kapoor 1985).

At the time of the Spanish conquest of the Inca Empire in 1532, quinoa, potatoes, and maize were the principal staple foods in Andean South America, with quinoa cultivation extending slightly beyond the region occupied by the Incas (Galwey 1995; Cusack 1984; Risi and Galwey 1984). Following the conquest of the Incas, quinoa cultivation precipitously declined with displacement by crops preferred by the conquistadores. According to Cusack (1984), quinoa held such a high position in Inca culture and religious ceremonies that the Spanish conquistadores may have actively suppressed its cultivation in an effort to eradicate traditional Inca religious rites. Furthermore, quinoa was not adopted as a crop by European settlers in South America or in Europe, as were the new world crops of maize and potatoes. Quinoa cultivation continued to decline into modern times as rural farmers migrated to urban centers, incentives were paid to farmers to plant barley, faba beans, and oats instead of quinoa, and increased dependence on imported food discouraged quinoa cultivation. According to Galwey (1995), the area of world quinoa cultivation had declined to ca. 39,000 ha by 1975.

In the mid-1970s, the exceptional nutritional characteristics of quinoa were discovered and its popularity began to increase. Andean countries established small, but effective, breeding programs, and several new varieties were released. Efforts to collect diverse landraces to prevent genetic erosion resulted in national quinoa germplasm banks in many Andean countries, the largest being in Bolivia and Peru. Quinoa's major use is still as a staple crop for subsistence farmers in rural regions of the Andes. However, a new international market for organically grown quinoa is increasing, creating a demand for export quinoa production in South America and some commercial production outside of South America.

9.2 Botanical Description

Wilson (1990) described in detail the botanical classification of cultivated *Chenopodium* species. The genus *Chenopodium* is in the family Amaranthaceae, although it was formerly in the family Chenopodiaceae. Phylogenic revision has merged the Amaranthaceae and Chenopodiaceae under the name Amaranthaceae (Angiosperm Phylogeny Group 1998). The domesticated *Chenopodium* species are classified in two subsections: Cellulata and Leiosperma. Leiosperma includes the South American species cañihua (*C. pallidculale*) and the Eurasian species group *C. album*. Quinoa (*C. quinoa*) and huazontle (*C. berlandieri* subsp. *nuttalliae*) are members of the subsection Cellulata. Most significantly for scientists interested in quinoa improvement, there is a dearth of published information on the breeding value of related *Chenopodium* species. Another serious problem concerns the taxonomy of this complex genus; for example, *C. album* has been used as a "convenient taxonomic receptacle" (Wilson 1980), although this species might actually form a complex of diploids, tetraploids, and hexaploids (Table 1). It should be noted that a few of quinoa's close relatives have been domesticated, such that they could potentially be developed into productive cultivars in their own right, either as vegetable or as seed crops. These include *C. album* (fat hen or lambsquarters) in Eurasia, hexaploid *C. giganteum* (khan or bithua) in the highlands of South and East Asia (Partap et al. 1998),

Table 1. *Chenopodium* species with their 2n chromosome number, 45S and 5S rRNA locus number (as determined by FISH to laminar meristem chromosomes) and orgins. Adapted with permission from Kolano 2004

Species	Chromosome #	Origin	rRNA loci		Ref.
			45S	5S	
C. album L.	18	Eurasia (wide spread)	1	2	1, 3, 4, 6
	36		1	3	
	54		2	4	
C. ambrosioides L.	32	N. America	1	1	1, 4, 6
C. aristatum L.	18	Eurasia	1	1	1, 4, 6
C. berlandieri subsp *nuttalliae* 'Huauzontle'	36	N. America	2	3	1, 4
C. berlandieri subsp *nuttalliae* 'Quelite'	36	N. America	1	3	1, 4
C. berlandieri Moq.	36	N. America	1	2	1, 4
C. bonus-henricus L.	36	Eurasia	2	2	1, 4
C. botrys L.	18, 36	Eurasia	1	1	1, 4, 6
C. bushianum Aellen	36, 54	N. America (midwest)	2	4	1, 3, 4
C. capitatum (L.) Ascher	18	N. America	1	1	1, 4, 6
C. ficifolium Smith	18	Asia	2	2	1, 3, 4
C. foliosum (Moench) Asch.	18	Eurasia	2	1	1, 4
C. giganteum D. Don	54	Eurasia	2	4	2, 3, 4
C. glaucum L.	18, 36	Asia/N. America	1	1	1, 3, 4, 6
C. hybridum L.	18	Eurasia/N. America	1	1	1, 3, 4
C. murale L.	18	Eurasia	1	1	1, 3, 4, 6
C. neomexicanum Stand.	18	N. America (southwest)	1	1	1
C. pallidicaule Aellen	18	S. America (Andes)	1	1	4
C. petiolare Kunth	36, 54	S. America (Andes)	2	4	4, 5
C. polyspermum L.	18	Eurasia	1	1	1, 3, 4
C. quinoa Willd.	36	S. America (Andes)	1	2	3, 4
C. rubrum L.	18, 36	Europe/N. America	1	2	1, 4, 6
C. schraderianum Schult.	18	Africa/Asia (southwest)	1	1	1, 4
C. sp. 'Silvestre Salinas'	18	S. America (Andes)	1	1	4
C. strictum Roth.	36	Eurasia	1	2	1, 3
C. vulvaria L.	18	Europe/N. America	1	1	1, 3, 4

References: (1) Clemants and Mosyakin 2003, (2) Zhu et al. 2003, (3) Rahiminejad and Gornall 2004, (4) Kolano 2004, (5) Wilson 1980, (6) Welsh et al. 2003

tetraploid *C. berlandieri* subsp. *nuttalliae* (huauzontle, chia, or quelite) in Mesoamerica (Wilson and Heiser 1979), and diploid *C. pallidicaule* (cañahua) in the Andes.

Quinoa is apparently an allotetraploid, although most genetic markers, both morphological and molecular, segregate in a typical disomic Mendelian fashion (Maughan et al. 2004; Ward 2000, 2001; Bonifacio 1990; Simmonds 1971; Gandarillas 1968). The chromosome number is $2n=36$. The basic chromosome number for all known species in the genus and most species in this family is $x=9$, and quinoa has numerous wild relatives with chromosome numbers of $2n=18$, 36, and 54, indicative of its apparent tetraploid origin (Table 1). The haploid genome of quinoa ($n=18$) is ca. 967 million nucleotide pairs, as determined by flow cytometry, and is thus relatively small compared to most plant species (Maughan et al. 2004).

Genetically and ecologically, quinoa varieties can be separated into two economically important subgroups: Andean ecotypes adapted to the high-altitude environments of the Altiplano regions of Bolivia and Peru, and coastal ecotypes adapted to the humid, disease-prone coastal lowlands of Chile and Ecuador (Wilson 1988a). Recently, several modern, improved varieties have been released, predominantly within the Andean subgroup. These varieties are genetically uniform inbred lines, and they display substantially higher yields than their landrace counterparts (Bonifacio and Gandarillas 1992; Bonifacio et al. 2003).

There is great morphological diversity among quinoa varieties and landraces. Leaf color is typically grayish-green, often with red betalin coloration at the bases of the petioles and in the leaf veins. Some varieties may have significant betalin coloration throughout the plant. Seed colors include coffee, black, white, red, yellow, cream, and various shades of white. Bitter saponins are present in the seed pericarp of many varieties.

9.3 Breeding Value of Other *Chenopodium* Species for Quinoa Improvement

Intrageneric hybridization experiments involving a limited number of related species have documented the breeding value of other tetraploid species of the subsection Cellulata for direct, sexual transfer of genes to improve quinoa (Nelson 1968; Heiser and Nelson 1974; Wilson and Heiser 1979; Wilson 1980). Wilson and Heiser (1979) and Wilson (1980) performed a series of interspecific hybridization experiments to assess relatedness of *C. quinoa* with other *Chenopodium* species. In the former study, the authors reported limited self-fertility in hybrids between quinoa and *C. berlandieri* subsp. *nuttalliae* or its wild North American relative *C. berlandieri*. In the latter study, crosses with *C. murale*, *C. gigantospermum*, *C. album*, *C. missouriense*, *C. strictum*, *C. standleyanum*, *C. dessicatum*, and *C. fremontii* failed to produce hybrids. Single hybrids produced from the *C. quinoa*×*C. petiolare* and *C. quinoa*×*C. neomexicanum* crosses did not survive to flowering. However, hybrids between quinoa and other tetraploid members of subsection Cellulata (*C. berlandieri*, *C. berlandieri* subsp. *nuttalliae*, and *C. bushianum*) produced partially fertile hybrids that could be backcrossed to quinoa to produce fruits. Viable, though sterile, hybrids have also been reported from crosses between quinoa and *Atriplex hortensis* (Bonifacio 1995). Breeding strategies involving bridging crosses, colchicine-mediated chromosome doubling, and/or embryo rescue may prove necessary to expand the breeding gene pool of quinoa to include Cellulata diploids and more distantly related *Chenopodium* and *Atriplex* species.

In a study analyzing the differential composition of flavonoids of 16 *Chenopodium* species, Rahiminejad and Gornall (2004) found identical profiles among the hexaploid "*C. album* aggregate" (Graebner 1919) consisting of *C. album*, *C. giganteum*, *C. opulifolium*, and *C. probstii*. This profile was identical in tetraploid *C. novopokrovskyanum* while two putative diploid ancestors, *C. ficifolium* and *C. suecicum*, possessed component flavonoids of the higher-ploidy species. Other diploids and tetraploids included in the study had unique flavonoid compositions, which indicates that they are not likely candidate-ancestors of the *C. album* hexaploid aggregate.

9.4 Economic Importance

Although quinoa is a minor crop in terms of worldwide importance, it has great economic significance for the Andean region of South America, particularly for impoverished subsistence farmers with small land holdings. For these people, quinoa is an important source of food as well as income. As the international market for quinoa grows, the quantity of quinoa harvested for export increases. The increased demand has led to higher market prices, which is both an advantage and disadvantage for Andean subsistence farmers. Many families now sell the majority of their quinoa to generate much needed income and purchase rice as a replacement grain in their diet, thus losing the nutritional benefits of quinoa.

Quinoa is produced principally in Bolivia and Peru with minor production in Ecuador, Chile, Argentina, and Colombia. The FAO tabulates data on quinoa production in Bolivia and Peru and estimates production in Ecuador (Statistical Databases at www.fao.org). In 2000 (the most recent year with actual data), quinoa was harvested on 36,847 ha producing 23,785 metric tons in Bolivia and on 28,889 ha producing 28,382 metric tons in Peru. Estimates for Ecuador in 2000 were 1,300 ha producing 650 metric tons. Because many of the people who produce quinoa directly consume much of what they produce and many live in remote regions, a significant proportion of the quinoa harvest does not enter local, national, or international markets. For this reason, quinoa production data likely underestimate actual production.

The Altiplano covers 255,000 km^2 at an altitude of 3,500 to 3,850 m above sea level. It is predominantly cold and arid with annual precipitation in the southern Altiplano of less than 250 mm and more than 200 d with frost. In spite of such harsh conditions, more than two million people reside there, half in rural areas. In these areas, quinoa production accounts for ca. 50 to 60% of income for farmers, and 60% of the crop is destined for export (Bonifacio 2004).

9.5 Breeding Objectives

Quinoa breeding programs have been established in Bolivia, Peru, Ecuador, and Chile. Quinoa is a self-pollinated species with predominantly perfect flowers (a few flowers are pistillate or staminate), and most improved varieties are homogeneous, pure-line genotypes. Landrace varieties typically consist of mixtures of pure-line, homozygous types. Breeding methods include mass selection, pedigree, and single-seed-descent approaches. Major objectives include grain yield, maturity, plant habit (amaranthiform or glomerulate), seed size and color, presence or absence of seed saponins, and resistance to downy mildew (*Peronospora farinosa*), cutworms, and armyworms (*Feltia* spp., *Agrotis* spp. and *Spodoptera* spp.) and to the quinoa moth (*Eurysacca melanocampta*).

Grain yields are inherently low, averaging less than one metric ton per hectare. Improved varieties often double grain yields and are being widely adopted. Most improved varieties have large, white seeds that meet high quality and processing characteristics required by the export market. However, improved varieties with dark-colored seeds, which are used in traditional recipes and for popping, have also just recently been released (Bonifacio et al. 2003).

The saponins, an antinutritional triterpenoid compound found in the seed pericarp, are bitter and must be removed by washing or through milling when seeds are processed for local and export markets. Saponin-free quinoa lines have been developed and have been shown to be controlled by a recessive allele at a single genetic locus (Ward 2001). Saponin-free types are advantageous in that they eliminate the cost associated with saponin removal during quinoa processing. However, the presence of saponins in the seed coat plays an important role in deterring avian predation in areas where birds are abundant (northern and central Altiplano). In the southern Altiplano, where birds are scarce and where most of the quinoa produced is destined for export markets, the development of saponin-free varieties is consequently a major breeding goal and one where marker-assisted selection (MAS) could be highly effective for breeding programs.

9.6 Quinoa Molecular Marker Studies: Introduction

Although quinoa is emerging as an important alternative crop species, only a few molecular investigations have been reported for quinoa to date. Initial molecular investigations focused on isozyme, seed protein variation, and morphological marker to study phylogenetic relationships within *C. quinoa* (Wilson 1988 a, b; Fairbanks et al. 1990). Wilson (1988 a) reported the detection of 26 allelic variants at 12 polymorphic isozyme loci that, when combined with morphological data, clearly differentiated *C. quinoa* into two distinct elements, specifically into Andean ecotypes (distributed at elevations above 1800 m) and Coastal/Valley (mainly Chilean) ecotypes. Moreover, the Andean ecotypes showed a weak but discernible differentiation into northern and southern phylogenetic groups. These data also support the hypothesis of the Altiplano as the center of origin for the species.

The first use of DNA-based molecular markers was reported by Fairbanks et al. (1993), who showed that random amplified polymorphic DNA (RAPD) analysis could be used to identify DNA polymorphisms among *C. quinoa* accessions. Bonifacio (1995) reported the use of RAPD markers to identify true hybrids from intergeneric crosses between *C. quinoa* and three species of the closely related genus *Atriplex*, specifically *A. hortenisis, A. joaquiniana,* and *A. serenana*. Ruas et al. (1999) used the RAPD technique to investigate the genetic relationship among 19 *Chenopodium* taxa, including 10 *C. quinoa* (8 domesticated and 2 wild) accessions and 6 related *Chenopodium* species (i.e., *C. berlandieri* subsp. *nuttalliae, C. ambrosioides, C. album, C. berlandieri* subsp. *joneianum,* and *C. pallidicale*). A total of 33 10-mer oligonucleotide primers amplified 399 DNA polymorphisms with an average of 12 polymorphisms per RAPD primer. The RAPD data clearly clustered the accessions relative to species classification. Within the *C. quinoa* group, the two wild accessions (termed "ajara" ecotypes) clustered closely among the eight cultivated varieties, suggesting little differentiation between sympatric domesticated cultivars and wild *C. quinoa* accessions. According to Wilson (1990), the origin and differentiation of *C. quinoa* includes placement of the Andean crop weed complex as a monophyletic, possibly coevolving unit.

9.6.1 First-Generation Genetic Linkage Map

Maughan et al. (2004) developed the first genetic linkage map of quinoa. This linkage map is based on 80 F_2 individuals from the cross KU-2 and 0654, a Chilean lowland type and a Peruvian Altiplano type, respectively. The linkage map consists of 230 amplified fragment length polymorphisms (AFLPs), 19 microsatellite or simple sequence repeats (SSRs), and 6 RAPD markers. It spans 1020 cM and contains 35 linkages with an average density of 4.0 cM per marker (Fig. 2). The number of linkage groups (LGs) reported in this study is higher than the haploid chromosome number of *C. quinoa* ($n=18$). The excess of LGs when compared to the haploid chromosome number suggests that several areas of the genome remain undetected. Employing the method of Hulbert et al. (1988), Maughan et al. (2004) predicted the total length of the quinoa map to be 1700 cM.

Sugar beet (*Beta vulgaris* L., $2n=2x=18$) is the best-studied member of the Amaranthaceae family and is a true diploid with a haploid chromosome number of nine (half that of quinoa). The total length of the sugar beet high-density genetic map ranges from 621 cM (Halldén et al. 1996) to 815 cM (Barzen et al. 1995), approximately half the size of the predicted genetic map for quinoa. Considering the likely allotetraploid nature of quinoa ($2n=4x=36$), a quinoa genetic map twice the size of sugar beet is not unexpected.

9.6.2 Simple Sequence Repeat (SSR) Markers

Unfortunately, the AFLP, RAPD and isozyme markers developed for the initial molecular investigations of quinoa are of only limited utility to laboratories based in less-developed countries (such as those in the Andean region) due to the requirement of technical expertise (AFLP), expensive instrumentation or detection chemistry (sequencing polyacrylamide gel electrophoresis and radioactivity for AFLP) or problems of reproducibility (RAPD), marker abundance (isozyme) and technology transfer (RAPD and AFLP). SSR-based markers are codominant, multiallelic, and generally more informative (based on heterozygosity measures) than RAPD or AFLP markers (Powell et al. 1996). The ease of use (simple PCR and horizontal agarose electrophoresis), high reproducibility, and low cost make them ideal marker systems for genetic analysis. Moreover, once developed they are easily maintained and shared among laboratories by publishing primer-pair sequence information.

In an attempt to make marker technology more widely available for quinoa researchers, Mason et al. (2005) sequenced 1276 clones from three SSR (CA, ATT, ATG) enriched genomic libraries, of which 397 contained potential SSRs. Polymorphism analysis using a panel of 31 diverse cultivated quinoa accessions showed that 208 (52%) of the SSRs were polymorphic among the accessions. The number of observed alleles ranged from 2 to 13, with an average of 4 alleles detected per locus. Heterozygosity values ranged from 0.2 to 0.9 with a mean value of 0.57. Sixty-seven SSRs were highly polymorphic with heterozygosity values greater than 0.7.

Transferability and potential utility of these SSRs as genetic markers in related crop species were assessed by cross-species amplification using two accessions each of three cultivated *Chenopodium* species (*C. pallidicaule*, *C. giganteum*, *C. berlandieri* subsp. *nuttalliae*). Sixty-seven percent of the SSRs amplified successfully in all species in the panel. The most notable PCR conservation was seen in *C. berlandieri* subsp. *nuttalliae*, where 99.5% of the SSR markers amplified specific PCR products. These results confirm the close genetic relationship between quinoa and *C. berlandieri* subsp. *nuttalliae* (see also Ruas et al. 1999) and suggest that these new quinoa sequence-based SSRs may serve as new molecular tools for genetic mapping and characterization across the genus.

9.6.3 Expressed Sequence Tags and Single Nucleotide Polymorphism Markers

Additional genomic resources for the improvement of quinoa are beginning to emerge, and the utility of these resources for cloning genes of interest has been demonstrated. For example, a set of 424 expressed sequence tags (ESTs) from two cDNA libraries, developing seed and floral tissue, were obtained and have been deposited in GenBank (Coles et al. 2005). Of these sequences, 349 have significant

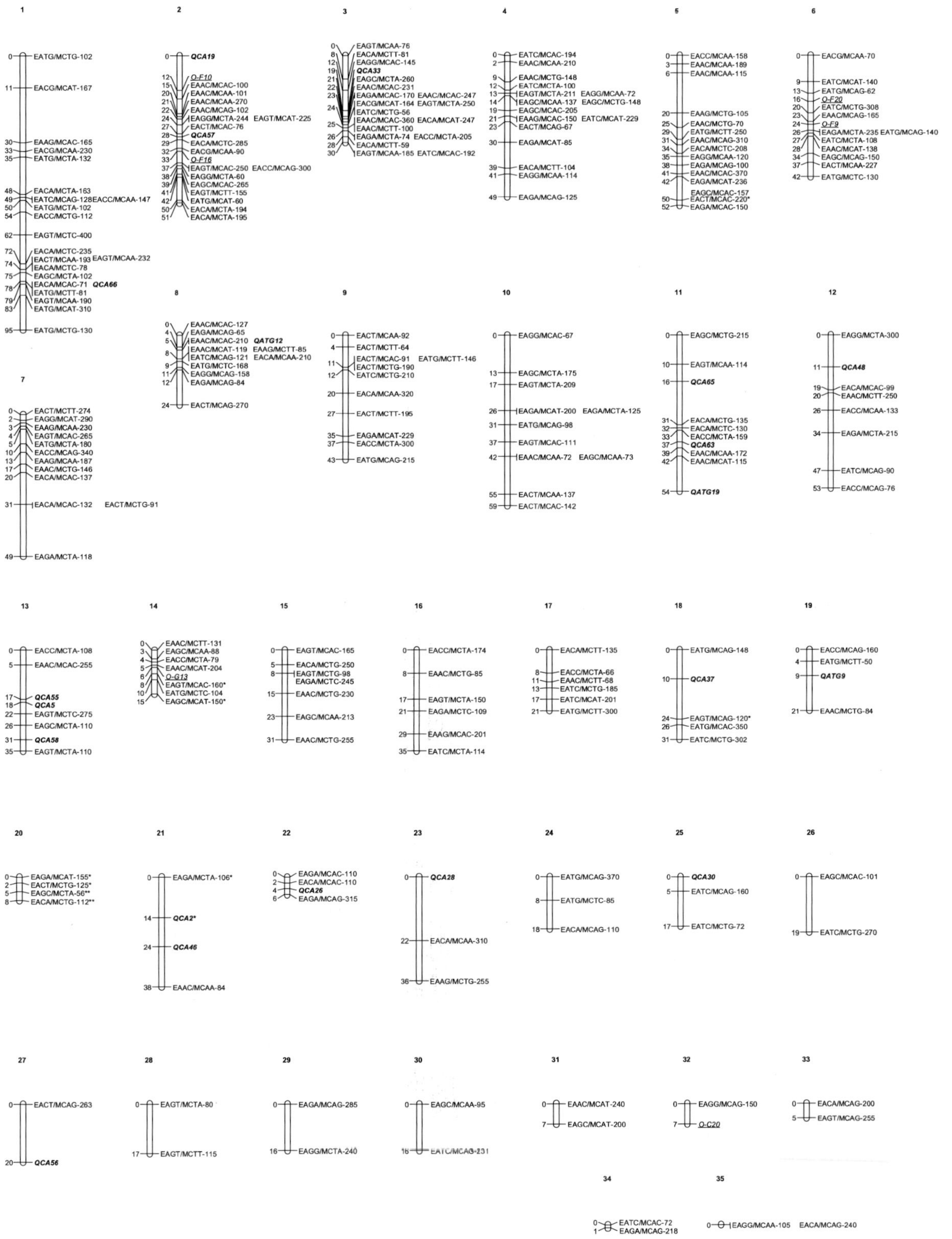

Fig. 2. Linkage map of *Chenopodium quinoa*. Cumulative distances are in centiMorgans (cM) and indicated on *left side* of LGs. Microsatellite markers are *bolded*. RAPD markers are *underlined*. All other markers are AFLP markers. AFLP markers are named as described in material and methods section. Loci marked with * or ** showed significant distortion: $p<0.05$ and $p<0.01$, respectively. Originally published in Theoretical and Applied Genetics (Maughan et al. 1994)

homology to protein-encoding genes from other plant species. Putative functions related to metabolism, protein synthesis, development, and so forth have been assigned to many of these EST sequences, although the majority of them have functions that are unclassified or not yet clear cut. Interestingly, the most abundant ESTs from the two cDNA libraries have putative functions related to plant defense.

Identification of the 11S globulin seed storage protein is an excellent example of the utility of EST databases for cloning of specific genes. Identification and characterization of the genes encoding storage proteins are likely to provide important clues to understanding the high protein content and excellent balance of amino acids in quinoa grain. Primers designed to match sequences from the *Amaranthus hypochondriacus* 11S seed storage globulin gene were used to amplify a portion of the 11S globulin gene from quinoa (C. Coleman, unpubl. data). The resulting PCR fragment was then used to screen the developing quinoa seed cDNA library (Coles et al. 2005), resulting in the identification of two candidate cDNA clones. Sequencing of the clones revealed two variants of a gene that encodes a protein with 72% homology to the 11S globulin protein from *A. hypochondriacus* (Fig. 3).

```
CQ1   MAKSTTTLFLLSCSIALVLLNGCMGQGRMREMQ-GNECQIDRLTALEPTYRIQAEGGLTEVWDTQDQQFQC
CQ2   MAKSTTTLFLLSCSIALVLLNGCMGQGRMREMQ-GNECQIDRLTALEPTNRIQAEGGLTEVWDTQDQQFQC
AH    MAKSTN-YFLISC-LLFVLFNGCMGEGRFREFQQGNECQIDRLTALEPTNRIQAERGLTEVWDSNEQEFRC

CQ1   SGVSVIRRTIEPNGLLLPSFTSGPELIYIEQGNGISGLMIPGCPETFESMSQ-------ESWREGMER--G
CQ2   SGVSVIRRTIEPNGLLLPSFTSGPELIYIEQGNGISGLMIPGCPETFESMSQ-------ESWREGMKR--G
AH    AGVSVIRRTIEPHGLLLPSFTSAPELIYIEQGNGITGMMIPGCPETYESGSQQFQGGEDERIREQGSRKFG

CQ1   MRGGRFQDQHQKIRHLRQGHIFAMPAGVAHWAYNSGNEPLVAVILIDTSNHANQLDKDYPKRFYLAGKPQQ
CQ2   MRGGRFQDQHQKIRHLRQGHIFAMPAGVAHWAYNTGNEPLVAVILIDTSNHANQLDKDYPKRFYLAGKPQQ
AH    MRGDRFQDQHQKIRHLRE DIFAMPAGVSHWAYNNGDQPLVAVILIDTANHANQLDKNFPTRFYLAGKPQQ

CQ1   EHSRHHHRGGESQRGEHGSDGNVFSGLDTKSVVQSFGVSEDIAEKLQAKQDERGNIVLVQEGLHVIKPPSS
CQ2   EHSRHQHRGGESQRGERGSGGNVFSGLGTKTIAQSFGVSEDIAEKLQAEQDERGNIVLVQEGLHVIKPPSS
AH    EHSGEHQFSRESRRGERNT-GNIFRGFETRLLAESFGVSEEIAQKLQAEQDDRGNIVRVQEGLHVIKPPS-

                       *********************
CQ1   RSYDDEREQ-RHRSPR--SNGLEETICSARLSENIDDPSKADVYSPEAGRLTTLNSFNLPILSNLRLSAEK
CQ2   RSYDDEREQRRHRSPR--SNGLEETICSARLSENIDEPSKADVYSPEAGRLTTLNSFNLPILSNLRLSAEK
AH    RAWE-EREQ-GSRGSRYLPNGVEETICSARLAVNVDDPSKADVYTPEAGRLTTVNSFNLPILRHLRLSAAK

CQ1   GVLYRNAIMAPHYNLNAHSIIYGVRGRGRIQIVNAQGNSVFDDELRQGQLVVVPQNFAVVKQAGEEGFEWI
CQ2   GVLYRNAIMAPHYNLNAHSIIYGVRGRGRIQIVNAQGNSVFDDELRQGQLVVVPQNFAVVKQAGEEGFEWI
AH    GVLYRNAMMAPHYNLNAHNIMYCVRGRGRIQIVNDQGQSVFDEELSRGQLVVVPQNFAIVKQAFEDGFEWV

CQ1   AFKTCENALFQTLAGRTSAIRAMPVEVISNIYQISREQAYRLKFSRSETTLFRPENQGRQRREMA-A
CQ2   AFKTCENALFQTLAGRTSAIRAMPLEVISNIYQISREQAYRLKFSRSETTLFRPENQGRQRRDLA-A
AH    SFKTSENAMFQSLAGRTSAIRSLPIDVVSNIYQISREEAFGLKFNRPETTLFRSSGQGEYRRKISIA
```

Fig. 3. Comparison of deduced 11S seed storage globulin amino acid sequences from two *Chenopodium quinoa* clones (CQ1 – GenBank Accession AAS67037 and CQ2 – GenBank Accession AAS67036), one *Amaranthus hypochondriacus* clone (AH – GenBank Accession CAA57633). The N-terminal sequence of the basic B chain as determined by Brinegar and Goundan (1993) is indicated with *asterisks*

The deduced amino acid sequence of the putative quinoa 11S globulin gene includes the 21-amino-acid N-terminal sequence of the B chain (indicated with asterisks in Fig. 3) as determined with direct protein sequencing by Brinegar and Goundan (1993), suggesting that the gene encodes a precursor similar to legumin A from pea (Domoney et al. 1986; Rerie et al. 1990).

EST databases are also a rich source of genetic information that can be exploited for genetic marker development. For instance, Coles et al. (2005) utilized the EST database described above as a starting point for the identification of single nucleotide polymorphisms (SNPs). By employing a resequencing strategy with five quinoa accessions, 51 SNPs were identified from 20 of the ESTs described above. This included 38 single-base changes and 13 insertion/deletions (indels) for an average of one SNP per 462 base pair (bp) and one indel per 1812 bp. When the sequence information from *C. berlandieri* subsp. *jonesianum* was included in the analysis, an additional 81 SNPs were identified, bringing the total number of SNPs identified to 132 (1 per 179 bp). The relatively low SNP frequency detected within quinoa reflects a narrow genetic base and demonstrates a need to collect and preserve primitive landraces in germplasm banks to serve as sources of potential genetic variation for improvement programs. The development of SNP markers is likely to be an important next step in the development of MAS protocols in quinoa, especially as DNA sequence information for quinoa becomes more available. Indeed, SNP-based markers are quickly becoming the marker system of choice because they are easily developed from sequence data, highly reproducible, and amenable to laboratory automation (Rafalski 2002).

9.6.4 Quinoa Bacterial Artificial Chromosome Library

Another important genomic tool that is available for quinoa is a 74,880 clone bacterial artificial chromosome (BAC) library that is available for public use at the Arizona Genomic Institute at the University of Arizona, Tucson, AZ, USA (Stevens et al. 2006). The library was constructed using *Bam*HI (26,880 clones) and *Eco*RI (48,000 clones), with an average clone insert size of 113 kb per insert and 130 kb, respectively for the two libraries. The estimated coverage of the library is 9.0 haploid nuclear genome equivalents, based on a calculated genome size of 967 Mbp. Stevens et al. (2006) characterize the BAC library probing with a set of 13 single-copy sequences developed from the aforementioned quinoa EST database (see above). These single-copy EST probes yielded an average of 12.2 positive hits per probe, suggesting that the estimation of 9.0 haploid genome equivalents coverage may be conservative. The utility of this BAC library for gene cloning was demonstrated by screening it with a probe consisting of a fragment of the quinoa 11S seed storage globulin cDNA. Positive clones have been identified that represent two different 11S seed storage globulin genetic loci (Fig. 4). Undoubtedly, the BAC library will be an important resource for efforts to identify specific genes and the continued characterization of the quinoa genome. Experiments are under way to identify additional genes controlling protein deposition and saponin accumulation in quinoa seeds. Furthermore, the development of a BAC library is essential for the construction of a quinoa physical map and map-based cloning efforts.

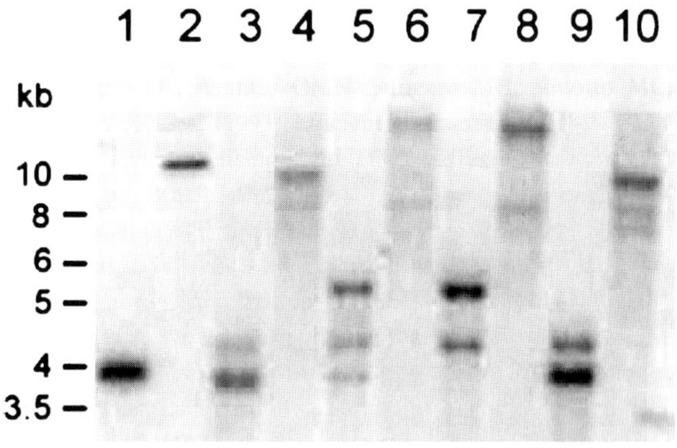

Fig. 4. Southern blot of quinoa BAC DNA containing an 11S seed storage protein gene compared to genomic DNA from four quinoa cultivars and breeding lines. Lanes 1–2: BAC clone 82M10; lanes 3–4: quinoa cultivar Chucapaca; lanes 5–6: quinoa breeding line KU-2; lanes 7–8: quinoa breeding line NL-6; lanes 9–10: quinoa breeding line 0654. DNA in lanes 1, 3, 5, 7, and 9 were digested with *Eco*RI. DNA in lanes 2, 4, 6, 8, and 10 were digested with *Hind*III. The blot was probed with a fragment of DNA containing the quinoa 11S seed storage protein gene amplified from BAC clone 77L9. The sizes of DNA markers given in kilobases (kb) are shown on the *left*

A high-density RFLP linkage map of sugar beet. Genome 39:634–645

Heiser CB, Nelson DC (1974) On the origin of the cultivated chenopods (*Chenopodium*). Genetics 78:503–505

Hulbert SH, Ilott TW, Legg EJ, Lincoln SE, Lander ES, Michelmore RW (1988) Genetic analysis of the fungus, *Bremia lactucae*, using restriction fragment length polymorphisms. Genetics 120:947–958

Kolano B, Pando LG, Maluszynska J (2001) Molecular cytogenetic studies in *Chenopodium quinoa* and *Amaranthus caudatus*. Acta Societ Botan Polon 70:85–90

Kolano BA (2004) Genome analysis of a few *Chenopodium* species. Ph.D. thesis, University of Silesia, Katowice, Poland

León J (1964) Plantas alimenticias andinas. Boletín Técnico No. 6, IICA Zona Andina, Lima, Peru

Mason SL, Stevens MR, Jellen EN, Bonifacio A, Fairbanks DJ, McCarty RR, Rasmussen AG, Maughan PJ (2005) Development and use of microsatellite markers for germplasm characterization in quinoa (*Chenopodium quinoa* Willd.). Crop Science 45:1618–1630

Maughan PJ, Bonifacio A, Jellen EN, Stevens MR, Coleman CE, Ricks M, Mason SL, Jarvis DE, Gardunia BW, Fairbanks DJ (2004) A genetic linkage map of quinoa (Chenopodium quinoa) based on AFLP, RAPD, and SSR markers. Theor Appl Genet 109:1188–1195

National Research Council (1989) Lost crops of the Incas. National Academy Press, Washington, DC

Nelson DC (1968) Taxonomy and origins of *Chenopodium quinoa* and *Chenopodium nuttaliae*. Ph.D. thesis, Indiana University, Bloomington, IN

Partap T, Kapoor P (1985) The Himalayan grain chenopods. I. Distribution and ethnobotany. Agric Ecosyst Environ 14:185–199

Partap T, Joshi BD, Galwey NW (1998) Chenopods. *Chenopodium* spp. Promoting the conservation and use of underutilized and neglected crops. 22. Institute of Plant Genetics and Crop Plant Research, Gatersleben/International Plant Genetic Resources Institute, Rome

Powell W, Morgante M, Andre C, Hanafey M, Vogel J, Tingey S, Rafalski A (1996) The comparison of RFLP, RAPD, AFLP, and SSR (microsatellite) markers for germplasm analysis. Mol Breed 2:225–238

Pulgar-Vidal J (1954) La quinua o suba, alimento básico de los Chibchas. Economía Colomiana 1:549–560

Rafalski A (2002) Applications of single nucleotide polymorphisms in crop genetics. Curr Opin Plant Biol 5:94–100

Rahiminejad MR, Gornall RJ (2004) Flavonoid evidence for allopolyploidy in the *Chenopodium album* aggregate (Amaranthaceae). Plant Syst Evol 246:77–87

Rerie WG, Whitecross MI, Higgins TJV (1990) Nucleotide sequence of an A-type legumin gene from pea. Nucleic Acids Res 18:655

Risi J, Galwey NW (1984) The *Chenopodium* grains of the Andes: Inca crops for modern agriculture. Adv Appl Biol 10:145–216

Ruas PM, Bonifacio A, Ruas CF, Fairbanks DJ, Andersen WR (1999) Genetic relationship among 19 accessions of six species of *Chenopodium* L., by Random Amplified Polymorphic DNA fragments (RAPD). Euphytica 105:25–32

Simmonds NW (1971) The breeding system of *Chenopodium quinoa*. I. Male sterility. Heredity 27:73–82

Stevens MR, Coleman CE, Parkinson SE, Maughan PJ, Zhang H-B, Balzotti MR, Kooyman DL, Arumuganathan K, Bonifacio A, Fairbanks DJ, Jellen EN, Stevens JJ (2006) Construction of a quinoa (*Chenopodium quinoa* Willd.) BAC library and its use in identifying genes encoding seed storage proteins. Theor Appl Genet 112:1593–1600

Tapia ME (1979) Historia y distribuición geográfica. In Tapia ME (ed) Quinua y Kañihua. Cultivos Andinos. Serie Libros y Materiales Educativos No. 49. Instituto Interamericano de Ciencias Agrícolas, Bogotá, Colombia, pp 11–15

Ward SM (2000) Allotetraploid segregation for single-gene morphological characters in quinoa (*Chenopodium quinoa* Willd.). Euphytica 116:11–16

Ward SM (2001) A recessive allele inhibiting saponin synthesis in two lines of Bolivian quinoa (*Chenopodium quinoa* Willd.). J Hered 92:83–86

Welsh SL, Atwood ND, Goodrich S, Higgins LC (2003) A Utah Flora, 3rd edn. ME Jones Endowment Fund, ML Bean Life Sci Mus, Brigham Young University, Provo, UT

Wilson HD (1980) Artificial hybridization among species of *Chenopodium* sect. *Chenopodium*. Syst Bot 5:252–263

Wilson HD (1988 a) Quinoa biosystematics. I. Domesticated populations. Econ Bot 42:461–477

Wilson HD (1988 b) Allozyme variation and morphological relationships of *Chenopodium hircinium* (s.l.). Syst Bot 13:215–228

Wilson HD (1990) Quinua and relatives (*Chenopodium* sect. *Chenopodium* subsect. *Cellulata*). Econ Bot 44(3 Suppl): 92–110

Wilson HD, Heiser CB (1979) The origin and evolutionary relationships of 'huauzontle' (*Chenopodium nuttaliae* Safford), domesticated chenopod of Mexico. Am J Bot 66:198–206

Zhu G, Mosyakin SL, Clemants SE (2003) Flora of China. http://www.efloras.org

10 Bambara Groundnut

S. Basu[1], J.A. Roberts[2], S.N. Azam-Ali[1], S. Mayes[1]

[1] Division of Agricultural and Environmental Sciences, School of Biosciences, University of Nottingham, Sutton Bonington Campus, Loughborough, Leicestershire, LE12 5RD, UK, *e-mail*: Sayed.Azam-ali@nottingham.ac.uk
[2] Division of Plant Sciences, School of Biosciences, University of Nottingham, Sutton Bonington Campus, Loughborough, Leicestershire, LE12 5RD, UK

10.1 Introduction

10.1.1 The Crop Bambara Groundnut

The semi-arid tropics (SAT) comprise parts of 48 countries in the developing world, including vast areas of India, South-east Asia, sub-Saharan Africa and Latin America. These environments are characterized by unpredictable weather, long dry seasons, inconsistent rainfall and nutrient-poor soils. Here several biotic and abiotic factors limit crop productivity and growth.

Bambara groundnut (*Vigna subterranea* L. Verdc., $2n=2x=22$) is grown primarily by subsistence farmers in sub-Saharan Africa and is used as a secondary food crop. It is a legume and is often grown alone or as an intercrop with drought-resistant crops like millet. The landraces consist of several genotypes, which result in their capacity to tolerate biotic and abiotic stress under a low-input agricultural system (Zeven 1998).

Bambara groundnut can be cultivated up to 1600 m above sea level. An average day temperature of 20 to 28 °C is ideal for the crop. A growth period of 110 to 150 d is required for the crop to develop, depending upon environment and landrace. Bambara beans will grow on any well-drained soil, but light, sandy, loams with a pH of 5.0 to 6.5 are most suitable. The crop does well on poor soil which is low in nutrients.

Although it produces a nutritious food and is cultivated throughout Africa the bambara groundnut remains one of the crops most neglected by science. Yet empirical evidence and research results suggest that it is a crop with great potential.

Bambara groundnut is an annual, which resembles groundnut (*Arachis hypogaea*) in both cultivation and habit. It is one of the five most important protein sources for many Africans (Vietmeyer 1978). The reported approximate chemical composition of bambara groundnut seeds is water 14.7%, ash 3.24%, crude protein 22.2%, fat 6.6%, cellulose 4.4% and carbohydrates 63.56%. The fatty acid composition for seeds with cream and dark red seed coats was found to be similar. However, the levels of palmitic and linoleic acids in bambara groundnut were higher than in groundnut (Ferrao et al. 1987).

Evans and Boulter (1974) estimated the amino acid and amino acid N, S and cysteine contents for various legumes. Total nitrogen in bambara groundnut was found to be 4.48 g and amino acid nitrogen 3.77 g per 100 g of meal. The amino acid composition was (in g/16 g N): aspartic acid 11.90, threonine 3.20, serine 5.40, glutamic acid 17.00, proline 4.10, glycine 3.80, alanine 3.60, valine 4.80, methionine 1.18, isoleucine 3.50, leucine 7.40, tyrosine 3.20, phenylalanine 5.80, histidine 3.10, lysine 7.70, arginine 7.70, cystine 1.14. Estimated total sulphur was 0.377% and sulphur in sulphur amino acids 0.154% (Evans and Boulter 1974).

In studies published on 16 indigenous plant species of Burkina Faso that serve to supplement the nutrients provided by cereals such as millets and sorghum, the nutritional quality of the protein in those plant specimens that contained greater than 10% protein on a dry-weight basis was assessed by comparing the amino acid composition of each of these specimens to that of a World Health Organization standard protein (WHO 1973). According to the WHO reference protein, the highest quality plant proteins were those of *Vigna subterranea*, *Pennisetum americanum*, and *Bixa orellana*; each of these

Genome Mapping and Molecular Breeding in Plants, Volume 3
Pulses, Sugar and Tuber Crops
C. Kole (Ed.)

scored at or above the score of the WHO standard for six of eight amino acids or amino acid pairs (Glew et al. 1997). However, the seeds are also reported to contain trypsin inhibitors and a chymotrypsin inhibitor, which act as anti-nutritional factors. Roasting the nuts helps greatly reduce the trypsin inhibitors and subsequently improves their nutritional value (Aregheore 1992).

Bambara groundnut also contains an appreciable quantity of the anti-beriberi vitamin B1. In the raw state, it contains about half that of mungbean (*V. radiata*), the pre-eminent source of B1 vitamin. However, when cooked, both legumes hold about the same vitamin percentage. The food value of bambara groundnut is higher than that of mungbean because of the better ratio of protein, fat and carbohydrates. Based on this information, increased cultivation of the crop was advocated by Donath and Spruyt in 1933.

The bambara groundnut has become less important in many parts of Africa because of the expansion of groundnut production. In recent years there has, however, been renewed interest in the crop for cultivation in the arid savannah zones.

10.1.2
The Plant

The species *subterranea* is further differentiated at the subspecific level into two groups: var. *spontanea*, comprising the wild forms, restricted to Cameroon in an area from Nigeria to Sudan, and var. *subterranea* comprising the cultivated forms found in many parts of the tropics, particularly sub-Saharan Africa. It is a typical short-day plant.

The cultivated bambara groundnut (which exists as landraces) is a self-pollinated legume with a compact well-developed tap root with many short lateral stems on which the leaves are borne. The leaves are trifoliate, the petiole is long, stiff and grooved, and the base is green or purple in color (Fig. 1 a). In contrast, the wild accessions have a spreading growth habit with few lateral stems and many pentafoliate leaves (Fig. 1 b).

The flowers of bambara groundnut of both wild and cultivated types are typically papilionaceous and are borne in a raceme on long, hairy peduncles which arise from the nodes on the stem (Fig. 2). The peduncles reach their maximum length at the initiation of pod filling. Anthers dehisce and stigma becomes receptive even before

Fig. 1. a Cultivated bambara groundnut (erect growth habit). **b** Wild bambara groundnut (spreading growth habit)

flowers open, and fertilization takes place on the same day as anthesis (Linnemann 1994). When fertilization takes place, the ovary covered by the calyx folds back at the flower base pointing towards the glandular apex. While the fertilized ovary stays unchanged for several days, the peduncle elongates to bring the fertilized ovary to or below the soil surface. The peduncle reaches its maximum length at the initiation of pod formation. The apical end, which bears the ovary with the fertilized ovules, expands into a pod. After 10 to 18 d (depending on the landrace), pod development commences, although there is a photoperiod requirement for this process in at least some landraces.

The pods of bambara groundnut develop underground or just above the ground and are spherical or oval in shape and may contain only one seed, although pods with two seeds are also common in

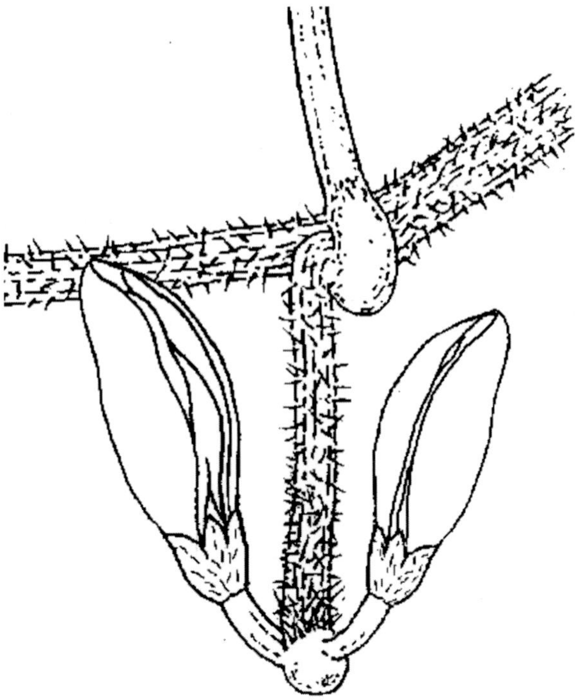

Fig. 2. Bambara groundnut peduncle with two flower buds and a glandular apex

some landraces. Pods with more than two seeds have also been reported (Pasquet and Fotso 1997). Mature pods are indehiscent, ranging from cream/yellow to reddish to dark brown in color (Massawe et al. 2004).

The alteration of photoperiod and temperature may affect flowering in some landraces (Linnemann 1994). Drought reduces the number of flowers per plant and the percentage of fertile pollen (Elia and Mwandemele 1986; Massawe et al. 2004).

Cultivated bambara groundnut seeds take 7 to 15 d to germinate. Flowering starts 30 to 35 d after sowing and may continue until the end of the plant's life. Pod and seed development take place ca. 30 to 40 d after fertilisation.

10.1.3 Intraspecific Hybridization for Development of a Mapping Population

Bambara groundnut occurs wild in West Africa and is distributed throughout Africa from where it later spread to America and Asia (Zeven and Zhukovsky 1975).

Species and genotypes of underexploited plants are threatened with disappearance through the rapid loss of native habitats. A combination of economic botany and biotechnology could help to prevent this. Although little known, bambara groundnut, which is grown by villagers throughout most of Africa south of the Sahara, is one of the major foods of the world. The seeds are a well-balanced food and the taste is acceptable. The crop is high yielding if managed well, but more research attention needs to be given to this crop (Vietmeyer 1986).

Despite its potential as crop, no coordinated plant-breeding programmes have been established for bambara groundnut. One major limitation to the establishment of breeding programmes is the failure to produce controlled hybrids from promising parental genotypes. Assessing the composition of landraces at the genetic level also makes the evaluation and selection of individual genotypes within the landrace mix more complex. To date, there are no established varieties of bambara groundnut, and for many centuries farmers have grown locally adapted landraces consisting of a mixture of genotypes. This often results in low and unpredictable yields with variable responses to agronomic interventions such as the application of fertilizers and irrigation. Artificial hybridization and selection will be a valuable strategy for the generation of genotypes with new combinations of desired traits through genetic recombination.

10.1.4 Distribution and Domestication

Cowpea (*V. unguiculata*) and bambara groundnut (*V. subterranea*) are the main African grain legume species. Although wild *V. unguiculata* is widespread from Sahara to Kalahari, wild *V. subterranea* is found in a smaller area, i.e. from Nigeria to Sudan. Both species were domesticated in the savannah belt extending from the Atlantic Ocean to the Red Sea, and while more than 500 herbarium specimens exist for *V. unguiculata* var. *spontanea*, there are less than 20 for *V. subterranea* var. *spontanea* (Pasquet 2004). The probable area of domestication of bambara groundnut is northern Cameroon (Harlen 1977) (Fig. 3).

The domestication of most major crop plants occurred during a brief period in human history about 10,000 years ago. During this time, ancient agriculturalists selected seed of preferred forms and culled out seed of undesirable types to produce each subsequent generation. Consequently, favoured alleles at genes controlling traits of interest increased in frequency, ultimately reaching fixation.

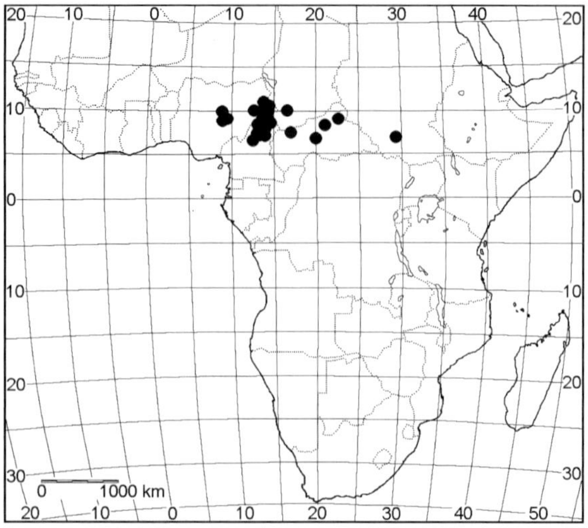

Fig. 3. Areas of Africa where wild bambara groundnut material has been identified (after Hepper 1968)

Domesticated bambara groundnut is characterized by larger seed size, larger leaves and slightly larger flowers as is the case with domesticated cowpea and most grain legumes. Unlike cowpea, however, domesticated bambara groundnut is characterized by longer leaf petiole and shorter stem internodes, two unusual traits in domesticated legumes. The thin shell wall of the wild pod is thickened in the domesticated form, which gives more protection from humidity and early germination of the seed. Therefore, the pods of the wild plant do not wrinkle upon drying, while the thick fleshy pods of the freshly dug domesticated fruit wrinkle on drying (Hepper 1963; Pasquet 2004).

Wild bambara groundnut plants are not as homogeneous as suggested by Hepper (1963). Within domesticated forms, it is possible to distinguish bunch, semi-bunch and open or spreading cultivars, mainly based on internode length (Doku 1969; Karikari 1972; Begemann 1988). A survey of Cameroonian landraces (Pasquet and Fotso 1997), however, pointed to a new morphological character. In Cameroon, domesticated accessions could be split into two distinct groups: a northern group characterized by one-seeded pods and a southern group characterized by two- to four- seeded pods, as well as an open habit. This can be correlated with a similar phenomenon observed with Cameroonian cowpea landraces. In the northern area, the cultigroups Biflora and Melanophthalmus display a low ovule number, and in the southern area, the cultigroup Unguiculata displays a high ovule number (Pasquet 1999, 2004). However, Pasquet and Fosto (1997) were not able to check if the change in the number of ovules in bambara groundnut was accompanied by a switch from photoperiod-sensitive to photoperiod-independent physiology as it was in cowpea. The ovule number and sensitivity to photoperiod splits domesticated cowpea into two opposite groups, and since Rassel (1960) reported a longer cropping cycle, a similar situation is suspected in domesticated bambara groundnut (Pasquet 2004), but this remains to be confirmed.

Many cultivated bambara groundnut plants have a similar upright growth habit. In contrast, wild forms exhibit spreading growth. The latter characteristic may be useful in many agronomic situations where limited soil moisture means that crop stands are established at low planting densities. In addition, a low spreading habit may have advantages for light capture and complementary growth when bambara groundnut is included as part of an intercropping system. The development of spreading forms in bambara groundnut provides a range of agronomic options analogous to those for groundnut (*Arachis hypogaea*) where erect, bunch and runner types are available for differing situations.

Additionally, there are likely to be other traits such as resistance to pests and diseases within wild forms as yet unidentified that may also be useful and worth introgressing into the cultivated gene pool. One such example is pubescence. Damage to cowpea pods caused by insect pests is a major problem for cowpea seed production. Pubescence found in wild relatives of cowpea, such as *V. vexillata*, could deter attack from some postflowering insect pests of cowpea. However, numerous attempts to make crosses between this wild species and cowpea were not successful. The close species *V. rhomboidea* has hairy pods, leaves and stems and is intercrossable with cowpea. The pubescence of this species is quite dense and long, though less pronounced than those found in *V. vexillata*. Hybrids derived from crosses made between cowpea and *V. rhombodiea* showed partial resistance to pod-sucking insects. It was concluded that pod hairs of this wild cowpea might play a role in resistance against some insect pests of cowpea (IITA 2000). Pubescence is also observed in VSSP18B, a wild accession of bambara groundnut collected in Cameroon and may have similar utility in cultivated bambara groundnut.

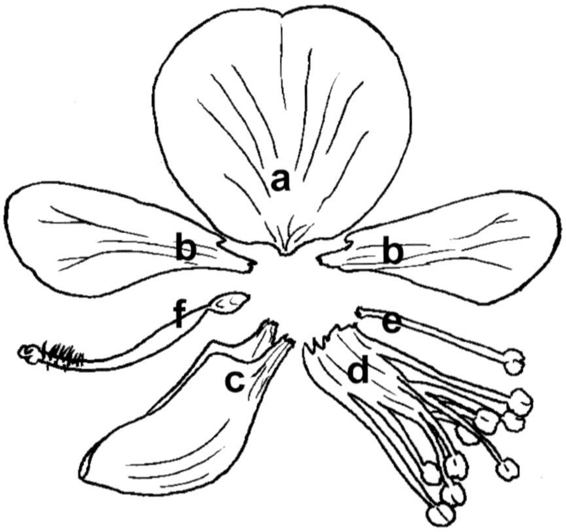

Fig. 4. Bambara groundnut flower. **a** Flag. **b** Wing petals. **c** Keel petal. **d** Fused stamen. **e** Vexillary stamen. **f** Style and stigma

10.1.5
Floral Biology

Bambara groundnut is predominantly a self-pollinating species. The flowers are typically papilionaceous, the term that describes a corolla having a standard [flag], wings and keel petals (Fig. 4). The flower has a pair of hairy epicalyces, and the calyx consists of five hairy sepals, four on the upper side and one on the lower side. The four upper sepals are completely joined, while the lower sepal is largely free. The standard petal encloses the wing and keel petals until the flower opens. While the standard petal is usually bright golden-yellowish in color, the wing petals are generally yellow and wrap around the keel. The keel is pale yellow and encloses the stamens, style and stigma. The stamens are diadelphous, nine with partly fused filament and one isolated vexillary stamen. The stigma, which is located at the end of the style, is spherical in shape with a curved lobe at the very tip. In young flower buds, the stigma is slightly above the anthers while in a mature flower the filaments elongate to place the anthers at the level of the stigma (Massawe et al. 2004).

Self- and cross-pollination, both brought about by ants, occur to varying extents depending on the variety or strain. Cross-pollination may be greater in varieties or strains with an open habit resembling that of the wild ancestor, but the bunched types are more likely to be self-pollinated (Doku 1968).

10.2
Evaluation of Genetic Diversity

10.2.1
Genetic Diversity in *Vigna*

Genetic variability is not uniformly distributed in plant populations. Some regions are particularly polymorphic, especially centers of origin of cultivated species. In these areas, wild populations are known to be a source of genes of interest, including those responsible for adaptation to different climatic conditions, resistance to stresses (e.g. drought, salinity) and resistance to pathogens (Harlen 1977; Cattan-Toupance et al. 1998).

In general, results of analyses of genetic diversity in a wild progenitor and its domesticated descendant crop species have shown that the latter has only a limited part of the diversity present in the progenitor (Doebley 1992; Gepts 1993). Three major phases in the reduction in genetic diversity can be identified: (1) the domestication process itself, (2) cultivation in the centre of domestication and dispersal in and from the centre of domestication, and (3) modern breeding practices (Gepts and Papa 2002). In studies on the characterization of genetic variation in three forms of *V. angularis*, namely cultivated, wild and weedy forms, it was found that the genetic variation was highest in the wild form, followed by the weedy form and the lowest in the cultivated form. Intrapopulation genetic variation was high in the weedy and wild populations. However, interpopulation variation was greater than intrapopulation genetic variation for all groups of populations studied in the *V. angularis* complex (Xu et al. 2000). Biochemical studies have shown that the cultivated bean (*Phaseolus vulgaris*) germplasm presents a low level of genetic diversity as compared to their wild ancestors (Gepts et al. 1986; Koenig et al. 1990; Sonnante et al. 1994; Johnson and Gepts 1998). Progress in bean breeding around the world has been slow, probably because of a limited genetic variability of the parents, which have been selected from the same gene pool (Alzate-Marin et al. 2003).

A survey of allozyme variation in cultivar groups of cowpea undertaken by examining 21 enzyme systems encoded by 36 loci across 271 accessions (representing the five cultivar groups) revealed very low levels of variation within acceshave occurred to reduce disequilibrium between moderately linked markers. When these advanced generation lines are created by selfing, the reduction in disequilibrium is not nearly as great as that under random mating.

Segregation for seed weight, pod length and days to flowering was studied in the F_3 (early generation) and F_6 (late generation) of intersubspecific crosses in cowpea. A wide range of segregants showed continuous distributions for these traits in both generations, confirming the quantitative nature of inheritance for these traits. The F_3 and F_6 generations were not significantly different in all three of the agronomic traits. Intergeneration correlations revealed strong associations between traits measured in the two generations. The results of this study also indicated that selection in early generations for superior types is feasible (Ubi et al. 2001).

10.3.6 Genetic Linkage Map of Bambara Groundnut

The construction of the first framework genetic linkage map of bambara groundnut, which has a genome size of 1C = 0.9 pg/haploid genome (Lakhanpaul and Babu 1991), is currently under way. This intraspecific map is based on an F_2 population derived by crossing *V. subterranea* var. *subterranea* (wild) × *V. subterranea* var. *spontanea* (cultivated). Sixty-seven AFLP and one SSR markers have been mapped to 20 linkage groups (LGs) spanning a total length of 516 cM. The minimum and maximum distances between markers varies from 4.7 to 32 cM.

The Mapping Population. Bambara groundnut is a self-pollinated diploid annual legume characterized with very low intralandrace diversity. The wild populations of this species, on the other hand, are characterized by higher genetic diversity, and yet a high genetic identity between wild and domesticated forms tends to suggest that wild bambara groundnut is the true progenitor of domesticated bambara groundnut. The potential of the wild germplasm in bambara groundnut breeding was highlighted by Pasquet and co-workers (1999).

Extensive polymorphism between the progenitors of the mapping population is a prerequisite for the construction of a comprehensive linkage map. This cannot be overemphasised, for in the absence of DNA polymorphism, segregation analysis and linkage mapping are impossible. The use of an F_2 mapping population of a wide intraspecific cross between a wild and cultivated accession for genome mapping in bambara groundnut was primarily due to a low level of genetic polymorphism detected within the cultivated gene pool.

Levels of DNA sequence variation are generally lower in naturally inbreeding species, which include most legumes. Low levels of polymorphisms based on RFLPs have also been detected within the cultivated gene pools of common bean (Vallejos et al. 1992).

The problems associated with fewer polymorphisms in genome mapping have been dealt with in many species by constructing linkage maps based on interspecific crosses. The map of bambara groundnut discussed here is based on a subintraspecific cross.

Population Size. Resolution of a map and the ability to determine marker order is largely dependent on population size and data quality. However, population size in most cases is technically limited by the number of seeds available or by the number of DNA samples that can reasonably be prepared.

In the present research a set of 98 F_2 progenies was used for map construction and QTL mapping of eight agronomic traits, which showed polygenic inheritance. This number was higher than the 85 F_2 progenies used in the construction of an intraspecific linkage map of chickpea (Flandez-Galvez et al. 2003), 75 RILs used in common bean (Freyre et al. 1998) and 94 RILs used in map construction in cowpea (Ouédraogo et al. 2002), although RIL populations have the practical advantage of being immortal.

Choice of Markers. Due to the large number of polymorphisms generated per primer combination, the AFLP marker system was primarily exploited in constructing the first core map of the bambara groundnut genome. AFLP analysis is very sensitive to polymorphism in the genome, with as little as 1-bp length differences in relatively short DNA fragments (50 to 1000 kb) being detectable (Qi et al. 1998).

AFLPs have been the marker of choice in map construction and map saturation in many plant and animal species including the model system *Arabidopsis thaliana* (Alonso-Blanco et al. 1998). Three hundred ninety-five AFLP markers were integrated into the previous Arabidopsis molecular map of 122 RFLPs, cleaved amplified polymorphic sequences and simple sequence length polymorphisms using an RIL population. This enabled the evaluation of the efficiency and robustness of AFLP technology for linkage analyses in Arabidopsis. AFLP markers were found throughout the linkage map. Two AFLP-based RIL maps were also integrated through 49 common markers, all of which mapped at similar positions. Comparison of both maps led to the conclusion that segregating bands from a common parent can be compared between different populations, and that AFLP bands of similar molecular size, amplified with the same primer combination in two different ecotypes, are likely to correspond to the same locus. AFLPs were found clustering around the centromeric regions, and the authors have established the map position of the centromere of chromosome 3 by a quantitative analysis of AFLP bands using trisomic plants (Alonso-Blanco et al. 1998).

The major disadvantage of AFLPs is that they are dominant markers, so it is not possible to distinguish heterozygotes from homozygotes. This is a limitation for an F_2 population and for a QTL analysis based on such a map. However, in the absence of bambara-specific markers, AFLPs were adopted for initial map construction.

Segregation Distortion. The present AFLP-based map of bambara groundnut revealed a high proportion of distorted loci (48.5%) that did not segregate in the expected Mendelian ratio of 3:1.

Skewed segregation ratios have been reported frequently in interspecific cross for all types of markers (morphological, isozyme, RFLP or AFLP). The percentage of loci showing segregation distortions was found to be highly variable: 69% in *Cryptomeria japonica*, 37% in *Citrus*, 36% in *Oryza*, 30% in coffee, 33% in *Prunus*, 23% in *Helianthus*, 8.4% in *Lens* sp. and only 1.4% in *Hevea* species. In the genus *Lycopersicum*, based on a wild species crossed with the cultivated *L. esculentum*, interspecific crosses led to distortion rates varying from 51 to 80% (Ky et al. 2000).

During the construction of an intraspecific map of chickpea based primarily on sequence tagged microsatellite site (STMS) and resistance gene analog (RGA) markers Flandez-Galvez and co-workers (2003) reported segregation distortion with all the marker types used in its construction (chickpea-STMS, ISSR, RAPD, RGA and field pea-STMS). However, in the extreme case, the field pea-STMS and RAPD markers showed 100% aberrant segregations. PCR-derived marker distortion levels as high as 36.9% and 40.6% have also been reported in F_2 intraspecific populations of *Medicago truncatula* and *M. tornata*, respectively (Jenczewski et al. 1997).

Therefore, compared to interspecific maps in most species and the intraspecific map in *Medicago*, a segregation distortion of 48.5% in the present case is not overly abnormal. The explanations given for such distortions are mostly gametic, zygotic or/and postzygotic selection.

10.4 QTL Mapping in Bambara Groundnut

The majority of biological traits are genetically complex. Mapping the QTLs that determine these phenotypes is a powerful means of estimating many parameters of the genetic architecture for a trait and potentially identifying the genes responsible for natural variation.

A direct application of genetic linkage maps has been in tagging genes of economic importance with molecular markers (Kumar 1999). The intraspecific F_2 population used for map construction was derived from a wide cross, which clearly segregated for a number of agronomic traits. Eight such traits, which either showed a continuous distribution or no known (Mendelian) segregation pattern and included emergence, flowering, internode length, stems per plant, leaf area, specific leaf area, carbon isotope discrimination and 100-seed weight, were subjected to QTL analysis.

It has been proved by many field experiments and QTL mapping results that among genes affecting some quantitative traits there are some major genes with larger genetic effect and some polygenes with smaller genetic effect. For such traits, the distribution of segregating population demonstrates multimodality, and this is the characteristic of the mixture of more than one distribution.

For example, Wang and Gai (1997) analyzed the inheritance of soybean flowering date in an F_2 population of a cross and reported the presence of one major gene and tested the existence of polygenes using the likelihood ratio test in controlling this trait.

Non-parametric (Kruskal Wallis analysis) and parametric (interval mapping, IM) mapping revealed associations of these traits to AFLPs and a single SSR marker across 20 different LGs on the genetic linkage map. Using IM it was possible to declare one to four major QTLs along with their positions on the map for traits like seed weight (one QTL), SLA (two QTLs), CID (three QTLs) and stems/plant (four QTLs).

Typically, QTL mapping is conducted in a single mapping population and, therefore, only has the potential to reveal genomic regions that are polymorphic between the progenitors of the population. What remains unclear is how well the QTL identified in any one mapping experiment characterise the genetics that underlie natural variation in traits. Symonds and co-workers (2005) aligned four RIL-based linkage maps of *Arabidopsis thaliana* used for QTL mapping for trichome density onto a common physical map. Seven of the nine QTLs identified were population-specific, while two were mapped in all four populations. Their results have shown that many lineage-specific alleles that either increase or decrease trichome density persist in natural populations or that most of this genetic variation is additive. More generally, these findings suggest that the use of multiple populations holds great promise for better understanding the genetic architecture of natural variation (Symonds et al. 2005).

This also has implications in bambara groundnut, where in the future, using mapping populations derived by crossing between landraces, it may still be possible to extend the results of the present study by exploiting linkage conservation and collinearity.

10.5 The Future Direction

10.5.1 Further Application of AFLPs in Saturation

A prerequisite for mapping and marker-assisted selection (MAS) of QTLs is the development of a genetic map with a high density of informative markers. AFLPs have been widely used to saturate existing genetic maps in a number of plant, animal and microbial genomes. In the present project only 67 AFLP markers could be mapped to 20 LGs of bambara groundnut. This number will be increased three- to four-fold by screening more AFLP primer combinations and mapping polymorphic markers to the existing core map.

10.5.2 Development of Microsatellite-Enriched Genomic Libraries

To perform a refined QTL mapping analysis for important agronomic traits, microsatellite markers must be developed in bambara groundnut. Unlike AFLPs and RAPDs, SSRs (as co-dominant and locus-specific markers) are useful as anchoring loci between unrelated crosses. This not only offers the potential for greater application in plant breeding but would also enable the map-based assessment of linkage conservation and collinearity between intraspecific and interspecific hybrid maps.

While one approach to generating species-specific SSRs for bambara groundnut could involve the screening of EST databases of related and model legumes like soybean and Medicago, another approach would be to develop microsatellite-enriched genomic DNA libraries. The argument against the latter is that it is a highly time-consuming and expensive technique. But recent studies in soybean by Song and co-workers (2004) have shown that, in contrast to SSRs derived from EST sequences, those derived from genomic libraries were a superior source of polymorphic markers, given that the mean number of tandem repeats in the former was significantly less than that of the latter ($p < 0.01$) (Song et al. 2004).

10.5.3 Oligonucleotide DNA Microarrays

Changes in gene expression underlie many biological phenomena. The use of DNA microarrays provides insight into tissue- and developmental-specific patterns of gene expression and the impact of environmental stimuli on transcript accumulation. To accelerate both gene discovery as well as hypothesis-driven research in bambara groundnut, the applications of microarray are unlimited.

cDNAs and ESTs representing a wide array of developmental stages and physiological conditions of the soybean plant are available. It has been demonstrated that the quality of the data from the soybean cDNA microarrays is sufficiently reliable to examine isogenic lines that differ with respect to a mutant phenotype and thereby to define a small list of candidate genes potentially encoding or modulated by the mutant phenotype (Vodkin et al. 2004).

The *Medicago* Genome Oligo Set (v1.0) (QIAGEN), which contains 16,086 70-mer oligonucleotides designed from the TIGR *Medicago truncatula* Gene Index (MtGI) Release 5.0 (May 2002), is now available to the research community. This has potential utility for work in bambara groundnut, although significant prescreening of response using bambara groundnut mRNA would be needed before it could be used routinely.

10.5.4 QTL Mapping

Initial QTL analysis in bambara groundnut was based on the scored trait data for the F_2 single plants. F_3 seeds derived by permitting the F_2 plants to self-fertilize will be bulked and used for multilocation field trialing and evaluation of phenological performance and ecophysiological responses in different environments. Composite analysis of QTL stability and relevance to different environments could be estimated. A linked QTL analysis could also be performed by combining the data from all local analyses into a global analysis for consistent and across-environment traits of value in this wide cross material developed in the present study, which may have potential use in future breeding work and relevance to end-user requirements.

10.5.5 The Potential of Marker-Assisted Selection in Bambara Groundnut

The AFLP assay is an efficient method for the identification of molecular markers, useful in the improvement of numerous crop species through MAS. BSA could be used to identify AFLP markers associated with important agronomic traits in bambara groundnut, as this would permit rapid selection of trait-specific genotypes for use in future breeding programmes.

The further development of selected F_3 and F_4 lines derived from the intraspecific cross as breeding lines and continuation of line development towards RILs will allow conclusions to be drawn concerning the genetics behind some of the key traits for bambara groundnut. This will allow a clearer understanding of end-user desirable traits and of how conventional or MAS may be employed to assist development of varieties or admixtures of varieties suitable for local conditions.

Thanks to the European Union-funded BAMFOOD (EUFP5 INCO-DEV) project on bambara groundnut, this crop has already come a long way from obscurity to popularity. With some more conscious and concentrated efforts in the present direction it could well be an important phenomenon in the legume circle.

References

Alonso-Blanco C, Peeters AJ, Koornneef M, Lister C, Dean C, van den Bosch B, Pot J, Kuiper MT (1998) Development of an AFLP based linkage map of *Ler*, Col and Cvi *Arabidopsis thaliana* ecotypes and construction of a *Ler*/Cvi recombinant inbred line population. Plant J 14:259–271

Alzate-Marin AL, Costa MR, Sortorato A, Del Peloso MJ, de Barros EG, Moreira MA (2003) Genetic variability and pedigree analysis of Brazilian common bean elite genotypes. Scientia Agricola 60:283–290

Amadou HI, Bebelli PJ, Kaltsikes PJ (2001) Genetic diversity in bambara groundnut (*Vigna subterranea* L.) germplasm revealed by RAPD markers. Genome 45:995–999

Aregheore EM (1992) A review of toxicity factors in some food and feedingstuffs in the nutrition of man and livestock in Nigeria. Vet Hum Toxicol 34:71–73

Begemann F (1988) Ecogeographic differentiation of bambara groundnut (*Vigna subterranea*) in the collection of the International Institute of Tropical Agriculture (I.I.T.A.). Wissenschaftlicher Fachverlag, Giessen

Brücher H (1998) The wild ancestor of *Phaseolus vulgaris* in South America. In: Gepts P (ed) Genetic Resources of *Phaseolus* Beans. Kluwer, Dordrecht, pp 185–214

Cattan-Toupance I, Michalakis Y, Neema C (1998) Genetic structure of wild bean populations in their South-Andean centre of origin. Theor Appl Genet 96:844–851

Darvasi A, Soller M (1995) Advanced intercross lines, an experimental population for fine genetic mapping. Genetics 141:1199–1207

Doebley J (1992) Molecular systematics and crop evolution. In: Soltis PS, Soltis DE, Doyle JJ (eds) Molecular Systematics of Plants. Chapman Hall, New York, pp 202–222

Doku EV (1968) Flowering, pollination and pod formation in bambara groundnut (*Voandzeia subterranea*) in Ghana. Exp Agric 4:41–48

Doku EV (1969) Growth habit and pod production in bambara groundnut (*Voandzeia subterranea*). Ghana J Agric Sci 2:91–95

Donath WF, Spruyt JP (1933) Het anti-beri-beri vitaminegehalte van katjang bogor (*Voandzeia subterranea* (L)

Thouars). [The content of anti-beriberi vitamin in ka-chang bogor (*Voandzeia subterranea* (L.) Thouars)]. Landbouw (Buitenzorg, Java) 9:133–144. Dutch (English summary)

Elia FM, Mwandemele OD (1986) The effect of water deficit droughts on some plant characters in bambara groundnut (*Vigna subterranea* Thouars). Trop Grain Legume Bull 33:45–50

EU (2004) Increasing the productivity of bambara groundnut (*Vigna subterannea* (L.) Verdc.) for sustainable food production in semi-arid Africa. Final Report of the EU BAMFOOD project. Contract number ICA4-CT-2000-30002

Evans IM, Boulter D (1974) Chemical methods suitable for screening for protein content and quality in cowpea (*Vigna unguiculata*) meals. J Sci Food Agric 25:311–322

Ferrao JEM, Ferrao AMBC, Antunes AMG (1987) A mancarra dos Bijagós (Vigna subterranea). Aspectos do seu valor nutricional. [Bambara groundnut (Vigna subterranea). Aspects of its nutritional value]. Garcia de Orta, Série de Estudos Agronómicos 14:35–39

Flandez-Galvez H, Ades PK, Ford R, Pang ECK, Taylor PWJ (2003) QTL analysis for ascochyta blight resistance in an intraspecific population of chickpea (*Cicer arietinum* L.). Theor Appl Genet 107:1257–1265

Freyre R, Skroch PW, Geffroy V, Adam-Blondon A-F, Shirmohamadali A, Johnson WC, Llaca V, Nodari RO, Pereira PA, Tsai SM, Tohme J, Dron M, Nienhuis J, Vallejos CE, Gepts P (1998) Towards an integrated linkage map of common bean. 4. Development of a core linkage map and alignment of RFLP maps. Theor Appl Genet 97:847–856

Gepts P (1993) The use of molecular and biochemical markers in crop evolution studies. Evol Biol 27:51–94

Gepts P, Osborn TC, Rashka K, Bliss FA (1986) Phaseolin-protein variability in wild forms and landraces of the common bean (*Phaseolus vulgaris*): evidence for multiple centers of domestication. Econ Bot 40:451–468

Gepts P, Papa R (2002) Evolution during domestication. In: Encyclopedia of Life Sciences. Nature Publishing, London

Glew RH, Vanderjagt DJ, Lockett C, Grivetti LE, Smith GC, Pastuszyn A, Millson M (1997) Amino acid, fatty acid, and mineral composition of 24 indigenous plants of Burkina Faso. J Food Comp Anal 10:205–217

Harlen JR (1977) The origins of cereal agriculture in the old world. In: Reed A (ed) Origins of Agriculture. Mouton, The Hague, The Netherlands, pp 357–383

Hepper FN (1963) The bambara groundnut (*Voandzeia subterranea*) and Kersting's groundnut (*Kerstingiella geocarpa*) wild in west Africa. Kew Bull 16:395–407

Jenczewski E, Gherardi M, Bonnin I, Prosperi JM, Olivieri I, Huguet T (1997) Insight on segregation distortions in two intraspecific crosses between annual species of *Medicago* (Leguminosae). Theor Appl Genet 94:682–691

Johnson WC, Gepts P (1998) Influence of epistatis and diverse evolution on wide crosses of common bean. In: Int Conf on the Status of Plant and Animal Genome Res, San Diego, Summary W 83

Karikari SK (1972) Correlation studies between yield and some agronomic characters in bambara groundnut (*Voandzeia subterranea* Thouars). Ghana J Agric Sci 5:79–83

Koenig R, Singh SP, Gepts P (1990) Novel phaseolin types in wild and cultivated common bean (*Phaseolus vulgaris*, Fabaceae). Econ Bot 44:50–60

Kumar LS (1999) DNA markers in plant improvement: an overview. Biotechnol Adv 17:143–182

Ky CL, Barre P, Lorieux M, Trouslot P, Akaffou S, Louarn J, Charrier A, Hamon S, Noirot M (2000) Interspecific genetic linkage map, segregation distortion and genetic conversion in coffee (*Coffea* sp.). Theor Appl Genet 101:669–676

Lakhanpaul S, Babu CR (1991) Genome size and evolution in Asiatic *Vigna savi*. In: Symp on Grain Legumes, New Delhi, pp 47–57

Linnemann AR (1994) Photothermal regulation of phenological development and growth in bambara groundnut (*Vigna subterranea* (L.) Verdc.). PhD thesis, Wageningen Agricultural University, The Netherlands

Mackay IJ, Caligari PDS (2000) Efficiencies of F2 and backcross generations for bulked segregant analysis using dominant markers. Crop Sci 40:626–630

Massawe FJ, Dickinson M, Roberts JA, Azam-Ali SN (2002) Genetic diversity in Bambara groundnut (*Vigna subterranea* (L) Verdc) landraces revealed by AFLP markers. Genome 45:1175–1180

Massawe FJ, Schenkel W, Basu S, Temba EM (2004) Artificial hybridization in bambara groundnut (*Vigna subterranea* (L.) Verdc.). In: Proc Int Bambara Groundnut Symp, 5–12 September 2003, Gaborone, Botswana, pp 193–209

Ouédraogo JT, Gowda BS, Jean M, Close TJ, Ehlers JD, Hall AE, Gillespie AG, Roberts PA, Ismail AM, Bruening G, Gepts P, Timko MP, Belzile FJ (2002) An improved genetic linkage map for cowpea (*Vigna unguiculata* L.) combining AFLP, RFLP, RAPD, biochemical markers, and biological resistance traits. Genome 45:175–188

Pasquet RS (1999) Genetic relationships among subspecies of *Vigna unguiculata* (L.) Walp. based on allozyme variation. Theor Appl Genet 98:1104–1119

Pasquet RS (2000) Allozyme diversity of cultivated cowpea *Vigna unguiculata* (L.) Walp. Theor Appl Genet 101:211–219

Pasquet RS (2004) Bambara groundnut and cowpea genepool organization and domestication. In: Conf Proc of the Intl Bambara Groundnut Symp, 5–12 September 2003, Gaborone, Botswana, pp 265–274

Pasquet RS, Fotso M (1997) The ORSTOM bambara groundnut collection. In: Heller J, Begemann F, Mushonga J (ed) Bambara Groundnut, *Vigna subterranea* (L.) Verdc. IPGRI. Rome, Italy, pp 119–123

Pasquet RS, Schwedes S, Gepts P (1999) Isozyme Diversity in Bambara Groundnut. Crop Sci 39:1228–1236

Peakall R, Gilmore S, Keys W, Morgante M, Rafalski A (1998) Cross-species amplification of soybean (*Glycine max*) simple sequence repeats (SSRs) within the genus and other legume genera: implications for the transferability of SSRs in plants. Mol Biol Evol 15:1275–1287

Qi X, Stam P, Lindhout P (1998) Use of locus specific AFLP markers to construct a high-density molecular map in barley. Theor Appl Genet 96:376–384

Rassel A (1960) Le voandzou Voandzeia subterranea Thouars et sa culture au Kwango. Bull Agr Congo Belge Ruanda-Urundi 51:1–26

Santalla Marta, Menendez-Sevillano MC, Monteagudo AB, De Ron AM (2004) Genetic diversity of Argentinean common bean and its evolution during domestication. Euphytica 135:75–87

Smartt J (1988) Morphological, physiological, and biochemical changes in *Phaseolus* beans under domestication. In: Gepts P (ed) Genetic Resources of *Phaseolus* Beans. Kluwer, Dordrecht, pp 143–161

Song QJ, Marek LF, Shoemaker RC, Lark KG, Concibido VC, Delannay X, Specht JE, Cregan PB (2004) A new integrated genetic linkage map of the soybean. Theor Appl Genet 109:122–128

Sonnante G, Stockton T, Nodari RO, Becerra Velasquez VL, Gepts P (1994) Evolution of genetic diversity during the domestication of common-bean (*Phaseolus vulgaris* L.). Theor Appl Genet 89:629–635

Symonds VV, Godoy AV, Alconada T, Botto JF, Juenger TE, Casal JJ, Lloyd AM (2005) Mapping quantitative trait loci in multiple populations of *Arabidopsis thaliana* identifies natural allelic variation for trichome density. Genetics, preprint, 16 January 2005

Tanksley SD (1993) Mapping Polygenes. Annu Rev Genet 27:205–233

Ubi BE, Mignouna H, Obigbesan G (2001) Segregation for seed weight, pod length and days to flowering following a cowpea cross. African Crop Sci J 9:463–470

Vallejos CE, Sakiyama NS, Chase CD (1992) A molecular marker-based linkage map of *Phaseolus vulgaris* L. Genetics 131:733–740

Vietmeyer ND (1978) Poor people's crops. World Farm 21:6–9

Vietmeyer ND (1986) Lesser-known plants of potential use in agriculture and forestry. Science 232:1379–1384

Vodkin LO, Khanna A, Shealy R, Clough SJ, Gonzalez DO, Philip R, Zabala G, Thibaud-Nissen F, Sidarous M, Strömvik MV, Shoop E, Schmidt C, Retzel E, Erpelding J, Shoemaker RC, Rodriguez-Huete AM, Polacco JC, Coryell V, Keim P, Gong G, Liu L, Pardinas J, Schweitzer P (2004) Microarrays for global expression constructed with a low redundancy set of 27,500 sequenced cDNAs representing an array of developmental stages and physiological conditions of the soybean plant. BMC Genomics 5:73

Wang J, Gai J (1997) Identification of major gene and polygene mixed inheritance model and estimation of genetic parameters of a quantitative trait from F_2 progeny. Yi Chuan Xue Bao 24:432–440

Weber J, May P (1989) Abundant class of human DNA polymorphisms which can be types using the polymerase chain reaction. Am J Hum Genet 44:388–396

Xu RQ, Tomooka N, Vaughan DA, Doi K (2000) The *Vigna angularis* complex: Genetic variation and relationships revealed by RAPD analysis, and their implications for *in situ* conservation and domestication. Genetic Resour Crop Evol 47:123–134

Zeven AC (1998) Landraces: a review of definitions and classifications. Euphytica 104:127–139

Zeven AC, Zhukovsky PM (1975) Dictionary of cultivated plants and their centers of diversity, excluding ornamentals, forest trees and lower plants. Centre for Agricultural Publishing and Documentation, Wageningen, The Netherlands, p 120

11 Sugarcane

Giovanni Cordeiro, Ouzi Amouyal, Frances Eliott, Robert Henry

Centre for Plant Conservation Genetics, Southern Cross University, PO Box 157, Lismore, NSW 2480, Australia,
e-mail: giovanni.cordeir@scu.edu.au

11.1 Introduction

Sugar (sucrose) is the major product of photosynthesis and a carbohydrate that occurs naturally in every fruit and vegetable. Used to alter the flavor of beverages and foods, it occurs in greatest quantities in sugar beet and sugar cane (sugarcane) and appears as a white crystalline substance in its pure extracted form. Sugarcane is the main sugar-producing crop in the world and has long been grown in tropical and subtropical regions.

11.1.1 History of Sugarcane

Whilst *Saccharum spontaneum* occurs in the wild from eastern and northern Africa, through the Middle East to India, China, Taiwan and Malaysia and through the Pacific to New Guinea, it is believed that New Guinea is the home of cultivated cane. High-fiber forms were used in construction whilst the softer and juicer forms were propagated in gardens for chewing. The migrations of people to the Pacific islands helped the spread of the plant. It later reached Indonesia, the Philippines and northern India.

Records of sugarcane in history have been in existence since 510 BC where 'reeds which produce honey without bees' were first indicated by the soldiers of the Emperor Darius near the river Indus. However, not until Alexander the Great's conquest of India in 327 BC did sugar start its spread westward (Purseglove 1972).

Possibly the first mention of 'sugar' as a commercial product occurs in 95 AD in 'Periplus Maris Erythraei' (or 'Guide Book to the Red Sea'), where we find the quote 'Exported commonly ... honey reeds which is called sakchar' made by a merchant. The Indians started pressing and boiling cane to crystallise sugar around 300 AD, and by 540 AD the Persians had learnt the art of sugar making. The cultivation of cane spread from Persia when it was invaded by the Arabs in 641 AD. The Arabs spread this knowledge to Egypt during the Islamic Holy War, but it was not until 710 AD that mention is made of the Egyptians developing clarification, crystallisation and refining of sugar. On the island of Crete, or Qandi (meaning 'crystallised sugar' in Arab), the Arabs installed the first sugar refinery around the year 1000 (Toussant 1987).

Through the Crusades in the 11th century, cane was brought to Europe, and in 1319 the first large shipment of sugar reached England. Sugar spread rapidly in the 1400s and in 1420 reached the Canary Islands, from where it was introduced to the New World by Columbus in 1493. Up until then, sugar was considered a luxury item. The colonisation of the New World brought sugar onto the world market in sufficiently large quantities at acceptable prices.

11.1.2 Morphology and Taxonomy

The sugarcane plant is a tall, thick-stemmed perennial grass (*Saccharum officinarum*; family Poaceae) (Fig. 1). Its parts and general appearance are similar to those of other grasses (Metcalfe 1960) and it is one of the most photosynthetically efficient of plants. The growth habit of the plant is characterized by a clumping of solid stalks bearing regularly spaced nodes or joints 10 to 25 cm apart, each with a single bud capable of asexual propagation. Leaves are sword-shaped, similar to that of a corn plant, and fold in a sheath with overlapping edges around the stem at the node. The sheath from one node encircles the stem up to the next node above and may overlap the base of the leaf on the next higher node. Mature cane stems extend 1.5 to 3 m in

Genome Mapping and Molecular Breeding in Plants, Volume 3
Pulses, Sugar and Tuber Crops
C. Kole (Ed.)

Fig. 1. Morphology of sugarcane plant. **A** Growth habit. **B** Stem morphology

height with a diameter from 2.5 to 7.5 cm (Britannica Concise Encyclopedia 2005). Its sap is high in sugar content. It is tender to cold, the tops being killed by temperatures slightly below freezing. The plant does not have a rest period and can be planted throughout the year in tropical countries where it grows continually until harvest. The derivation of the species name is of some interest: Up until the 17th century, sugar was a rare foodstuff and was attributed with numerous medicinal virtues, with some doctors in the habit of prescribing doses of sugar to their patients. It is from this official use that the name *Saccharum officinarum* was derived.

Sugarcane belongs to the genus *Saccharum* L., first established by Linnaeus in Species plantarum in 1753 with two species: *S. officinarum* and *S. spicalum*. The genus falls in the tribe Andropogoneae in the grass family, Poaceae. The tribe includes other tropical grasses such as *Sorghum* and *Zea* (maize). Very closely related to *Saccharum* are another four genera that readily interbreed, forming what is now commonly referred to as the *Saccharum* complex (Daniels and Roach 1987). The other four genera are *Erianthus* section *Ripidum*, *Miscanthus* section *Diandra*, *Narenga* and *Sclerostachya* (Table 1). They have in common a high level of polyploidy, aneuploidy (unbalanced number of chromosomes) that creates a challenge for the taxonomist (Daniels and Roach 1987; Sreenivasan et al. 1987). Linnaeus' original classification has since been revised to contain six species: *S. officinarum*, known as the noble cane; *S. spontaneum*, *S. robus-*

Table 1. Classification of the *Saccharum* complex.

Family:	Poaceae
Sub-family:	Panicoideae
Tribe:	Andropogoneae
Sub-tribe:	Saccharastrae
Genera:	*Erianthus* Michx Sect. Ripidium Henrard
	Miscanthus Anderss Sect. Diandra Keng
	Narenga Bor
	Saccharum Linn.
	Sclerostachya

tum, and *S. edule,* classified as wild species; and *S. sinense* and *S. barberi,* classified as ancient hybrids (Buzacott 1965; Daniels and Roach 1987; D'Hont and Layssac 1998).

Saccharum officinarum

SYNONYM(S): *Saccharum chinense* Roxb., *Saccharatum hybridum* hort., *Saccharum officinale* Salisb., *Saccharatum violaceum* Tussac
ENGLISH: Sugar cane, noble cane, cultivated sugarcane
S. officinarum, being high in sucrose content, has been used primarily for sugar production. It is characterized by being moderately tall, of various colors, thick stalks and low in fiber. Over 400 clones of *S. officinarum* have been recorded. *S. officinarum* is generally characterized by having a chromosome number of $2n=80$, with a basic chromosome number of $x=10$ (D'Hont et al. 1998a). It is thought to be derived from $2n=80$ forms of the wild species *S. robustum* in New Guinea, the center of origin (Brandes 1958). From there, clones of *S. officinarum* would have first been dispersed to the surrounding islands and later to various regions that became the current sugar-producing countries (Jannoo et al. 1999). In the early 19th century, to produce interspecific hybrids, clones of *S. officinarum* were crossed with *S. spontaneum* and backcrossed twice with *S. officinarum* as the recurrent parent. *S. officinarum* transmitted its somatic chromosome number during the first hybridization and the first backcross contrarily to the *S. spontaneum* clones. The chromosome number after the second backcross became normal (Bremer 1961). This process, referred to as 'nobilisation', has resulted in modern varieties with genomes that comprise multiple sets of homologous chromosomes derived from a single species (autopolyploid); possessing two or more unlike sets of chromosomes (allopolyploid) (Sreenivasan et al. 1987); and having a high total chromosome number ($2n=100$ to 130) with ca. 80% of the genome derived from *S. officinarum* and the rest from *S. spontaneum.*

Saccharum spontaneum L.

SYNONYM(S): *Saccharum chinense* Nees
ENGLISH: Wild cane, wild sugar cane, fodder cane, thatch grass
S. spontaneum is a wild species that has thinner canes, shows great variability in its morphology, is high in fiber and produces little sugar. Characteristics of the spikelets at the end of tertiary branches of the inflorescence are used to distinguish the species from other *Saccharum* spp. (Purseglove 1972). It, however, is highly adaptable to cold, drought and other difficult growing conditions; is highly vigorous and has resistance to several diseases. It too is a complex polyploid with a chromosome number varying from $2n=40$ to 128. Its basic chromosome number was initially suggested as being either 8 or 10 (Sreenivasan et al. 1987) but has now been confirmed by genomic in situ hybridization (GISH) as $x=8$ (D'Hont et al. 1998a). *S. spontaneum* is widely used in hybridization with *S. officinarum* and is widely distributed from Japan and New Guinea to the Mediterranean and Africa, with India thought to be the center of origin (Daniels et al. 1975). The vigour and adaptability of *S. spontaneum* has, however, earned it a reputation for being weedy and as such has been classified as a weed species in 33 countries. In Thailand, India, the Philippines and Indonesia, it is a serious agricultural weed competing on disturbed sites (Holm et al. 1997).

Saccharum robustum

This species was discovered in New Guinea and is thought to be the intermediate form in the evolutionary pathway from *S. spontaneum* to *S. officinarum.* In 1928, Brandes and Jeswiet, on expedition to New Guinea, collected 154 clones (Daniels and Daniels 1975). Its stems are large in diameter, tall (up to 10 m) and vigorous but, like *S. officinarum,* is susceptible to similar diseases, such as sugarcane mosaic virus. Due to its growth form, *S. robustum* is often used as either wind breaks or fences in New Guinea.

Saccharum edule

S. edule is thought to be an intergeneric hybrid of *Miscanthus floridus* and *S. robustum* or *S. officinarum.* It is similar in appearance to *S. robustum* except that its inflorescence is compacted and it is cultivated as a vegetable on the islands of the Pacific and New Guinea.

Saccharum sinense and Saccharum barberi

Saccharum sinense Roxb.
ENGLISH: Chinese cane, Chinese sugar cane, Indian sugar cane, cultivated sugarcane
S. sinense ($2n=81$ to 124) and S. barberi ($2n=11$ to 120) are thought to be wild species that have

been in cultivation since prehistory in northern India and China (Price 1968). The two species are thought to be derived from natural hybridizations. *S. sinense* is believed to be a result of natural hybridization between *S. officinarum*×*Erianthus* (sect. *Ripidium*), whilst *S. barberi* is native to India and thought to have been derived from *S. officinarum*× *Miscanthus* introgression. Each species also contains chromosomes homologous to *S. officinarum* and *S. spontaneum* as well as those from the genera *Erianthus* and *Miscanthus*, giving evidence to the complex relationships between members of the *Saccharum* complex (Daniels and Roach 1987; D'Hont et al. 1996). *S. sinense* has been used for sugar production in areas of India, China, Japan, the Philippines and Hawaii. It typically has narrow leaves and its stalks are generally thin, but the plant is resistant to disease, drought and flooding.

11.1.3 Karyotype

The *Saccharum* karyotype varies between species and cultivars. Chromosome numbers within the different species were determined early in sugarcane research. However, it is only recently that greater understanding of the karyotype varieties has been achieved. Cytogenetic studies had been used largely to determine chromosome behavior in sugarcane during meiosis. It had also been applied to studies in determining chromosome number of cultivars and interspecific and intergeneric hybrids (Sreenivasan et al. 1987). With the development of GISH, a method developed in the 1990s and involving the direct labelling of DNA from two sources with SpectrumRed and SpectrumGreen fluorescent dyes to act as probes that are then hybridized to the DNA of interest (Jiang and Gill 1994; Bennett 1995; Gill and Friebe 1998), the technique paved the way for studies of chromosomal composition in sugarcane, and in the mid-1990s it was shown that distinguishing the chromosome contributions made by *S. officinarum* and *S. spontaneum* was possible using GISH (D'Hont et al. 1995, 1996). In the cultivar R570, where the chromosome count of chromosome preparations varied between 107 and 115, about 10% of chromosomes were identified as having been contributed by *S. spontaneum* and about 80% from *S. officinarum*. The remaining 10% are recombinants of both species (D'Hont et al. 1995). Whilst these figures indicate that recombination of chromosomes does occur in contrast to previous assumptions to the contrary (Price 1963), the fact that 10% of the *S. spontaneum* chromosomes did not recombine indicates that recombination is not a frequent event (D'Hont et al. 1996).

The basic genome size (1x) estimated by flow cytometry is ca. 930 Mbp (0.98 pg DNA) for *S. officinarum* and 750 Mbp (0.78 pg) for *S. spontaneum*, whilst the total genome size estimated for the cultivar R570 (2n = 115) is ca. 10,000 Mbp. This is significantly larger than related grass species such as rice (860 Mbp, 2n = 24), sorghum (1600 Mbp, 2n = 20) or maize (5500 Mbp, 2n = 20) and highlights the high ploidy level of sugarcane cultivars (D'Hont & Glaszmann 2001).

Bremer (1924) established the chromosome number of *S. officinarum* as 2n = 80, but he believed that clones with chromosome numbers other than 2n = 80 were hybrids. Whilst Bremer provided a range of 2n = 60–200 for *S. robustum*, Price gave a more detailed breakdown of the range as 2n = 60, 63-70, 70, 80, 86, 92, 100–112, 157, 164 and 194 (Price 1957, 1965). The probable basic number was thought to be x = 10. This has now been confirmed through physical mapping of the rDNA sites, where eight copies of the locus was present in the species, with further support coming from studies on the interspecific hybrid R570 (D'Hont et al. 1996).

The chromosome number of *S. spontaneum* clones also range widely from 2n = 40 to 128, with the five major cytotypes for the species being 2n = 64, 2n = 80, 2n = 96, 2n = 112, 2n = 128 (Panje and Babu 1960; Sreenivasan et al. 1987). Based on this series of cytotypes, it was believed that the most probable basic number would be x = 8. This has been confirmed again through in situ hybridi-

Table 2. Summary of members of genus *Saccharum* (Buzacott 1965; Daniels et al. 1987)

Species	Classification	Sugar content	Chromosome number
S. barbari	Ancient hybrid	Low	2n = 11–120
S. edule	Wild species	Used as a vegetable	2n = 60–80 with aneuploid forms
S. officinarum	Noble canes	High	2n = 80
S. robustum	Wild species	Nil	2n = 60–~200
S. sinense	Ancient hybrid	Low	2n = 80–140
S. spontaneum	Wild species	Nil	2n = 40–128

zation of rDNA sites on a *S. spontaneum* clone with 2n = 64 (D'Hont et al. 1995).

Amongst the remaining *Saccharum* species, *S. barberi* and *S. sinense* have been separated into five morphologically different groups with chromosome numbers ranging between 2n = 82 and 124. *S. robustum* clones range from 2n = 60 to 200.

11.1.4 Economic Importance

The sugar market includes beet, cane, high fructose syrup and other sweeteners and farmers, processors, food manufacturers and consumers; in addition, there is extensive government involvement around the world in this market. Over three-fourths of sugar produced is from sugarcane, with the remainder produced from sugar beet.

Sugarcane is the main sugar-producing crop and is the source of sugar in all tropical and subtropical countries of the world. Sugarcane today is cultivated in 127 countries in both the tropics and subtropics covering a land area of over 130,000 km^2 (32 million acres). The top 20 producing countries harvested 125 million metric tonnes of sugarcane in 2002 (over 6 times that of sugar beets produced) with the largest producers being Brazil, India and the European Union (sugar beet) and the United States. However, the top five largest exporters are Brazil, Thailand, the European Union (sugar beet), Australia and Cuba. However, it is the developing nations such as Latin America, the Caribbean, southern Africa, Asia and the Pacific that have economies which are reliant on sugar as a significant share of their export earnings.

Of the approximately 125 million metric tonnes harvested, about 71% of this is sold profitably within its country of origin, or at a profit to other countries under agreement. What remains unsold is traded internationally (or dumped) at below production cost on the 'world sugar market'. Since 1985, this dump price has averaged about half the world average cost of producing sugar and is well below that of even the most efficient producers (Carter and Wenner 2003). At early-21st-century prices, most sugar industries in the developing nations have little hope of competing on the world market without the support of tariffs, and between early 2001 and late 2004, sugar prices went through at least two rounds of fluctuations between US$ 0.096 per pound and US$ 0.05 cents per pound.

The importance of achieving reasonable price stability in the sugar market is a result of the need for rational planning in the industry. Sugar is a special agricultural product that is highly capital intensive because it requires mills and factories for final processing. The capital intensity means that long-term planning is essential. This situation is aggravated due to sugarcane being a multiyear crop. It is therefore extremely difficult to match production with price conditions, and there is an in-built tendency to overproduce, driving prices downwards in surplus years (Mulherin 1986).

Changes to world sugar prices can be partially attributed to changes in sugar production by Southern Hemisphere countries taking a greater role as primary suppliers since 1985. Brazil is now the leading producer and exporter of sugarcane, and sustained production in Australia, South Africa, Zimbabwe and Swaziland have contributed to the predominance of Southern Hemisphere producers. Over the 10 years to 2003, Brazil has increased its exports ten-fold, to over 10 million tonnes. This trend is expected to continue, with Brazil targeting ca. 50% of the world sugar market. Brazil's ability to expand its exports has been helped largely by the repeated devaluation of its currency, thereby artificially reducing its production costs. In addition, the cross subsidy by the Brazilian government-supported bioethanol industry has also assisted in the reduction of these costs. Expansion programs in Brazil's central, south and north-north-east regions are expected to continue due to rising investment in the industry through mergers, joint ventures and direct acquisitions. In addition, an estimated 40 mills expected to be operational by 2007 will strengthen Brazil's supply response capabilities (Knapp 2003).

11.1.5 Breeding Objectives

Increasing sugar production has been the aim of improving cultivars for the sugarcane industries since the 19th century. Whilst molecular diversity in *S. officinarum* and commercial clones has been shown to be greater than previously expected (D'Hont et al. 1993; Al-Janabi et al. 1994; Harvey and Botha 1996; Jannoo et al. 1999), there are still many agriculturally desirable traits lacking in commercial cane. Traits such as ratooning ability, vigour and disease resistance are found in closely related members of the Andropogoneae tribe, and multiple attempts have been made to introgress these into commercial cane (D'Hont et al. 1995; Besse et al. 1997; Alix et al. 1999; Piperidis et al. 2000) though various breeding efforts.

11.1.6 Classical Breeding Achievements

Sugarcane breeding consists of intercrossing superior parental varieties with subsequent selection of the best F_1 hybrids, a process that may take anywhere from 8 to 20 years. This proven technique has produced all the current elite commercial cultivars, but it is limited by the extreme ploidy and heterozygosity of the parental breeding lines, the diversity in the gene pool, and the expense of selection of commercial traits such as ratooning ability. The result is often a compromise between what is technically feasible for the breeder and the needs of the farmers, processor and consumer, often leading to commercial varieties with serious faults (Skinner et al. 1987).

Prior to sugarcane farming and breeding programs, disease control relied mainly on naturally occurring resistant clones. The introduction of breeding programs and genetic improvement have contributed significantly to the sugarcane industry by rasing yield as a consequence of disease-resistant and stress-tolerant varieties that changed the sugarcane farming industry. Genetic improvement and breeding exercises developed the varieties of modern sugarcane that can be grown in a wide range of environmental conditions and agronomic practices.

The ability to produce seedlings in the early 20th century was a breakthrough in classical sugarcane breeding. Obtaining vital and fertile seeds is a necessity for breeding programs and provides a wide genetic variation in the hybrids. Hybrid sugarcane fertility was doubtful until 1858, when it was reported in Barbados, and not until 1885 when Soltwedel in Java obtained seeds from *S. spontaneum* was fertility in sugarcane finally recognised. Intercrossing of the noble cane in the early years of sugarcane breeding was the only means of developing disease resistance and other commercial targets. However, the continuous intercrossing program aimed at commercial characteristics in cultivated sugarcane resulted in a genetic drift that made the improvement of clones difficult. Nonetheless, intensive interspecific crossing programs around the world have resulted in a relative resistance to gumming disease as well as improved sugar yield. These late interspecific crosses have provided a valuable basis for today's germplasm in breeding programs based upon interspecific hybridizations.

S. officinarum as we know it today evolved in 1893 with the first successful crossing between *S. officinarum* Black Cheribon and the wild *S. spontaneum* Kassoer (Bremer 1961). Stimulated by the viral Sereh disease, Javan and Indian breeders selectively introduced desired traits from *S. spontaneum* whilst still preserving the *S. officinarum* characteristics through a double backcrossing with *S. officinarum*. This process was later given the term 'nobilisation' by Jeswiet (Jeswiet 1929) and Stevenson (Stevenson 1965), which means developing a cane that carries 'noble' characters of high productivity and disease resistance (Roach 1972).

The current commercial hybrid varieties are a consequence of years of selection. This selection for improved sugar-producing varieties commenced early in China and India where the main varieties used were *S. sinense* and *S. barberi* respectively. The outcome of sugar trading and the movement of sugar from these areas to Europe through the Mediterranean and to the rest of the world saw an international growth in the use of sugarcane for sugar that eventually evolved into the modern sugarcane industry we see today. The needs and demands of the new world for more sugar forced the search for more efficient sugar-producing clones to fit these needs, thereby leading to the development of modern-day cultivars.

In breeding for disease resistance, the strategy adopted by the Australian sugar industry is to preferentially use disease-resistant parent varieties when the trait is known to show high heritability but to limit the selection pressure so that parents with superior agronomic performance can still be used (Hogarth 1987). Breeding of resistant varieties is now fundamental to the control of some of the major diseases of sugarcane, including Fiji disease, leaf scald, mosaic, red rot and smut. However, breeding for resistance has not been successfully developed for the control of other major diseases such as ratoon stunting disease (RSD), controlled with hot water treatment; pineapple disease, controlled with fungicides; or for the control of major insect pests such as cane grub.

11.1.7 Limitations of Classical Activities

The complexity and size of the sugarcane genome is a major limitation in classical breeding methods. Linkage between different traits is a major factor contributing to this complexity, and whilst continued selective breeding for enhanced sucrose accumulation has been responsible for over half of the yield increase in the past 50 years, it has probably reached the limit of its capabilities. Nevertheless, the employment of new technologies can aid in this respect. The genetic selection process in sugarcane can now be assisted by the ability to associate specific traits with genetic markers and genetic maps, and the use of these molecular tools, mapping and gene-identification techniques will be crucial if yield increases through breeding are to be continued.

11.2 Genetic Maps

Due to the high ploidy level, complexity of aneuploidy and the highly heterozygous nature of the sugarcane genome, the application of molecular markers for mapping purposes presents challenges not commonly associated with diploid plants. Marker systems such as amplified fragment length polymorphisms (AFLPs), restricted fragment length polymorphisms (RFLPs) and simple sequence repeats (SSRs) reveal several alleles for a single locus. These markers may be single dose (SD) and present as a single copy, double dose (DD) and present as two copies, or multidose. Ideally, a mapping population would show a high level of SD markers. SD markers are present in one parent, absent in the other and segregate in 1:1 ratio in the progenies.

11.2.1 First-Generation Maps

RFLPs were amongst the first DNA marker systems used to develop genetic maps in higher organisms (Botstein et al. 1980). Initial efforts in developing genetic maps of *Saccharum* were carried out simultaneously by da Silva using RFLP markers and by Al-Janabi using arbitrarily primed polymerase chain reaction (PCR) (Welsh and McLelland 1990), a method similar to random amplified polymorphic DNA (RAPD) markers (Al-Janabi et al. 1993; da Silva et al. 1993). They based their initial maps on a proposal that suggested direct mapping of any polyploid was possible if it had strict bivalent pairing at meiosis (Wu et al. 1992). This premise allowed mapping based on the segregation of SD markers (Bonierbale et al. 1988). The RFLP map consisted of 216 SD loci comprising 44 linkage groups (LGs), whilst the arbitrarily primed PCR map had 208 SD loci comprising 41 LGs (Al-Janabi et al. 1993; da Silva et al. 1993). Further efforts by da Silva extended the RFLP map to 319 loci through the development of a method to map DD and triple-dose (TD) markers (da Silva et al. 1993). Estimates of genome size and coverage of these two studies were in close agreement, and both concluded that sugarcane presented polysomic segregation, typical of autopolyploids. Both authors had also developed their maps on the *S. spontaneum* clone SES 208, which allowed for the combination of both maps into a single genetic linkage map comprising 527 markers randomly distributed across the genome and bearing an average interval of 6 cM between the markers (da Silva et al. 1995). The results also suggested this clone to be an autopolyploid. Maps were later created for *S. officinarum* using RAPDs (Mudge et al. 1996) and RFLPs (Ming et al. 1998), and for *S. robustum* with RFLPs (Ming et al. 1998; Hoarau et al. 2001). The RFLP maps by Ming were developed from interspecific crosses of *S. officinarum*×*S. spontaneum*, providing a total of four maps, two for *S. officinarum* (clones Green German and Muntok Java) and a further two for *S. spontaneum* (clones IND81-146 and PIN84-1). On Green German (2n=97–117), 418 markers were mapped of which 270 were distributed over 72 LGs and covered 2304 cM; with Muntok Java (2n=140), 355 markers were mapped and 206 markers also distributed over 72 LGs covering 1443 cM. Of the 385 markers mapped on IND81-146 (2n=52–56), 248 were distributed over 69 groups covering 2063 cM, and with the 297 markers on PIN84-1 (2n=96), 182 were also assembled in 69 LGs over 1103 cM (Ming et al. 1998).

11.2.2 Second-Generation Maps

The first maps of a cultivar were initiated on the selfed progeny of the cultivar SP70-1006 (D'Hont

1994). This map was later transferred and further developed on the cultivar R570 (Grivet et al. 1996 a) using RFLP probes from maize and sugarcane. By 2001 the R570 map, as it had become commonly known, contained some 600 RFLP markers derived from a number of grass (Poaceae) species (D'Hont and Glaszmann 2001). The markers on this map distribute over 98 cosegregation groups covering a total length of 2008 cM. A parallel mapping effort was also carried out to place 939 SD AFLP markers on R570, of which 887 were distributed into 120 cosegregation or linkage groups (Hoarau et al. 2001). A more recent map has been developed on a cross between the Australian commercial variety Q165 (2n = 115) with the *S. officinarum* clone IJ76-514 (2n = 80) using a combination of AFLP and SSR markers. A total of 967 SD markers were generated from the two marker systems, and 910 were distributed across 116 LGs covering a total map length of 9058.3 cM (Fig. 2) (Aitken et al. 2005).

11.2.3 Comparative Mapping

Despite the complexities, parallel work has been initiated and continues to be carried out in comparative mapping using diploid members of the Adropogoneae tribe as a reference. The choice of maize as a diploid reference was motivated by the plausible synteny (Beckmann and Heun 1989) among members of the Andropogoneae tribe that was suggested by Hulbert (Hulbert and Richter 1990). Strong hybridization of maize DNA probes to DNA of sorghum, foxtail millet, Johnsongrass and sugarcane was observed and revealed the existence of large homosequential portions between the

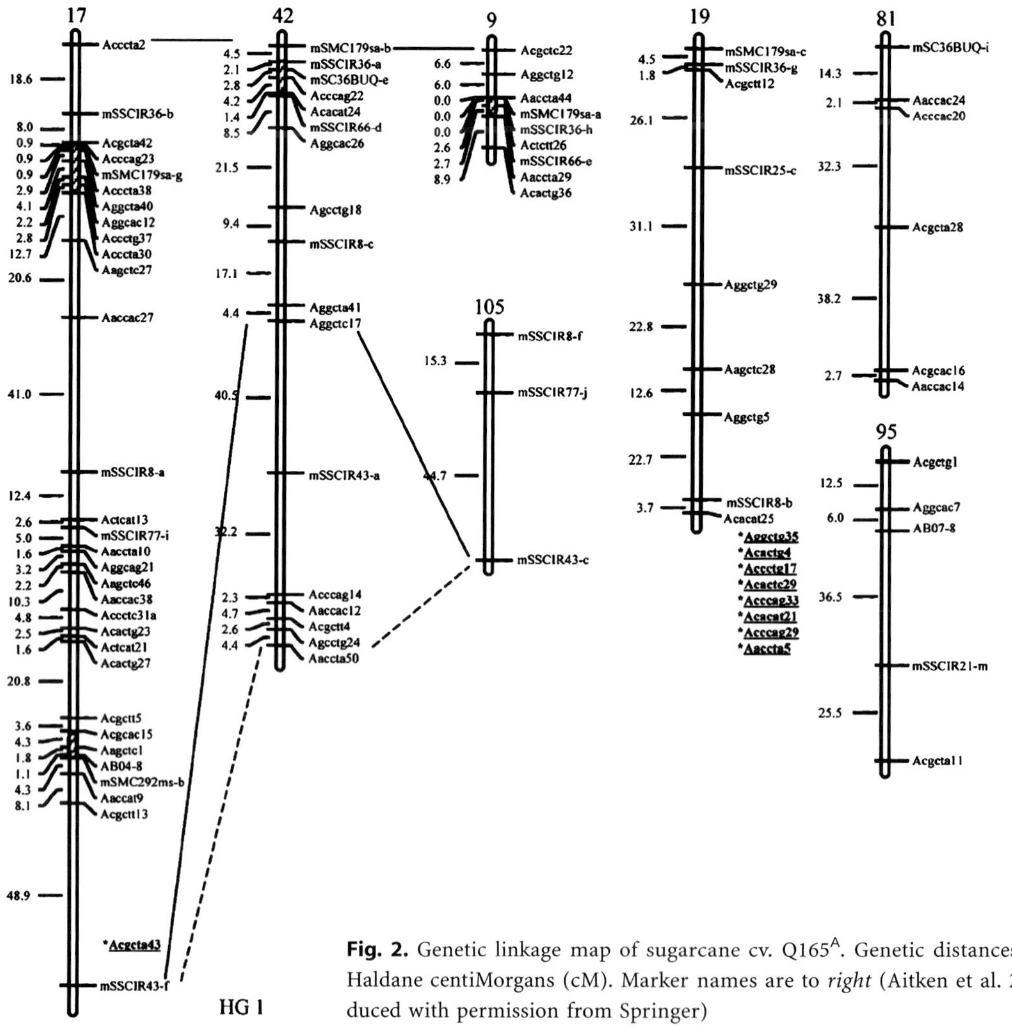

Fig. 2. Genetic linkage map of sugarcane cv. Q165[A]. Genetic distances are to *left* and in Haldane centiMorgans (cM). Marker names are to *right* (Aitken et al. 2005) (Maps reproduced with permission from Springer)

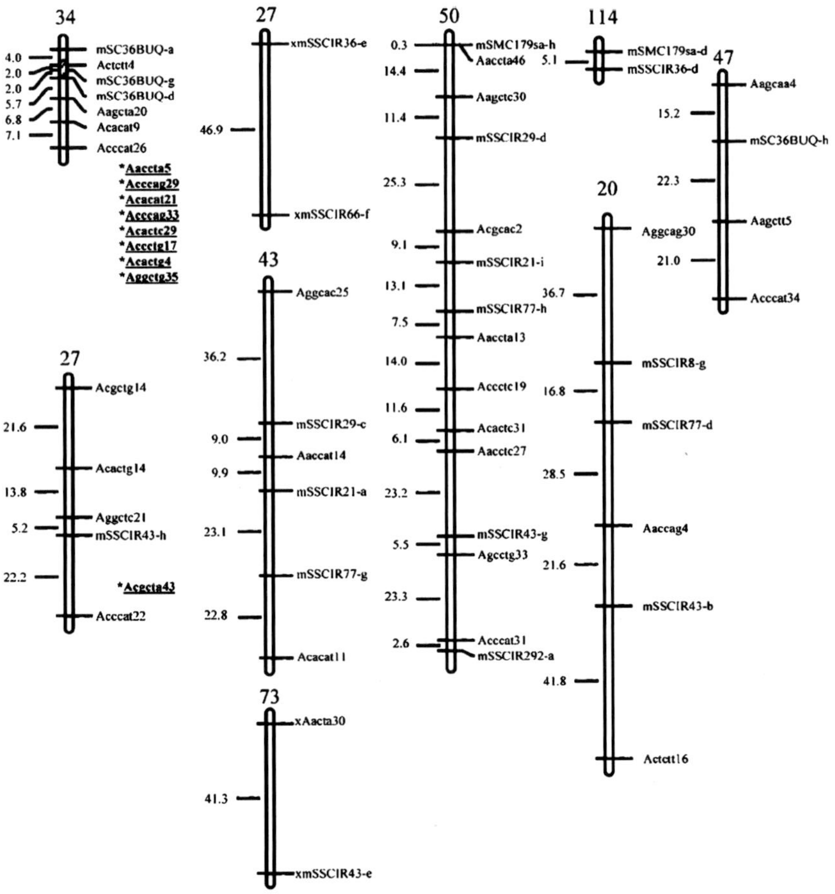

Fig. 2 (continued)

genomes of maize and sorghum, both of which show synteny with sugarcane. Initial efforts in comparative mapping were carried out only with maize due to its more advanced map at that stage (Grivet et al. 1994; Dufour et al. 1996). However, these first comparisons of LGs between sorghum and maize revealed the presence of large chromosome rearrangements, i.e. translocations, centric fusions or fissions (Grivet et al. 1994). Later efforts showed sorghum and sugarcane as having a high degree of colinearity between their chromosomes and they appeared more closely related than either is to maize (Fig. 3) (Ming et al. 2002b). The chromosome count of 2n = 10 to 40 in sorghum makes it a relatively simple genome as compared to that of sugarcane (2n = 107 to 115 in cultivar R570), making sorghum the ideal candidate for comparative mapping (Grivet et al. 1994; Glaszmann et al. 1997; Guimaraes et al. 1997; Ming et al. 1998). Comparative mapping should lead to a saturation of the sugarcane map. Linked loci can be predicted based on their location on the complex sorghum map, emphasizing the usefulness of using synteny to diploids as a means of accelerating the mapping efforts of the complex sugarcane genome.

11.3 Detection of QTLs

Quantitative trait loci (QTLs), or groups of genes or alleles controlling complex traits, can be used for locating genomic regions associated with quantitative traits. The prerequisites for QTL detection include having a mapping population that displays the genetic variation for the trait of interest, sufficient phenotypic data on the population and being able to establish genetic LGs. The identification of QTLs requires knowledge of the genome of the species under investigation, and this can be acquired by utilizing existing genetic maps, comparative mapping and association mapping.

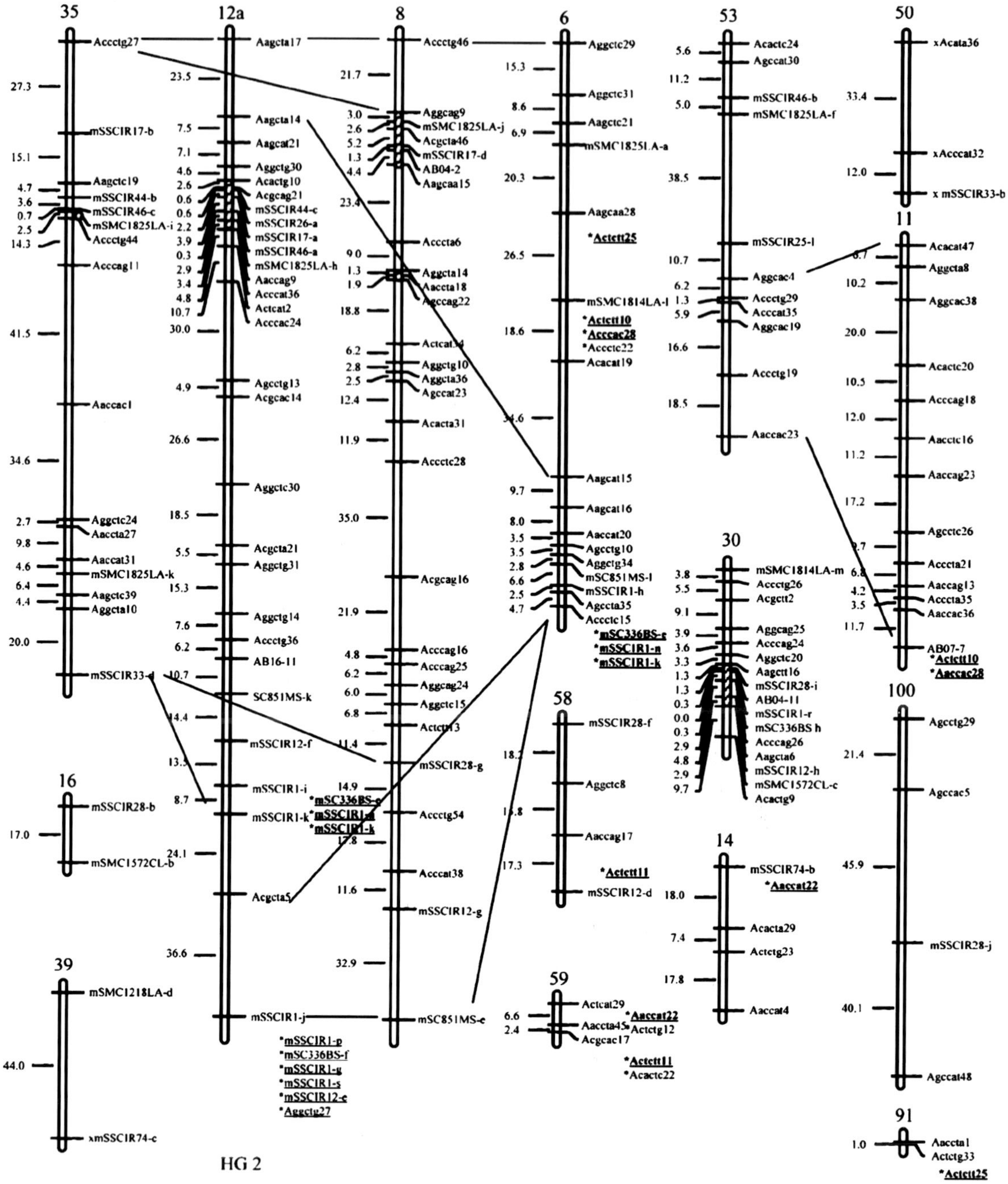

Fig. 2 (continued)

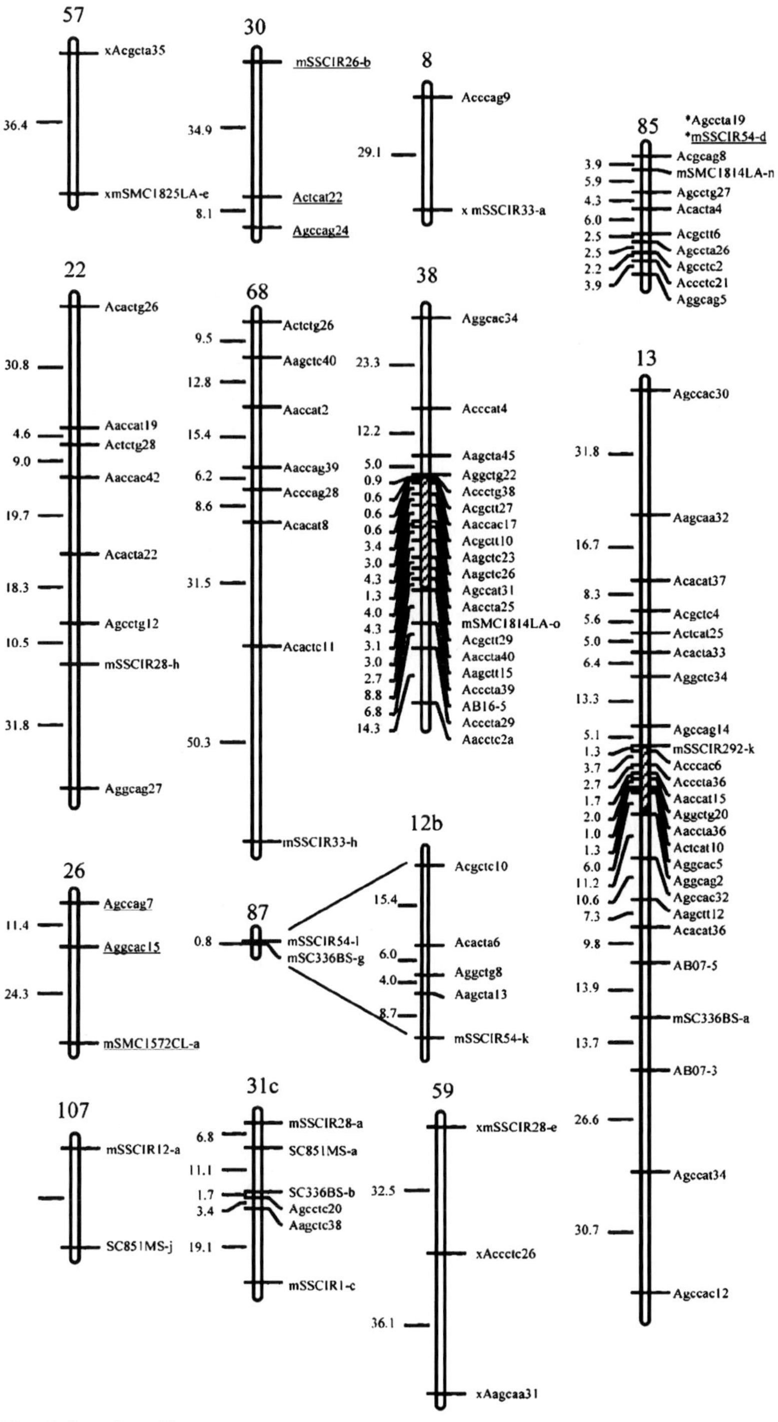

Fig. 2 (continued)

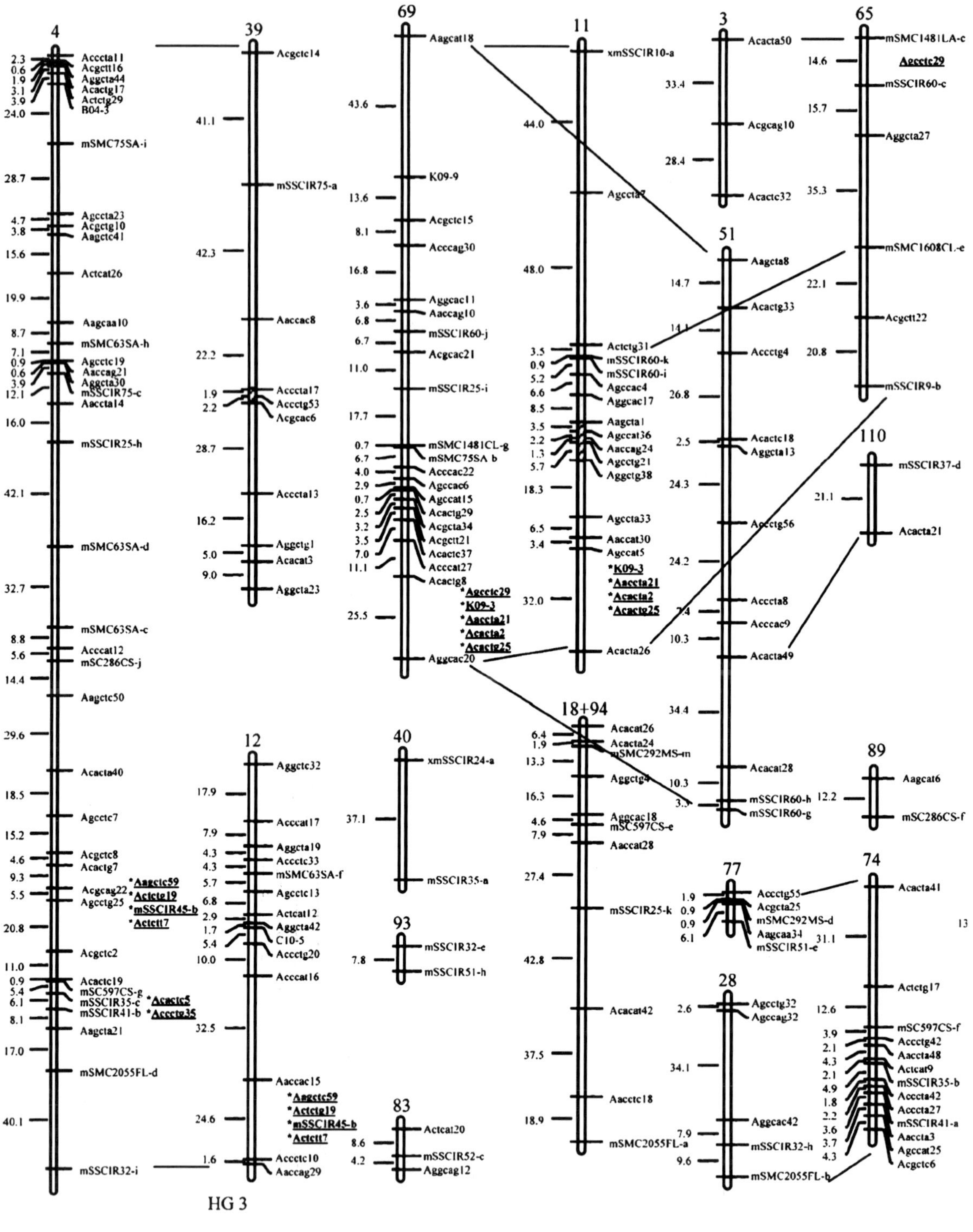

Fig. 2 (continued)

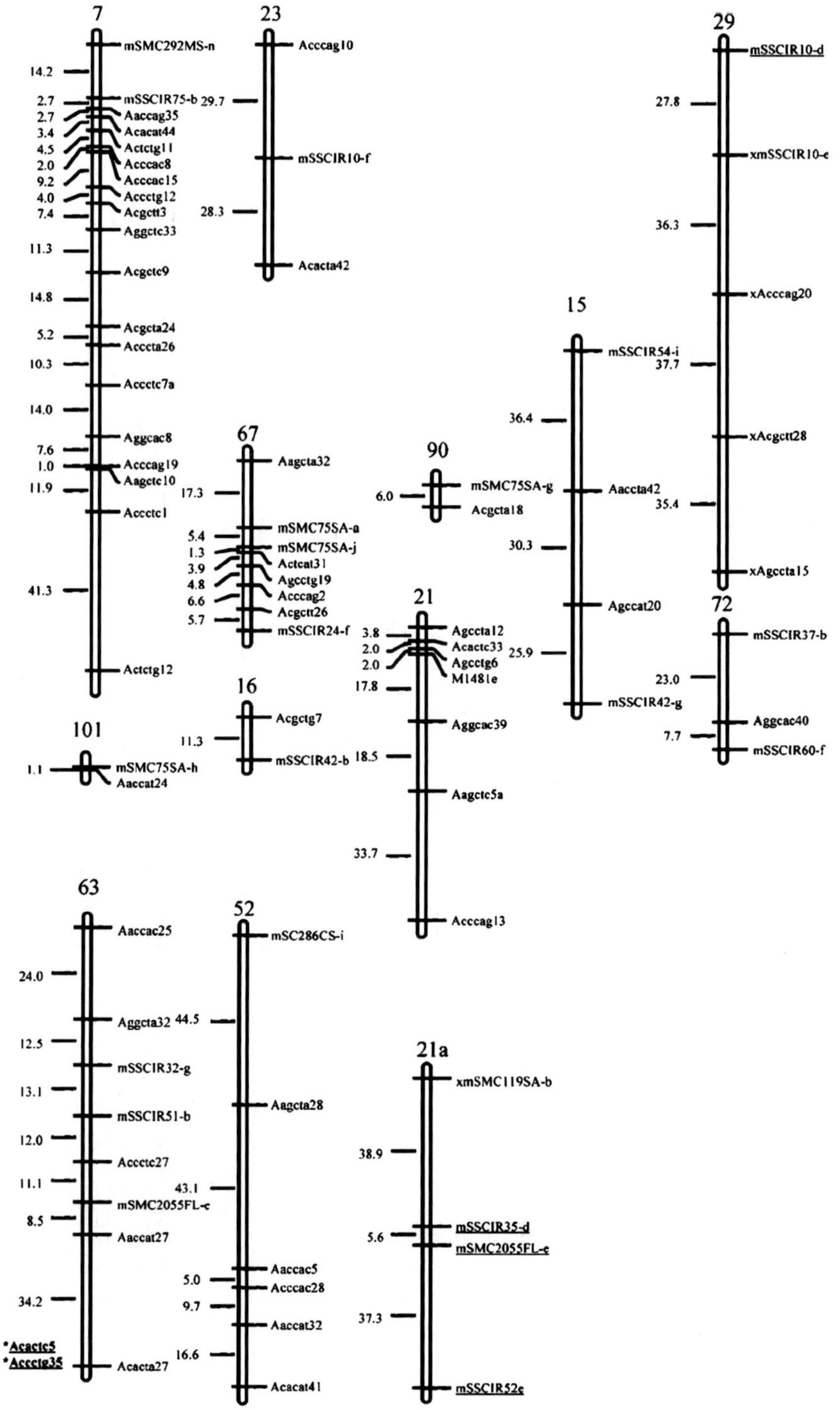

Fig. 2 (continued)

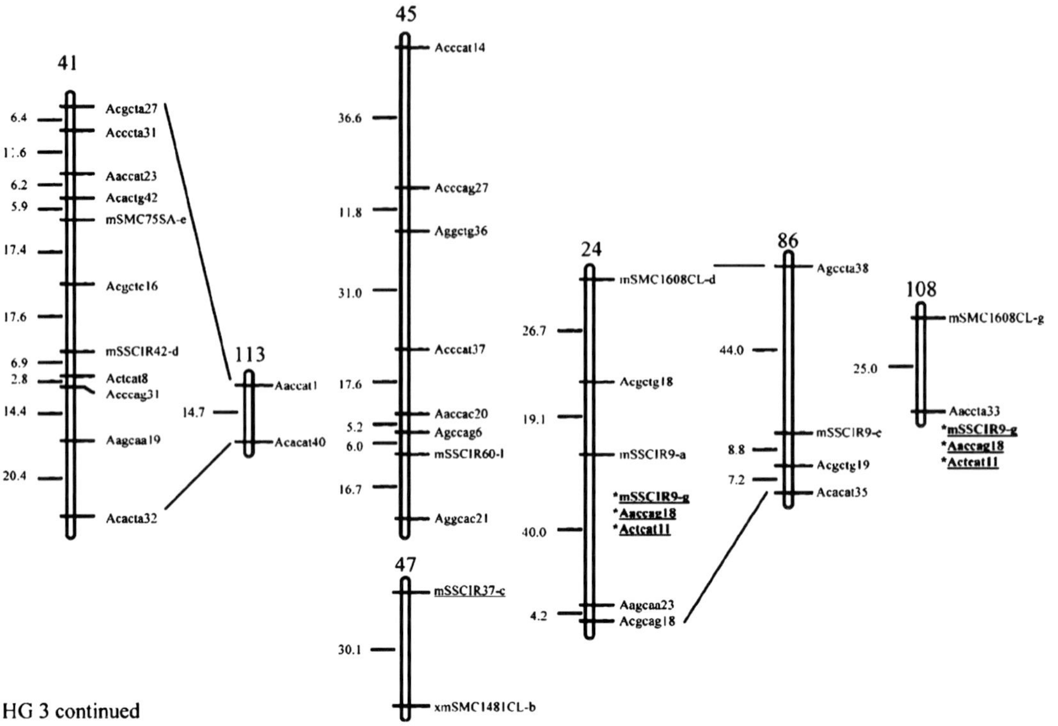

HG 3 continued

Fig. 2 (continued)

There are inherent difficulties in the detection of QTLs in the complex polyploid genome of sugarcane. The majority of QTL analyses and the statistical methods used to identify QTLs have been based on diploid species that have homologous pairs of chromosomes arranged in sets and, therefore, have predictable meiotic outcomes, single-copy molecular markers and progeny that segregate in expected patterns. In contrast, the high ploidy nature of sugarcane has more homologous chromosomes per set, increasing the number of possible genotypes per marker and/or QTL. Added to this, the number of each marker and/or QTL in the parents and progeny is not obvious and may not be observable. Added to this, the additional marker copies can mask recombination information, and the outcome of meiosis is usually unknown (Doerge and Craig 2000).

Despite these difficulties, comparative mapping has been particularly useful for QTL analysis in sugarcane. Synteny with diploid relatives, in particular sorghum, has been beneficial in constructing genetic maps and for identifying QTLs for homologous agronomic traits in sugarcane (Paterson et al. 1995; Dufour et al. 1996, 1997; Grivet et al. 1996b; Guimaraes et al. 1997; Ming et al. 1998; 2002a; Bowers et al. 2003; Jordan et al. 2004)

While considerable effort has been put into creating various sugarcane genetic maps, there are relatively few published QTL mapping studies in sugarcane. This, no doubt, is due in part to the difficulties associated with QTL analysis in polyploid species. The earliest QTL studies in *Saccharum* species used small mapping populations with limited phenotypic data. D'Hont et al. (1999) recognized the 'bottleneck' for identifying QTLs in sugarcane as the limited availability of robust field data with insufficient replications. To achieve an adequately dense map for QTL analysis, particularly in polyploid species, a mapping population of 200 to 300 individuals and a reasonably high number of markers are required (Hoarau et al. 2002; McMullen 2003).

In the first published QTL analysis in sugarcane, RAPD analysis was used to evaluate a small mapping population consisting of 44 F_1 progeny of a cross between *S. officinarum* and *S. robustum* (Sills et al. 1995). The study included QTLs for stalk number, tasseled stalks and stalks with smut (*Ustilago scitaminea*). In addition, a RAPD marker linked to eyespot resistance in a population of 84 F_1 progeny derived from a *S. officinarum*×*S. spontaneum* cross (Mudge et al. 1996) has been identified, as well as an RFLP marker associated with

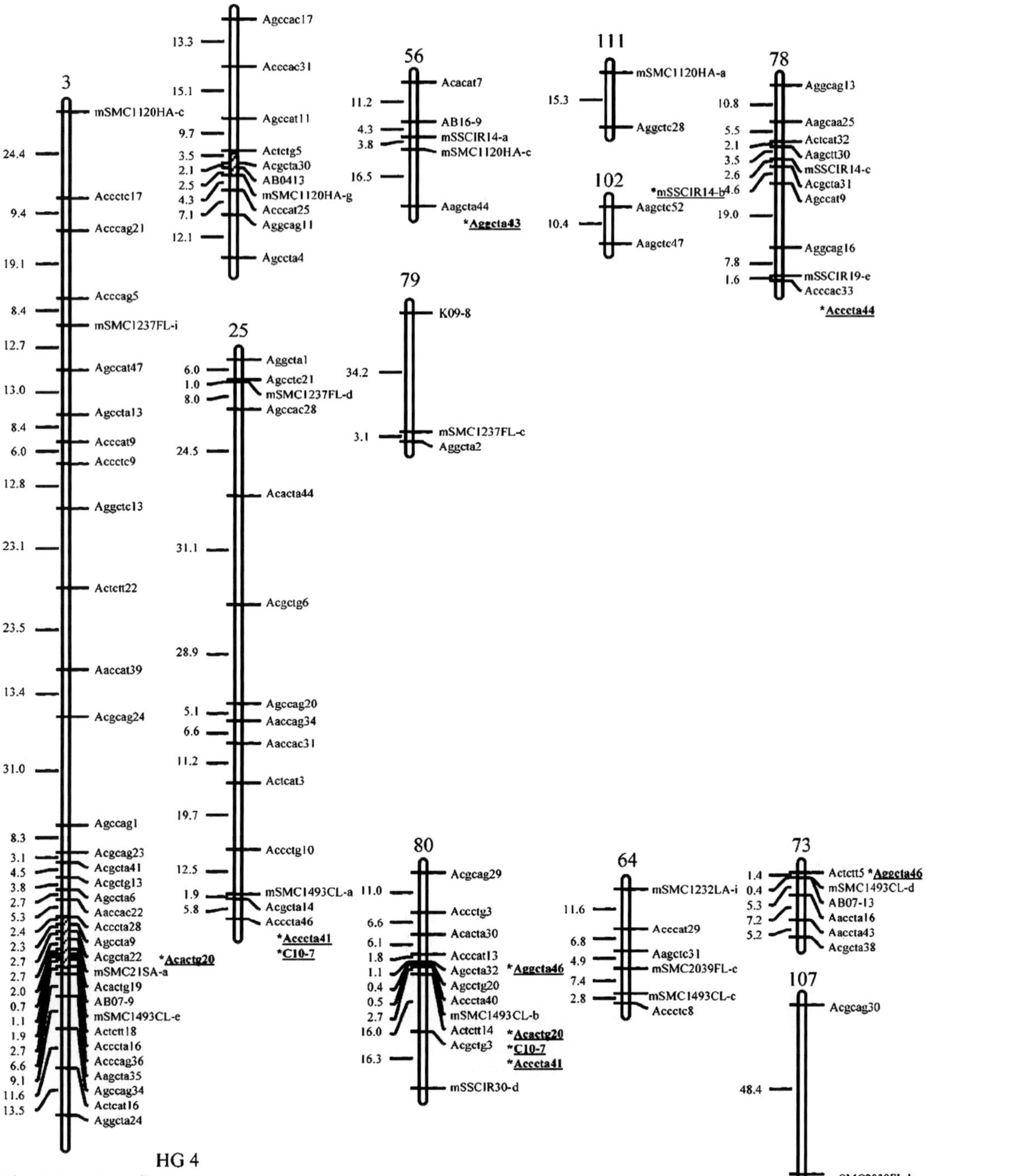

Fig. 2 (continued)

short-day flowering in a mapping population of 100 individuals (Guimaraes et al. 1997).

11.3.1 QTLs Associated with Resistance

Daugrois (Daugrois et al. 1996) used RFLP markers to investigate the inheritance of rust resistance in a population of 141 progenies from the rust resistant cultivar R570. Rust resistance was evaluated with field experiments and artificial inoculation in greenhouse tests. The overall data revealed a 3:1 segregation (resistant/susceptible), suggesting the presence of a major rust resistance gene in R570.

In an attempt to saturate the region surrounding the major rust gene, Asnaghi (Asnaghi et al. 2000)

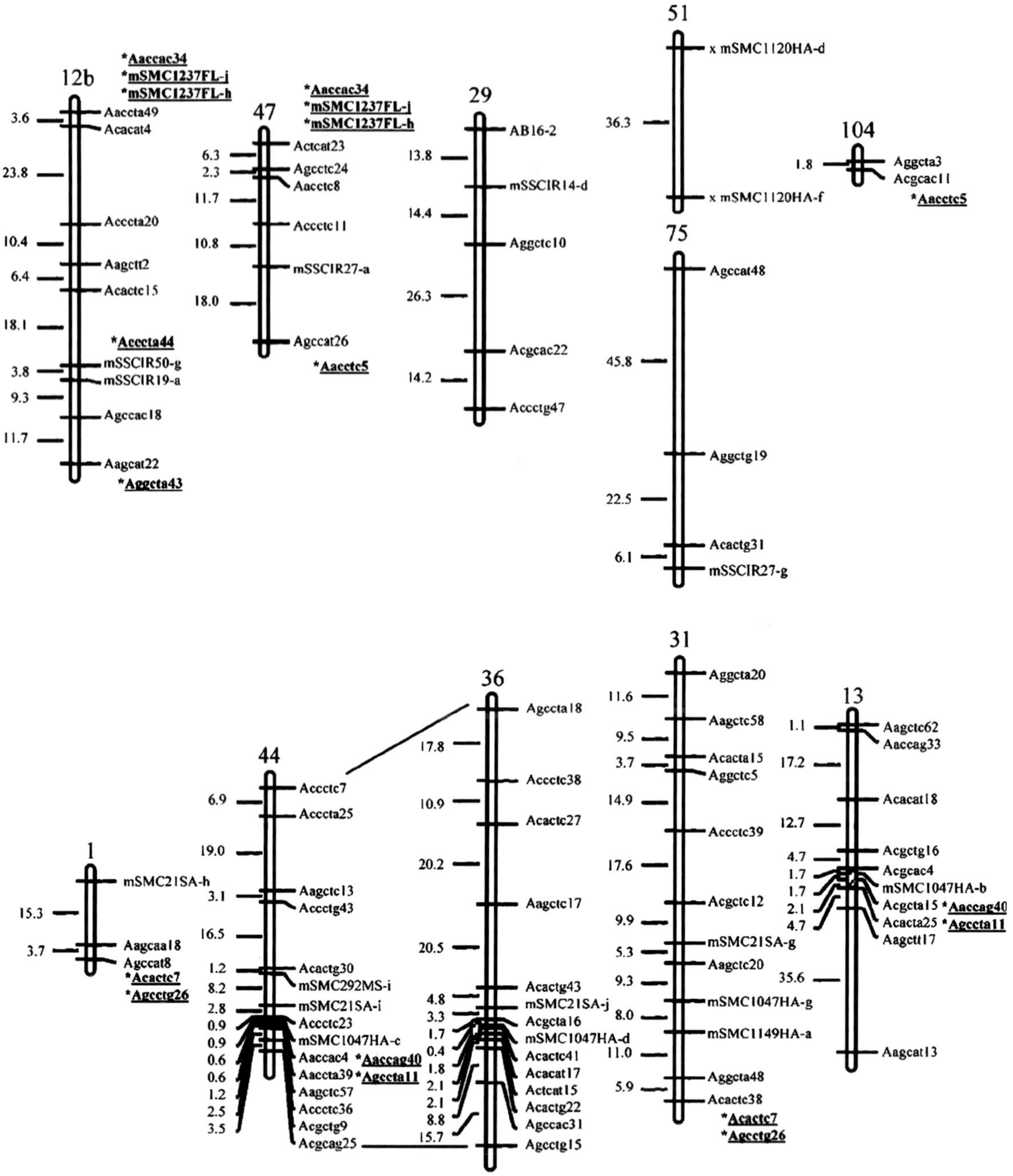

Fig. 2 (continued)

utilised comparative mapping between sugarcane, sorghum, maize and rice. Using RFLPs on a mapping population of 88 selfed progeny of R570, another 217 SD markers were added to the R570 genetic map of which 66% mapped to the rust-resistant gene.

In 2003, a large number of sugarcane expressed sequence tags (ESTs) were released into the public domain by the Brazil-based SUCEST project. Rossi et al. (2003) exploited the SUCEST EST database to identify candidate genes for pathogen resistance and map their distribution on the sugarcane genome. A total of 261,609 EST sequences were analysed, resulting in the identification of 88 resistance gene analogs (RGAs). Of these, 50 RGAs were located on the sugarcane reference genetic map constructed on the cultivar R570 (Grivet et al. 1996b; Hoarau et al. 2001). Using SD markers, 148 polymorphic RFLP markers and 55 SSR loci were

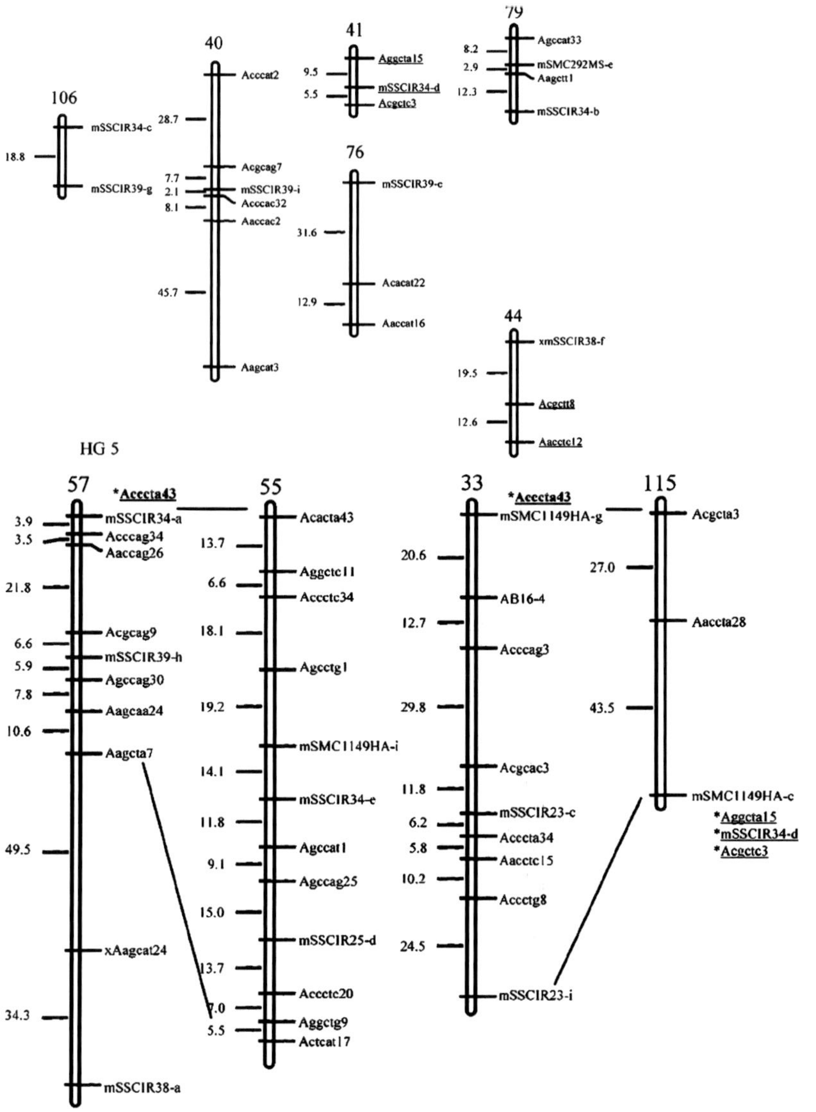

Fig. 2 (continued)

mapped. The mapping population used for this study consisted of 112 selfed progeny from R570 and was a subset of the population used by Hoarau (Hoarau et al. 2001) to build an AFLP genetic map. Prior to this study, the only pathogen resistance locus mapped in sugarcane was the common rust gene (Daugrois et al. 1996; Asnaghi et al. 2000).

In the first well-characterized Mendelian trait described in sugarcane, Asnaghi et al. (2004) built upon the work of Daugrois et al. (1996), who first demonstrated the rust gene in R570, and Grivet et al. (1996 a), who developed the R570 RFLP genetic map. In this extensive study, Asnaghi et al. (2004) used bulked segregated analysis (BSA) and AFLPs to map the rust resistance gene (*Bru*1). A large population of 658 individuals, derived from selfing of clone R570, were analysed for segregation of rust resistance. Phenotypic data were collected from two geographically distinct trials over 2 years. In addition, to confirm the Mendelian inheritance of the R570 resistance gene, over 2000 clones were studied. These clones were derived from four biparental crosses using R570 as the male parent and four susceptible cultivars as the female parents.

11.3.2 QTLs Associated with Sugar Yield and Content

Initiating a systematic candidate gene approach, Ming et al. (2001) used linkage analysis in two dif-

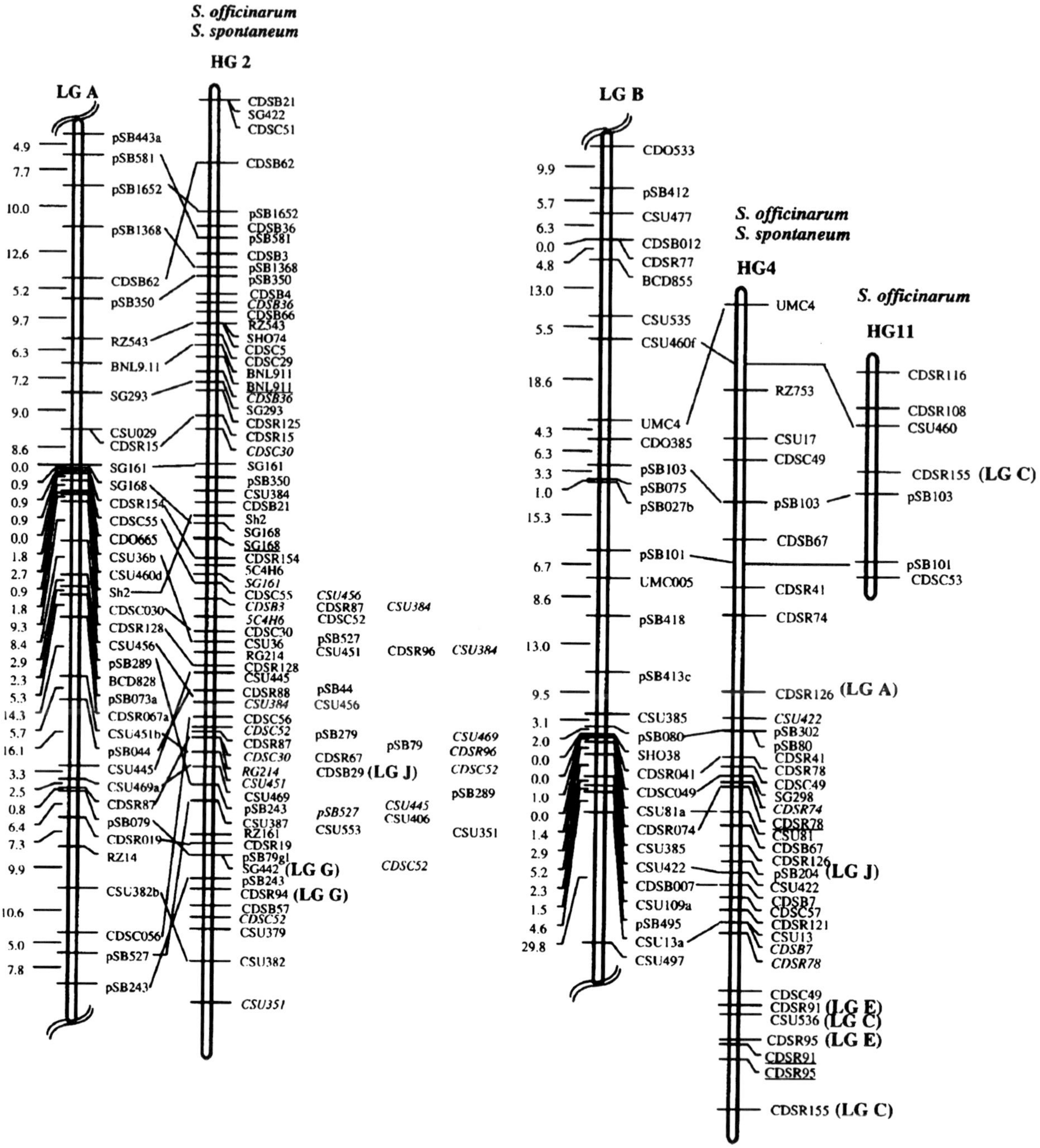

Fig. 3. Saccharum consensus linkage map and corresponding sorghum LGs. Loci connected by a *line* are detected by the same probe in both genomes. This correspondence of 86% of 982 loci/alleles on the consensus map to sorghum illustrates how a high-density linkage map of sorghum can be used to facilitate mapping of the complex sugarcane genome (Ming et al. 2002b). Maps reproduced with permission from Crop Science

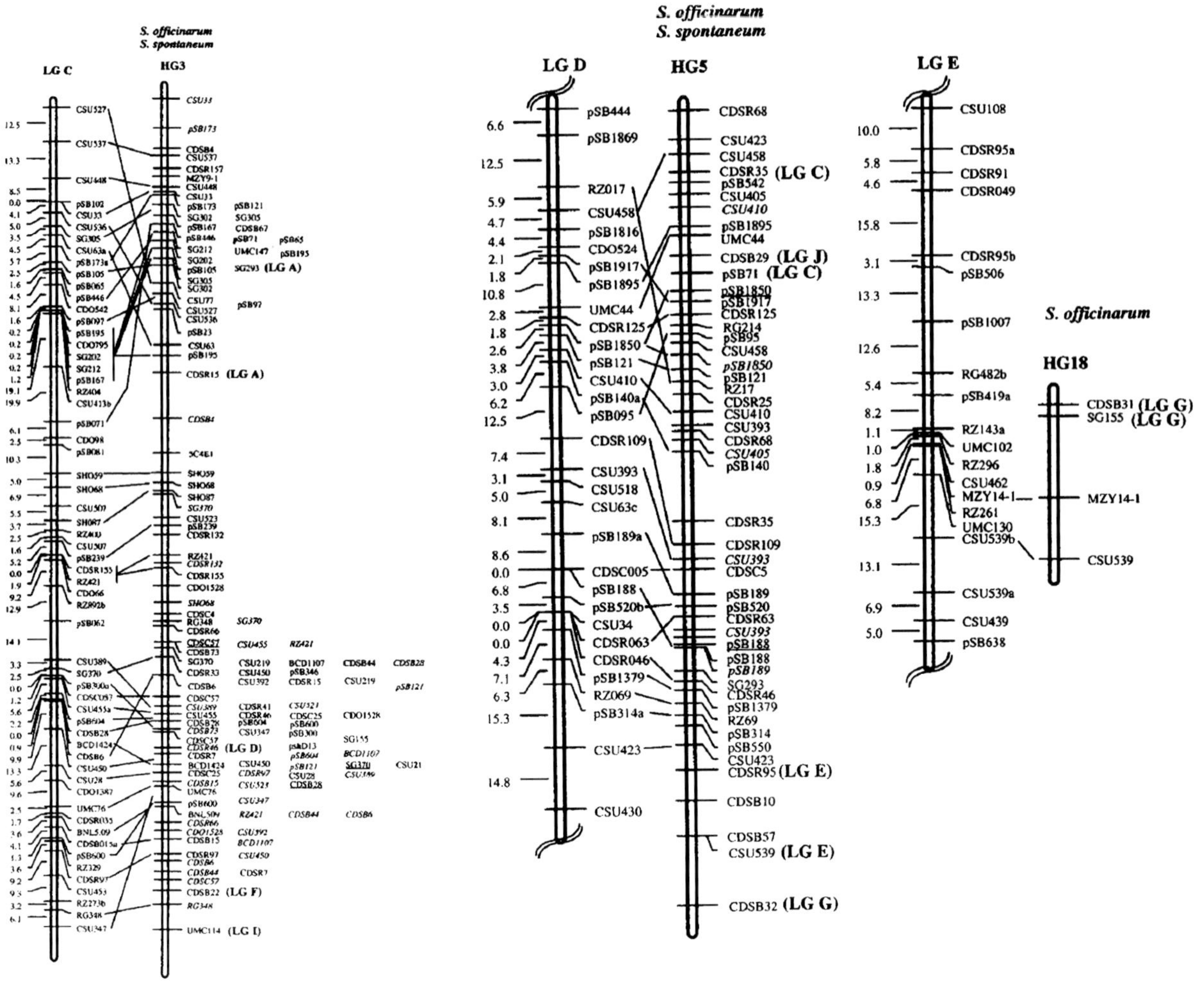

Fig. 3 (continued)

were identified and were consistent over both years. This study is still in progress and will involve further field trial data for traits and the repeatability of trait-marker associations in populations derived from the progeny used in this cross.

Jordan et al. (2004) conducted a preliminary mapping study scoring 258 RFLP and RAF (radio-labelled amplified fragments) markers over a population of 108 F_1 progeny from a cross between two elite Australian sugarcane clones (Q117×74C42). The population was evaluated at two sites over 2 years for stalk number and suckering, both important factors influencing sugar yield. Sixteen (7 RFLP and 9 RAF) markers were significantly associated with stalk number and 14 (6 RFLP and 8 RAF) markers with suckering across both sites and years. The 13 RFLP markers were generated by 8 different RFLP probes, 7 of which had previously been mapped in sorghum (Paterson et al. 1995) and/or sugarcane. The researchers observed that these seven probes were located within or near QTLs for tillering and rhizomatousness in sorghum (Paterson et al. 1995). Thus their study highlighted the benefits of comparative mapping between sorghum and sugarcane and suggests useful markers for sugarcane can be identified from sorghum QTLs associated with related traits.

The opportunity to detect QTLs for traits of interest in sugarcane will be greatly enhanced with the construction of maps with an extensive genome coverage and a large number of markers. Aitkin et al. (2005) constructed a map using 40 AFLP, 5 RAF and 72 SSR markers generating 1365 polymorphic markers of which 967 (71%) were single dose. Aitkin et al.'s map is probably the largest sugarcane map to date, with the second largest constructed rust resistance target area was saturated. Markers were placed at between 0.14 and 0.3 cM on either side of the gene, and a physical map built with sugarcane BACs. An initial attempt at cloning the gene identified three different BACs with markers located at one or two recombinations respectively on each side of the gene. Unfortunately, none of these three BACs corresponded to the chromosome segment bearing the rust resistance gene, which may have been a result of the low coverage of the BAC library. A second repeat attempt at recreating the BAC library identified a clone containing the rust resistance gene (D'Hont 2004).

New PCR based marker systems have been and continue to be developed. In 1997, a consortium was developed by the International Consortium for Sugarcane Biotechnology to look into and develop simple sequence repeats (SSRs), or microsatellite sequences, into a marker system for sugarcane. These markers were developed from an enriched microsatellite library and shown to have the capacity to distinguish between sugarcane genotypes due to their ability to detect large numbers of alleles (Cordeiro et al. 2000). Microsatellite markers have been used in genome mapping, fingerprinting and population and evolutionary studies in many plant species (Dje et al. 2000; Hokanson and Lamboy 2001). Whilst the use of various marker types in mapping and in understanding relationships between members of the sugarcane tribe have now been occurring for over 10 years, the use of markers in sugarcane breeding is still a relatively recent event. Anecdotal evidence is available on the use of microsatellite markers in confirming mislabelled genotypes or identifying progeny derived from contaminating pollen. Microsatellite markers linked to agronomic traits of interest are also known to have been identified. A number of reports are now available on the use of markers in breeding, and these have largely been for the verification of hybrid progeny.

The Hainan Sugarcane Breeding Station (HSBS) of the Guangzhou Sugarcane Industry Research Institute (GSIRI) in China have applied these markers to the confirmation of fertile intergeneric F_1 hybrids of S. *officinarum* and *E. arundinaceus* as well as backcross (BC_1) progeny from the F_1 to hybrid sugarcane (*Saccharum* sp.) (Cai et al. 2005). RAPD and SSR markers were used by the USDA-ARS Sugarcane Research Unit to validate introgression of genes into F_1 hybrids of crosses made between *S. spontaneum* and elite commercial clones (Pan et al. 2004); and a combination of 5S rDNA and microsatellite markers has been used to verify first-generation backcross progeny from an intergeneric F_1 parent (YC96-66 derived from *S. officinarum* clone Badilla×*Erianthus arundinaceus* clone HN92-105) and a commercial clone produced by the USDA Canal Point facility (CP84-1198) (Cai et al. 2005). A number of reports using 5S rDNA as a marker to verify successful introgression in F_1 hybrids between *E. arundinaceus* and *S. officinarum* (or hybrids) are also available. However, these hybrids have largely been sterile (D'Hont et al. 1995; Besse et al. 1997; Alix et al. 1999; Piperidis et al. 2000).

11.4.1 Advanced Works

SSR markers have been widely adopted for use in sugarcane genetics and breeding. However, high-throughput genotyping technologies based on single nucleotide polymorphisms (SNP) or small-scale insertion/deletions (indel) could become efficient alternative tools for traditional markers because of their greater abundance in the genome and ease of measurement. A further major advantage of SNP markers is that they allow easy and unambiguous identification of alleles or haplotypes. Adoption of SNP markers will require large efforts in SNP discovery. An efficient strategy, which has been extensively tested in humans, relies on the comparison of the sequences of ESTs derived from different genotypes or different chromosomes in a polyploid genome like that in sugarcane (Grivet et al. 2002). The SUCEST project in Brazil has resulted in the public release of 300,000 sugarcane EST sequences. SUCEST was a large-scale EST program sequencing clones from cDNA libraries prepared from calli, root, stalk, etiolated leaves, flowers, seed and pathogen-induced plants. The aim was to identify around 50,000 sugarcane genes. The information provided by SUCEST can be exploited by the research community, as a source of markers for agriculturally significant characteristics, to understand the molecular basis of plant growth and development in plant physiology, biochemistry, cell biology, pathology and, ultimately, in plant breeding. ESTs are being used today as a major source of DNA sequence data for identifying SNPs that could potentially be used as genetic markers for mapping and breeding purposes. An important issue to consider in this research is the possible confusion between orthologous alleles and paralogous genes (Grivet et al. 2002). The complete genome sequence of sugarcane remains a distant goal, and thus the large collection of sugarcane ESTs will remain a highly valuable resource for a long time to come. This collection will be of primary interest for gene discovery and functional analysis in sugarcane and possibly in related grasses, such as maize and sorghum (Grivet and Arruda 2001).

11.5 Future Scope of Works

Improving the genetic performance of the sugarcane plant, for increased sugar production in diverse environments and for the generation of new products, is the ultimate target of sugarcane genetic research. Achieving this target requires understanding the biology molecular and cellular level of processes that underpin sucrose accumulation and candidate genes that may provide the basis for improved selection and development of high sugar genotypes in breeding programs.

The complexities of the sugarcane genome have left its genetic peculiarities a mystery to science for a long time, and sugarcane research has mainly been carried out in an exploratory manner in order to unravel its genetic intricacy. Current research is focused on achieving more functional goals – to develop genomics-based knowledge and experimental platforms for the discovery of new gene technologies that can be applied in sugarcane improvement.

To circumnavigate the issue of genetic bottlenecks, genetic manipulation methods were developed in the early 1990s for the production of transgenic sugarcane (Birch and Franks 1991; Birch and Maretzki 1992; Franks and Birch 1991; Bower and Birch 1992). The technology is capable of delivering its intended goals and has been adopted by various sugarcane biotechnology groups. However, in countries such as Australia, where the negative attitude of the community in regard to sugar from genetically modified cane is high, the industry has formed the view that this source of sugar cannot be marketed successfully (Grice et al. 2003). The industry has, however, seen the potential for genetic engineering to diversify and develop cane plants with the ability to store novel products with wider markets than sugar (Grice et al. 2003).

The development of agronomic traits such as reduction in use of nitrogen and pesticide or shoot architecture in sugarcane farming systems is also being investigated. Application of molecular tools for the analysis of the sugarcane genome makes use of DNA marker technology to clarify phylogenetic relationships, document varieties, assess genetic diversity in the germplasm, detect major genes and resolve complex genetic traits.

One of the main research goals of this 'knowledge gathering' and a very important tool in gaining a whole-genome perspective is the generation of genetic maps that indicate the location and function of genes and the relationship between genes and provide the markers for their identification. A result of the high polyploidy nature of sugarcane, development of genetic marker technology, is possibly more difficult in sugarcane than it would be in any other plant. However, the object remains highly desirable. SNPs is a very promising genetic marker technology for sugarcane currently under development in Australia. A major source for SNP detection is the recently established EST database. ESTs have proven to be a valuable tool in the discovery of SNPs in human genes, but their use for this purpose is still limited in higher plants (Grivet et al. 2002). There are two reasons why ESTs might provide a large number of SNPs for genotyping. First, the sequencing of ESTs provides sequence data for expressed genes. It is reasonable to expect that at least some of these ESTs will be responsible for the observed agronomic traits. Therefore, unlike anonymous markers such as SSRs, SNPs derived from ESTs could underline the traits being examined. Secondly, the independent sequencing of ESTs derived from specific and different genotypes will allow the in silico search for SNPs to take place. However, while this latter approach is likely to be the method of choice for the discovery of plant SNPs, it is not without problems. For instance, in sugarcane the occurrence of multigene families and the presence of large introns may interfere with the sequence analysis. Furthermore it is clear that not all genotypes will be the subject of large-scale EST sequencing, and therefore further targeted sequencing will be necessary to cover all useful genotypes (Edwards and Mogg 2001).

Nevertheless, SNPs are well on their way to becoming the dominant marker system in commercial plant breeding. Given their significant practical advantages and the ability that they confer to examine polymorphisms within ESTs associated with key underlying traits, there is no doubt that SNPs will become the method of choice for DNA fingerprinting

(Edwards and Mogg 2001). One of the most direct implications of SNPs is when an SNP in a coding region (cSNP) has functional significance resulting in an amino acid change in the encoded protein and causing an altered phenotype. Among the different applications of information obtained with such markers is the construction of genetic linkage maps and saturation of genetic maps in polyploids. Once a dense map has been constructed, it allows linkage detection between markers and genes. Such linkages are useful for marker-assisted selection in breeding programs (Burr et al. 1983). Furthermore, high-density maps can provide starting points for map-based gene cloning (Wicking and Williamson 1991). Dense genetic maps (up to one marker per kilobase) are possible with SNPs allowing genome scans for linkage disequilibrium to screen directly for coding polymorphisms of genes derived from complex populations to directly identify candidates for agronomic traits under investigation (Brookes 1999).

A considerable amount of work is dedicated to the improvement of agronomic practices for environmental and biological applications. Applications of biotechnology for environmental sustainability address the control of weeds, pests and nutrient utilisation efficiency in sugarcane by a list of research directions such as reduction of nitrogen input, development of insect-resistant plants and enzymatic degradation of pesticide residues.

Sugarcane breeders use quantitative inheritance knowledge in breeding programs to breed superior varieties. Applications of DNA-based markers and QTL mapping technology in sugarcane will enable breeders to recombine loci, which account for major variation in key traits in a more targeted way by providing knowledge of genetic loci. Ignoring the 'popularity' of genetically modified crops, molecular mapping and gene identification techniques are definitely to be the basis of sugarcane breeding programs in the future.

The competitive sugar world market is pushing sugarcane research to identify new markets for other sugarcane products. Sugarcane as a biofactory is the new term that brings together the technologies for production of high-value materials such as functional foods and nutraceuticals , biopolymers and enzymes in sugarcane.

References

Aitken K, Jackson P, Piperidis G, McIntyre L (2004) QTL identified for yield components in a cross between a sugarcane (*Saccharum* spp.) cultivar Q165^{A} and a *S. officinarum* clone IJ76-514. In: Proceedings for the 4th International Crop Science Congress, Brisbane, Australia, 26 Sept – 1 Oct 2004. www.cropscience.org.au

Aitken KS, Jackson PA, McIntyre CL (2005) A combination of AFLP and SSR markers provides extensive map coverage and identification of homo(eo)logous linkage groups in a sugarcane cultivar. Theor Appl Genet 110:789–801

Alix K, Paulet F, Glaszmann JC, D'Hont A (1999) InterAlu like species-specific sequences in the *Saccharum* complex. Theor Appl Genet 6:854–864

Al-Janabi SM, Honeycutt RJ, McClelland M, Sobral BWS (1993) A genetic linkage map of *Sacchrum spontaneum* (L.) 'SES 208'. Genetics 134:1249–1260

Al-Janabi SM, McClelland M, Petersen C, Sobral BWS (1994) Phylogenetic analysis of organellar DNA sequences in the Andropogoneae: Saccharinae. Theor Appl Genet 88:933–944

Asnaghi C, Paulet F, Kaye C, Grivet L, Deu M, Glaszmann J-C, D'Hont A, Rao P, Feldmann P, Grivet L, Islam-Faridi N, Berding N (2000) Application of synténie across Poaceae to determine the map location of a rust resistance gene of sugarcane. Theor Appl Genet 101:962–969

Asnaghi C, Roques D, Ruffel S, Kaye C, Hoarau JY, Telismart H, Girard JC, Raboin LM, Risterucci AM, Grivet L, D'Hont A (2004) Targeted mapping of a sugarcane rust resistance gene (*Bru*1) using bulked segregant analysis and AFLP markers. Theor Appl Genet 108:759–764

Beckmann JS, Soller M (1989) Genomic genetics in plant breeding. In: P. Congress. EUCARPIA, Berlin, Germany, pp 91–106

Bennett MD (1995) The development and use of genomic in situ hybridisation (GISH) as a new tool in plant biosystematics. In: Brandham PE, Bennett D (eds) Kew Chromosome Conference IV, Royal Botanic Gardens, Kew, pp 167–183

Besse P, McIntyre CL, Burner DM, Dealmeida CG (1997) Using genomic slot blot hybridization to assess intergeneric *Saccharum*×*Erianthus* hybrids (Andropogoneae-Saccharinae). Genome 40:428–432

Birch RG, Franks T (1991) Development and optimisation of microprojectile systems for plant genetic transformation. Aust J Plant Physiol 18:453–469

Birch RG, Maretzki A (1992) Transformation of sugarcane. In: Bajaj YPS (ed) Biotechnology in Agriculture and Forestry. Springer, Berlin Heidelberg New York, pp 348–360

Bonierbale MW, Plaisted RL, Tanksely SD (1988) RFLP maps based on common set of clones reveal modes of chromosomal evolution in potato and tomato. Genetics 120:1095–1103

Botstein D, White RL, Skolnick M, Davis RW (1980) Construction of a genetic linkage map in man using restriction fragment length polymorphisms. Am J Hum Genet 32:314–331

Bower R, Birch RG (1992) Transgenic sugarcane plants via microprojectile bombardment. Plant J 2:409–416

Bowers JE, Abbey C, Anderson S, Chang C, Draye X, Hoppe AH, Jessup R, Lemke C, Lennington J, Li ZK, Lin YR, Liu SC, Luo LJ, Marler BS, Ming RG, Mitchell SE, Qiang D, Reischmann K, Schulze SR, Skinner DN, Wang YW, Kresovich S, Schertz KF, Paterson AH (2003) A high-density genetic recombination map of sequence-tagged sites for *Sorghum*, as a framework for comparative structural and evolutionary genomics of tropical grains and grasses. Genetics 165:367–386

Brandes E (1958) Origin, classification and characteristics. USDA Agriculture Handbook

Bremer G (1924) The cytology of sugarcane. A cytological investigation of some cultivated kinds and their parents. Genetica 6:497–525

Bremer G (1961) Problems in breeding and cytology of sugarcane. Euphytica 10:59–78

Britannica Concise Encyclopedia (2005) Sugarcane. Encyclopedia Britannica

Brookes AJ (1999) The essence of SNPs. Gene 234:177–186

Burr BJ, Evola SV, Burr FA, Beckmann JS (1983) The application of restriction length polymorphism to plant breeding. In: Setlow JK, Hollaender A (eds) Genetic Engineering vol 5. Plenum, New York, pp 45–59

Buzacott JH (1965) Cane varieties and breeding. In: King NJ, Mungomery RW, Hughes CG (eds) Manual of Cane Growing. Angus & Robertson, Sydney, Australia, pp 220–253

Cai Q, Aitken K, Deng HH, Chen XW, Cheng F, Jackson PA, McIntyre CL (2005) Verification of the introgression of *Erianthus Arundinaceus* germplasm into sugarcane. Plant Breed (in press)

Coleman RE (1968) Physiology of flowering in sugarcane. In: Proceedings of the XIII Congress of the International Society of Sugar Cane Technologists XIII, pp 992–1000

Cordeiro GM, Taylor GO, Henry RJ (2000) Characterisation of microsatellite markers from sugarcane (*Saccharum* sp.), a highly polyploid species. Plant Sci 155:161–168

Daniels J, Daniels C (1975) Geographical, historical and cultural aspects of the origin of the Indian and Chinese Sugarcanes *S. barberi* and *S. sinense.* Sugarcane Breed Newslett 36:4–23

Daniels J, Roach BT (1987) Taxonomy and evolution in sugarcane. In: Heinz DJ (ed) Sugarcane Improvement Through Breeding. Elsevier, Amsterdam, pp 7–84

Daniels J, Smith P, Paton N, Williams CA (1975) The origin of the genus *Saccharum.* Sugarcane Breed Newslett 36:24–39

da Silva J, Sorrells ME, Burnquist WL, Tanksley SD (1993) RFLP linkage map and genome analysis of *S. spontaneum.* Genome 36:782–791

da Silva J, Honeycutt RJ, Burnquist W, Al-Janabi S, Sorrells ME, Tanksley SD, Sobral B (1995) *Saccharum spontaneum* L. 'SES 208' genetic linkage map combining RFLP- and PCR-based markers. Mol Breed 1:165–179

Daugrois JH, Grivet L, Roques D, Hoarau JY, Lombard H, Glaszmann JC, D'Hont A (1996) A putative major gene for rust resistance linked with a RFLP marker in sugarcane cultivar 'R570'. Theor Appl Genet 92:1059–1064

D'Hont A (1994) A molecular approach to unraveling the genetics of sugarcane, a complex polyploid of the Andropogoneae tribe. Genome 37:222–230

D'Hont A (2004) Unraveling the genome structure of polyploids using FISH and GISH; examples of sugarcane and banana. In: Puertas MJ, Naranjo T (eds) Plant Cytogenetics. CIRAD, UMR1096, Montpellier, France

D'Hont A, Glaszmann JC (2001) Sugarcane genome analysis with molecular markers: A first decade of research. In: Hogarth DM (ed) Proceedings of the International Society of Sugarcane Technologists XXVI Congress. The Australian Society of Sugar Cane Technologists, Brisbane, Australia, pp 556–559

D'Hont A, Layssac M (1998) Analysis of cultivars genome structure by molecular cytogenetics and the study of introgression mechanisms. Centre de coopération internationale en recherche agronomique pour le développement. Annual Crops Department CIRAD-CA, pp 9–10

D'Hont A, Lu YH, Feldmann P, Glaszmann JC (1993) Cytoplasmic diversity in sugar cane revealed by heterologous probes. Sugar Cane 1:12–15

D'Hont A, Rao PS, Feldmann P, Grivet L, Islam-Faridi N, Taylor P, Glaszmann JC (1995) Identification and characterisation of sugarcane intergeneric hybrids, *Saccharum officinarum*×*Erianthus arundinaceus*, with molecular markers and DNA in situ hybridisation. Theor Appl Genet 91:320–326

D'Hont A, Grivet L, Feldmann P, Rao P, Berding N, Glaszmann JC (1996) Characterisation of the double genome structure of modern sugarcane cultivars (*Saccharum* spp.) by molecular cytogenetics. Mol Gen Genet 250:405–413

D'Hont A, Ison D, Alix K, Roux C, Glaszmann JC (1998) Determination of basic chromosome numbers in the genus Saccharum by physical mapping of ribosomal RNA genes. Genome 41:221–225

Dje Y, Heuertz M, Lefebvre C, Vekemans X (2000) Assessment of genetic diversity within and among germplasm accessions in cultivated sorghum using microsatellite markers. Theor Appl Genet 100:918–925

Doerge RW, Craig BA (2000) Model selection for quantitative trait locus analysis in polyploids. Proc Natl Acad Sci USA 97:7951–7956

Dufour P, Grivet L, D'Hont A, Deu M, Trouche G, Glaszmann JC, Hamon P (1996) Comparative genetic mapping between duplicated segments on maize chromosomes 3 and 8 and homeologous regions in sorghum and sugarcane. Theor Appl Genet 92:1024–1030

Dufour P, Deu M, Grivet L, D'Hont A, Paulet F, Bouet A, Lanaud C, Glaszmann JC, Hamon P (1997) Construction of a composite sorghum genome map and comparison with sugarcane, a related complex polyploid. Theor Appl Genet 94:409–418

Edwards KJ, Mogg R (2001) Single Nucleotide Polymorphisms. CABI, Lismore, Australia

Franks T, Birch RG (1991) Gene transfer into intact sugarcane cells using microprojectile bombardment. Aust J Plant Physiol 18:471–480

Gill B, Friebe B (1998) Plant cytogenetics at the dawn of the 21st century. Curr Opin Plant Biol 1:109–115

Glaszmann JC, Dufour P, Grivet L, D'Hont A, Deu M, Paulet F, Hamon P (1997) Comparative genome analysis between several tropical grasses. Euphytica 96:13–21

Grice J, Wegener MK, Romanach LM, Paton S, Bonaventura P, Garrad S (2003) Genetically modified sugarcane: a case for alternate products. AgBioForum 64:162–168

Grivet L, Arruda P (2001) Sugarcane genomics: depicting the complex genome of an important tropical crop. Curr Opin Plant Biol 5:122–127

Grivet L, D'Hont A, Dufour P, Hamon P, Roques D, Glaszmann JC (1994) Comparative genome mapping of sugarcane with other species within the andropogoneae tribe. Heredity 73:500–508

Grivet L, D'Hont A, Roques D, Feldmann P, Lanaud C, Glaszmann J-C (1996 a) RFLP mapping in a highly polyploid and aneuploid interspecific hybrid. Genetics 142:987–1000

Grivet L, D'Hont A, Roques D, Feldmann P, Lanaud C, Glaszmann JC (1996 b) RFLP mapping in cultivated sugarcane (*Saccharum spp.*): Genome organization in a highly polyploid and aneuploid interspecific hybrid. Genetics 142:987–1000

Grivet L, Glaszmann J-C, Vincentz M, Da Silva FA, Arruda P (2002) ESTs as a source for sequence polymorphism discovery in sugarcane: example of the Adh genes. Theor Appl Genet 106:190–197

Guimaraes CT, Sills GR, Sobral BWS (1997) Comparative mapping of Andropogoneae: *Saccharum* L. (sugarcane) and its relation to sorghum and maize. Proc Natl Acad Sci USA 94:14261–14266

Harvey M, Botha FC (1996) Use of PCR based methodologies for the determination of DNA diversity between *Saccharum* varieties. Euphytica 89:257–265

Hoarau JY, Offmann B, D'Hont A, Risterucci AM, Roques D, Glaszmann JC, Grivet L (2001) Genetic dissection of a modern sugarcane cultivar (*Saccharum* spp.). I. Genome mapping with AFLP markers. Theor Appl Genet 103:84–97

Hoarau JY, Grivet L, Offmann B, Raboin LM, Diorflar JP, Payet J, Hellmann M, D'Hont A, Glaszmann JC (2002) Genetic dissection of a modern sugarcane cultivar (Saccharum spp.). II. Detection of QTLs for yield components. Theor Appl Genet 105:1027–1037

Hogarth M (1987) Genetics of Sugarcane. In: Heinz DJ (ed) Developments in crop science. II. Sugarcane improvement through breeding. Elsevier, New York, pp 255–271

Hokanson SC, Lamboy WF (2001) Microsatellite (SSR) variation in a collection of *Malus* (apple) species and hybrids. Euphytica 118:281–294

Holm L, Doll J, Holm E, Pancho J, Herberger J (1997) World weeds. Natural Histories and Distribution. Wiley, New York

Hulbert SH, Richter TE (1990) Genetic mapping and characterization of sorghum and related crops by means of maize DNA probes. Science 87:4251–4255

Jannoo N, Grivet L, Seguiin M, Paulet F, Domaingue R, Rao PS, Dookun A, D'Hont A, Glaszmann JC (1999) Molecular investigation of the genetic base of sugarcane cultivars. Theor Appl Genet 99:171–184

Jeswiet J (1929) The development of selection and breeding of the sugarcane in Java. Proceedings of the International Society of Sugarcane Technologists III Congress. The Executive Committee, Soerabaia, pp 44–57

Jiang J, Gill B (1994) Nonisotopic *in situ* hybridisation and plant genome mapping: the first 10 years. Genome 37:717–725

Jordan DR, Casu RE, Besse P, Carroll BC, Berding N, McIntyre CL (2004) Markers associated with stalk number and suckering in sugarcane colocate with tillering and rhizomatousness QTLs in sorghum. Genome 47:988–993

Knapp R (2003) Changes in the world sugar situation. USDA Foreign Agricultural Service. http://www.fas.usda.gov/htp/sugar/2003/Article.pdf

Lin YR, Schertz KF, Paterson AH (1995) Comparative analysis of QTLs affecting plant height and maturity across the poaceae, in reference to an interspecific sorghum population. Genetics 141:391–411

McIntyre L, Aitken K, Berding N, Casu R, Drenth J, Jackson J, Jordan J, Piperidis G, Reffay N, Smith G, Tao Y, Whan V (2001) Identification of DNA markers linked to agronomic traits in sugarcane in Australia. In: Proceedings of the International Society of Sugarcane Technologies, 24:560–562

McMullen MD (2003) Quantitative trait locus analysis as a gene discovery tool. In: Grotewold E (ed) Plant Functional Genomics, vol 236. Humana, Totowa, NJ (Methods in Molecular Biology), pp 141–154

Metcalfe CR (1960) Anatomy of the Monocotyledons. I. Gramineae. Clarendon, Oxford, UK

Ming R, Liu S-C, Lin Y-R, da Silva J, Wilson W, Braga D, van Deynze A, Wenslaff TF, Wu KK, Moore PH, Burnquist W, Sorrels ME, Irvine JE, Paterson AH (1998) Detailed alignment of Saccharum and sorghum chromosomes: comparative organization of closely related diploid and polyploid genomes. Genetics 150:1663–1682

Ming R, Liu S-C, Moore PH, Irvine JE, Paterson AH (2001) QTL analysis in a complex autopolyploid: genetic control of sugar content in sugarcane. Genome Res 11:2075–2084

Ming R, Del Monte TA, Hernandez E, Moore PH, Irvine JE, Paterson AH (2002 a) Comparative analysis of QTLs affecting plant height and flowering among closely-related diploid and polypoild genomes. Genome 45:794–803

Ming R, Liu S-C, Bowers JE, Moore PH, Irvine JE, Paterson AH (2002 b) Construction of a *Saccharum* consensus genetic map from two interspecific crosses. Crop Sci 42:570–583

Ming R, Wang Y-W, Draye X, Moore PH, Irvine JE, Paterson AH (2002 c) Molecular dissection of complex traits in autopolyploids: mapping QTLs affecting sugar yield and related traits in sugarcane. Theor Appl Genet 105:332–345

Mudge J, Andersen W, Kehrer R, Fairbanks D (1996) A RAPD genetic map of *Saccharum officinarum*. Crop Sci 36:1362–1363

Mulherin KS (1986) The economic importance of sugar and sugar cane – problems and perspectives. In: Sansoucy R, Aarts G, Preston TR (eds) Proceedings of an FAO Expert Consultation, Santo Domingo, Dominican Republic

Pan YB, Burner DM, Wei Q, Cordeiro GM, Legendre BL, Henry RJ (2004) New *Saccharum* hybrids in *S. spontaneum* cytoplasm developed through a combination of conventional and molecular breeding approaches. Plant Genet Resour 2:131–139

Panje R, Babu C (1960) Studies in Saccharum spontaneum. Distribution and geographical association of chromosome numbers. Cytologia 25:152–172

Paterson AH (1996) Mapping genes responsible for differences in phenotype. In: Unit BI (ed) Genome Mapping in Plants. Academic, San Diego, pp 41–54

Paterson AH, Schertz KF, Lin YR, Liu SC, Chang YL (1995) The Weediness of Wild Plants – Molecular Analysis of Genes Influencing Dispersal and Persistence of Johnsongrass, Sorghum-Halepense (L) Pers. Proc Natl Acad Sci USA 92:6127–6131

Piperidis G, Christopher MJ, Carroll BJ, Berding N, D'Hont A (2000) Molecular contribution to selection of intergeneric hybrids between sugarcane and the wild species *Erianthus arundinaceus*. Genome 43:1033–1037

Price S (1957) Cytological studies in *Saccharum* and allied genera. III. Chromosome numbers in interspecific hybrids. Bot Gaz 118:146–159

Price S (1963) Cytogenetics of modern sugarcane. Econ Bot 17:97–106

Price S (1965) Interspecific hybridization in sugarcane breeding. Proceedings of the International Society of Sugarcane Technologists XII Congress, Puerto Rico, pp 1021–1026

Price S (1968) Cytology of Chinese and North Indian sugarcanes. Econ Bot 22:155–164

Purseglove JW (1972) Tropical crops: monocotyledons. Longman, New York

Rao PS (1977) Effects of flowering in yield and quality of sugarcane. Exp Agric 13:381–387

Roach BT (1972) Nobilisation of sugarcane. Proceedings of the International Society of Sugarcane Technologists XIV Congress, pp 206–216

Rossi M, Araujo PG, Paulet F, Garsmeur O, Dias VM, Chen H, Van Sluys M-A, D'Hont A (2003) Genomic distribution and characterization of EST-derived resistance gene analogs (RGAs) in sugarcane. Mol Gen Genet 269:406–419

Sills G, Bridges W, Al-Janabi S, Sobral BWS (1995) Genetic analysis of agronomic traits in a cross between sugarcane (*Saccharum officinarum* L.) and its presumed progenitor (*S. robustum* Brandes & Jesw. Ex. Grassl). Mol Breed 1:355–363

Skinner JC, Hogarth DM, Wu KK (1987) Selection methods, criteria, and indices. In: Heinz DJ (ed) Developments in Crop Science II: Sugarcane Improvement Through Breeding. Elsevier, New York, pp 409–454

Sreenivasan TV, Ahloowalia BS, Heinz DJ (1987) Cytogenetics. In: Heinz DJ (ed) Sugarcane Improvement Through Breeding. Elsevier, Amsterdam, pp 211–253

Stevenson GC (1965) Genetics and Breeding of Sugarcane. Longmans, Green, London, UK

Tomkins JP, Yu Y, Miller-Smith, Frisch H, Woo DASS, Wing RA (1999) A bacterial artificial chromosome library for sugarcane. Theor Appl Genet 99:419–424

Toussant M (1987) A History of Food (Trans Anthea Bell 1992). Blackwell, Cambridge, MA

Welsh J, McClelland M (1990) Fingerprinting genomes using PCR with arbitrary primers. Nucleic Acids Res 18:7213–7218

Wicking C, Williamson B (1991) From linkage markers to genes. Trends Genet 7:288–293

Wu K, Burnquist W, Sorrels M, Tew TL, Moore P, Tanksley S (1992) The detection and estimation of linkage in polyploids using single-dose restriction fragments. Theor Appl Genet 83:294–300

12 Potato

Dan Milbourne[1], Barnaly Pande[2], Glenn J. Bryan[3]

[1] Plant Biotechnology Unit, Teagasc, Crops Research Centre, Oak Park, Carlow, Eire, *e-mail*: d.milbourne@oakpark.teagasc.ie
[2] Department of Environmental Horticulture, University of California, Davis, One Shields Avenue, Davis, CA 95616, USA
[3] Genetics Programme, Scottish Crop Research Institute, Invergowrie, Dundee, DD2 5DA, UK

12.1 Introduction

The cultivated potato has undergone significant biological evolution over the past four centuries that has led to its position as the fourth most important cultivated crop worldwide after wheat, maize and rice, with a total global production of ca. 311 million metric tonnes on an area of ca. 18.9 million ha (*http://faostat.fao.org*). Table 1 gives a summary of the top six potato-producing countries/areas in the world, which account for over two thirds of total global potato production. In developing economies, the majority of the crop is used for direct consumption. Potatoes are a good source of carbohydrate, protein, vitamins, e.g. vitamin C or ascorbic acid, thiamine, niacin, pyroxidine, vitamin B6, riboflavin, folic acid, and minerals, e.g. calcium, potassium, phosphorus (Smith 1984). In recent years, particularly in developed economies, there has been a shift in the end use of potatoes, with production for direct consumption being replaced by a trend towards processing potatoes for the production of convenience foods such as french fries and potato chips (*http://faostat.fao.org*). In addition, diverse non-food uses of potato are emerging, for example, potato as a source of starch for the production of biodegradable plastics (Doane 1994).

12.1.1 History, Taxonomy and Distribution

Despite the fact that it has only been known in Europe for the past 400 years, the potato is one of humanity's most ancient cultivated plant species. It originated in South America and was probably first seen by Europeans when the Spanish arrived on the coast of what is modern-day Peru in 1532. Published accounts describing potato as a staple food crop of the peoples of the Andes of southern Peru and Bolivia exist from this period (Lopez de Gomara 1552), and the antiquity of potato as a food crop is supported by archaeological finds such as ceramic pots of the Moche (1–600 AD) and Chimu (900–1450 AD) tribes of the northern Peruvian coast, which are illustrated with potato plants (Hawkes 1990). Evidence predating the Christian era has been found in plant remains excavated from graves, food stores and rubbish dumps. In Peru, potato remains, frequently in the form of

Table 1. Top six potato-producing countries/areas in the world (2003)

	Production (MT)	Area (ha)	Average yield (MT/ha)	Seed production (MT)
World	310,810,336	18,896,832	16.45	36,201,982
China	66,813,331	4,501,667	14.84	3,001,500
EU 25 states, May 2004	57,224,856	2,171,259	24.78 *(42.4 in Denmark to 13.1 in Latvia)	5,455,314
Russian Federation	36,746,512	3,171,990	11.58	10,000,000
India	23,161,400	1,337,200	17.32	2,100,000
USA	20,821,930	505,980	41.15	1,250,000
Ukraine	18,500,000	1,600,000	11.56	5,000,000

Genome Mapping and Molecular Breeding in Plants, Volume 3
Pulses, Sugar and Tuber Crops
C. Kole (Ed.)

chuno (freeze dried tubers), have been found dating from 4500 years ago for coastal sites (Mosely 1975; Martins-Farias 1976) and 7000 years ago for one Andean site (Engel 1984). At one site in Chile, potato remains dating from 13,000 years ago have been found, predating the development of agriculture (Ugent et al. 1987).

The taxonomy, evolution and route of domestication of potato are all the subject of debate amongst experts in these respective fields. Taxonomically, it is generally agreed that potato and its wild relatives occupy section Petota within subgenus Potatoe of the large and very diverse genus *Solanum*. Section Petota is divided into subsections Estolonifera (non-tuber-bearing) and Potatoe (tuber-bearing), and these subsections are further divided into series in which species are placed. However, beyond this level there exist at least two major schools of thought as to the further treatment of section Petota. Hawkes (1994) recognizes 228 wild and seven cultivated species (*S. tuberosum*, *S. stenotonum*, *S. ajanhuiri*, *S. chaucha*, *S. phureja*, *S. curtilobum* and *S. juzepczukii*) of potato. Within this classification system, *S. tuberosum* is subdivided into two subspecies, *S. tuberosum* ssp. *tuberosum* (to which the modern cultivated potato belongs) and *S. tuberosum* ssp. *andigena*. Spooner and colleagues, however, describe only 199 wild species (Spooner and Hijmans 2001) and recognize all cultivated potatoes as a single species, *S. tuberosum* (a classification originally proposed by Dodds 1962), which they further subdivide into eight cultivar groups: Chilotanum Group, Andigenum Group, Stenotonum Group, Ajanhuiri Group, Chaucha Group, Phureja Group, Curtilobum Group, and Juzepcukii Group (Huaman and Spooner 2002). The eight cultivar groups in this system are representative of the seven cultivated species in the Hawkes system with the exception that the two subspecies of *S. tuberosum* (ssp. *tuberosum* and ssp. *andigena*) are individually represented by the Chilotanum and Andigenum Groups respectively.

The geographic distribution of wild potato species is extensive, ranging from the southwestern states of the USA, through the countries of Central America, to South America, where they are found along the entire length of the Andes from Venezuela to northwest Argentina, as well as lowland areas of countries occupying the southernmost half of South America (Hawkes 1994). Accompanying this wide geographic distribution is a wide range of environmental conditions including cold, high-alti-

Table 2. Major potato germplasm collections worldwide. Information derived from Web site of Dutch-German Potato Collection. Web addresses were active at time of writing, some are not in English

Collection name	Host institute	URL
The Dutch-German Potato Collection	Centre for Genetic Resources, The Netherlands (CGN)	www.genebank.nl/potato/
The Commonwealth Potato Collection (CPC)	Scottish Crop Research Institute (SCRI), UK	http://www.external.scri.sari.ac.uk/SCRI/Web/Site/home/ResearchAreas/CommonwealthPotatoCollection/cpc.asp
The Groß Lüsewitz Potato Collection (GLKS)	Institute of Plant Genetics and Crop Plant Research (IPK), Germany	pgrc.ipk-gatersleben.de/potato/
The Vavilov Institute Potato Collection	N. I. Vavilov All-Russian Scientific Research Institute of Plant Industry (VIR)	http://vir.nw.ru/
The US Potato Genebank (NRSP-6)	USDA-ARS Potato Introduction Station, USA	http://www.ars-grin.gov/ars/MidWest/NR6/
The CIP Potato Collection	International Potato Center (CIP), Peru	www.cipotato.org/
The University of Cusco Potato Collection	University of Cusco, Peru	
The PROINPA Potato Collection	Proyecto de Investigacion de la Papa (PROINPA), Bolivia	http://www.proinpa.org/
The INTA-Balcarce Potato Collection	INTA-Balcarce, Argentina	www.inta.gov.ar/balcarce/index.htm
The CORPOICA Potato Collection	Corporacion Colombiana de Investigacion Agropecuaria (CORPOICA), Colombia	http://www.corpoica.org.co/

tude regions, near-desert regions, temperate rain forests and coastal plains. Conversely, the cultivated species/groups were originally confined to the South American Andes and southern Chilean lowlands, both of which experience cool temperate climates. Thus, while cultivated species, because of their relatively narrow ecogeographical origin, are adapted to relatively few environments and diseases, their wild relatives, by contrast, have had to adapt to a whole range of biotic and abiotic stresses. Because of this, wild potato species represent a valuable resource which potato breeders can exploit to broaden the originally narrow genetic base of the modern cultivated potato. In order to maximise the utility of these resources, several extensive potato germplasm collections have been set up around the world (Table 2) in which many of the species of potential benefit to potato breeding are held and characterized with respect to a variety of potentially important agronomic traits.

Cultivated and wild potato species exist in a polyploid series from diploid to hexaploid, with a generally accepted basic haploid chromosome number of x=12. The modern cultivated potato, *Solanum tuberosum* ssp. *tuberosum*, is tetraploid (2n=4x=48) and displays autotetrasomic inheritance, with random pairing amongst its four sets of homologous chromosomes during meiosis. The South American cultivated species that bear the greatest resemblance to the modern cultivated potato are the Chilean *S. tuberosum* ssp. *tuberosum* (Chilotanum Group) and the Andean *S. tuberosum* ssp. *andigena* (Andigenum Group), and several hypotheses for the evolution of these species have been proposed. In the case of *S. tuberosum* ssp. *andigena*, three main possibilities have been hypothesised, unified by their assumption of the involvement of the diploid cultivated species *S. stenotonum*. An early hypothesis suggested the autopolyploidisation of *S. stenotonum* (Swaminathan and Magoon 1961), while two alternative hypotheses involve the hybridization of *S. stenotonum* with another diploid species, in one case the weedy diploid wild species *S. sparsipilum* (Cribb and Hawkes 1986), and in the other case the diploid cultivated species *S. phureja* (Matsubayashi 1991). Similarly, there are several hypotheses about the origin of the Chilean populations of *S. tuberosum* ssp. *tuberosum*. Perhaps the most widely accepted theory is that *S. tuberosum* ssp. *andigena*, which is adapted to the short 12-h days of the high Andes, was transported to Chile via trade or human population movement, where it adapted to the longer day length of that region, evolving into subspecies *tuberosum* (Hawkes 1990). There are alternative theories; for instance, Ugent et al. (1987) proposed the wild species *S. maglia* as a progenitor of subspecies *tuberosum* on the basis of starch grains, and Grun (1990) suggested that subspecies *tuberosum* evolved from a cross between subspecies *andigena* and an unidentified wild species. There is some agreement by various authors, however, that all of the above hypotheses are oversimplifications, and that South American cultivated potatoes are likely of multiple origins.

The introduction of potato from South America into Europe, and its development into what we now refer to as cultivated potato, is also a subject of some controversy. It is generally accepted that the first introductions occurred somewhere towards the end of the 16th century. Historical records listing the purchase of potatoes by the Hospital de la Sangre in Seville in 1576 suggest an introduction into Spain around 1570 (Hawkes 1990). The aforementioned controversy surrounds the point of origin and subspecies (or cultivar group) of the earliest introductions. Juzepcuk and Bukasov (1929) originally proposed that the earliest introductions were of the Chilean *Solanum tuberosum* ssp. *tuberosum* (Chilotanum Group) because of the fact that both modern potato and these Chilean landraces are morphologically similar and adapted to tuberise under long days. Salaman (1949), however, proposed that historical evidence indicated that the earliest introductions were of the short-day-length-adapted Andean *Solanum tuberosum* ssp. *andigena* (Andigenum Group) and that their introduction was followed by a rapid convergent selection towards the day length and morphological characteristics of the Chilean landraces because of the similar environments of the Chilean lowlands and continental Europe. Whatever the source of the original introduction, modern potato cultivars certainly derive much of their genetic background from the Chilean *tuberosum* subspecies, as evidenced by the fact that nearly all of these cultivars have a chloroplast genome possessing a 241-bp deletion characteristic of Chilean *tuberosum* landraces (Hosaka 1995). Proponents of a sole original introduction of *S. tuberosum* ssp. *andigena* attribute this fact to the widespread decimation of potato crops during the pan-European potato famine beginning in Ireland in 1845, followed by widespread breeding with Chilean germplasm in an effort to increase levels of resistance (Hawkes 1990). However, Spooner and colleagues have advanced an alternative hypothesis that there were multiple early introductions of both *S. tuberosum* subspecies into Europe, most of which are simply not recorded in historical literature. In this scenario it is likely that that the Chilean races, which were adapted to European day-length conditions, would have rapidly been selected over the Andean landraces, explaining the possession of the Chilean Tuberosum-type chloroplast genome by most modern cultivars (D. Spooner, personal communication).

12.1.2 Potato Breeding

Potato breeding began in earnest after the global blight epidemics of the mid-19th century. The fact that European cultivated potato was founded on relatively few introductions of non-diverse landrace material from South America resulted in a crop with a narrow genetic base, and genetic diversity was further eroded by the widespread decimation of the species caused by disease epidemics. Early attempts at introducing new germplasm failed to widen the gene pool significantly, since many initial attempts involved crosses with descendants of a single Chilean *S. tuberosum* ssp. *tuberosum* variety, Rough Purple Chili (Hawkes 1990). These efforts were also not very effective in increasing the level of blight resistance in potato, and it was not until the beginning of the 20th century, with the widespread use of the Mexican hexaploid species *Solanum demissum*, that effective blight-resistant cultivars began to be developed (Hawkes 1990). Successes such as this saw the realisation of the potential, in potato breeding, of wild and exotic cultivated germplasm of Central and South America, which resulted in numerous collecting expeditions that led to the establishment of several potato germplasm collections worldwide.

While upwards of 130 wild and cultivated exotic species of potato are available in potato germplasm collections (Hawkes 1994), relatively few (of the wild species) have been used extensively in breeding programs. This is in part due to the existence of interspecific crossing barriers such as different nuclear ploidy levels, gametophytic self-incompatibility, endosperm instability and sterility of F_1 progeny. Although many of these barriers can be overcome by various techniques, their existence has limited the widespread use of many wild species in potato breeding. Some of the evolutionarily more advanced tuber-bearing diploid wild species can be crossed with *S. tuberosum* ssp. *tuberosum* relatively easily. The success of such crosses is facilitated by the fact that the embryos that develop after fertilization have a viable endosperm. Potato and its relatives have been empirically assigned a hypothetical value called an endosperm balance number (EBN) and species with the same EBN (and nuclear ploidy) can be crossed successfully (Hermsen 1994). Crossing is facilitated by the ease with which ploidy and EBN can be manipulated in both cultivated potato (4×, EBN=4) and these wild species (2×, EBN=2). Tetraploid cultivated potato can be artificially haploidised to produce functional diploids (with an EBN=2) through prickle pollination using specific *S. phureja* inducer genotypes or through anther culture ('androgenesis'), and many diploid wild species (and potato dihaploids) have a tendency to produce unreduced, diploid gametes, effectively analogous to the gametes of tetraploids (with an EBN=4). Strategies such as these allow both individual traits and genetic diversity in general to be transferred from these wild species to germplasm that can be used in commercial breeding programs, although several rounds of backcrossing may be required to eliminate undesirable traits from the donor species.

Use of wild species material has not been the sole strategy employed in attempts to both widen the genetic base of potato and introgress new traits. Another strategy, developed by Simmonds (1969) and subsequently adopted by breeders in the USA, Canada and the Netherlands, was based on establishing populations of unadapted *S. tuberosum* ssp. *andigena* and essentially rapidly 'evolving' them towards the phenotype of ssp. *tuberosum* by recurrent mass selection over two decades. Material from these programs is referred to as Neotuberosum. Simmonds' program resulted in the generation of clones with comparable yield and maturity, good cooking quality and better blight resistance compared to Tuberosum, although they were inferior in the regularity of tuber shape. Some of the clones were also found to have an improved resistance to low-temperature sweetening. Neotuberosum material from such programs has been used successfully for cultivar development, for example the cultivars Shelagh and Rosa both resulted from crosses between Tuberosum and Neotuberosum material (Bradshaw and Mackay 1994).

The majority of potato-breeding programs worldwide aimed at commercial cultivar development (as opposed to the germplasm development schemes described above, which might be referred to as prebreeding) follow the same basic framework. Generally, each cycle of such a program is initiated by a large number of pair crosses involving well-adapted tetraploid material such as existing cultivars and advanced breeding material from previous cycles. This is followed by several years of recurrent phenotypic selection during which up to 60 traits are assessed. It can take up to 11 years from the initial selection of seedlings to official trials before commercialisation of the best clones can take place (a process that takes several more years). A schematic outlining the structure of the current format of the potato-breeding program at Oak Park Research Centre in Ireland is presented in Fig. 1. This could be said to represent a typical example of a 'traditional' breeding program. In this scheme, ca. 150 pair crosses are performed in the first year and 90,000 individual seedlings are raised in pots during the second year. Apart from loss due to events such as lack of tuberisation, all material is carried forward as single plants in the field in the third year, when the first round of selection occurs, based on a non-destructive visual assessment of tuber characteristics, including number, size, shape, appearance, growth cracking, etc. Selection in subsequent years is carried out in clonal plots to approximate commercial production and

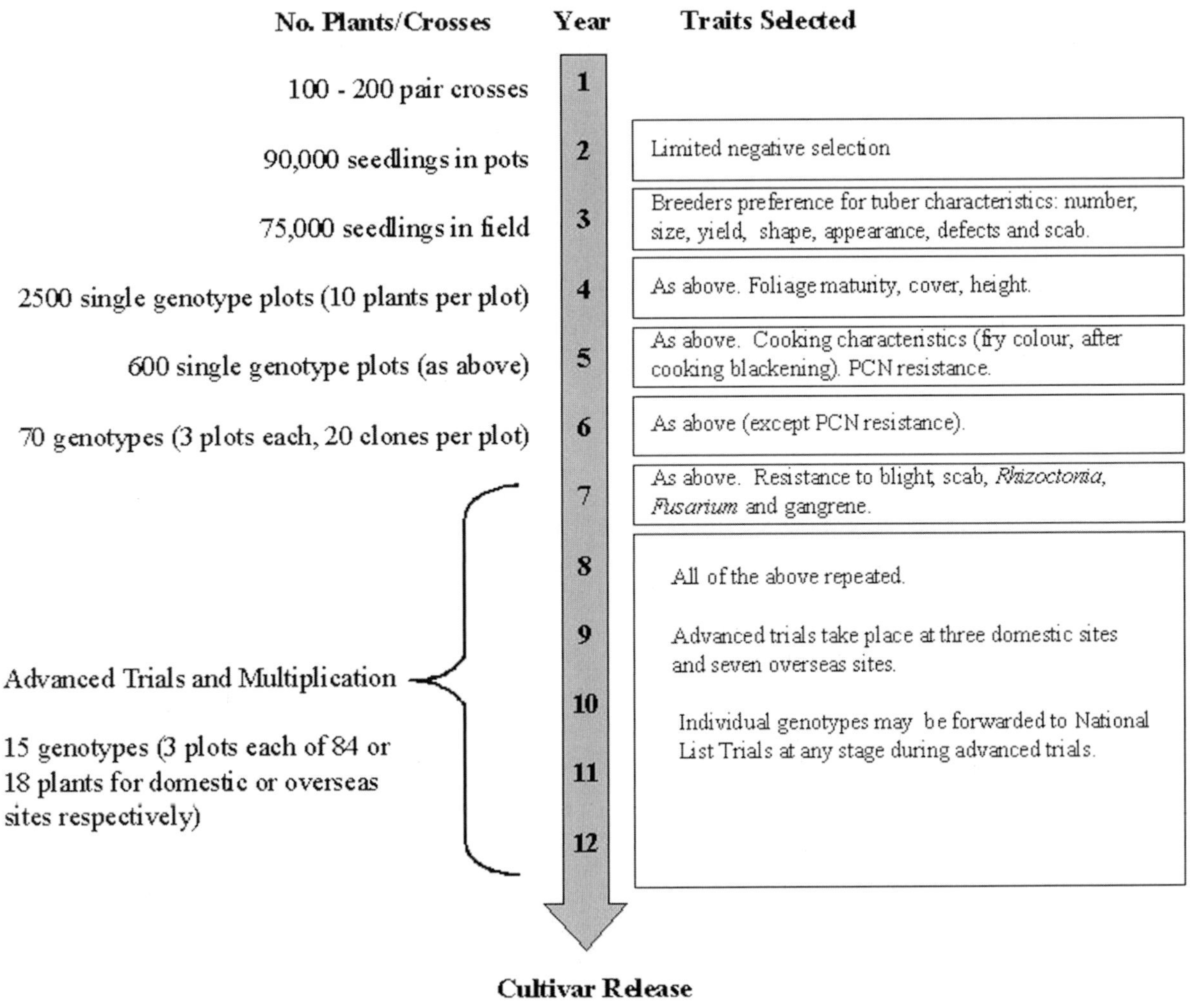

Fig. 1. Potato breeding program as practiced at Oak Park Research Centre at the time of writing. Information courtesy of D. Griffin, potato breeder

enable an accurate assessment of foliage characteristics. Assessment of tuber quality traits such as dry matter content, after-cooking blackening and tendency to accumulate reducing sugars during cold storage (leading to oxidative browning during frying, an undesirable trait in potato chip production) begins in the fifth year, when sufficient tubers become available for such tests. Assessment of resistance to potato cyst nematodes also takes place at this point. Resistance levels to most other diseases, e.g. late blight, scab, *Rhizoctonia*, *Fusarium* and gangrene are assessed early in the seventh year. The material subsequently moves on to advanced trials at three sites in Ireland and sites in North Africa, the UK and the Netherlands. Material performing sufficiently well in these trials may be forwarded to National List Trials to obtain cultivar status.

The above strategy is representative of many breeding programs, which may differ in specifics such as numbers of clones and traits targeted (e.g. different diseases, different end uses, different regional preferences). Programs similar to this have successfully produced the majority of commercial cultivars available today. There is a drive, however, to increase the efficiency and speed of potato breeding. Arguably the single most useful means of driving this increase in efficiency would be an increased ability to predict the extent of transmission of desired parental characteristics to the next generation. The goal of this concept is the reduction of the total number of crosses (and thus progeny) required for cultivar development by identifying the most useful parental sources of desired characteristics and the best crosses to combine complementary characteristics. One method of achieving this goal is to examine the segregation of traits of interest in progeny populations derived from pairwise crosses, in order to estimate the extent to which particular parents will stably transmit desirable traits to their offspring. This concept, known as progeny testing, represents an attempt to define the heritable, genetic component (and possible interactions) of the traits under examination. For the process to be truly efficient, it generally requires a multifactorial scheme involving the crossing of several parents of interest in different combinations (e.g. diallels, selfs, reciprocals) and testing the progeny of all of these crosses for the traits in question (Bradshaw and Mackay 1994). The amount of information gained depends largely on the genetic architecture of the trait in question. Traits controlled by dominant major genes may exhibit the Mendelian segregation ratios expected in an autotetraploid, and these can be used to predict gene dosage in the parent (tetrasomic inheritance is discussed in greater depth later). Inferences (such as narrow-sense heritability and general and specific combining ability) about the heritable component of quantitative traits can also be made, allowing the identification of the best parents and parental combinations to achieve particular breeding goals. A breeding program based around a progeny testing approach may differ significantly in structure from the mass selection program described above. The recurrent genotypic multitrait selection scheme employed at SCRI makes use of these strategies within the framework of a 3-year cycle of recurrent selection. In the first year, crosses are performed between chosen parents, and seed is obtained. In the second year, seedling progeny tests are carried out for traits such as disease resistance, and visual assessment of glasshouse grown tubers is performed. In year 3, selected tuber progenies are field planted, and selected clones from amongst these are either advanced directly to the cultivar selection stage or serve as material for subsequent rounds of hybridization and selection (the reader is referred to Bradshaw et al. 1994, 1995, 2000 and 2003 for a description of the various elements of this program).

While progeny tests can give some insights into the inheritance patterns of traits, their main limitation is the fact that they generally provide only an indirect indication of the genetic component of the character being assessed. Only with the development of molecular-marker-based genetic linkage mapping has it been possible to move towards identifying and quantifying the genetic component of quantitative traits through the identification of quantitative trait loci (QTLs) for these traits. Not only does genetic mapping provide a more direct indication of the genetic component of quantitative traits, but the association of QTLs with closely linked molecular markers offers the possibility of practising selection directly on the genetic components underlying quantitative traits. The potential for marker-assisted selection (MAS) of both quantitative and qualitative traits in potato breeding will be discussed later in the chapter after a thorough review of the current state of the art in genetic mapping of potato.

12.2 First-Generation Genetic Linkage Maps of Potato

Despite the fact that the modern cultivated potato is autotetraploid, the vast majority of molecular-marker-based genetic linkage mapping in potato to date has been carried out at the diploid level in order to circumvent the complexities associated with tetrasomic inheritance. This section will examine several diploid maps that could be considered as 'first-generation reference maps' of potato by virtue of the fact that they have been constructed using markers whose chromosomal location, once established in one mapping experiment, will remain consistent in subsequent mapping studies using different mapping populations. The ability to align genetic maps relative to each other by virtue of the inclusion of markers from these maps is obviously of central importance if the genetic mapping of traits of interest is going to have any significance beyond the specific germplasm in which the mapping study has been undertaken.

The first published molecular-marker-based genetic linkage maps of potato were based on restriction fragment length polymorphisms (RFLPs). Bonierbale et al. (1988) took advantage of the synteny between the genomes of tomato and potato by employing tomato-derived cDNA and gDNA probes that had been used previously to construct an RFLP-based map in tomato (Bernatzky and Tanksley 1986). One hundred thirty-five markers were mapped to 12 linkage groups (LGs) in a population of 65 F_1 individuals in an interspecific cross: *Solanum phureja*×(*Solanum tuberosum*×*Solanum chacoense*). The map was produced by following the segregation of alleles from the hybrid parent only. The similarity of the genomes of potato and tomato was confirmed by the fact that all but two of the tomato-derived probes tested hybridized in potato. Furthermore, almost all probes mapped to the same LGs in both species, and the linear order of markers was found to be the same on all but three chromosomes. The order differences were all due to paracentric inversions with breakpoints at or near the centromere. The resulting map length of 606 cM was almost half the length of the tomato map produced using the same markers, with potato LGs exhibiting a 1.4- to 3.6-fold reduction in length compared to their tomato homeologues. Comparison of recombination fractions between tomato and the two potato parents suggested that the reduced level of recombination was restricted to the interspecific hybrid parent. Significant deviations from the expected marker segregation ratios for a diploid cross were observed for 28% of mapped markers, and these loci occurred in clusters on five LGs. The map was extended in a subsequent study by analysing the segregation of a larger set (approx. 170) of RFLP markers in an interspecific backcross (BC): (*Solanum tuberosum*×*Solanum berthaultii*)× *Solanum berthaultii* (Tanksley et al. 1992). The resulting map length increased to 684 cM, and the increased marker density allowed the observation of two further paracentric inversions differentiating the tomato and potato genomes, bringing the total to five inversions located on chromosomes V, IX, X, XI and XII. An interesting feature of this study was the fact that the potato genetic map was constructed in parallel with the construction of a high-density genetic map of the tomato genome comprising over 1000 molecular markers, of which the markers used to construct the potato map were a subset. The high level of gene-order conservation observed between the two species, and the fact that the majority of probes used in tomato also hybridize to potato DNA, means that the order of those markers not mapped directly in potato can be approximately inferred by their positions on the tomato map and that these markers can be applied to the construction of genetic maps in other potato crosses.

In many plant systems, interspecific crosses are used to generate mapping populations in order to maximise the genetic diversity, and thus the level of marker polymorphism between mapping parents. Gebhardt et al. (1989 a) screened 38 diploid *S. tuberosum* breeding lines for RFLPs using 168 potato-derived gDNA and cDNA probes. The lines chosen exhibited variability for various morphological and agronomically important characters. Eighty percent of the probes tested were polymorphic in the germplasm examined (three probe/enzyme combinations), with a minimum of 49% and a maximum of 95% polymorphism between any two lines. The observed intraspecific variation in potato is very high in comparison to the closely related tomato, probably due to a mixture of the outbreeding nature of potato and the introgression of wild germplasm into many potato breeding lines. The level of variability observed implied that the F_1

progeny of the majority of *S. tuberosum* intraspecific crosses would provide sufficient polymorphism for the creation of genetic maps in potato, and subsequent mapping experiments have confirmed this.

Based on the preliminary data a BC population created from two of the lines was used to generate an RFLP-based linkage map with 141 loci covering all 12 chromosomes (Gebhardt et al. 1989b). The map was 14% longer than that of Bonierbale et al. (1988) at 690 cM, and showed similar levels of segregation distortion. In contrast to the previous studies involving interspecific BCs, markers segregating from both parents were used for map construction. Individual maps were constructed for each of the parents using RFLP alleles derived from the individual parents (expected segregation ratio 1:1, presence:absence), and a third map comprised RFLP marker alleles present in both parents and segregating in the population (expected segregation ratio 3:1, presence:absence). These maps were aligned relative to each other based on assumed allelism between marker alleles derived from the same probe. The map was further extended by increasing the total number of markers analysed in the cross and by incorporating data from an interspecific BC (*S. tuberosum*×*S. spegazzinii*)×*S. spegazzinii* (Barone et al. 1990) indirectly onto the map on the basis of linkage to flanking markers (Gebhardt et al. 1991). Increasing the number of markers by 117% almost doubled the map length to 1034 cM, estimated at 80% genome coverage. In addition, mapping a number of markers from the study of Bonierbale et al. (1988) in the intraspecific *S. tuberosum* BC and a concurrent transfer of markers from this population to the tomato population of Bernatzky and Tanksley (1986) allowed the alignment of both sets of potato RFLP maps described previously. This allowed a comparison of potato RFLP maps derived from different genetic backgrounds. Significant reduction in map length was observed in the interspecific compared to the intraspecific crosses. While some segregation distortion could be attributed directly to the interspecific hybridity of parents, regions of distortion consistent across several mapping parents from this study and that of Bonierbale et al. (1988) were found on chromosomes I, II, III, VI, IX and X. The most notable of these was the interval surrounding the SI (self-incompatibility) locus on LG I.

Jacobs et al. (1995) subsequently used a *S. tuberosum* intraspecific BC to produce a map combining molecular markers (gDNA and cDNA derived RFLPs, and probes flanking T-DNA inserts or maize transposable element reintegration sites, morphological markers and isozyme markers). Maps were produced separately for the male and female parents, and these were subsequently combined on the basis of allelic bridges between the maps. The final map consisted of 175 molecular markers, 10 morphological markers and eight isozyme markers, with a total map length of 1120 cM.

As evidenced by the above examples, the codominant and highly transferable nature of RFLPs makes them a very powerful tool for map integration both within and across species boundaries. However, features of the assay (such as the requirement for microgramme quantities of DNA, the necessity to acquire and maintain large collections of probes, and the recalcitrance of the method to automation and other measures to increase throughput) have limited their widespread use in the face of the advent of polymerase chain reaction (PCR)-based marker systems that can overcome many of these limitations.

One of the early illustrations of the utility of PCR-based markers for map construction in potato was the use by van Eck et al. (1995) of the highly multiplex amplified fragment length polymorphism (AFLP) technique (Vos et al. 1995) to add a further 264 markers, using only six AFLP primer combinations, to the maps constructed by Jacobs et al. (1995). Subsequently, Rouppe van de Voort et al. (1997a) demonstrated that AFLP-based maps produced in fairly unrelated diploid *S. tuberosum* genotypes may be aligned relative to each other by virtue of the fact that AFLP markers of the same size (electrophoretic mobility) produced by the same primer combination in different genotypes are generally allelic and will map to homologous positions. Using the six AFLP primer combinations previously used by van Eck et al. (1995), they identified a total of 117 comigrating markers mapping to homologous positions the genetic maps of five genotypes. Because two of the genotypes were those used by Jacobs et al. (1995), the remaining three (entirely AFLP-based) maps could also be aligned relative to the RFLP-based maps of potato. In order to allow this approach to be used in other AFLP-based mapping studies, the results of this and further AFLP mapping experiments were used to assemble a catalogue of chromosome-specific AFLP markers to allow any new AFLP-based map of potato generated using the appropriate primer combinations to be aligned relative to the potato RFLP maps (Rouppe van der Voort et al. 1998a). The catalogue consists of a Web site (*http://www.dpw.wau.nl/pv/aflp/catalog.htm*) comprising 12 autoradiogram images of AFLP fingerprints generated with commonly used primer combinations on six mapping parents. AFLP markers are described by indicating on the autoradiogram images:

1. Their name (including enzyme combinations, primer extensions and mobility)
2. Their position on the potato genome
3. The parental genotype in which the mapping was carried out.

Inclusion of a size standard and one or several of the mapping parents from the catalogue in AFLP assays when producing a map in other *S. tuberosum* genotypes allows the identification of chromosome-specific AFLPs in the resulting maps. While the AFLP catalogue described here has proven to be a very useful tool in constructing genetic linkage maps in potato, it does suffer from some limitations. On a practical level, minor differences in assay conditions can occasionally make it difficult to relate AFLP patterns to the catalogue images, even when the appropriate mapping parents and size standard are present. There is also an increasing trend towards use of fluorescence-based genotyping/sequencing systems for AFLP mapping studies in plants, causing issues of back-compatibility to autoradiogram images as well as intersystem compatibility between different platforms. Finally, the catalogue was assembled from genotypes that are derived largely from a *S. tuberosum* background, with introgressions from a limited number of other cultivated and wild species. The number of markers (per primer combination) that comigrate with those from the catalogue parents can drop radically when using wild species mapping parents, reducing the utility of this method for chromosomal identification and orientation in these situations.

Microsatellite- or simple sequence repeat (SSR)-based markers combine the convenience of a PCR-based assay with the codominant and locus-specific nature of RFLPs and have become the system of choice for linkage mapping in most well-characterized plant species. Milbourne et al. (1998) characterized SSR loci in potato, from a mixture of database searches, cDNA libraries and selectively enriched small insert libraries. Primer pairs flanking 112 SSRs were designed, and 65 of these were mapped to 89 loci on two separate populations (Fig. 2). One of the mapping populations used was the *S. tuberosum* BC population of Gebhardt et al. (1991), allowing the alignment of the SSR-based map relative to the potato RFLP maps. To date, the utility of SSR-based markers in potato has been limited by the relatively low numbers that have been mapped. Their role is frequently limited to providing chromosomal identification and orientation on maps constructed largely of other markers such as AFLPs. Originally, the low number of SSRs available was due to the fact that the isolation and characterization of SSR loci required extensive library construction and screening, followed by sequencing and primer testing. Development of SSRs from databases was hampered by the low number of entries for potato – only 451 *S. tuberosum* sequences were present in the EMBL and GenBank databases at the time of the study of Milbourne et al. (1998), but these yielded a total of 38 primer pairs that successfully amplified SSR loci. However, the advent of large-scale expressed sequence tag (EST) programs has led to a radical increase in the amount of gene-derived sequence currently available, and at the time of writing over 120,000 EST sequences are available for potato. The Institute for Genomic Research (TIGR) has developed a pipeline for the identification of potato ESTs containing SSR motifs, and to date over 6000 SSR-containing ESTs have been identified (a summary of these is available at *http://www.tigr.org*). Development of successful SSR markers is a process with a relatively high attrition rate, and it is not expected that all of the SSR sequences identified in ESTs can successfully be developed as markers. For instance, approx. 50% of SSRs in ESTs are mononucleotide repeats (after removal of poly-A tails), which are less desirable as SSR markers because the segregation of single-base-pair-length polymorphisms can be difficult to follow over entire populations. In addition, coding-sequence-derived SSRs, many of which are codon-derived trinucleotide repeats, generally exhibit lower levels of polymorphism than those derived from non-coding sequence, and this has been confirmed specifically for EST-derived SSRs in potato by Ghislain et al. (2004). Forty percent of the SSRs in potato ESTs are trinucleotide repeats, the majority of which are likely to be codon derived. Despite these factors, ESTs represent a pro-

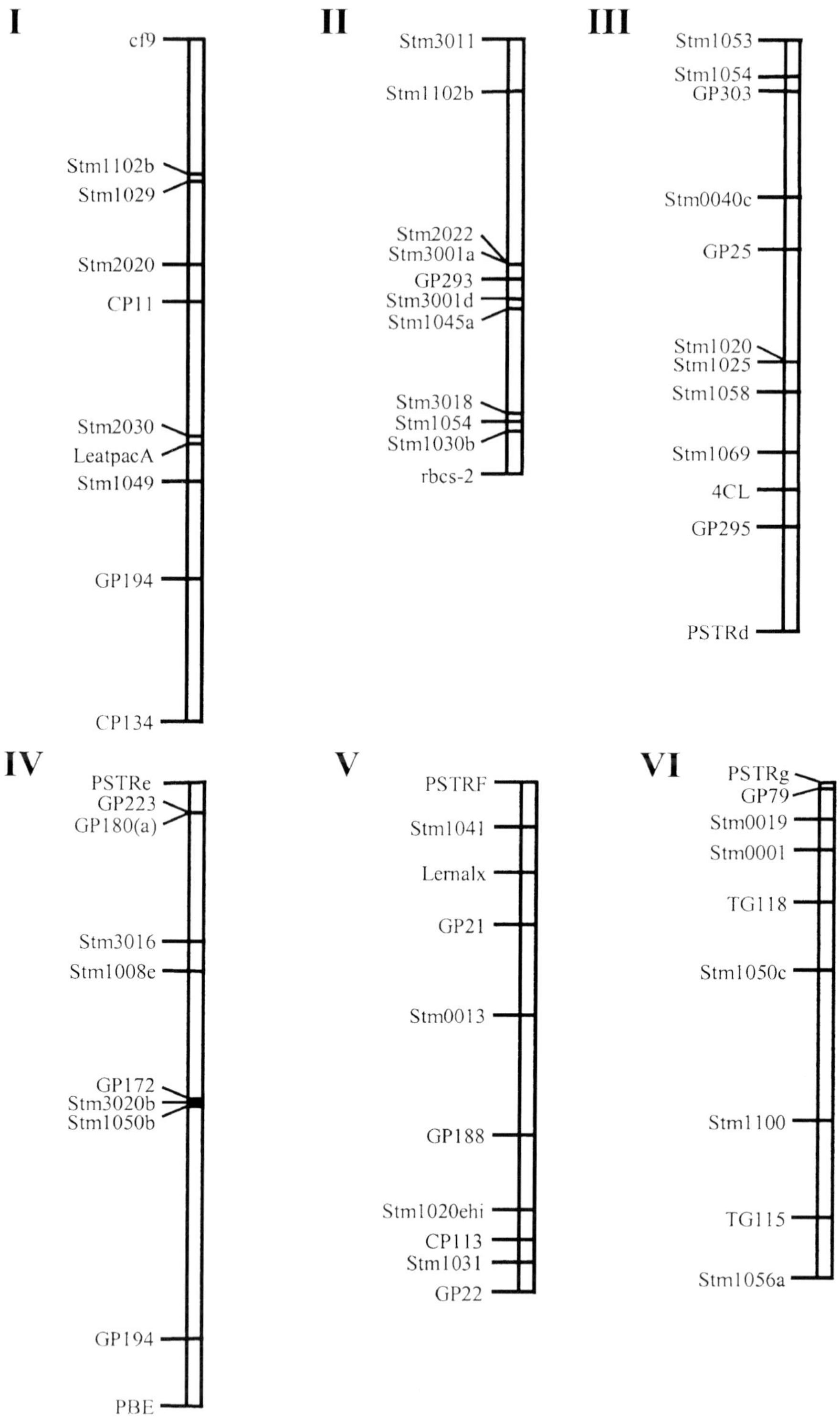

Fig. 2. Representation of microsatellite-based map of Milbourne et al. (1998). Information from two separate mapping populations have been amalgamated in a single representative map. Some marker positions have been approximated because of different recombination rates observed for the two maps, and the reader is directed to the original publication for more accurate positions of individual markers

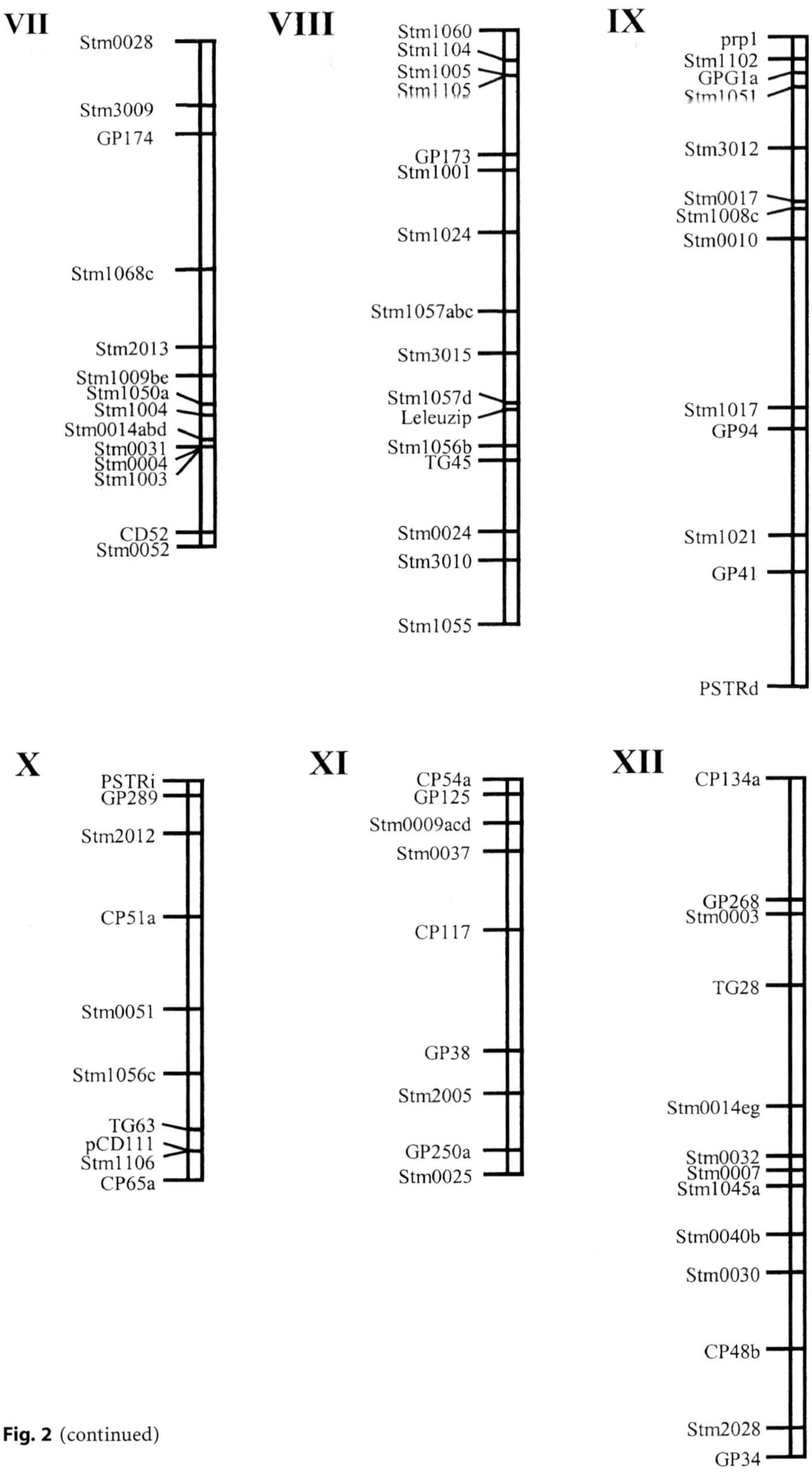

Fig. 2 (continued)

mising source for the development of SSR markers in potato, and work is currently under way in several laboratories worldwide towards this goal.

Knowledge of the sequence of any stretch of DNA (preferably single-copy and more than 100 nucleotides) for an organism offers the potential to develop a PCR-based marker for that piece of DNA (referred to as a sequence-tagged site or STS) by simply designing PCR primers that amplify the sequence. The PCR products generated can then be examined for differences in restriction-endonuclease-site content, either by sequencing or screening with multiple restriction enzymes, in order to develop cleaved amplified polymorphic site (CAPS) markers. The earliest applications of this approach in potato involved attempts to circumvent some of the aforementioned technical complexities associated with RFLP markers by sequencing probes mapped to loci of specific interest and developing CAPS markers that cosegregated with the original RFLP marker, for example by Niewöhner et al. (1995), who successfully converted RFLPs linked to nematode resistance to CAPS markers. The high level of heterozygosity present in most potato genotypes can make it practical, in certain circumstances, to use CAPS markers in a more high-throughput fashion, to map multiple loci for which sequence information is available. In an experiment to map functional genes involved in carbohydrate metabolism in potato, Chen et al. (2001) successfully mapped 45 genes relative to markers from the potato RFLP maps using a CAPS-based approach. In most cases, primer pairs were designed to include introns within the amplification product, since these are more likely to exhibit variation between genotypes. Genes were amplified and tested for polymorphism in the parents of four diploid potato populations, for which RFLP segregation data also existed, using 4-bp-recognition-site restriction enzymes (generally *Alu*I, *Taq*I and *Rsa*I). CAPS markers were then mapped in one or both populations and the resulting maps integrated on the basis of common markers. This study demonstrated that, given due attention to primer design, and with the availability of multiple populations with common reference markers, it is possible to map reasonable numbers of specific gene sequences in potato using CAPS.

12.3 Genetic Mapping of Agronomically Significant Traits

Since the publication of the first genetic linkage map of potato in 1988, about 60 papers have been published in the scientific literature describing experiments in mapping both quantitative and qualitative traits in potato. While this represents a large number of individual studies, the types of traits mapped can, with a few exceptions, be placed into two broad categories – namely loci associated with disease resistance and loci associated with tuber traits.

12.3.1 Mapping of Loci Associated with Disease Resistance

Chemical control of disease represents one of the biggest costs in the production of potatoes. To control late blight, caused by the oomycete pathogen *Phytophthora infestans*, it is estimated that farmers in Europe apply fungicide to their crop an average of eight times a year at a cost of ca. € 200 (US$ 370) per hectare (Duncan 1999). In the UK alone, it is estimated that up to £30 million sterling (US$ 55 million) p.a. are spent on fungicides for the control of late blight and £9 million sterling (almost US$ 17 million) p.a. are spent on the application of granular nematicides for the control of potato cyst nematodes (Minnis et al. 2002). Thus, the development of cultivars with genetically derived resistance to disease is a high-priority target for both traditional breeding and marker development programs. Both qualitative (vertical) and quantitative (horizontal) resistance is evident for a broad range of diseases in potato germplasm. Indeed, attempts to breed late-blight-resistant cultivars of potato at the beginning of the century provided one of the earliest models of vertical, or qualitative, resistance in plants (Wastie 1991). The introgression of resistance from the wild Mexican hexaploid species *S. demissum* resulted in cultivars with complete (hypersensitive-response-based) resistance to *P. infestans*. However, strains of the pathogen capable of overcoming the resistance developed rapidly (Ross 1986). Breeding and classical genetics experiments in the 1950s and 1960s identified the source of the *S. demissum*-based resistance as a series of dominant genes, designated R genes (Black et al. 1953). Eleven *S. demissum*-derived R genes (*R1* to *R11*) are known, and these form the basis of the pathotype scheme by which races of *P. infestans* are characterized through their ability to overcome resistance conferred by the presence of these genes in a series of differential lines referred to as Black's differential series (Black et al. 1953; Malcomson and Black 1966). The following section contains a brief summary of genetic mapping of disease resistance loci in potato over the last 15 years. For a more in-depth analysis of the area, the reader is referred to the excellent review of Gebhardt and Valkonen (2001), who have summarised the results of many of the disease resistance mapping studies to date and constructed a potato function map for pathogen resistance on which the relative positions of all of these loci are illustrated.

To date, over 20 single dominant genes conferring resistance to fungi (including *P. infestans*), nematodes or viruses have been genetically mapped in potato. Five of the 11 aforementioned *S. demissum*-derived R genes have been mapped: *R1* to chromosome V (Leonards-Schippers et al. 1992), *R2* to chromosome IV (Li et al. 1998) and *R3*, *R6* and *R7* (El-Kharbotly et al. 1994, 1996) to chromosome XI. More recently, dominant resistance genes conferring apparently durable, non-race-specific resistance to *P. infestans* have also been mapped. The R_b gene derived from *S. bulbocastanum* is located on ChrVIII (Naess et al. 2000) and the R_{ber} gene from *S. berthaultii* on chromosome X (Ewing et al. 2000). A dominant resistance gene from *S. pinnatisectum*, named $R_{pi}1$ (which is possibly equivalent to R9) has been mapped to chromosome VII (Kuhl et al. 2001). The only other fungal pathogen of potato for which a major resistance gene has been mapped to date is *Synchytrium endobioticum*, the causal agent of potato wart. The dominant resistance gene *Sen1* (of unknown origin) is located on chromosome XI (Hehl et al. 1999). Major resistance genes for both the golden potato cyst nematode species *Globodera rostochiensis* and the white PCN species *G. pallida* have also been mapped. The gene *Gro1*, derived from *S. spegazzinii*, is located on chromosome VII (Barone et al. 1990). The *S. tuberosum* ssp. *andigena*-derived *H1* gene, which confers durable resistance to pathotype *Ro1*, is located on chromosome V (Gebhardt et al. 1993), as is the *GroVI* gene originating from *S. vernei* (Jacobs et al. 1996).

The gene *Gpa2*, conferring resistance to populations of *G. pallida* found in the Netherlands, has been mapped to the distal end of chromosome XII (Rouppe van der Voort et al. 1997b). The gene R_{Mc1}, conferring resistance to the root knot nematode *Meloidogyne chitwoodii*, has been mapped to chromosome XI (Brown et al. 1996). Several dominant genes conferring resistance to potato viruses have also been mapped. *Rx1* (derived from *S. tuberosum* ssp. *andigena*) and *Rx2* (*S. acaule*), which confer resistance to potato virus X (PVX), map to loci on chromosomes XII and V respectively (Ritter et al. 1991). The PVX resistance gene *Nb* also maps to chromosome V (De Jong et al. 1997), and the *S. phureja*-derived PVX resistance gene Nx_{phu} maps to chromosome IX (Tommiska et al. 1998). The Potato virus Y resistance genes Ry_{adg} (*S. tuberosum* ssp. *andigena*) and Ry_{sto} (*S. stoloniferum*) both map to chromosome XI (Hämäläinen et al. 1998; Brigneti et al. 1997), as does the Potato virus A resistance gene Na_{adg} (also derived from *S. tuberosum* ssp. *andigena*) (Hämäläinen et al. 2000).

As has been observed in other plant species, dominant resistance genes in potato tend to occur in clustered loci, and indeed, approximately three-quarters of all resistance genes mapped in potato to date have been found in only five 'resistance-gene hotspots' located on chromosomes V, XI and XII (Fig. 3). Resistance-gene clusters in potato comprise both genes conferring specificity to different races of the same pathogen and/or genes conferring resistance to completely different pathogens. Both types of cluster are evident, for example, on chromosome XI (Fig. 3). In separate studies, genes conferring resistance to Potato virus A (Na_{adg}), Potato virus Y (Ry_{adg} and Ry_{sto}), the wart fungus *Synchytrium endobioticum* (*Sen1*) and the root knot nematode *Meloidogyne chitwoodii* (R_{Mc1}) were mapped to the distal end of chromosome XI, defined by RFLP anchor marker CP58. On the proximal end of the same chromosome, three of the five mapped *Solanum demissum*-derived late blight resistance genes (R3, R6 and R7) were mapped (in two separate studies) to the same genomic interval near RFLP marker GP185.

QTLs for resistance to many potato pathogens, including *P. infestans*, both PCN species, *Erwinia carotovora* ssp. *atroseptica* and Colorado potato beetle (CPB), have also been mapped in potato. Quantitative, or field, resistance to late blight, which tends to be partial, but race non-specific, is

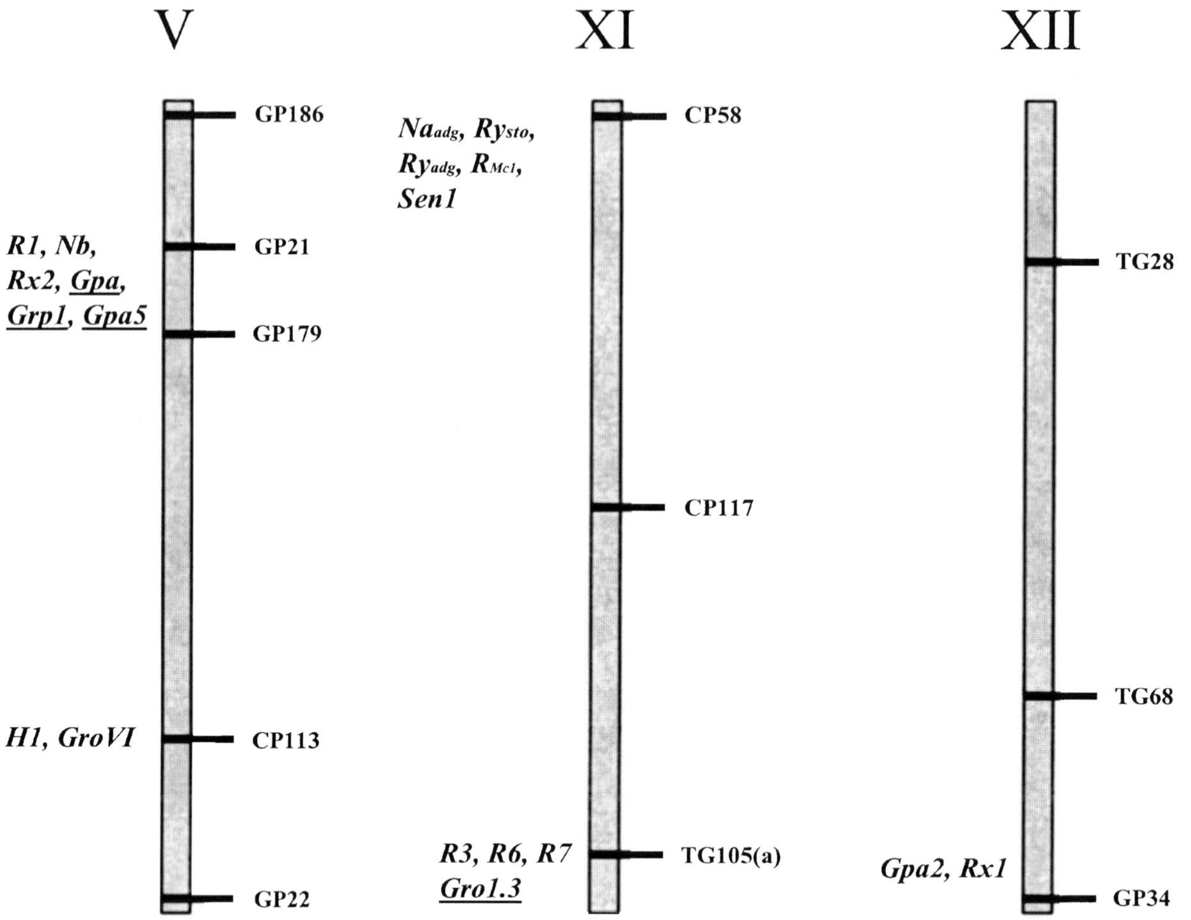

Fig. 3. A schematic representation of the five most extensive resistance gene hotspots in the potato genome. *Italics*: Loci mapped as major dominant genes; *underline*: loci mapped as QTLs

of greater interest to breeders than *S. demissum*-derived R-gene-mediated resistance because of the rapidity with which *P. infestans* can overcome the latter. QTLs conferring resistance to late blight have been mapped on almost every potato chromosome in a number of different diploid and tetraploid populations (Leonards-Schippers et al. 1994; Collins et al. 1999; Oberhagemann et al. 1999; Ewing et al. 2000; Costanzo et al. 2002; Trognitz et al. 2002; Bradshaw et al. 2004). The use of some common RFLP anchor markers in many of the studies has allowed observation of the fact that some of the blight resistance QTLs found in different populations map to similar regions, most notably on chromosomes III, IV, V and VI. The consistent observation of QTLs for blight resistance in homologous positions in several mapping populations is obviously indicative of the possibility that the effects observed at these loci are mediated by the same sets of genes in the different populations. Such consistency of effect is hopeful for the prospect of marker-assisted breeding for quantitative resistance to late blight. However, the genomic intervals encompassed by most of the individual QTLs for blight resistance in the various studies are significant, and caution must be used when inferring common underlying causal genetic components for mapped traits in intervals which potentially harbour hundreds of genes.

Despite this caveat, there is at least one region which consistently displays a high level of association with blight resistance across most of the populations analysed to date. This QTL is located on chromosome V near the RFLP markers GP21 and GP179. This QTL is particularly interesting because it coincides with a major QTL for maturity in many populations. The association between maturity and blight resistance is well known to potato breeders, who have long noted a positive correlation between field resistance to late blight and a late-maturing phenotype in long-day-adapted material (Toxopeus 1958). Of further interest is the fact that this QTL also coincides with the most extensive disease resistance hotspot in the potato genome (Fig. 3), where the *R1* gene, amongst others, has also been located. The first study in which large-effect QTLs for both field resistance to blight and maturity were mapped to chromosome V in the same (diploid) population was carried out by Collins et al. (1999). No major gene for resistance to blight was present in this population, and the authors of the study hypothesised a direct causal link between the maturity phenotype and resistance. In brief, they proposed that late-maturing varieties are in a period of active growth when levels of blight inoculum are at their highest and, as a result, are naturally more resistant to infection at this point. Conversely, early varieties are approaching flowering at this point and are in a more blight-susceptible stage as the blight epidemic peaks. However, as the season progresses, and late-maturing varieties enter the blight-susceptible flowering stage, the pathogen population has decreased to a level where infection is less problematic.

Although this pleiotropic-effect hypothesis is attractive, it cannot be the sole mechanism underlying field resistance to late blight for several reasons. Firstly, there exist QTLs on other chromosomes at which no effects on maturity are detected. Furthermore, in a more recent experiment where a maturity-corrected blight resistance scoring system was used for QTL analysis (Visker et al. 2003), the effect on blight resistance was reduced (by about half) but not eliminated, indicating that causal agents other than a pleiotropic effect of the maturity phenotype may also be involved in resistance at this complex locus. Visker et al. (2003) tentatively acknowledged the possibility that the existence of a resistance-gene cluster/hotspot in this interval may contribute to the remaining effect observed at the locus. The concept that members of resistance-gene clusters may also play a role in quantitative disease resistance in potato has been proposed (Gebhardt and Valkonen 2001). This is supported by the observation (Stewart et al. 2003) that potato genotypes possessing *S. demissum*-derived R genes exhibit greater resistance to complex races of blight than those not possessing R genes, even when the blight races used were known to overcome the particular R genes present. This, combined with genetic mapping evidence, supports a role for 'defeated' R genes in quantitative resistance to late blight.

As mentioned above, the majority of blight resistance QTLs mapped to date do not coincide with QTLs for maturity. In the course of several blight resistance QTL mapping studies in potato (Leonards-Schippers et al. 1994; Oberhagemann et al. 1999; Trognitz et al. 2002) it was observed that genes and gene families known to be involved in pathogen defence response map to similar positions to blight-resistant QTLs. These genes, therefore, are also candidates for the effect observed at QTL loci for resistance to blight, and, given their non-specific nature, to other pathogens as well.

QTLs for resistance to the PCN species *G. pallida* and *G. rostochiensis* have also been mapped in a number of diploid and tetraploid crosses. However, unlike quantitative resistance to late blight, which in most cases seems to be controlled by numerous loci of relatively small effect (after maturity is taken into account), quantitative resistance to PCN seems to be controlled by relatively few loci of large effect. Indeed, the number of loci are so few and the effects sufficiently large to warrant actually naming many of the loci in a manner similar to major disease resistance genes. Much evidence supports the idea that the type of genes underlying these large-effect QTLs are the same resistance genes that are responsible for 'monogenic' resistance to PCN. Firstly, in many cases, quantitative resistance to PCN is not only species specific but pathotype specific, in the same manner as monogenic resistance. Secondly, unlike truly polygenic resistance to late blight, where, in most experiments, QTLs (for resistance and susceptibility) are observed from both parents in the mapping population, resistance to PCN is generally derived solely from the resistant parent. Finally, and perhaps most significantly, in the vast majority of cases, QTLs for PCN resistance map to well-known disease resistance hotspots or to genetic intervals to which at least one major disease resistance gene has been mapped. For instance, the *G. pallida*-specific QTLs *Gpa* and *Gpa5*, the broad-spectrum QTL *Grp1* (which confers resistance to pathotypes of both species), and unnamed QTLs from *S. vernei* and *S. spegazzinii* conferring resistance to *G. pallida* are all located in the largest resistance-gene hotspot on LGV (Fig. 3) (Kreike et al. 1994; Rouppe van der Voort et al. 1998b, 2000; Bryan et al. 2002;

studies. However, van den Berg et al. (1996a) discovered no QTL on chromosome VII, which harboured the largest-effect QTL in the other study. In addition, while Freyre et al. found that their *S. phureja* parent contributed dominant alleles for short dormancy, van den Berg et al. found that *S. berthaultii* contributed recessive alleles for longer dormancy. The differences between these studies highlight the complexity of the trait under analysis and illustrate that it is unwise to assume the control of complex traits in different genetic backgrounds will always be mediated by the same loci. Three of the QTLs for dormancy detected by van den Berg et al. (1996a) coincided with QTLs for tuberisation detected in the same population (van den Berg et al. 1996b), prompting the authors to suggest that these loci might be related to plant hormones or other substances that are involved in the regulation of both traits. In an attempt to test this hypothesis, they followed the segregation of abscisic acid (ABA) content in the tubers of their mapping population (Simko et al. 1997) and mapped three QTLs for this trait directly to chromosomes 4 and 7, and a further QTL, through the epistatic interaction of one of these loci (on chromosome 4), to the distal arm of chromosome 2. This epistatic interaction mirrored one of the epistatic interactions found in the dormancy mapping experiment, indicating a possible role for ABA level in dormancy for this population.

Tuber starch content (sometimes measured as specific gravity) is an important character in potato because it is a component of many quality traits, including suitability for processing, texture and the ability to resist mechanical damage. QTLs for starch content were detected during three separate studies, in a wide range of material and, in some cases, over multiple environments (Bonierbale et al. 1993; Freyre and Douches 1994; Schäfer-Pregl et al. 1998). Numerous QTLs have been identified for the trait on all 12 chromosomes of potato, indicating that starch content is a complex trait under the influence of many genes, and analysis of the same trait over multiple years and environments has shown that many of the QTLs associated with the trait are detected inconsistently, indicating both population specific and environmental influences. As in many other situations, comparison of the relative positions of QTLs over the three studies is rendered difficult by a varying number of shared markers between the experiments. Nevertheless, taking the results of all of these studies as a whole, there is some evidence for reproducible QTLs affecting tuber starch content/specific gravity on chromosomes I, II, III, V and VII.

Schäfer-Pregl et al. (1998) found coincidence of QTLs for tuber starch content and several QTLs for tuber yield, which they ascribed to the fact that tuber starch content has an effect on tuber weight (a component of tuber yield). This study also found significant QTLs for both tuber starch content and yield at the locus on chromosome V that has been shown to harbour a major QTL for maturity in other studies, although no maturity data were available in the populations under study at the time. Physiologically, the accumulation of starch in potato tubers is the net result of several processes involved in carbohydrate metabolism and transport within the plant. Many of the genes involved in these processes have been identified and isolated in potato and other solanaceous species. It is a reasonable assumption that allelic variants of these genes are likely to be the source of the natural variation observed for these traits in potato genotypes, and this idea is certainly supported by some studies in which such genes have been repressed or overexpressed in transgenic systems. If this is the case, then genetic linkage should be evident between such genes and QTLs for tuber starch content in populations segregating for this character.

In order to test this hypothesis, Chen et al. (2001) created a molecular-function map for carbohydrate metabolism and transport by creating a genetic map comprising 85 loci from 69 genes involved in starch synthesis and degradation, sucrose metabolism, membrane transport, the Calvin cycle and photorespiration, glycolysis and the oxidative pentose phosphate pathway and the TCA cycle. This map (which was mentioned earlier with specific reference to CAPS-based mapping) was a synthesis of CAPS and RFLP mapping data for the gene-based markers in four populations, two of which had been used in the study of Schäfer-Pregl et al. (1998) and two of which had been used to construct previous RFLP-based maps. A single unified map showing the relative positions of the 85 functional gene loci was constructed on the basis of shared markers. When the positions of 17 QTLs from the study of Schäfer-Pregl et al. (1998) were overlaid onto the function map, only three of them were not linked to any candidate-gene loci. The remaining QTL intervals coincided with at least one and, in some cases, several functional gene loci, reinforcing the candidacy of these genes for a role in tuber starch accumulation.

As previously mentioned, another aspect of carbohydrate metabolism that is of significance in potato production is the phenomenon of cold sweetening, whereby storage of tubers at low temperatures results in the accumulation of free sucrose, which is subsequently converted into the reducing sugars glucose and fructose. At high frying temperatures, these reducing sugars interact with amino acids in a non-enzymatic Maillard browning reaction, resulting in unacceptably dark and bitter tasting crisps/chips. Menéndez et al. (2002) mapped QTLs for sucrose, fructose and glucose content of potato tubers after a period of cold storage in two diploid potato populations using AFLPs, RFLPs and CAPS. The populations were grown in several different locations over several years, resulting in a total of six separate environments from which sugar content data were scored on both populations. Between the two populations, a total of 24 QTLs with an effect on sugar content were detected on all 12 chromosomes. One advantage of using multiple environments is the ability to assess QTL stability, and for glucose and fructose content, seven and eight QTLs respectively were detected in three to six of the environments. Furthermore, six of these QTLs were coincident for the two traits, conforming to the observation of a high phenotypic correlation between the two sugars. In contrast, of 17 QTLs detected for sucrose content in the two populations, only 3 were reproducible in more than two environments. Six of the 17 sucrose content QTLs corresponded to QTLs exhibiting an effect of glucose and/or fructose content. A total of 6 of the 24 QTLs for sugar content were shown to be repeatably detectable in four or more environments. Statistical analysis showed that the likelihood that this was a chance detection of different QTLs in similar positions was low; suggesting that the same group of genes is responsible for the effect observed at these loci across all environments. Loci such as these, which give a consistent effect across environments, are an obvious first target for MAS.

Ten genes involved in starch and hexose metabolism or transport were mapped to 14 loci in the two populations. Eight of these candidate-gene loci showed an effect on two (and in some cases all three) of the traits. The likelihood that at least some of the candidate-gene loci were responsible for effects observed at QTLs was reinforced by the fact that, for several loci, the candidate genes were the best predictors of the phenotypic variance observed at the QTL. In an older QTL mapping study relating to the cold sweetening phenomenon, Douches and Freyre (1994) mapped six QTLs for variation in chip color (after a period of cold storage) in an interspecific diploid cross. As chip color is mostly a function of reducing sugar content, the authors of the above study attempted to compare the positions of sugar-content QTLs in their maps to chip-color QTLs in the map of Douches and Freyre (1994). As in previous examples, a lack of common markers made comparisons difficult, but tentative coincidences were proposed for QTLs on chromosomes V and X.

As can be seen from the above, the map position of many major genes and QTLs for resistance to major pathogens and various tuber-related characters has been established in potato. This not only lays the groundwork for the possibility of performing MAS for these traits in potato breeding programs, but has also given insights into the mechanisms underlying both disease resistance and the tuber traits examined. In particular, it is apparent that the functional distinction between qualitative and quantitative traits in potato is becoming increasingly blurred. A consistent problem with a genetic mapping approach to resolving the genes underlying QTLs (for any trait) in potato is a lack of the appropriate genetic tools for fine-scale mapping of quantitative traits in this species. The outcrossing nature of potato largely prevents the production of recombinant inbred and near-isogenic lines (RILs and NILs) and mutant populations that have made the candidate-gene approach to QTL analysis in self-compatible species such as tomato so powerful. Approaches such as allele mining of EST databases, linkage disequilibrium mapping and physical mapping/genome sequencing will play an important role resolving many of the questions arising as a result of the genetic mapping studies described here.

12.4 Second-Generation Maps: Linkage and QTL Analysis in Tetraploid Potato

Originally, most molecular-marker-based linkage analyses were performed in populations derived from the F_1 of crosses between two inbred (homozygous) diploid parents (e.g. BCs, F_2, RILs and doubled haploids). This is due to the fact that many agronomically important crop species are fully self-fertile and that in such populations linkage analysis is simplified because only two alleles segregate per locus and the linkage phase (coupling or repulsion) of all markers is known. Linkage analysis is more complicated in populations derived from the progeny of crosses between non-inbred parents of an outbreeding species. Markers vary in the number of segregating alleles, one or both parents may be heterozygous at a locus, and frequently the linkage phase of the markers is unknown (Maliepaard et al. 1997). This is the case in diploid potato crosses, which have been used to avoid the complicating factor of tetrasomic inheritance in most mapping studies to date. Diploid potatoes are highly heterozygous and generally self-incompatible, precluding the possibility of obtaining inbred lines. Formulae for the calculation of recombination fractions and methods for the construction of genetic linkage maps in crosses between heterozygous parents have been presented by Ritter et al. (1990) and Ritter and Salamini (1996). In addition, most computer packages used to construct genetic linkage maps can now deal with segregation data obtained from crosses between heterozygous parents for both dominant and codominant markers, eg. JoinMap (Stam and Van Ooijen 1995). However, most potato breeding is carried out at the tetraploid level, and much variation for traits of interest to breeders is present in cultivars and advanced tetraploid breeding lines. Thus, the ability to perform linkage and QTL analysis at the tetraploid level in potato would be extremely useful in a practical breeding context.

The autotetraploid nature of potato implies the random pairing of four homologous chromosomes at meiosis. This, in conjunction with its high level of heterozygosity, results in a large number of possible allelic combinations at a single locus. In the most extreme case, with four independent alleles contributed by each parent, 36 possible genotypic classes can be found in the progeny. It is the computational difficulty in resolving this number of possible genotypic classes, and the subsequent problem of following recombination events between two such loci (as well as the occurrence of quadrivalents, and double reduction), that has discouraged linkage analysis at this ploidy level.

Several approaches to performing linkage analysis in polyploids were originally devised in species other than potato. The simplest of these was proposed by Wu et al. (1992) and involved using markers present in only one parent and segregating in a 1:1 proportion in the progeny. These 'single-dose' (or 'simplex') markers are found in their simplest form in the F_1 generation of a diploid cross between a dominant heterozygote and a recessive homozygote (AO×OO, where A indicates allele presence and O indicates allele absence). The fact that their inheritance pattern is the same in the F_1 progeny of diploids and polyploids (1:1 presence: absence) means that the methodology for calculating recombination fractions between simplex markers in diploids (Ritter et al. 1990) can also be applied to autopolyploids (this applies to coupling-phase linkages only). This strategy was employed by Al-Janabi et al. (1993, 1994) to construct linkage maps using AP-PCR-based markers in wild *Saccharum* (sugarcane), a complex polyploid of unknown genetic constitution. The disadvantage to this approach is that, using only linkages in coupling, each resulting cosegregation group represents only one homologue of N homologous chromosomes (N = ploidy level). Applying this technique to potato could result in 48 (4×12) LGs per parent. Merging of these homologous groups via the use of repulsion linkages between simplex markers alone is hampered by the reduced sensitivity of their detection and by high associated standard errors in polyploids.

Two approaches were originally devised to overcome this problem. One method involves the use of markers present in 'double dose' (duplex markers) in one parent and absent in the other (AAOO×OOOO). This strategy was originally adopted by Yu and Pauls (1993) to construct a very limited linkage map in tetraploid alfalfa using RAPD markers. A second strategy involves the use of codominant, multi-allelic markers revealing several bands/peaks (depending on detection method) segregating in the expected ratio for simplex markers. Bands from the same assay are assumed to be allelic, and this acts as an anchor point to align homologous groups. This approach was originally deployed in mapping both wild (Da Silva et al. 1993) and cultivated sugarcane (Grivet et al. 1995) using RFLPs.

Both approaches were used in the construction of the first genetic linkage map of tetraploid potato (Meyer et al. 1998). In this study an F_1 population of 78 individuals derived from a cross between the cultivar Stirling and the advanced SCRI breeding clone 12601ab1 was genotyped with both AFLP and SSR markers. Linkages were calculated between simplex (AOOO×OOOO), duplex (AAOO×OOOO) and double simplex AFLP markers (AOOO×AOOO, equivalent to a diploid cross between two dominant heterozygotes; AO×AO, 3:1 segregation ratio). Parallel simulation studies by Hackett et al. (1998) showed that a population of this size allowed detection of only the most reliable linkages (generally coupling phase) and that population sizes of between 150 and 250 would be necessary to detect repulsion-phase linkages, which would allow greater use of available marker information. This inability to make full use of the marker segregation data meant that, despite the availability of over 500 segregating AFLP markers for the population, the resulting maps covered only 12% and 25% of the paternal and maternal genomes respectively. The use of SSR markers allowed the identification of some of the LGs in the tetraploid map relative to diploid maps but was not very successful in identifying homologous groups within each parent. Theoretically, if an SSR had a different simplex allele on each homologous chromosome of both parents, all homologous chromosomes could be identified. However, in practice, the number of alleles detected per SSR locus in the study was quite low (an average of just over three alleles out of a possible eight). This complicates the determination of the parental marker genotype because the gel band phenotype may be compatible with more than one genotype. The impact on linkage mapping is significant since map construction relies on distinguishing parental and recombinant genotypes and thus knowledge of the parental genotype is essential. Manual reconstruction of the parental genotype is possible but very time consuming, particularly where each possible genotype must be tested for linkage with a large number of loci.

In the mapping studies in tetraploid potato described above, SSR alleles had to be assessed as separate dominant markers, reducing the inherent advantage of this marker system. Luo et al. (2000) showed how the codominance of SSRs could be exploited by using the joint segregation ratios of the SSR bands present in both parents and progeny and using this to infer all possible configurations of the parental genotypes for each SSR marker, and their associated probabilities, using Bayes' theorem. Then, for every predicted parental genotypic configuration, the expected number of progeny phenotypes and their frequencies are calculated and compared with the actual frequencies of phenotypes observed in the progeny. Linkage analysis can then be carried out using the most probable parental marker genotype(s). Knowing the phase relationship between pairs of markers is important not only for linkage analysis but in the application of map information to locate QTL affecting traits of importance and to optimise MAS strategies. The challenge of determining linkage phase was approached in further work by Luo et al. (2001), where the EM algorithm was applied to calculate the recombination frequencies and likelihoods between pairs of dominant and/or codominant markers testing up to eight alleles in any phase to determine the most likely configuration(s). Since tightly linked markers are less subject to separation by recombination, several similar map orders can result from these computational analyses with equal likelihood estimates. To examine such marker placements in greater detail, Hackett and Luo (2003) used a 'simulated annealing' approach, to optimise the least-squares criterion for segregating markers, whereby orders close to the optimum are compared to identify markers that remain fixed with respect to map position and markers whose placings are less certain.

The earliest QTL mapping studies in tetraploid potato (Bradshaw et al. 1998; Meyer et al. 1998) were based on testing for associations at single markers using regression-based approaches. However, unless there is sufficiently dense genome coverage by markers, or unless the marker and QTL are tightly linked in coupling, single-marker analysis will still result in a low resolution of QTL location. Hackett (2001) was able to apply interval mapping theory to reanalyse the study of Meyer et al. (1998) and examine the intervals between marker locations. All possible chromosome configurations were constructed based on the observed marker phenotypes for each offspring, and configurations with the minimum number of crossovers consistent with the observed data were identified. It was found that marker types differed in their reliability to reconstruct chromosomal configurations. Not surprisingly, and as mentioned earlier, reconstruction of individual LGs using codominant markers with multiple simplex alleles was more reliable than when mixtures of dominant and codominant markers were used.

For simplicity, the methodology developed by Hackett (2001), Luo et al. (2001) and Hackett and Luo (2003) assumed only bivalent formation during meiosis, and that all gametes have an equal likelihood and viability. This is unlikely to represent the situation in real data with the possibility of multivalent formation, double reduction, departure from random pairing of chromosomes due to incomplete homology, distorted segregation and therefore differential fertility and viability, cytological anomalies and, finally, errors in data collection. To test the above methodology on actual data, therefore, Bradshaw et al. (2004) and Bryan et al. (2004) used an expanded 227-individual family resulting from the same cross as used by Meyer at al. (1998) to more precisely locate QTL for resistance to late blight and *G. pallida*, plant height and maturity using interval mapping. These studies demonstrated that the methodology developed in the previous practical and simulation studies could effectively be deployed for the effective dissection of QTLs using interval mapping in tetraploid potato. It was also evident that a bulked segregant analysis could be very effective in a tetraploid cross to develop markers tightly linked to a large effect PCN resistance QTL. The widespread utility of the approaches described here for linkage and QTL analysis depends on the ability of other groups to deploy them in tetraploid potato. To that end, user-friendly computer programs based on the principles described above are currently under development and will hopefully become available in the near future (J. Bradshaw, personal communication). Further development of the software to allow incorporation of other types of marker data (e.g. SNP), and especially codominant marker data where allele dosage is known, would greatly enhance its utility, particularly for the generation of 'local' linkage maps.

12.5 Second-Generation Maps: The Ultra-High Density Map of Potato

The ultra-high-density (UHD) map of potato has been constructed by a consortium of European potato groups and is based on the application of over 400 AFLP primer combinations to a population derived from two heterozygous diploid parents (SH×RH). The objective was to develop a framework for marker saturation of the potato genome by the construction of a genetic map comprising ca. 10,000 AFLP markers. This project has involved developing a new approach to mapping very large numbers of markers, which conventional mapping software could not accommodate. This approach is demonstrated using chromosome I as a model (Isidore et al. 2003) and relies on the use of a 'bin concept', due to the relatively small size (in terms of possible recombination events) of the UHD mapping population (130 individuals). All marker and map information is available on the Web (http://www.dpw.wau.nl/uhd/). The UHD map has become a very useful resource for all of those performing mapping using AFLP markers in potato. Previous work by van Eck et al. (1995) and Rouppe van der Voort et al. (1997a) looked at the question of 'transferability' of AFLPs in potato, and it is evident that AFLP-based linkage maps derived from different crosses can be aligned, based on the observation that 'comigrating' fragments in different populations are generally allelic. Thus the UHD map functions as an invaluable genetic tool for both map alignment and for identifying chromosomes from linkage map data from new mapping studies. Moreover, the map of the male parent of the UHD cross has become the means for genetically 'anchoring' the potato physical map, currently under construction, and will ultimately form the template for the potato-genome-sequencing effort.

12.6 Map-Based Gene Cloning in Potato

To date most map-based cloning in potato has been directed towards isolation of major genes for resistance to the aforementioned pests and pathogens of cultivated potato. These activities have necessitated the development of dense genetic maps around the targeted resistance loci, as well as a concomitant generation of genomic resources, chiefly large-insert genomic (e.g. BAC) DNA libraries. Prior to these relatively recent efforts, a small number of potato genes (mostly 'housekeeping' and genes involved in well-studied metabolic processes) were cloned from potato, either by homology to other plant species or by random sequencing of cDNA clones from EST libraries. Similarly, many of the RFLP probes used to make the early potato maps are derived from potato cDNA libraries. These will not be discussed further here.

The first successful map-based cloning of a potato resistance gene was isolation of the PVX resistance gene *Rx1*, introgressed into cultivated potato from *S. tuberosum* ssp. *andigena* accession CPC1673 (Bendahmane et al. 1999). A population derived from a self of the cultivar Cara (which carries *Rx1* in a single dose) was used to fine map the gene, and a BAC library of 160,000 clones was constructed from a progeny plant carrying the gene in the duplex condition (Kanyuka et al. 1999). This library was used to isolate a BAC clone carrying the *Rx1* gene. The *Rx2* gene on chromosome V was cloned by homology to *Rx1* (Bendahmane et al. 2000). This work also aided the isolation of *Gpa2*, which resides with *Rx1* in a resistance gene 'hotspot' on chromosome XII (van der Vossen et al. 2000). These genes share certain structural similarities and form part of a cluster of four resistance-gene 'homologues'. The work on the *Gpa2/Rx1* 'cluster' has yielded a lot of important information concerning the evolution and structure of R gene loci and has shown beyond any doubt that resistances to different pests/pathogens can be coded by structurally similar genes from the same cluster.

The *R1* gene confers race-specific resistance to *P. infestans* and has been isolated using a map-based approach (Ballvora et al. 2002). This gene has been found to be a member of the leucine zipper/NBS/LRR class of plant resistance genes, being most similar to the *Prf* gene, which confers resistance to *Pseudomonas syringae* in tomato. *R1* maps to a hotspot for resistance in the GP21-GP179 region of chromosome V and was introgressed into cultivated potato from the wild hexaploid Mexican species *S. demissum*. The resistance allele was observed to represent an insertion of a functional resistance gene with respect to the susceptible allele.

The *RB* gene, which confers broad spectrum resistance to *P. infestans*, was introduced into *S. tuberosum* from the wild Mexican diploid *S. bulbocastanum* via somatic hybridization (Helgeson et al. 1998). Advanced BCs have been used to develop large populations, which have enabled the cloning of the gene using a map-based approach allied to long-range PCR (Song et al. 2003). The unusual PCR approach was used to address the fact that all of 11 BACs identified that spanned the *Rb* locus appeared to be from the susceptible haplotype. This gene is of the CC-NBS-LRR class and is located within a cluster of four very similar full-length genes. This gene holds great promise as a source of broad-spectrum resistance against late blight that can be deployed via transgenesis or though conventional breeding. Another allele of this gene has also been cloned by another group (van der Vossen et al. 2003), and this latter group have recently reported the isolation of another broad-spectrum resistance gene derived from *S. bulbocastanum*, *Rpi-blb2*, which resides on chromosome VI (van der Vossen et al. 2004).

Cloning of the *Gro1-4* gene, which confers resistance to pathotype Ro1 of the cyst nematode *G. rostochiensis*, has been achieved using a candidate-gene approach (Paal et al. 2004). The gene was found to colocalise in a large segregating population with a marker derived from a 'resistance-gene-like' sequence. The marker was used to isolate 15 members of a closely related gene family from genomic libraries. By taking into account different types of information (inheritance patterns in resistant and susceptible germplasm, mapping data, DNA sequence information) it was possible to reduce the number of candidates to three genes, which were subsequently tested for complementation of a susceptible phenotype by stable transformation. The identified functional gene, a member of the TIR-NBS-LRR class, differs from susceptible members of the same family by 29 amino acid changes. This study may pave the way for other resistance-gene-cloning efforts in potato using a similar approach. It is hoped that this type of approach may also lead to the isolation of QTLs conferring partial and durable resistance to the major potato pests and pathogens, which to now has remained an elusive goal.

Isolation of the *R3* locus, which maps to a cluster of *P. infestans* and other resistance genes at the bottom of LG XI, is very close to completion (Huang et al. 2004). This 'gene' turns out to be comprised of two very tightly linked genes with different specificities, *R3a* and *R3b*. This locus is found to be syntenic with the *I2* locus of tomato. It is highly likely that several other potato resistance genes will be cloned in the near future. Efforts are under way to isolate the *H1* gene on chromosome V (Bakker et al. 2004), around which a fine-scale map has been constructed. The genes isolated to date from potato conform to the same types of R-gene structure seen in other plants, further information on which can be found in the excellent review by Hulbert et al. (2001).

12.7 Physical Mapping and Sequencing of Potato Genome

Physical mapping of the potato genome is being performed using a strategy based on the use of the UHD genetic map. Approximately 73,000 BAC clones from a BAC library of the male parent of the UHD male parent (RH) have been fingerprinted using an AFLP-based method. The resultant fingerprints are being analysed using standard software tools for making physical maps. At the time of writing there were 7134 BAC contigs and 7580 'singletons' from 70,988 good quality fingerprint traces. Subsequent contiging will use a reduced stringency alignment approach that will reduce the number of contigs yet further. A sophisticated BAC pool mapping method has been developed for anchoring BACs and BAC contigs to the UHD genetic map (de Boer et al. 2004). This should lead to the development of an integrated genetic and physical map in 2005. The RH physical map will form the basis for the current objective of sequencing the potato genome, proposed by the Potato Genome Sequencing Consortium (*http://www.potatogenome.net*), by the year 2008. In common with the already-in-progress tomato-sequencing project, this recent and ongoing development will have huge implications for all those working in the area of potato genetics and genomics.

12.8 Association Genetics

Trait analysis in crop plants is undergoing a radical shift away from the use of single-pair crosses towards an approach based on analysing trait variation in large populations of unrelated individuals. This new approach relies on the presence of linkage disequilibrium (LD) between trait and marker and is also known as association genetics. For potato, there exist considerable genetic resources in the form of elite cultivars, landraces, breeding lines, primitive cultivars and wild species, and a cross between two such parents captures only a fraction of the potential variation that exists. To date, there have been three published studies in potato (Gebhardt et al. 2004; Simko et al. 2004a,b). In the former study, the authors investigated the association between late blight resistance and plant maturity and markers often found to be linked to QTL for these traits, using a germplasm collection comprising 600 European, American and Asian cultivars. This collection includes cultivars that are related by descent, but with an interesting situation of admixture between related genome segments accrued through 100 years of breeding and selection in many different geographical regions. Earlier studies had located a major QTL for maturity and blight resistance and also the *R1* gene for blight resistance to a 1-cM region on potato chromosome V. Gebhardt et al. (2004) scored the presence or absence of four linked PCR-based markers within this region in the cultivar collection and tested association with phenotypic data for resistance and maturity. LD was found to extend to three of the markers (BA47f2, CosA and the *R1* resistance gene), which are linked within 0.3 cM on an introgressed segment from the wild species *S. demissum*. Interestingly, the marker GP179, previously found in linkage analysis studies to be significantly linked to both traits in an experimental population (Oberhagemann et al. 1999), was not significantly associated in the cultivar collection and was also outside of the introgressed segment, mapping 0.6 to 0.9 cM away from the three markers. Since Ballvora et al. (2002) have made a physical map of the region between BA47f2 and the *R1* gene, this segment being at least 300 kb, this reduces the region in which to search for a candidate gene. This example highlights an important distinction between linkage of a marker to a trait over large genetic distances and the tight linkage required for detecting LD in this study. This latter property can be exploited in MAS experiments where if a marker is tightly linked, markers would be diagnostic for selecting individuals carrying the trait of interest.

The two other examples of association mapping in potato are related studies from the same research group investigating resistance to two species of *Verticillium* fungus (*V. dahliae* and *V. albo-atrum*), which causes vascular wilt disease. Often, resistance to both species is found to be strongly correlated. Simko et al. (2004 a) employed a candidate-gene approach and derived a probe from a *Verticillium* resistance gene in tomato (*Ve1*), using this to identify homologous potato sequences. The sequences were found to map to a single locus, named *StVe1*, and a closely linked SSR marker, STM1051, located 1.5 cM from *StVe1*, was used for LD tests. The association test was carried out using 137 tetraploid cultivars, thought to be closely related due to the narrow genetic base of North American potato germplasm. Association mapping relies on the measurement of any population structure in the sampled germplasm. To do this, Simko et al. (2004 a) constructed a genetic similarity matrix based on the similarity of the STM1051 banding pattern among the genotypes as an indication of their interrelatedness. The difference in the frequency of an allele of STM1051, STM1051-193 successfully divided the population into two subpopulations, with the allele explaining around 10 and 25% in each subpopulation. The *StVe1* locus was further found to comprise an 11-member family of plant-specific leucine-rich repeat containing proteins highly similar to the *Ve* proteins in tomato. The map position of the potato homologues was both found to correspond with the mapped position of the tomato *Ve1* gene on chromosome IX. These pieces of evidence appear to suggest that the potato *StVe1* homologues could be orthologues of the tomato *Ve* genes, although further work would be needed to state this conclusively (Simko et al. 2004 a).

In a second study, Simko et al. (2004 b) tested the association between a haplotype at the *StVe1* locus and resistance to *V. albo-atrum*. They amplified an 839-bp fragment of an *StVe1* homologue from a set of 30 North American tetraploid cultivars. The analysis revealed 56 SNPs, forming 3 distinct haplotypes (A, B and C) with frequencies of 97%, 10% and 33% in the population respectively. Only the C haplotype of the StVe1-839 amplicon was significantly associated with disease status, its presence associated with susceptibility. Since one practical application of this work would be to use this information to improve MAS, Simko et al. (2004 b) subjected the data to a Bayesian analysis to establish the occurrence of false positive and false negative predictions of resistance status of the plants. They showed that using this haplotype alone to predict disease susceptibility was correct in calling truly susceptible and truly resistant individuals only 72% of the time. The authors concluded that in this case, a single haplotype would not provide a diagnostic test and that MAS would be made more efficient by tagging additional *Verticillium* resistance genes.

A major issue in using LD approaches in highly heterozygous species such as potato is that of determining haplotype phase. Several methods have been developed for diploids, but none of them can be utilised for tetraploids, and so phase must be determined experimentally. This can be achieved by cloning multiple PCR products per locus to 'capture' all possible haplotypes, or by generating individuals of lower ploidy to simplify the analysis. Simko et al. (2004 b) used the latter approach and generated monoploids using *Solanum phureja* in their study.

Another important consideration is the extent of LD, which is currently not known across the potato genome. In cultivated species such as potato, where the initial introductory gene pool was limited, LD is likely to be extensive, and this is suggested by the studies in potato where few distinct haplotypes are shared among many individuals. Gebhardt et al. (2004) estimated a frequency of one SNP every 21 bp and an InDel every 243 bp. The study by Simko et al. (2004 b) found a slightly higher frequency of one SNP per 15 bp, which is plausible for the leucine rich repeat region they were studying due to its involvement in resistance-gene evolution. However, to establish the overall frequency of SNPs in the potato genome, multiple loci from several populations need to be analysed.

The use of LD mapping is becoming established as a tool in mapping plant genes and complex traits. By utilising existing populations of germplasm, time and resources involved in generating new populations are saved. LD mapping also offers a more realistic hope for MAS by identifying markers that are truly closely linked and therefore diagnostic for a phenotype of interest, and it furthermore defines the genomic region that should be searched for candidate genes.

12.9 Functional Genomics Resources

Our knowledge of the relationship of genes and traits in cultivated potato has expanded greatly in the last two decades, and this expansion looks set to continue over the coming decade. This chapter has focused on advancements in genetic and physical mapping, map-based cloning and genome sequencing, all of which could be described under the heading of structural genetics/genomics. Parallel to developments in these areas are exciting developments in functional genetics/genomics tools and resources for potato. As previously mentioned, at the time of writing, about 120,000 ESTs are available in potato (*http://www.tigr.org*), and these represent a large (but unknown) portion of the total gene complement of the species. The fact that the ESTs are from a variety of genotypes, tissues, environmental influences and developmental changes is allowing researchers to begin to examine differences in global gene expression patterns for various states and potentially identify key genes active in these states. A non-redundant set of 10,000 of these ESTs has been used by the Institute for Genomic Research (TIGR) to develop a potato microarray that is available to the potato genetics research community at minimal cost. This, and future microarrays, in combination with other functional genomics tools such as virus-induced gene silencing (VIGS) and activation tagged lines, has the potential to facilitate the identification of the role of thousands of potato genes over the next several years. Furthermore, combining structural genetics approaches (such as QTL and candidate-gene mapping) with functional genomics information (such as microarray-derived gene expression data for candidate genes) has great potential for the dissection of many of the complex, polygenic traits mentioned here.

12.10 MAS in Potato Breeding

One of the most significant challenges to be faced over the next decade is how to deploy all of this information in potato cultivar development. The stated goal of most genetic mapping studies in plant species has been to provide a framework for MAS, in order to increase the rapidity and decrease the cost of breeding new varieties. Despite the fact that genetic mapping studies in potato have been going on for 15 years, reported incidences of the actual use of molecular markers in potato breeding programs have been few. One of the greatest impediments to the deployment of MAS in potato breeding to date has undoubtedly been the difficulty associated with developing markers that are truly diagnostic for the trait in question. Association of a particular marker allele with a gene/QTL for a trait in the appropriate parent of a specific mapping population does not guarantee that the same marker allele will be diagnostic for this genetic component in a broad selection of germplasma (Niewöhner et al. 1995). However, with recent advances such as the actual cloning of many disease resistance genes, the demonstration of the potential of association genetics and the availability of the genomics tools and resources described above, this situation is gradually improving, especially for major genes and large-effect QTLs. However, MAS for many of the polygenic traits described in this chapter remains a truly challenging proposition, since it requires the development of diagnostic markers for many loci whose effect varies with environment, genetic background and the presence/absence of other loci. Identifying the genes (and the specific alleles thereof) underlying each individual QTL for such traits, although challenging in itself, may be the most prudent route to truly effective MAS for QTL in potato breeding. The candidate-gene approach described earlier in the chapter represents the first step in this process.

Another challenge in the application of MAS to potato breeding lies in the question of how best to actually deploy the technology in potato breeding programs. It would be difficult, for instance, to simply deploy a MAS scheme directly in a program such as that illustrated in Fig. 2 because of the sheer numbers of genotypes involved. MAS in the context of a program like this would probably best be carried out at the 'prebreeding' stage discussed earlier in order to speed the introgression of target genes from wild species donors into a cultivated potato genetic background (via some form of recurrent BC selection process) and for the development of multiplex tetraploid parents. This might largely obviate the necessity for any marker-based selection in the subsequent stages, since parental material that was triplex or quadruplex for the target genes would definitely transfer these genes to the subsequent generation. Breeding programs that employ extensive progeny testing as part of the selection process (such as SCRI's recurrent genotypic multitrait selection scheme, described earlier) might be more suitable for the direct application of MAS with relatively few modifications, because of the use of fewer parental clones, and the fact that segregating populations suitable for mapping are generated, and phenotyped, as a natural consequence of the program. In the long term, it is likely that a realisation of the full potential of MAS in potato breeding may require a redesign of breeding programs and an adoption of a marker-led ethos in potato breeding schemes rather than adaptation of marker technology to present phenotype-based breeding strategies.

References

Al-Janabi S, Honeycutt R, McClelland M, Sobral B (1993) A genetic linkage map of *Saccharum spontaneum* L. 'SES 208'. Genetics 134:1249–1260

Al-Janabi S, Honeycutt R, Sobral B (1994) Chromosome assortment in *Saccharum*. Theor Appl Genet 89:959–963

Bakker E, Achenbach U, Bakker J, van Vliet J, Peleman J, Segers B, van der Heijden S, van der Linde P, Graveland R, Hutten R, van Eck H, Coppoolse E, van der Vossen E, Bakker J, Goverse A (2004) A high-resolution map of the *H1* locus harbouring resistance to the potato cyst nematode *Globodera rostochiensis*. Theor Appl Genet 109:146–152

Ballvora A, Ercolano MR, Weiss J, Meksem K, Bormann CA, Oberhagemann P, Salamini F, Gebhardt C (2002) The *R1* gene for potato resistance to late blight (*Phytophthora infestans*) belongs to the leucine zipper/NBS/LRR class of plant resistance genes. Plant J 30:361–371

Barone A, Ritter E, Schachtshabel U, Debener T, Salamini F, Gebhardt C (1990) Localisation by restriction fragment length polymorphism mapping in potato of a major dominant gene conferring resistance to the potato cyst nematode *Globodera rostochiensis*. Mol Gen Genet 244:177–182

Bendahmane A, Kanyuka K, Baulcombe DC (1999) The *Rx* gene from potato controls separate virus resistance and cell death responses. Plant Cell 11:781–791

Bendahmane A, Querci M, Kanyuka K, Baulcombe DC (2000) *Agrobacterium* transient expression system as a tool for the isolation of disease resistance genes: application to the *Rx2* locus in potato. Plant J 21:73–81

Bernatzky R, Tanksley S (1986) Toward a saturated linkage map in tomato based on isozymes and random cDNA sequences. Genetics 112:887–898

Black W, Mastenbroek C, Millsand W, Peterson L (1953) A proposal for an international nomenclature of races of *Phytophthora infestans* and of genes controlling immunity in *Solanum demissum* derivatives. Euphytica 2:173–178

Bonierbale MW, Plaisted RL, Tanksley S (1988) RFLP maps based on a common set of clones reveal modes of chromosomal evolution in potato and tomato. Genetics 120:1095–1103

Bonierbale MW, Plaisted RL, Tanksley SD (1993) A test of the maximum heterozygosity hypothesis using molecular markers in tetraploid potatoes. Theor Appl Genet 86:481–491

Bonierbale MW, Plaisted RL, Pineda O, Tanksley SD (1994) QTL analysis of trichome-mediated insect resistance in potato. Theor Appl Genet 87:973–987

Bradshaw JE, Mackay GR (1994) Breeding strategies for clonally propagated potatoes. In: Bradshaw JE, Mackay GR (eds) Potato Genetics. CAB, Wallingford, UK, pp 3–42

Bradshaw JE, Dale MFB, Mackay GR, Wilkinson MJ (1994) Potato Breeding at SCRI: from wild species to finished cultivars. SCRI Annual Report, pp 36–39

Bradshaw JE, Stewart HE, Wastie RL, Dale MFB, Phillips MS (1995) Use of seedling progeny tests for genetical studies as part of a potato (*Solanum tuberosum* subsp. *tuberosum*) breeding programme. Theor Appl Genet 90:899–905

Bradshaw JE, Meyer RC, Milbourne D, Mcnicol JW, Phillips MS, Waugh R (1998) Identification of AFLP and SSR markers associated with quantitative resistance to *Globodera pallida* (Stone) in tetraploid potato (*Solanum tuberosum* subsp. *tuberosum*) with a view to marker-assisted selection. Theor Appl Genet 97:202–210

Bradshaw JE, Todd D, Wilson RN (2000) Use of tuber progeny tests for genetical studies as part of a potato (*Solanum tuberosum* subsp. *tuberosum*) breeding programme. Theor Appl Genet 100:772–781

Bradshaw JE, Dale MF, Mackay GR (2003) Use of mid-parent values and progeny tests to increase the efficiency of potato breeding for combined processing quality and disease and pest resistance. Theor Appl Genet 107(1):36–42

Bradshaw JE, Pande B, Bryan GJ, Hackett CA, McKlean K, Stewart HE, Waugh R (2004) Interval mapping of quantitative trait loci for resistance to late blight (*Phytophthora infestans* (Mont.) de Bary), height and maturity in a tetraploid population of potato (*Solanum tuberosum* subsp. *tuberosum*). Genetics 168:983–995

Brigneti G, Garciamas J, Baulcombe DC (1997) Molecular mapping of the potato virus Y resistance gene $Ry_{(sto)}$ in potato. Theor Appl Genet 94:198–203

Brown CR, Yang C-P, Mojahedi H, Santo S (1996) RFLP analysis of resistance to Columbia root-knot nematode derived from *Solanum bulbocastanum* in a BC2 population. Theor Appl Genet 92:572–576

Bryan GJ, McLean K, Bradshaw JE, De Jong WS, Phillips M, Castelli L, Waugh R (2002) Mapping QTLs for resistance to the cyst nematode *Globodera pallida* derived from the

Kuhl JC, Hanneman RE Jr, Havey MJ (2001) Characterization and mapping of $R_{pi}1$, a late-blight resistance locus from diploid (1EBN) Mexican *Solanum pinnatisectum*. Mol Genet Genom 265(6):977–985

Leonards-Schippers C, Gieffers W, Salamini F, Gebhardt C (1992) The R1 gene conferring race specific resistance to *Phytophthora infestans* in potato is located on potato chromosome V. Mol Gen Genet 233:278–283

Leonards-Schippers C, Gieffers W, Schäfer-Pregl R, Ritter E, Knapp SJ, Salamini F, Gebhardt C (1994) Quantitative Resistance to *Phytophthora infestans* in potato – a case-study for QTL mapping in an allogamous plant species. Genetics 137:67–77

Li X, Van Eck HJ, Van der Voort J, Huigen DJ, Stam P, Jacobsen E (1998) Autotetraploids and genetic mapping using common AFLP markers: the *R2* allele conferring resistance to *Phytophthora infestans* mapped on potato chromosome 4. Theor Appl Genet 96:1121–1128

Lopez de Gomara F (1552) Historia General de las Indias. Zaragoza. (Madrid Edition 1922)

Luo ZW, Hackett CA, Bradshaw JE, McNicol JW, Milbourne D (2000) Predicting parental genotypes and gene segregation for tetrasomic inheritance. Theor Appl Genet 100(7):1067–1073

Luo ZW, Hackett CA, Bradshaw JE, McNicol JW, Milbourne D (2001) Construction of a genetic linkage map in tetraploid species using molecular markers. Genetics 157(3):1369–1385

Malcomson J, Black W (1966) New R-genes in *Solanum demissum* Lindl. and their complementary races of *Phytophthora infestans* (Mont.) de Bary. Euphytica 15:199–203

Maliepaard C, Jansen J, Van Ooijen J (1997) Linkage analysis in a full sib family of an outbreeding plant species: overview and consequences for applications. Genet Res 70:237–250

Martins-Farias R (1976) New Archaeological Techniques for the Study of Ancient Root Crops in Peru. PhD Thesis, University of Birmingham, UK

Masson MF (1985) Mapping, combining abilities, heritabilities and heterosis with 4x×2x crosses in potato. PhD Thesis, University of Wisconsin, Madison

Matsubayashi M (1991) Phylogenetic relationships in the potato and its related species. In: Tsuchiya T, Gupta PK (eds) Chromosome Engineering in Plants: Genetics, Breeding and Evolution, Part B. Elsevier, Amsterdam, pp 93–118

Menéndez CM, Ritter E, Schäfer-Pregl R, Walkemeier B, Kalde A, Salimini F, Gebhardt C (2002) Cold sweetening in diploid potato: Mapping quantitative trait loc and candidate genes. Genetics 162:1423–1434

Meyer R, Milbourne D, Hackett C, Bradshaw J, McNichol J, Waugh R (1998) Linkage analysis in tetraploid potato and association of markers with quantitative resistance to late blight (*Phytophthora infestans*). Mol Gen Genet 259:150–160

Milbourne D, Meyer R, Collins A, Ramsay L, Gebhardt C, Waugh R (1998) Isolation, characterisation and mapping of simple sequence repeat loci in potato. Mol Gen Genet 259:233–245

Minnis ST, Haydock PPJ, Ibrahim SK, Grove IG, Evans K, Russell MD (2002) Potato cyst nematodes in England and Wales – occurrence and distribution. Ann Appl Biol 140(2):187–195

Moseley M (1975) The Maritime Foundations of Andean Civilisation. Cummings Archaeology Series, Menlo Park, CA

Naess SK, Braden JM, Wielgus SM, Haberlach GT, McGrath JM, Helgeson JP (2000) Resistance to late blight in *Solanum bulbocastanum* is mapped to chromosome 8. Theor Appl Genet 101:697–704

Niewöhner J, Samamini F, Gebhardt C (1995) Development of PCR assays diagnostic for RFLP marker alleles closely linked to alleles *Gro1* and *H1*, conferring resistance to the root cyst nematode *Globodera rostochiensis* in potato. Mol Breed 1:65–78

Oberhagemann P, Chatot-Balandras C, Schäfer-Pregl R, Wegener D, Palomino C, Salamini F, Bonnel E, Gebhardt C (1999) A genetic analysis of quantitative resistance to late blight in potato: towards marker-assisted selection. Mol Breed 5:399–415

Paal J, Henselewski H, Muth J, Meksem K, Menendez CM, Salamini F, Ballvora A, Gebhardt C (2004) Molecular cloning of the potato *Gro1-4* gene conferring resistance to pathotype Ro1 of the root cyst nematode *Globodera rostochiensis*, based on a candidate gene approach. Plant J 38:285–297

Ritter E, Salamini F (1996) The calculation of recombination frequencies in crosses of allogamous plant species with applications to linkage mapping. Genet Res 67:55–67

Ritter E, Gebhardt C, Salamini F (1990) Estimation of recombination frequencies and construction of RFLP linkage maps in plants from crosses between heterozygous parents. Genetics 125:645–654

Ritter E, Debener T, Barone A, Salamini F, Gebhardt C (1991) RFLP mapping on potato chromosomes of two genes controlling extreme resistance to potato virus X (PVX). Mol Gen Genet 227:81–85

Ross H (1986) Potato breeding – problems and perspectives. Adv Plant Breed Suppl 13:82–86

Rouppe van der Voort J, Van Zandvoort P, Van Eck H, Folkertsma R, Hutten R, Draaistra J, Gommers F, Jacobsen E, Helder J, Bakker J (1997 a) Use of allele specificity of comigrating AFLP markers to align genetic maps from different potato genotypes. Mol Gen Genet 255:438–447

Rouppe van der Voort J, Wolters P, Folkertsma R, Hutten R, van Zandvoort P, Vink H, Kanyuka K, Bendahmane A, Jacobsen E, Jansen R, Bakker J (1997 b) Mapping the cyst nematode locus *Gpa2* in potato using a strategy based on comigrating AFLP markers. Theor Appl Genet 95: 874–880

Rouppe van der Voort J, Van Eck HJ, Draaistra J, Van Zandvoort PM, Jacobsen E, Bakker J (1998a) An online catalogue of AFLP markers covering the potato genome. Mol Breed 4(1):73–77

Rouppe van der Voort J, Lindeman W, Folkertsma R, Hutten R, Overmars H, Van der Vossen E, Jacobsen E, Bakker J (1998b) A QTL for broad-spectrum resistance to cyst nematode species (*Globodera* spp.) maps to a resistance gene cluster in potato. Theor Appl Genet 96:654–661

Rouppe van der Voort J, Van der Vossen R, Bakker E, Overmars H, Van Zandvoort P, Hutten R, Lankhorst RK, Bakker J (2000) Two additive QTLs conferring broad-spectrum resistance in potato to *Globodera pallida* are localized on resistance gene clusters. Theor Appl Genet 101:1122–1130

Salaman R (1949) The character of the early european potato. Proc Linn Soc 161:71–84

Schäfer-Pregl R, Ritter E, Concilio L, Hesselbach J, Lovatti L, Walkemeier B, Thelen H, Salamini F, Gebhardt C (1998) Analysis of quantitative trait loci (QTLs) and quantitative trait alleles (QTAs) for potato tuber yield and starch content. Theor Appl Genet 97:834–846

Simko I, McMurray, Yang H, Manschot A, Davies P, Ewing E (1997) Evidence from polygene mapping for a causal relationship between potato tuber dormancy and abscisic acid content. Plant Physiol 115:1453–1459

Simko I, Costanzo S, Haynes KG, Christ BJ, Jones RW (2004a) Linkage disequilibrium mapping of a *Verticillium dahliae* resistance quantitative trait locus in tetraploid potato (*Solanum tuberosum*) through a candidate gene approach. Theor Appl Genet 108:217–224

Simko I, Haynes KG, Ewing EE, Costanzo S, Christ BJ, Jones RW (2004b) Mapping genes for resistance to *Verticillium albo-atrum* in tetraploid and diploid potato populations using haplotype association tests and genetic linkage analysis. Mol Genet Genom 271:522–531

Simmonds NW (1969) Prospects for potato improvement. Scottish Plant Breed Stn 48th Annual Report, 1968–69, pp 18–38

Smith MAE (1984) Encyclopedia Americana. Groller, Danbury, CT

Song JQ, Bradeen JM, Naess SK, Raasch JA, Wielgus SM, Haberlach GT, Liu J, Kuang HH, Austin-Phillips S, Buell CR, Helgeson JP, Jiang JM (2003) Gene *RB* cloned from *Solanum bulbocastanum* confers broad spectrum resistance to potato late blight. Proc Natl Acad Sci USA 100:9128–9133

Spooner DM, Hijmans R (2001) Potato systematics and germplasm collecting, 1989–2000. Am J Potato Res 78:237–268, 395

Stam P, Van Ooijen J (1995) JoinMap version 2.0: Software for the calculation of genetic linkage maps. CPRO-DLO, Wageningen, The Netherlands

Stewart HE, Bradshaw JE, Pande B (2003) The effect of the presence of R-genes for resistance to late blight (*Phytophthora infestans*) of potato (*Solanum tuberosum*) on the underlying level of field resistance. Plant Pathol (Oxford) (2)52:193–198

Swaminathan MS, Magoon ML (1961) Origin and cytogenetics of the commercial potato. In: Advances in Genetics, vol 10. Academic, London, pp 217–256

Tanksley S, Ganal M, Prince J, de Vicente M, Bonierbale M, Brown P, Fulton T, Giovannoni J, Grandillo S, Martin G, Messegeur R, Miller J, Miller L, Paterson A, Pineda O, Röder M, Wing R, Wu W, Young N (1992) High density molecular linkage maps of the tomato and potato genomes. Genetics 132:1141–1160

Tommiska TJ, Hamalainen JH, Watanabe KN, Valkonen JPT (1998) Mapping of the gene $Nx_{(phu)}$ that controls hypersensitive resistance to potato virus X in *Solanum phureja* lvP35. Theor Appl Genet 96:840–843

Toxopeus H (1958) Some notes on the relations between field resistance to *Phytophthora infestans* in leaves and tubers and ripening time in *Solanum tuberosum* ssp *tuberosum*. Euphytica 7:123–130

Trognitz FC, Manosalva P, Giysin R, Nino-Liu D, Reinhard S, Herrera MR, Trognitz B, Ghislain M, Nelson R (2002) Plant defence genes associated with quantitative resistance to potato late blight in *Solanum phureja*×*S. tuberosum* hybrids. Mol Plant-Micr Interact 15(6):587–597

Ugent D, Dillehay T, Ramirez C (1987) Potato remains from a late Pleistocene settlement in south central Chile. Econ Bot 4(1):17–27

van den Berg J, Ewing E, Plaisted R, McMurray S, Bonierbale M (1996a) QTL analysis of potato tuberisation. Theor Appl Genet 93:307–316

van den Berg J, Ewing E, Plaisted R, McMurray S, Bonierbale M (1996b) QTL analysis of potato tuber dormancy. Theor Appl Genet 93:317–324

van der Vossen E, van der Voort J, Kanyuka K, Bendahmane A, Sandbrink H, Baulcombe D, Bakker J, Stiekema W, Klein-Lankhorst R (2000) Homologues of a single resistance-gene cluster in potato confer resistance to distinct pathogens: a virus and a nematode. Plant J 23:567–576

van der Vossen E, Sikkema A, Hekkert BTL, Gros J, Stevens P, Muskens M, Wouters D, Pereira A, Stiekema W, Allefs S (2003) An ancient R gene from the wild potato species *Solanum bulbocastanum* confers broad-spectrum resistance to *Phytophthora infestans* in cultivated potato and tomato. Plant J 36:867–882

van der Vossen E, Muskens M, Sikkema A, Gros BJ, Wouters D, Wolters P, Pereira A, Allefs S (2004) The *Rpi-blb2* gene from *Solanum bulbocastanum* is allelic to the *Mi-1* gene from tomato and confers broad-spectrum resistance to *Phytophthora infestans* both in cultivated potato and tomato. In: 1st Solanaceae Genome Workshop 2004, 19–21 Sept 2004, Wageningen, The Netherlands, p 121

van Eck H, Jacobs J, Van Dijk J, Stiekema W, Jacobsen E (1993) Identification and mapping of three flower colour et al. 1998). Many genotypes can produce sizable quantities of anthocyanin and carotenoid plant pigments. Sweetpotato with white- to cream-colored flesh is common in the Pacific, whereas US sweetpotato normally has yellow to orange flesh (Bradbury and Holloway 1988). Some of these are currently considered as a source of natural dye because of their unique stability characteristics, while others (e.g., beta-carotene, a vitamin A precursor) are known to be highly beneficial to human health and nutrition (Woolfe 1992). In developing countries, sweetpotato ranks third in value of production and fifth in caloric contribution to human diet (Plucknett 1991). Sweetpotato offers the highest solar energy fixation capacity per unit area of any crop that can be grown in temperate climate (JRT 2000). Sweetpotato can be grown in areas that are not suitable for corn. Currently, corn is a major crop for ethanol, but as demand increases, it would be more sustainable to develop other bio-based sources of ethanol that can be used either separately or as a blended feedstock for the production of alcohol. Sweetpotato is highly adaptable, drought tolerant, and easy to grow and produces higher yields with minimal fertilizer inputs. There are mounting concerns about groundwater contamination from the gasoline additive methyl tertiary-butyl ether (MTBE), and its recent ban, coupled with increasing gasoline prices, has increased demand for the use of ethanol as an economical, more environment-friendly fuel additive. Currently, the ethanol industry adds more than $2 billion to the US economy each year, but its future contribution is potentially greater (*www.ethanolrfa.org*). Sweetpotato is capable of producing 30% more starch yield per unit area than corn. Indeed, sweetpotato produces more edible energy per hectare per day than any other major food crop, and the enormous potential of this crop as a food and carbohydrate source is widely recognized (Woolfe 1992). Sweetpotato is known for its high-grade starches for the food and pharmaceutical industries.

According to Kays and Kays (1998) and Yoshimoto (1998), Japan started using sweetpotato to produce carbohydrate-based products during the World War II period. The value-added byproducts captured from the ethanol production process are CO_2, distiller's dried grain soluble that can also be used to supplement animal feeds, and high-quality starch and carbohydrate-based polymers for pharmaceuticals. The characteristics of the starch content of sweetpotato storage roots are well documented (Huang et al. 1999). Roughly 80% of the dry matter of a storage root is made up of carbohydrates, and the bulk of this consists primarily of type A starches (similar to most cereals) that are typically composed of amylopectin and amylose in 3:1 to 4:1 ratios (Bradbury et al. 1985). The enzymes α- and β-amylase present in the storage roots readily break down a portion of these starches into readily fermentable sugars.

13.1.2 *Ipomoea* Series *Batatas (Convolvulaceae)*: Phylogenetic Relationships

To date, *Ipomoea* species are A genome types that include *I. triloba* (2n = 30), *I. cordatotriloba* (2n = 30), *I. cynanchifolia* (2n = 30), *I. grandifolia* (2n = 30), *I. launosa* (2n = 30), *I. ramosissima* (2n = 30), *I. tenuissima* (2n = 30), *I. tiliaea* (2n = 60), *I. umbraticola* (2n = 30), and *I. xlecantha* (2n = 30). B genome is characterized in *I. littoralis* (2n = 30), *I. trifida* (2n = 30), *I. tabascana* (2n = 60), *I. batata* var. *apiculata* (2n = 60), *I. batatas* var. *batatas* (2n = 60), *I. batatas* var. *batatas* (sweetpotato) (2n = 90), and *I. setosa* and *I. nil* are out-group species, awaiting further characterization (Ting et al. 1957; Jones 1965, 1974; Jones and Deonier 1965; Martin and Jones 1973; Nishiyama et al. 1975; Oracion et al. 1990; Jarret et al. 1992). Intragenome types of *Ipomoea* are cross-compatible. Austin (1998) and Bohac et al. (1993) have shown the strong morphological relation between sweetpotato and *I. trifida*. Studies using molecular markers (Jarret et al. 1992; McDonald and Marby 1992; Jarret and Austin 1994; Huang and Sun 2000) have confirmed the similarity between *I. batatas* (6×) and *I. trifida*. Based on available information on β-amylase, a fairly conserved nuclear gene (Hattori and Nakamura 1988; Yoshida and Nakamura 1991; Yoshida et al. 1992), Rajapakse et al. (2004) cloned and sequenced β-amylase gene to study phylogenetic relationships in several species. In this study, *I. tabascana*, a tetraploid species with B genome, was shown to be closely related to *I. trifida*. The species *I. tabascana* is a recently discovered species in the series (McDonald and Austin 1990). Austin (unpubl. data) concluded that the *I. tabascana* is more similar to *I. batatas* than to *I. trifida*. Rajapakse et al. (2004) resolved the phylogenetic relationships of these three species to a terminal clade. Successful crossing has been reported between *I. trifida* and *I. ba-*

tatas (Oracion et al. 1990; Freyre et al. 1991; Mont et al. 1993). Exon analysis of *β*-amylase indicates (Rajapakse et al. 2004) that *I. ramosissima, I. umbraticola*, and *I. littoralis* are distantly related to the other species in the series.

13.1.3 Genetic Diversity Among Cultivars

Huaman et al. (1999) characterized 1939 Peruvian collections out of 5000 sweetpotato accessions, available at the International Potato Center (CIP), based on morphological traits and electrophoretic banding patterns of total proteins and esterase. A total of 21 morphological descriptors were used to support a clustering pattern based on UPGMA. Huang and Sun (2000) used inter simple sequence repeats (ISSRs) for restriction site variation in four noncoding regions of chloroplast DNA and scored 2071 bands in 40 accessions of *Ipomoea*. This study included *I. trifida, I. ramosissima, I. umbraticola*, and *I. triloba*. The study concluded that *I. triloba* could be an ancestor of *I. batatas*. He et al. (1995) used DNA amplification fingerprinting (DAF) on 73 plant introductions of sweetpotato drawn from the USA and New Guinea along with tetraploid *I. batatas* and *I. triloba*. In this study, US cultivars formed a single cluster, indicating less diversity, while accessions from New Guinea showed wide variation. To further extend their study, Prakash et al. (1996) used DAF on 30 heirlooms that also included Regal and Excel, lines that are developed using a population-based breeding approach. Regal and Excel have shown greater divergence from other heirlooms. In this study, a total of 144 bands were used to support a phenogram depicting molecular relationships among cultivars. Wang et al. (1998) used DAF on 42 sweetpotato accessions from Guangdong and Fujian provinces of China and from Japan to verify pedigree records. This study concluded that the DAF could resolve the domestication history of sweetpotato germplasm.

Sagredo (1998) used random amplified polymorphic DNA (RAPD) markers to understand the genetic diversity of 28 Chilean cultivars. This study was very informative for understanding historical records of sweetpotato breeding and cultivar collection. Bruckner et al. (2005) presented a comprehensive amplified fragment length polymorphisms (AFLP)-based genetic diversity study on 775 accessions from the Plant Genetic Resources Conservation Unit USDA-ARS in Griffin, GA, USA and the International Potato Center (CIP) in Lima, Peru. The data of 183 polymorphic bands were subjected to ANOVA and principal coordinate analysis to conclude that several clusters existed in the collection.

13.2 DNA Markers

DNA markers (AFLPs, DAF, RAPDs, and ISSRs) are used quite intensively, as reviewed in the previous section, the paucity of codominant markers such as RFLPs, microsatellites (or single sequence repeats, SSRs), and single nucleotide polymorphisms (SNPs) are currently hindering efforts of allele mining in sweetpotato germplasm and marker-assisted selection (MAS). DNA markers are known to greatly enhance breeding methodology by allowing breeders to reduce breeding cycles, establish early generation selections, decipher the genetic complexity of polygenic traits, and pyramid favorable allele combinations. The basic essence of molecular-marker technology is to resolve nucleotide-sequence differences within and between closely as well as distantly related species. The DNA differences could be insertion and deletion events, SNPs that are located in restriction sites gaining or losing classical RFLP sites, or SNPs that are not in restriction sites but can still be resolved based on hetero- and homoduplexes or other advanced polymerase chain reaction (PCR) technologies. In several crops RFLPs are used to detect genomewide homeologous relationships. Several reports of RFLP mapping have been published on many crops including cotton, peanut, rice, tomato, maize, and wheat. Extensive use of RFLPs has facilitated the identification of duplicate chromosomal regions and an understanding of the mechanisms of polyploid formation (Paterson 1995; Lan and Paterson 2000; Osterlund and Paterson 2002).

Many technologies resolve the above-mentioned DNA differences by classical hybridization methods like RFLPs or PCR-based technologies. Based on whether the priming site is intact or lost in a genome, we see dominant markers resolved in populations losing a lot of information. Dominant markers are largely from nontargeted and random primers that are used for single-directional PCRs. Codominant polymorphisms result from the bidirec-

13.4 Gene Identification and Functional Genomics

Tanaka et al. (2005) performed differential display on adventitious roots at different developmental stages of the storage root. The expression patterns were further confirmed by RT-PCR. In this study a total of ten genes were identified. A full-length clone of an SRF6-encoded receptor similar to kinase and structurally similar to the leucine-rich repeat II RLK family of *Arabidopsis thaliana*. Tanaka et al. (2004) cloned dihydroflavonol 4-reductase B(DFR-B) gene from sweetpotato and determined that exons and flanking regions were highly homologous to previously reported DFR-B genes of Japanese morning glory. Three transposable elements (MITEs) were noted, and four different alleles were characterized in different sweetpotato varieties.

Tahara et al. (2004) characterized an active member of *Ty1-copia* retrotransposon with an estimated 400 copies. Transcripts containing long terminal repeats (LTRs) were amplified from callus using the 3′RACE technique. Patterns of sequence-specific amplification polymorphism (S-SAP) of the LTR sequences in the genome were compared between different callus lines and normal plant. Jang et al. (2004) cloned ten POD genes, and specific activity and expression were investigated in four sweetpotato varieties after infection with *Pectobacterium chrysanthemi*. This study helped to understand peroxidase in terms of biotic stress and plant defense organization. Chen et al. (2004) cloned and characterized asparaginyl endopeptidase gene (SPAE) to understand posttranslational processing of pro-proteins. Phylogenic analysis of this gene (SPAE) displayed close association with vascular processing enzymes that function via cleavage for pro-protein maturation in the protein bodies during seed maturation and germination. In conclusion of this study, SPAE was implicated in leaf senescence.

Park et al. (2004) cloned cytosolic ascorbate peroxidase (APX), swAPX1 gene and expressed in cultured cells and mature leaves. These results indicate swAPX1 may be involved in hydrogen peroxide detoxification and thus help to overcome oxidative stress induced by abiotic and biotic stress. Burgmann et al. (2003) tested endophytic *nif*H gene diversity in African sweetpotato. In this study, nitrogenase reductase genes (*nif*H) were amplified by PCR and sequences were used for understanding gene diversity. Park et al. (2003) studied differential expression of six novel peroxidase (POD genes) cDNAs from cell cultures in response to stress. Based on differential expression of POD genes, they proposed that each POD may have different enzymatic properties and physiological stress during cell growth and development.

Kim et al. (2003) cloned strong oxidative-stress-inducible peroxidase (POD) from sweetpotato and characterized transgenic tobacco plants and cultured cells in terms of environmental stress. A POD genomic clone (referred as SWPA2) promoter contained several cis-element sequences implicated in oxidative stress. Employing a transient expression assay in tobacco protoplasts, with five different 5′-deletion mutants of SWPA2 promoter fused to the β-glucuronidase (GUS) reporter gene, the 1314-bp mutant deletion showed about 30 times higher GUS expression than CaMV35S promoter. This promoter is useful for its ability to produce transgenic plants with enhanced tolerance to environmental stress and cell lines engineered to produce key pharmaceutical proteins.

Chen et al. (2003) cloned the ipomoelin (IPO) gene from sweetpotato to investigate its regulation of hydrogen peroxide and nitric oxide upon wounding. This study revealed that the IPO gene is stimulated by hydrogen peroxide whether or not the plant is wounded, but its expression after wounding is totally suppressed by the presence of diphenylene iodonium, an inhibitor of NADPH oxidase. In conclusion, when sweetpotato was wounded, H_2O_2 and NO were produced to modulate the plant's defense system. Together H_2O_2 and NO regulate the expression of the IPO gene, and their interaction might further stimulate plants to protect themselves from invasions by pathogens and herbivores.

You et al. (2003) assembled 2859 sweetpotato cDNA clones into 483 clusters and 442 singletons and compared these sequences with existing nodule/tumor/tuber and development-related databases to identify homologs for 39 ESTs that are potentially involved in gene regulation, signal transduction, and development. Differential expression of 22 out of 39 sequences was identified in early storage root and fibrous root. Chen et al. (2002) characterized the expression and structure of a senescence-associated gene (SPG31) encoding a cystene proteinase precursor of sweetpotato. This study revealed the role of SPG31 in proteolysis and nitrogen remobilization during the leaf senescence process. Further, treatment of mature green leaves with ethylene for 3 d resulted in a high-level induction of SPG31 transcripts. Ethylene-regulated expression of SPG31 is consistent with the presence of a number of putative ethylene-responsive elements in the 899-bp SPG31 promoter region.

Kim et al. (2002) isolated MADS-box genes (IbMADS3 and IbMADS4) from sweetpotato that expressed preferentially in vegetative tissues, especially root tissues, and in white fibrous roots, pigmented roots, and developing tuberous roots. These genes shared sequence similarity with the STMADS group, and their transcripts were localized in a vascular cambium region, indicating their role in the high proliferation potential of vegetative tissues and morphogenesis of vegetative structures. Wang et al. (2002) studied the wound-response regulation of the sporamin gene promoter region. Characterization involved two wound-response-like elements, a G boxlike element and a GCC corelike sequence, that are in the promoter. A construct containing the sporamin promoter fused to a β-glucuronidase (GUS) gene was transferred to tobacco plants. The wound induced a high level of GUS activity in stems and leaves of sweetpotato and tobacco plants. Huang et al. (2001) cloned eight genes that are expressed during leaf senescence in sweetpotato. This senescence associated, ESTs were differentially expressed, and northern blot analysis indicated that all of them were upregulated during natural senescence.

Yao et al. (2001) studied site-directed mutagenesis for a negatively charged trypsin inhibitory loop in sweetpotato sporamin. Kim et al. (1999) characterized two anionic peroxidases (POD genes) that are isolated from suspension cultures and whose expression was investigated with a view to understanding their role in environmental stresses. Ishiguro and Nakamura (1994) characterized SPF1, a novel DNA-binding protein that recognizes SP8 sequences in the 5′ upstream regions of genes coding for sporamin and β-amylase from sweetpotato. This study indicated that the gene encoding SPF1 is present in several copies and could be detected in leaves, stem, and tuberous roots of sweetpotato, in addition to petioles. Takeda et al. (1994) characterized protein phosphatases 1 and 2A that block sugar-inducible gene expression in plants. Lin et al. (1993) cloned a full-length cDNA clone that encodes copper/zinc superoxide dismutase (SOD) of sweetpotato. Hattori et al. (1991) characterized genes coding for sporamin, the most abundant protein of the tuberous root of sweetpotato. These genes are expressed at high levels in the stems of plantlets cultured axenically in a sucrose-containing medium.

13.5 Sweetpotato Transformation and Transgene Stacking

Sweetpotato could be an ideal system for transformation as once a foreign gene is introduced, if the gene is functional in the T_0 generation, by taking advantage of subsequent vegetative propagation, it is less likely that that gene will undergo silencing. Transformation with single transgenes (or single T-DNAs) in sweetpotato can be carried out according to procedures commonly used (Allen 1996; Dickey et al. 1998; Ulker et al. 1999; Love 2000). For stable transformation, biolistic and *Agrobacterium*-mediated transformation procedures are optimized for sweetpotato (Prakash and Varadarajan 1992 a, b; Newell et al. 1995; Gama et al. 1996). Complex transgenics can be made by introducing into the same plant line several genes that stack simultaneously (Chen et al. 1998; Petracek et al. 1998; Love 2000).

13.6 Conclusions

The genes that are currently available in sweetpotato cDNA libraries represent only a small percentage of the total number of genes. Currently 4829 total nucleotide sequences and 466 protein sequences are available for public use in GenBank (http://ncbi.nln.nih.gov). Compared to crops like potato, rice, and corn, very little is known about molecular genetics of species (McGregor et al. 2004). In an attempt to make a global comparison with Arabidopsis arrays, McGregor et al. (2004) used sweetpotato genomic DNA for hybridization with Arabidopsis Genome Array ATH1 (Affymetrix, P/N 900385) and noted the limited amount of homology. Limie et al. (2001) constructed a BAC library with *Bam*HI sites in *pBECBAC1* vector with an insert size of 120 kb and cloned five resistant genes (*RPS2, RPM1, N, L6,* and *Cre*) using degenerate primers.

For sweetpotato, the identification of natural genetic variants can provide much information about gene function and can also be useful for association mapping and linkage disequilibrium analysis. Targeting induced local lesions in genomes (TILLING), a functional genomic technique that goes by the name ECOTILLING when applied to natural populations (Comai et al. 2004; Gilchrist and Haughn 2005), is a powerful reverse genetic technique that employs a mismatch-specific endonuclease to detect induced or natural DNA polymorphisms (from germplasm accessions). Henikoff et al. (2004) used ECOTILLING to detect small deletions, insertions, and microsatellite polymorphisms in addition to single-bp changes in Arabidopsis. Only the ends of the PCR products are labeled for TILLING, and as there are often multiple polymorphisms within a single gene, the conditions of nuclease digestion used are such that only partial cleavage of the DNA is achieved. This results in a collection of fragments that represent cleavage at each of the individual polymorphism sites within the amplicon.

The strength of ECOTILLING is its inherent economy with DNA sequencing. This approach eliminates unnecessary sequencing often required to obtain specific gene sequences from many genotypes in a population (Bendahmane 2005; Meksem et al. 2005). The sequence from one or two representative samples will indicate what the DNA polymorphism is at any particular site in each of the carriers. To extend this nontransgenic approach of high-throughput identification of natural gene variants in germplasm, a major project is currently under way (funded by USDA-CSREES) between West Virginia State University and the US Vegetable Laboratory (USDA-ARS).

References

Allen GC, Hall S, Michalowski W, Newman S, Spiker S, Weissinger AK, Thompson WF (1996) High-level transgene expression in plant cells: Effects of a strong scaffold attachment region from tobacco. Plant Cell 8:899–913

Almazan AM, Begum F, Johnson C (1997) Nutritional quality of sweetpotato greens from greenhouse plants. J Food Comp Anal 10:246–253

Arumuganathan K, Earle ED (1991) Nuclear DNA content of some important plant species. Plant Mol Biol Rep 9:208–218

Austin DF (1988) The taxonomy, evolution and genetic diversity of sweetpotato and related wild species. In: Exploration, Maintenance and Utilization of Sweet potato. Genetic Resources International Potato Centre (CIP), Lima, Peru, pp 27–60

Austin DF (1998) Parallel and convergent evolution in the Convolvulaceae. In: Mathews P, Sivadasan M (eds) Diversity and Taxonomy of Tropical Flowering Plants. Mentor Books, Calicut, India, pp 201–234

Bendahmane A, Tebbji F, Triques K, Sturbois B, Chauvin S, Rameau C, Caboche M, Jamai A, Guillum J, Meksem K (2005) Development of a TILLING Platform in pea and soybean, bridging the gap between structural and functional genomics. In: Plant, Animal and Microbe Genomes XIII Conf, San Diego, CA, USA

Berenyi M, Gichuki S, Schmidt J, Burg K (2002) Ty1-copia retrotransposon-based S-SAP (Sequence-Specific Amplified Polymorphism) for genetic analysis of sweetpotato. Theor Appl Genet 105(6/7):862–869

Bohac JR, Austin DF, Jones A (1993) Discovery of wild tetraploid sweetpotatoes. Econ Bot 47:193–201

Bradbury JH, Holloway WD (1988) Chemistry of tropical root crops: significance for nutrition and agriculture in the Pacific. Australian Centre for International Agricultural Research Monograph 6:201

Bradbury JH, Hammer B, Nguyen T, Tamate J, Anders M, Millar JS (1985) Analysis of vegetables from the highlands of Papua New Guinea. PNG Med J 28:127–130

Bruckner A, Sosinski B, Jarret R, Yencho GC (2005) AFLP-based genetic diversity assessment of global sweetpotato (Ipomoea batatas (L.) Lam.) germplasm resources: Progress toward the development of a sweetpotato core collection. In: Plant and Animal Genome XIII Conf, San Diego

Buteler MI, Jarret RL, LaBonte DR (1999) Sequence characterization of microsatellites in diploid and polyploid Ipomoea. Theor Appl Genet 99(1/2):123–132

Cervantes-Flores Jim C, Sosinski B, Roldan MA, Pecota K, Craig YG (2005) Development of a Genetic Linkage Map in Sweetpotato. In: Plant, Animal and Microbe Genomes XIII Conf, San Diego

Chen JM, Rawlings ND, Stevens RAE, Barrett AJ (1998) Identification of the active site of legumain links it to caspases, clostripain and gingipains in a new clad of cysteine endopeptidases. FEBS Lett 441:361–365

Chen GH, Huang LT, Yap MN, Lee RH, Huang YJ, Cheng MC, Chen SC (2002) Molecular characterisation of a senescence-associated gene encoding cysteine proteinase and its gene expression during leaf senescence in sweet potato. Plant Cell Physiol 43(9):984–991

Chen HJ, Hou WC, Yang CY, Huang DJ, Liu JS, Lin YH (2003) Molecular cloning of two metallothionein-like protein genes with differential expression patterns from sweet potato (*Ipomoea batatas* (L.) Lam.) leaves. J Plant Physiol 160:547–555

Chen HJ, Hou WC, Liu JS, Yang CY, Huang DJ, Lin YH (2004) Molecular cloning and characterisation of a cDNA encoding asparaginyl endopeptidase from sweetpotato (*Ipomoea batatas* (L.) Lam) senescent leaves. J Exp Bot 55(398):825–835

Choisy JD (1984) Convolvulaceae Orientalis. Mem Soc Phys Geneve 6:49–502

Comai L, Young K, Reynolds SH, Codomo C, Enns L, Johnson J, Burtner C, Henikoff JG, Greene EA, Till BJ, Henikoff S (2004) Efficient discovery of nucleotide polymorphisms in populations by ECOTILLING. Plant J 37:778–786

Dickey LF, Petracek ME, Nguyen TT, Hansen ER, Thompson WF (1998) Light regulation of Fed-1 mRNA requires an element in the 5′Untranslated region and correlates with differential polyribosome association. Plant Cell 10:475–484

FAO (1990) Food and Agriculture Organization of the United Nations, Rome. Production Yearbook, vol 43

FAOSTAT (2001) Food and Agriculture Organization of the United Nations, Production Statistics (*http://www.apps.-fao.org/*)

Freyre R, Iwanga M, Orjeda G (1991) Use of *Ipomoea trifida* (HBK) G. Don germplasm for sweetpotato improvement. 2. Fertility of synthetic hexaploids and triploids with 2n gametes of *I. trifida* and their interspecific crossability with sweetpotato. Genome 34:209–214

Gama M, Leite RP, Cordeiro AR, Cantliffe DJ (1996) Transgenic sweetpotato plants obtained by *Agrobacterium tumifaciens*-mediated transformation. Plant Cell Tiss Org Cult 46:237–244

Gilchrist EJ, Haughn GW (2005) TILLING without a plough: a new method with applications for reverse genetics. Curr Opin Plant Biol 8:211–215

Hattori T, Nakamura K (1988) Genes coding for the major tuberous root protein of sweet potato: Identification of putative regulatory sequence in the 5′upstream region. Plant Mol Biol 11:417–426

Hattori T, Nakagawa S, Nakamura K (1990) High-level expression of tuberous root storage protein genes of sweet potato in stems of plantlets grown in vitro on sucrose medium. Plant Mol Biol 14:595–604

He G, Prakash CS, Jarret RL (1995) Analysis of genetic diversity in a sweetpotato (*Ipomoea batatas*) germplasm collection using DNA amplification fingerprinting. Genome 38:938–945

Henikoff S, Till BJ, Comai L (2004) TILLING: traditional mutagenesis meets functional genomics. Plant Physiol 135:630–636

Hu J, Nakatani M, Lalusin AG, Fujimura T (2004) New microsatellite markers developed from reported *Ipomoea trifida* sequences and their application to sweetpotato and its related wild species. Sci Hort 102:375–386

Huaman Z, Aguilar C, Ortiz R (1999) Selecting a Peruvian sweetpotato core collection on the basis of morphological, eco-geographical, and disease and pest reaction data. Theor Appl Genet 98(5):840–844

Huang JC, Sun M (2000) Genetic diversity and relationships of sweetpotato and its wild relatives in Ipomoea series Batatas (Convolvulaceace) as revealed by inter-simple sequence repeat (ISSR) and restriction analysis of chloroplast DNA. Theor Appl Genet 100(7):1050–1060

Huang AS, Tanudjaja L, Lum D (1999) Contents of Alpha-, Beta-carotene, and dietary fiber in 18 sweetpotato varieties grown in Hawaii. J Food Comp Anal 12:147–151

Huang YJ, To KY, Yap MN, Chaing WJ, Suen DF, Chen SC (2001) Cloning and characterisation of leaf senescence up-regulated genes in sweetpotato. Physiol Plant 113(3): 384–391

Ishiguro S, Nakamura K (1994) Characterisation of a cDNA encoding a novel DNA-binding protein, SPF1, that recognises SP8 sequences in the 5′upstream regions of genes coding for sporamin and beta-amylase from sweetpotato. Mol Gen Genet 244(6):563–571

Jang IC, Park SY, Kim KY, Kwon SY, Kim JG, Kwak SS (2004) Differential expression of 10 sweetpotato peroxidase genes in response to bacterial pathogen, *Pectobacterium chrysanthemi*. Plant Physiol Biochem 42(5):451–455

Jarret RL, Austin DF (1994) Genetic diversity and systematic relationship in sweetpotato (*Ipomoea batatas* (L) Lam.) and related species as revealed by RAPD analysis. Genet Resour Crop Evol 41:165–173

Jarret RL, Gawel N, Whittensmore A (1992) Phylogenetic relationships of the sweetpotato [*Ipomoea batatas* (L.) Lam.]. J Am Soc Hort Sci 117:633–637

Jones A (1965) Cytological observations and fertility measurements of sweetpotato (*Ipomoea batatas* (L.) Lam.). Proc Am Soc Hort Sci 86:527–537

Jones A (1974) Chromosome numbers in genus Ipomoea. J Hered 55:216–219

Jones A, Deonier MT (1965) Interspecific crosses among *Ipomoea lacunosa*, *I. ramoni*, *I. trichocarpa* and *I. triloba*. Bot Gaz 126:226–232

Jones A, Dukes PD, Schalk JM (1986) Sweetpotato breeding. In: Bassett MJ (ed) Breeding Vegetable Crops. AVI, Westport, CT, pp 1–35

JRT (2000) Japan Soc for Root and Tuber Crops, Mini White paper, sweetpotato in Japan Vol 21, August 2000. The Foundation for Development of Tuber and Root Crops

Kays SJ, Kays SE (1998) Sweetpotato chemistry in relation to health. In: Proceedings of international workshop on sweetpotato production system toward the 21st Century, 9–10 Dec 1997, Miyakonojo, Japan, Kyushu National Experimental Experiment Station, Miyakonojo, Japan, pp 231–272

Kim KY, Huh GH, Lee HS, Kown SY, Hur Y, Kwak SS (1999) Molecular characterisation of cDNAs for two anionic peroxidases from suspension cultures of sweetpotato. Mol Gen Genet 261(6):941–947

Kim KY, Kwon SY, Lee HS, Hur Y, Bang JW, Kwak SS (2003) A novel oxidative stress-inducible peroxidase promoter from sweetpotato: molecular cloning and characterisation in transgenic tobacco plants and cultured cells. Plant Mol Biol 51(6):831–838

Kim SH, Mizuno K, Fujimura T (2002) Isolation of MADS-box genes from sweetpotato (*Ipomoea batatas* (L.) Lam.) expressed specifically in vegetative tissues. Plant Cell Physiol 43(3):314–322

Kriegner A, Cervantes JC, Burg K, Mwanga RO, Zhang D (2003) A genetic linkage map of sweetpotato (*Ipomoea batatas* (L.) Lam.) based on AFLP markers. Mol Breed 11(3):169–185

Lan TH, Paterson AH (2000) Comparative mapping of quantitative trait loci sculpting the curd of *Brassica oleracia*. Genetics 155(4):1927–1954

Limie H, Revonda P, Yencho CG, Sosinski B (2001) Construction of a Sweetpotato BAC library and its application for resistance gene isolation. In: Plant and Animal Genome IX Conf, San Diego

Lin CT, Yeh KW, Kao MC, Shaw JF (1993) Cloning and characterisation of a cDNA encoding the cytosolic copper/zinc-superoxide dismutase from sweetpotato tuberous root. Plant Mol Biol 23(4):911–913

Love J, Scott AC, Thompson WF (2000) Stringent control of transgene expression in *Arabidopsis thaliana* using the Top 10 promoter system. Plant J 21(6):579–588

Magoon ML, Krishnan R, Vijaya Bai K (1970) Cytological evidence on the origin of sweetpotato. Theor Appl Genet 40:360–366

Martin FW, Jones A (1973) The species of Ipomoea closely related to the sweetpotato. Econ Bot 26:201–215

McDonald JA, Mabry TJ (1992) Phylogenetic systematics of New World Ipomoea (Convolvulaceae) based on chloroplast DNA restriction site variation. Plant Syst Evol 180:243–259

McDonald JA, Austin DF (1990) Changes and additions in Ipomoea section Batatas (Convolvulaceae). Brittonia 42:116–120

McGregor EC, Moser EB, He L, Sosinske B, LaBonte DR (2004) Determining the usefulness of the Arabidopsis ATH1 Genome array for expression studies in Sweetpotato (*Ipomoea batatas* L.). In: Plant, Animal and Microbe Genomes XII Conf, San Diego

Meksem K, Tebbji F, Jamai A, Cooper J, Laport R, Henikoff S, Bendahmane A (2005) TILLING in soybean: A reverse genetic tool for functional gene analysis from two soybean populations. In: Plant, Animal and Microbe Genomes XIII Conf, San Diego

Mont J, Iwanga M, Orjeda G, Watanabe K (1993) Abortion and determination of stages for embryo rescue in crosses between sweetpotato, *Ipomoea batatas* Lam. (2n=6x=90) and its wild relative, *I. trifida* (H.B.K.) G. Don. (2n=2x=30). Sex Plant Reprod 6:176–182

Newell CA, Lowe JM, Merryweather A, Rooke LM, Hamilton WDO (1995) Transformation of sweetpotato (*Ipomoea batatas* (L.) Lam) with *Agrobacterium tumifaciens* and regeneration of plants expressing cowpea trypsin inhibitor and snowdrop lectin. Plant Sci 107:215–227

Nimmakayala P, Janice Bohac, Saha S, Reddy OUK (2004) Microsatellite Based Genetic Map of Sweetpotato. In: Plant, Animal & Microbe Genomes XII Conf, San Diego

Nishiyama I, Miyazaki T, Sakamoto S (1975) Evolutionary autoploidy in the sweetpotato (*Ipomoea batatas* (L.) Lam.) and its progenitors. Euphytica 24:197–208

Oracion MZ, Niwa K, Shiotani I (1990) Cytological analysis of tetraploid hybrids between sweetpotato and diploid *Ipomoea trifida* (H.B.K.) Don. Theor Appl Genet 80:617–624

Osterlund MT, Paterson AH (2002) Applied plant genomics: the secret is integration. Curr Opin Plant Biol 5:141–145

Park SY, Ryu SH, Jang IC, Kwan SY, Kim JG, Kwak SS (2004) Molecular cloning of a cytosolic ascorbate peroxidase cDNA from cell cultures of sweetpotato in response to stress. Mol Genet Genom 271(3):339–346

Park SY, Ryu SH, Kwon SY, Lee HS, Kim JG, Kwak SS (2003) Differential expression of six novel peroxidase cDNAs from cell cultures of sweetpotato in response to stress. Mol Genet Genom 269(4):542–552

Paterson AH (1995) Molecular dissection of quantitative traits: progress and prospects. Genome Res 5(4):321–333

Paterson AH, Bowers JE, Van de Peer Y, Van de Poele K (2005) Ancient duplication of cereal genomes. New Phytol 165(3):658–661

Paterson AH, Bowers JE, Chapman BA (2004) Ancient polyploidization predating divergence of the cereals, and its consequences for comparative genomics. Proc Natl Acad Sci USA 101:9903–9908

Petracek ME, Dickey LF, Nguyen C, Gatz D, Sowinski GC, Allen, Thompson WF (1998) Ferridoxin-1 mRNA is destabilised by changes in photosynthetic electron transport. Proc Natl Acad Sci USA 95:9009–9013

Plucknett DL (1991) Forward. In: Jansson RK, Raman KV (eds) Sweetpotato Pest Management: A Global Perspective. West View, Boulder, CO

Prakash CS, Varadarajan U (1992a) Genetic transformation of sweetpotato by particle bombardment. Plant Cell Rep 11:53–57

Prakash CS, Varadarajan U (1992b) Genetic transformation of sweetpotato. In: Hill WA, Bonsi CK, Loretan PA (eds) Sweetpotato Technology for the 21st Century. Tuskegee University Press, Tuskegee, AL, pp 27–37

Prakash CS, He G, Jarret RL (1996) DNA marker-based study of genetic relatedness in United States sweetpotato cultivars. J Am Soc Hort Sci 121(6):1059–1062

Rajapakse S, Sasanda DN, Matthew M, Robert EB, Austin DF, Bohac JR (2004) Phylogenetic relationships of the sweetpotato in Ipomoea series Batatas (Convolvulaceae)

based on nuclear Beta-amylase gene sequences. Mol Phylogenet Evol 30:623–632

Sagredo B, Hinrichsen P, Lopez H, Cubillos A, Munoz C (1998) Genetic variation of sweetpotato (*Ipomoea Batatas* L.) cultivated in chile determined by RAPDs. Euphytica 101:193–198

Shiotani I (1988) Genomic structure and the gene flow in sweet potato and related species. In: Exploration and maintenance and utilization of sweet potato genetic resources. First planning conference, Lima, Peru, International Potato Centre (CIP), pp 61–73

Tahara M, Aoki T, Suzuka S, Yamashita H, Tanaka M, Matsunaga S, Kokumai S (2004) Isolation of an active element from a high-copy-number family of retrotransposons in the sweetpotato genome. Mol Genet Genom 272(1):116–127

Takeda S, Mano S, Ohto M, Nakamura K (1994) Inhibitors of Protein Phosphates 1 and 2A Block the sugar-Inducible gene expression in plants. Plant Physiol 106(2):567–574

Tanaka M, Nakatani M, Nakazawa Y, Takahata Y (2004) Structural characterisation of the dihydroflavonol 4-reductase B (DFR-B) gene in the sweetpotato. DNA Seq 15(4):277–282

Tanaka M, Takahata Y, Nakatani M (2005) Analysis of genes developmentally regulated during storage root formation of sweetpotato. J Plant Physiol 162(1):91–102

Ting YC, Kehr AE (1953) Meiotic studies in the sweetpotato (*Ipomoea batatas* Lam.). J Heredity 44:207–211

Ting YC, Kehr AE, Miller JC (1957) A cytological study of the sweetpotato plant *Ipomoea batatas* (L.) Lam. and its related species. Am Nat 91:197–203

Ukoskit K, Thompson PG (1997) Autopolyploidy versus allopolyploidy and low-density randomly amplified polymorphic DNA linkage maps of sweet potato. J Am Soc Hort Sci 122:822–828

Ulker B, Allen GC, Thompson WF, Spiker S, Weissinger AK (1999) A tobacco matrix attachment region reduces the loss of transgene expression in the progeny of transgenic plants. Plant J 18(3):253–263

Wang J, He G, Prakash CS, Lu S (1998) Analysis of genetic diversity in Chinese sweetpotato [*Ipomoea batatas* (L.) Lam.] germplasm using DNA amplification fingerprinting. Plant Genet Resour Newslett 113:13–16

Wang SJ, Lan YC, Chen SF, Chen YM, Yeh KW (2002) Wound – response regulation of the sweetpotato sporamin gene promoter region. Plant Mol Biol 48(3):223–231

Woolfe JA (1992) Sweetpotato: An Untapped Resource. Cambridge University Press, New York

Wu KK, Burnquist W, Sorrells ME, Tew TL, Moore PH, Tanksley SD (1992) The detection and estimation of linkage in polyploids using single-dose restriction fragments. Theor Appl Genet 83:294–300

Yao PL, Hwang MJ, Chen YM, Yeh KW (2001) Site directed mutagenesis evidences for a negatively charged trypsin inhibitory loop in sweetpotato sporamin. FEBS Lett 496(2/3):134–138

Yoshida N, Nakamura K (1991) Molecular cloning and expression in *Escherichia coli* of cDNA encoding the subunit of sweetpotato Beta-amylase. J Biochem 110:196–201

Yoshida N, Hayashi K, Nakamura K (1992) A nuclear gene encoding Beta-amylase of sweetpotato. Gene 120:255–259

Yoshimoto M (1998) Sweetpotato as a multifunctional food. In: Proceedings of international workshop on sweetpotato production system toward the 21st century, 9–10 Dec 1997, Miyakonojo, Japan, Kyushu National Experimental Experiment Station, Miyazaki, Japan, pp 273–283

You MK, Hur CG, Ahn YS, Suh MC, Joeng BC, Shin JS, Bae JM (2003) Identification of genes possibly related to storage root induction in sweetpotato. FEBS Lett 536(1/3): 101–105

Zhang D, Hurtado O, Kriegner A, Benavides J, Ghislain M, Li X (2003) Mapping functional genes associated with carbohydrate metabolism in sweetpotato using cleaved amplified polymorphism sequences (CAPS). In: Plant and Animal Genome XI Conf, San Diego

Zhang D, Rossel G, Kriegner A, Hijmans R (2004) AFLP assessment of diversity in sweetpotato from Latin America and the Pacific region: its implications on the dispersal of the crop. Genet Resour Crop Evol 51:115–120

14 Cassava

Y. Lokko[1,4], E. Okogbenin[2], C. Mba[3], A. Dixon[1], A. Raji[1], M. Fregene[2]

[1] International Institute of Tropical Agriculture (IITA), Oyo Road PMB 5320, Ibadan, Nigeria
[2] Centro Internacional de Agricultura Tropical, Apartado Aéreo 6713, Cali, Colombia
[3] Plant Breeding Unit, Joint FAO/IAEA Agriculture and Biotechnology Laboratory, International Atomic Energy Agency Laboratories, A-2444 Seibersdorf, Austria
[4] Current adress: Plant Breeding & Genetics Section, International Atomic Energy Agency, Wagramer Straße 5, PO Box 200, A-1400, Vienna, Austria, *e-mail*: Official.Mail@iaea.org

14.1 Introduction

14.1.1 Brief History of Crop

Cassava (*Manihot esculenta* Crantz) belongs to the family Euphorbiaceae, which is characterized by lactiferous vessels composed of secretory cells. Its relatives in the Euphorbiaceae family include several commercially important plants, such as rubber trees (*Hevea brasiliensis*), castor oil plants (*Ricinus comunis*) and ornamental plants (*Euphorbia* spp.). Cassava grows in the tropics and subtropics and was introduced into Africa and Asia by Portuguese travellers in the 15th century (Jennings and Hershey 1985; Allen 1994).

The center of origin of cassava was first reported to be Central America, including Colombia, Venezuela, Guatemala and Southern Mexico, due to the large number of varieties present there (Sauer 1952; Rogers 1965). *Manihot* species found in the Central America region are only distantly related to cassava (Schaal et al. 1994; Roa et al. 1997). Central America was later referred to as the 'minor' center of origin and Brazil as the 'major' center of origin (Ekanayake et al. 1997).

Cassava is believed to have originated by hybridization between two wild *Manihot* species, followed by vegetative reproduction of the hybrid. Olsen and Schaal (1999) investigated the crop's domestication based on haplotypes of the single-copy nuclear gene glyceraldehyde 3-phosphate dehydrogenase (*G3pdh*) in cassava and its relatives, which confirmed that cassava was likely domesticated from wild *M. esculenta,* mainly from populations of subspecies *flabellifolia* along the southern border of the Amazon basin.

14.1.2 Botanical Descriptors

Cassava, *M. esculenta* Crantz, is a perennial woody shrub with edible tuberous root. It is also called yuca (Spanish), manioc (French) and mandioca (Portuguese). The height of a mature cassava plant usually ranges from 1 to 2 m, although some cultivars reach 4 m. Branching height can be as low as 20 cm, while some varieties never branch and, as a result, never flower. The leaves are simple, with between three and nine leaf lobes (usually odd numbers) arranged spirally around the stem. The plant is monoecious; male flowers develop near the tip, while female flowers develop closer to the base of the inflorescence (Ekanayake et al. 1997). Some varieties flower frequently and regularly, while others flower rarely or not at all. Environmental factors, such as photoperiod and temperature, influence flowering.

Cassava is generally propagated by stem cuttings, thereby maintaining a genotype. However, under natural conditions as well as in plant breeding, propagation by seed is common and farmers in Africa are known to occasionally use spontaneous seedlings for subsequent planting (Silvestre and Arraudeaus 1983). Cassava plants propagated from stem cuttings produce adventitious roots at the base of the cuttings within a week, while cassava plants propagated from seeds first develop a tap root system. The adventitious roots develop into a fibrous root system which within 30 to 60 d increase in diameter and become tuberous roots (also called 'storage roots'). These swell with time due to starch accumulation (Ekanayake et al. 1997).

Cassava is sometimes described as a segmental allotetraploid with basic chromosome number

Genome Mapping and Molecular Breeding in Plants, Volume 3
Pulses, Sugar and Tuber Crops
C. Kole (Ed.)

x=9. Studies on the meiotic behavior of several cassava genotypes have revealed regular 18 bivalent formation of the chromosomes typical of its diploid (2n=2x=36) chromosomes (Jos and Nair 1979). Studies on the inheritance of molecular markers have also further confirmed a diploid genome (Lefevre and Charrier 1993; Sanchez et al. 1999). Although most cassava genotypes studied are diploid, spontaneous polyploid such as triploids (3n) and tetraploids (4n) of some genotypes have been reported (IITA 1980). Triploid and tetraploid plants differ from diploid plants in plant vigour and leaf shape and size. Triploid plants usually grow and yield better than tetraploid and diploid plants. The nucleic acid content of diploid cassava is 1.67 pg per nucleus, which is 772 Mb base pairs in the haploid genome (Awoleye et al. 1994).

14.1.3 Economic Importance

Cassava accounts for approximately one-third of the total staple food production and provides over 50% of the energy for more than 200 million people in sub-Saharan Africa (IITA 1992). In Central Africa alone, cassava is estimated to provide over 1000 Kcal per day to 30 million people (Cock 1985). The storage roots form the basic carbohydrate component of the diet, and the leaves, which contain appreciable amounts of vitamins, minerals and proteins, are consumed as a preferred green vegetable in many parts of Africa, providing protein, minerals and vitamins (Lancaster and Brooks 1983; Hahn et al. 1989).

The crop is grown almost exclusively as food in 39 African countries stretching through a wide belt from the island of Madagascar in the south-east to Senegal and the Cape Verde Islands in the north-west (Hahn and Keyser 1985). It is grown in areas where rainfall exceeds 600 mm over a period of at least 2 to 3 months and altitudes range from sea level to 1800 m. It is particularly important in those areas where food supply is constantly threatened by environmental constraints such as drought and pest outbreaks because of its ability to grow under conditions considered as suboptimal for the majority of food crops. It can be harvested any time from 6 to 24 months after planting and can be left in the ground as a food reserve for household food security in times of famine, drought and war (Cock 1985; Best and Henry 1994; FAO 2000). Global cassava production projected for 2004 is expected to be around 192 million tonnes, as was recorded in 2003, with Africa contributing 103 million (FAO 2004).

14.1.4 Classical Breeding Objectives

Increased yield, multiple pest and disease resistance, desirable agronomic traits such as appropriate plant architecture, early bulking of storage roots, with high dry-matter content, low cyanide content and consumer preference traits, e.g. easy peeling, and early vigour in plant growth (for high foliage yield for leaf vegetable) have been the main breeding objectives. Recently breeding for improved micronutrient content has been emphasised (Gregorio 2002). Broad-based populations for different agroecologies are developed through recurrent selection (intra- and interspecific population improvement techniques) and backcrossing as well as multiple crossing schemes. Superior individuals are selected as parents, and segregating families are generated by multiple crossings among these elite clones, complementing one another for various agronomic, consumer-quality and major pest resistance traits. The breeding values of the parents are evaluated through progeny testing in seedling nurseries. Based on evaluations, selected parental clones or half-sib progenies are hybridized for further improvement in a recurrent selection scheme. Backcrossing has also been a useful procedure for the transfer of resistance into elite populations by providing resistant lines quickly to prevent the severe infestation of relevant pests.

14.1.5 Classical Breeding Achievements

Classical breeding has contributed to improvement in dry-matter and starch content as well as reduction in the cyanogenic potential in cassava. Achievement in cassava breeding in Africa is in the development of a range of elite genotypes, such as TMS 30572 and TMS 4(2)1425, that combine high stable yields, agronomic and consumer quality with acceptable levels of resistance to cassava mosaic disease (CMD) and cassava bacterial blight (CBB) for which these are widely cultivated in Africa (Hahn et al. 1989). The adoption of some of these multiple pest and disease resistance genotypes in Uganda, for instance, curtailed the devastating effect of CMD on cassava production in that country (Otim-Nape et al. 1994).

The introduction of Latin American germplasm into the breeding programs in Africa resulted in significant broadening of the genetic base of cassava in Africa. An IFAD-financed research project executed by the National Center for Research on Cassava and Tropical Fruits (CNMPF, the Brazilian acronym), Cruz das Almas, CIAT and IITA, beginning in 1991 led to the collection and evaluation at representative semi-arid sites of cassava germplasm in north-east Brazil. Genotypes with high efficiency for extraction of limited soil water, low levels of water loss through transpiration and resistance to severe mite attacks were selected and recombined by genetic crosses to form the basis of a breeding population for the semi-arid agroecosystem (CIAT 1996). Subsequently, a farmer participatory plant breeding scheme was employed to evaluate recombinant progenies and select several varieties adapted to the semi-arid region of north-eastern Brazil (Fukuda and Saad 2001). Improved cassava varieties developed in the project had 25 to 100% yield increases over the traditional cultivars in 41 communities of the states of Bahia and Pernambuco of north-east Brazil (CIAT 1999).

Recombinant seeds from the above breeding population for semi-arid ecosystems were also introduced to sub-Saharan Africa (hereafter Africa) through IITA and evaluated in the drier areas of northern Nigeria. Evaluations revealed the potential of the germplasm to increase productivity in the dry and hot conditions of northern Nigeria (IITA 1999). However, the presence in Africa of a major biotic constraint, CMD, a disease not found in Latin America, limited immediate use of the germplasm but required introgression of CMD resistance into the Latin American germplasm. Resistance to CMD was introduced by backcrosses, and thousands of recombinant seeds were produced and distributed to participating countries in West and Central Africa. Evaluation under typical farmer conditions and selection has led to the selection of 25, 32, 10, 28 and 122 improved genotypes that combine CMD resistance and adaptation to water stress for the semi-arid areas of Burkina Faso, Chad, Ghana, Niger and Nigeria respectively (IITA 2000a). The improved germplasm has extended considerably the range of cultivation of cassava beyond its traditional area in the humid and subhumid tropics into the semi-arid zone of West and Central Africa by more than 100,000 ha between 1989 and 1999 (IITA 2000b).

14.1.6 Limitations of Classical Breeding and Utility of Molecular Mapping

Most traits studied in cassava are polygenic (Bryne 1984; Hahn et al. 1989; Rajendran 1989; Amma et al. 1995). Variation in polygenic traits is attributed to quantitative trait loci (QTLs). Quantitative traits in plants and animals are studied using a variety of genetic models, and designs including the analysis of mating designs in segregating populations estimate effective factors using biometrical techniques or molecular markers; biometrical methods can also be used to estimate effective factors for quantitative traits (Zeng et al. 1990; Lynch and Walsh 1998). Most biometrical techniques estimate not the exact number of genes but the number of effective factors. Furthermore, the detection of multiple genes for a trait using segregation analysis alone is not efficient because of the differences due to genotype by environment interaction. Molecular markers, which are not affected by environmental conditions and are insensitive to gene interactions, are suitable for such studies.

Molecular markers occur in large numbers, and their expression is independent of phenotypic value; thus they are powerful tools of genetic research (Beckmann and Soller 1983, 1986). They allow geneticists and plant breeders to locate and follow the numerous interacting genes that determine a complex trait as well as to tag traits controlled by single genes. Molecular markers offer a spectacular improvement in the efficiency and sophistication of plant breeding. They are (1) developmentally stable, (2) detectable in all tissues, (3) unaffected by environmental conditions and (4) virtually insensitive to epistatic or pleiotropic effects. They provide a choice of codominant markers, which can be identified in heterozygotes, or dominant markers, which are identified as present or absent subclasses (Botstein et al. 1980; Williams et al. 1990).

The two main strategies used to identify molecular markers associated with traits of interest are genetic linkage mapping and bulked segregant analysis (BSA) (Tanksley et al. 1989; Giovannoni et al. 1991; Michelmore et al. 1991).

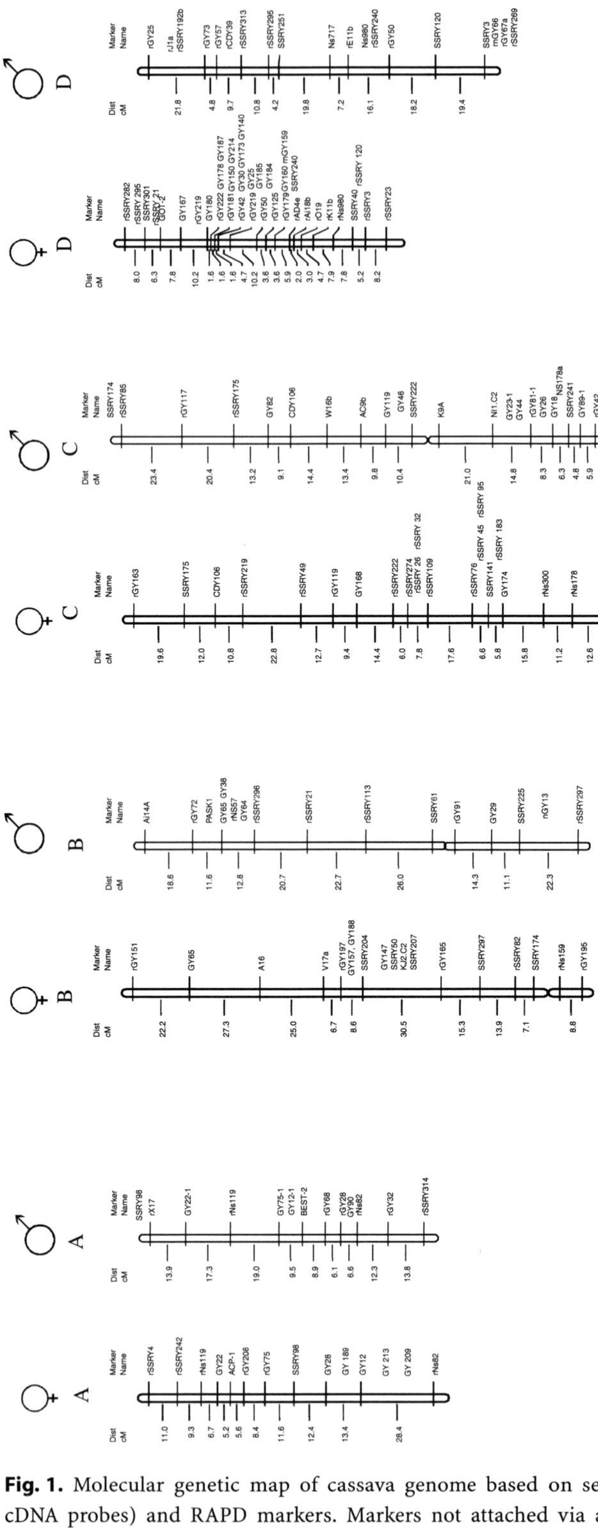

Fig. 1. Molecular genetic map of cassava genome based on segregation of SSR (prefix SSRY and NS), RFLP (genomic and cDNA probes) and RAPD markers. Markers not attached via a *horizontal line* to the LGs do not belong to the framework map (LOD >2.0); those with prefix *r* are linked in repulsion. Map distances on *left* are in Kosambi map units

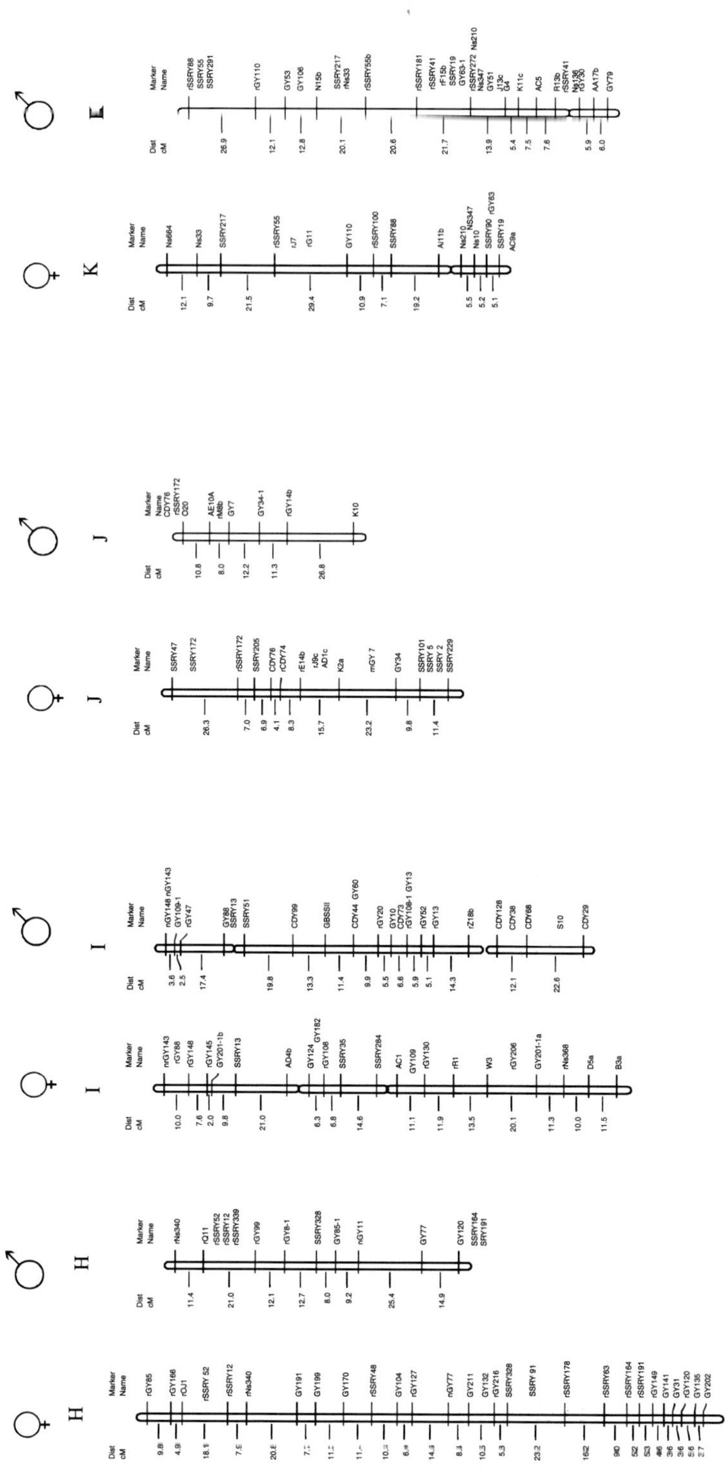

Fig. 1 (continued)

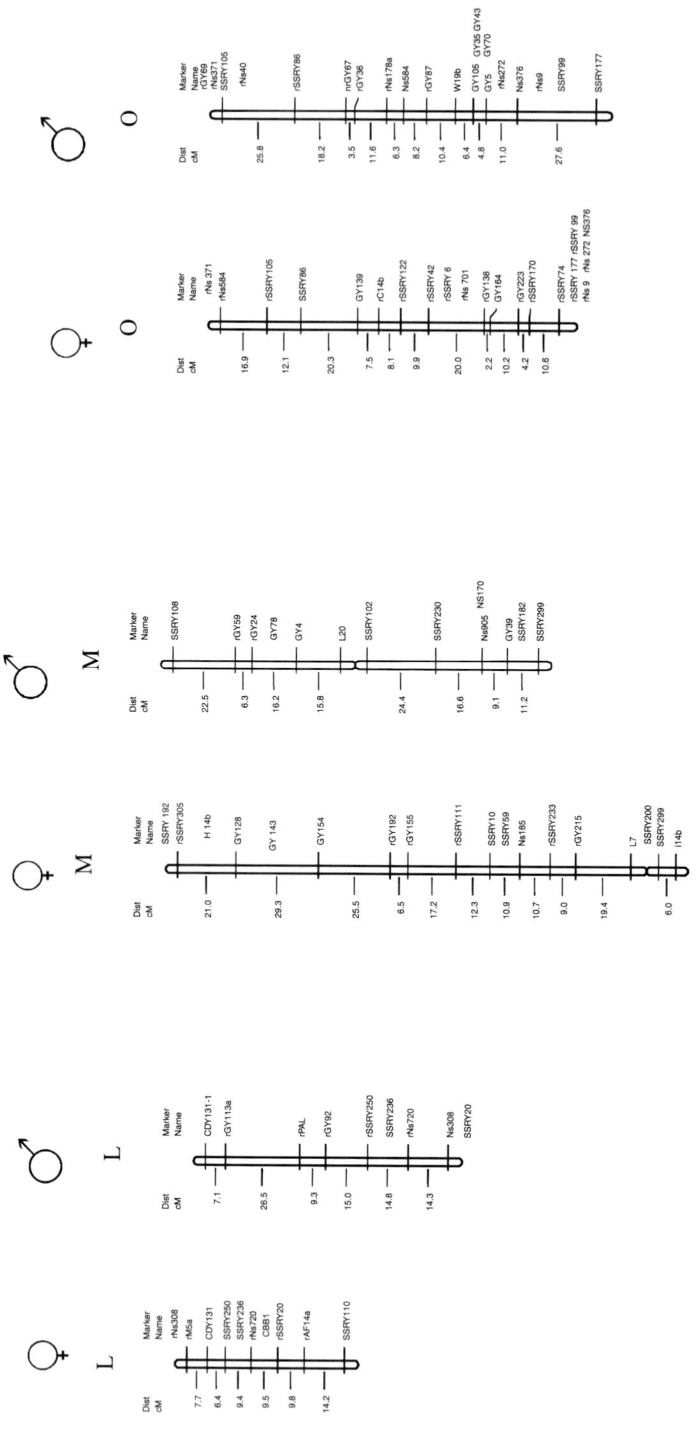

Fig. 1 (continued)

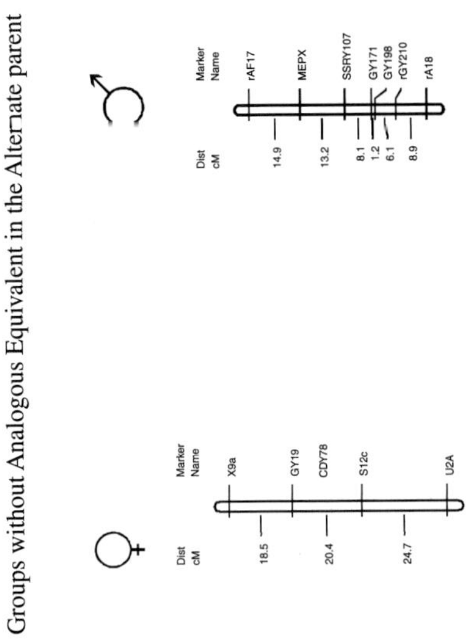

Fig. 1 (continued)

14.3 Gene Mapping

Using bulked segregant analysis (BSA), several SSR and AFLP markers were screened for their association with resistance to CMD in the mapping populations TMS 30555×TME3 and TMS 30555×TME7 (Akano et al. 2002; Lokko et al. 2003). The populations were characterized in multiple environments, and the CMD status of each genotype was used to classify them into highly resistant, resistant, susceptible and highly susceptible individuals. In the TMS 30555×TME3 cross, the CMD-resistant and CMD-susceptible bulks were made up of 40 individuals each. In the TMS 30555×TME7 cross, 10 individuals made up either the CMD-resistant or CMD-susceptible DNA bulks. In the two studies, the SSR marker SSRY28 was tightly linked to CMD resistance. Later on, the group at CIAT identified NS158 to be tightly linked to CMD resistance (Fig. 2).

Marker-trait association detected by regression analysis of TMS 30555×TME7 showed that the marker accounted for 57.41% of total phenotypic variation for resistance. The analysis further showed that another SSR marker, SSRY106, and AFLP marker E-ACC/M-CTC-225 accounted for 35.59% and 22.5%, respectively, of the total phenotypic variation for resistance. Correlation coefficients for SSRY28-180 and E-ACC/M-CTC-225 with CMD responses were significant and negative, while correlation between SSRY106-270 and the CMD responses was significant and positive. This suggests that the CMD resistance genes associated with E-ACC/M-CTC-225 and SRY28-180 are different from genes associated with SSRY106-270. Interval mapping analysis further revealed large QTL regions between SSRY28-180 and E-ACC/M-CTC-225 covering about 50% of the length of the LG and 3% of the length of genome, which explained most of the phenotypic variation.

To date five known genes have been placed on the genetic linkage map TMS 30572×CM 2177-2 developed at CIAT. They include *CMD1*, a recessive resistance gene from TMS 30572 (Fregene 2000), and *CMD2*, a dominant resistance gene from TMS 30555 (Akano et al. 2002). *CMD2* mapped onto LG R of the male cassava framework map 8 cM apart from the marker SSRY28.

14.4 Quantitative Trait Loci Analysis

Quantitative trait loci (QTLs) are genes whose variations contribute incrementally to a multifactorial trait. Molecular mapping and genomic approaches offer new opportunities to dissect major genes and QTLs of important traits. The identification of QTLs affecting important traits is a key step in the use of molecular markers for plant improvement and in understanding the genetic factors that determine these traits. Detection of markers flanking QTLs is the primary goal of QTL mapping. These markers are useful for marker-assisted breeding, prebreeding and map-based cloning of genes. QTL analysis not only provides DNA markers for effi-

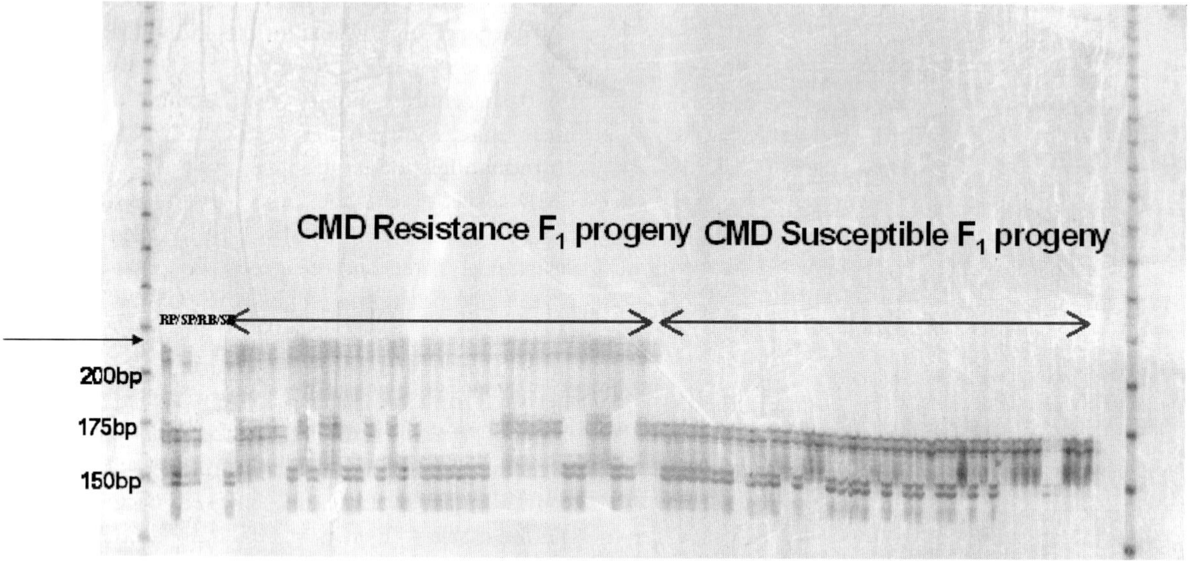

Fig. 2a. Bulked segregant analysis (BSA) in mapping population TMS 30555×TME3 with SSR marker SSRY28 in CMD-resistant parent (RP), susceptible parent (SB), resistant bulk (*RB*), susceptible bulk (*SB*) and 40 resistant and 40 susceptible genotypes used as bulks. The *arrow* points to the SSR allele that is associated with CMD resistance. The molecular-weight marker (*M*) is a 25-bp ladder

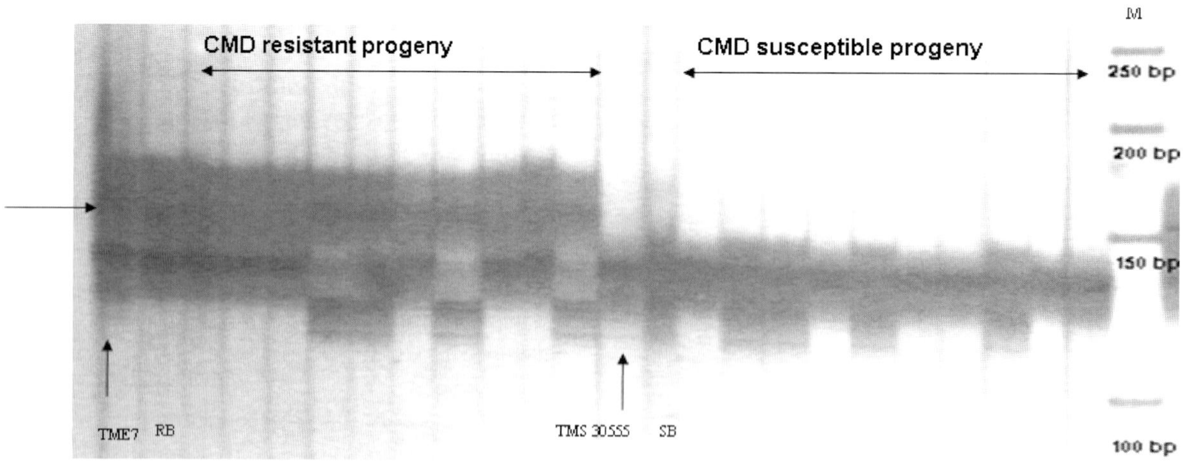

Fig. 2b. BSA in mapping population TMS 30555×TME7 with SSR marker SSRY28 in CMD-resistant parent (TME7), resistant bulk (*RB*), 10 resistant genotypes used as bulk, susceptible parent (TMS 30555), susceptible bulk (*SB*), and 10 susceptible genotypes used as bulk. The *arrow* points to the SSR allele that is associated with CMD resistance. The molecular-weight marker (*M*) is a 50-bp ladder

cient selection, but it is also of particular value in resolving interacting environmental and genetic effects that are common in agronomically important traits.

The TMS 30572×CM 2177-2 F_1 mapping population used for developing the first genetic map of cassava (Fregene et al. 1997) was used for QTL analysis of many traits (Jorge et al. 2000, 2001; Okogbenin and Fregene 2002, 2003). The F_1 population is a full-sib progeny from two non-inbred parents. QTL analysis was done for disease resistance, yield, morphological and quality traits. This study identified QTLs in both male- and female-derived maps with effects of opposite directions for traits studied; thus good alleles increasing the trait value were found in both parents. This result demonstrates the ability of marker analysis to uncover cryptic genetic variation that otherwise would have been masked between the parents. QTL analysis has revealed that the gene effects were not equally distributed among QTLs of each trait.

The genetic basis of early bulking (early yield at 7 MAP) in cassava was studied in sequential harvests to identify traits and genes strongly associated with early bulking. Analyses of regression revealed harvest index, dry shoot yield, and storage root diameter as the most important factors influencing early bulking. The phenotypic variance accounted for by individual QTLs in the early bulking study ranged from 7 to 33%. Some of the QTLs identified were stable throughout the crop growth cycle while others were only active at specific stages in the growth cycle. Of the 18 QTLs found for early bulking, 7 (39%) coincided with QTLs associated with one or more traits that exhibited significant influence on early yield. For traits associated with early bulking QTLs detected includes 2 for dry foliage, 8 for harvest index and 17 for root diameter. Three QTLs were also detected for bulking rate. The results indicate that sink and source capacities are very important in determining early yield in cassava (Okogbenin and Fregene 2002).

Attempts were also made to identify QTLs in the TMS 30572×CM 2177-2 F_1 population for several productivity and plant architecture traits at two locations (Palmira and Quilichao) in Colombia in 1998 and 1999. Architecture traits evaluated were plant height, branching height, branching levels, branching index, stem portion with leaves, and leaf area index. Productivity traits were those related to total dry-matter production and distribution, viz. fresh root yield, fresh shoot yield, harvest index and number of storage roots. A total of 30 primary QTLs (from a single QTL model) and 84 secondary QTLs (through a primary QTL interaction model) were detected. Okogbenin and Fregene (2003) identified 35% of the detected QTLs in two or more trials; the other QTLs were environment specific. This implies that the expression of QTLs can vary among environments. The results underscore the significant genotype (QTL)×environment interactions found for most of the traits. All QTLs identified for fresh root yield were found associated with component traits of productivity or architecture. The results indicate that QTLs for plant architecture can be used to improve productivity.

Genome locations of the wound-response genes that were expressed during postharvest physiological deterioration (PPD) of cassava, such as phenylalanine lysae, β-1.3 glucanase, and hydroxyproline-rich glycoprotein catalase, have been identified on the molecular genetic map of cassava. Molecular markers linked to putative QTLs (explaining between 5 and 13% of the phenotypic variance) influencing the PPD of cassava were mapped onto a TMS 30572×CM 2177-2 mapping population (Cortes et al. 2002). Quality-related traits analysed in the TMS 30572×CM 2177-2 F_1 population resulted in the detection of four QTLs for culinary quality, five for amylose contents and six for starch content.

Cassava bacterial blight is one of the most important diseases of cassava. Jorge et al. (2000, 2001) identified QTLs for cassava bacterial blight resistance, which explains 9 to 27% of phenotypic variance of response to five Xam strains. QTL analysis of cassava leaf morphology (the leaf width/leaf length ratio), a quantitative assessment of leaf shape (LS), indicates that this trait is controlled by a major gene in LG H. This gene accounts for 80% of the phenotypic variance. LS is thus simply inherited, with broad leaf shape being recessive to the narrow LS.

QTL mapping in the TMS 30572×CM 2177-2 population also showed that QTLs affecting a number of quantitative traits were distributed through out the genome. However, certain chromosomal regions appeared to contribute greater effects than others. The clustering of QTLs of several traits of agronomic importance was found in a single region in LGs of the genome. The low level of recombination of on LG D, relative to the rest of the genome, and its large number of markers suggest it might have been an introgressed segment from *M. glaziovii*. Gene effects for different allele combinations (heterozygote vs. homozygote) were also examined. The high proportion of QTLs with improved performance for the heterozygote condition suggests that the superiority of cassava performance for yield strongly derives from its high heterozygosity. Thus, heterosis can be fully exploited in the development of high-yielding cassava varieties.

QTLs associated with the polygenic and recessive source of resistance to CMD in the TMS 30572×TME117 population were studied based on the mean disease severity scores (MDSSs) of the progeny in two growing seasons in Nigeria (Lokko et al. 2004). Five highly significant ($P<0.001$) marker-associated QTL effects, explaining between 10.47 and 12.15% of the total phenotypic variation, were detected by regression. The Kruskal-Wallis analysis further detected four highly significant marker-associated QTL effects. Significant marker-trait associations were due to markers donated by both parents, which confirms the polygenic and recessive nature of this source of resistance.

An F_2 SSR-based genetic map was recently constructed (Okogbenin et al. 2006) to further test the stability of QTLs identified in F_1 under a different genetic background. Preliminary genetic analysis with the F_2 genetic map has revealed QTLs in similar genomic regions with the TMS 30572×CM 2177-2 F_1 genetic map for early bulking. Thirteen QTLs associated with traits linked to early bulking were found to be linked in both populations to the same markers, suggesting that allelic variation at the same loci may be responsible for the effects. It may be possible that effects associated with markers mapping to similar regions on homologous LGs resulted from the same QTL. Conversely, the possibility that effects linked to the same marker may result from different, closely linked QTLs cannot be ruled out. In an F_2 population one can determine the effect of different gene actions on the phenotype because all three possible gene dosages at the locus are represented. In the F_2 genetic mapping population QTLs associated to early bulking with pure additive gene action have been identified. Genetic advances can easily be predicted for accelerated improvement of quantitative traits in cassava with the identification of pure additive QTLs. If a significant amount of the additive variance associated with a QTL can be accounted for by molecular markers, then marker-assisted selection (MAS) can increase breeding efficiency (Edwards and Page 1994).

Identified QTLs hold great promise for introgression of genes into breeding programs. The localization of molecular markers closely linked to loci that condition agronomic traits (QTLs) can be used to search efficiently for new useful genes. More genetic analysis is still required to effectively maximise the application of identified QTLs for MAS in cassava improvement. Based on the initial mapping results, research efforts are under way aimed at MAS to validate the usefulness of identified QTLs.

14.5 Marker-Asisted Breeding

Genomic tools, particularly molecular markers, will permit an immediate increase in efficiency in cassava breeding pending when inbred lines are developed. Molecular markers are particularly useful in:

1. Using a large set of progenitors, especially when no information on combining ability is available, in breeding schemes and the subsequent elimination of a large part of the resulting progeny very early in the breeding cycle.
2. Increasing the number of replications and environments for testing breeding lines (thus achieving a more accurate performance of the genotypes), by working with reduced progeny sizes.
3. Increasing heritability by the elimination of the confounding effect of the environment; this is particularly true in cassava where the first stage of selection is conducted on a single plant per genotype or low/absence of pathogen/pest pressure in initial sites of selection (nurseries), or a trait that has a large environmental influence.
4. The pyramiding of different sources of resistance against a single pathogen or the combination of resistance to several pests and diseases and other traits without resort to time-consuming field evaluations, also known as 'express breeding'.

The success of conventional cassava breeding is inversely proportional to the number of traits it attempts to improve. Combining many genes controlling quantitative traits from diverse sources into a single variety is a long-term, high-risk venture that requires every available tool for success. MAS can be used to achieve these goals more efficiently. For example, CMD-resistant donor parents can be crossed to other parents with excellent resistance to CGM, and markers used to select recombinants that combine resistance to CMD and CGM in a single generation, without the need for field trials. Resulting selections can then be crossed to other genotypes that carry, for example, high β-carotene content to produce multitrait hybrids, again without the need for field evaluations. The best of these selections are then crossed to elite progenitors of the appropriate gene pool, to capture genes for yield and adaptation, and the resulting hybrids are selected with markers to eliminate those progenies that do not have resistance to CMD, CGM and high β-carotene content, leaving a smaller number of progeny to be thoroughly evaluated in the regular breeding scheme.

Molecular MAS for breeding resistance to CMD has been successfully implemented for introducing resistance into elite gene pools at CIAT (CIAT 2003; Fregene et al. 2006) and also to introgress resistance to CGM and CMD in local cassava varieties in Tanzania (Kullaya et al. 2004). With the successful development of markers for resistance to CMD and CGM, efforts have turned to developing markers for another major source of loss to cassava production, postharvest physiological deterioration (PPD). Dramatically delayed PPD was found in *Manihot walkerae*, a wild relative of cassava found in Mexico and the USA (Texas) (Ceballos et al. 2006). An accession of *M. walkerae* (MWal 001) was crossed extensively to elite cassava varieties. A single successful genotype was found with delayed PPD. The storage roots of the hybrid remained intact a month after harvest. Backcrosses of this hybrid to elite progenitors of the CIAT cassava gene pools and selfed (S_1) populations were made for genetic mapping of the delayed PPD traits. Genetic mapping of the delayed PPD genes is progressing, and, following the identification of genes involved in the regulation of PPD, MAS will be used to combine these genes with progenitors that already have combined CMD and CGM genes.

CIAT and a number of partners have initiated a project to genetically fortify cassava with the inherent ability to produce higher levels of β-carotene. This is one way of combating the deficiency of this key micronutrient in areas where cassava is a major staple. The experimental approach to increasing cassava β-carotene content includes conventional breeding and genetic transformation. The discovery of a wide segregation pattern of root color in two S_1 families from the Colombian landrace MCol 72 (cross code AM 273) and the Thai variety MTAI 8 (AM 320) was the basis for molecular genetic analysis of β-carotene content in cassava (CIAT 2003). Three SSR markers were found to be associated with β-carotene content (Fregene and Mba, unpublished results). One of these markers explained 30% of phenotypic variation for β-carotene content in the population used for this study. The homozygous state of certain alleles of these markers translates into higher β-carotene content, suggesting that breeding for β-carotene can benefit from molecular markers to assist in combining these favorable alleles in breeding populations. Few key traits in cassava hold greater potential for increasing cost-effectiveness via molecular MAS than root dry-matter content (DMC). This trait is usually measured at the end of the growth cycle. A number of genetic and environmental effects influence DMC. It is usually the highest before the onset of the rains but drops after the rains begin as the plant mobilises starch from the roots for re-growth of leaves (Bryne 1987). Defoliation from pest and disease attacks can lower DMC. Breeding programs have been quite successful in improving DMC, especially for industrial markets.

The entry point for developing markers associated with DMC was three diallel experiments carried out from 2000 to 2002 (Jaramillo et al. 2005). Diallels, in this case made up of 90 families, are an ideal method to identify genes controlling DMC that are useful in many genetic backgrounds. Estimates of general and specific combining ability (general combining ability (GCA) and specific combining ability (SCA) respectively) for many traits of agronomic interest were calculated, with an emphasis on DMC. Based on GCA estimates, parents were selected to generate larger sized progenies for DMC mapping. Sizes of families in the original diallel experiment were about 30 progenies, a rather small size for genetic mapping. Parallel to the development of mapping populations was the search for markers associated with DMC using two F_1 families GM 312 and GM 313 selected from the diallel experiment having parents with high GCA for DMC.

cross-hybridize (Anderson et al. 2001). Using DNA macroarrays developed using clones from a leafy spurge EST database (Anderson and Horvath 2001), we further demonstrated that at least 35% of the leafy spurge clones showed greater than 2× above-background hybridization with dehydration-stressed cassava target sample (Anderson et al. 2005). Based on this finding, clones from the leafy spurge EST database are being used as probes to monitor gene expression profiles in cassava tissues challenged with various environmental stresses as well as other Euphorbiaceae family members. Genes involved in the G1/S and G2/M cell cycle progression in cassava have also been identified (Anderson et al. 2005).

To date 20,000 ESTs have been generated for clones isolated from two normalised cDNA libraries of cassava. The libraries were constructed from drought-stressed and control tissues of the African accessions TME117 (Isunikankiyan). Preliminary analysis of a subset of these revealed that 73.3% of the sequences had coding potential, 3741 sequences were single unique genes and 1451 were unigene clusters with an average cluster size of 3.53. A wide range of gene functions could be assigned, and microsatellite and SNP sites, which could be used in marker development, were identified (Y. Lokko et al. 2004, unpubl. data).

14.8 Future Scope of Works

The three CMD linkage maps require further saturation with markers for a more extensive coverage of the genome. Addition of SRR markers, high-throughput markers such as EST and single nucleotide polymorphisms (SNPs), may be employed in these populations. SNPs could also facilitate mapping CMD resistance and other disease resistance genes of interest. Due to the high frequency of SNPs in the genome, there is a high probability of their being closely associated with resistance (Lohmann et al. 2000). SNPs also offer the opportunity to uncover allelic variation directly within expressed sequences of candidate genes to develop haplotypes based on gamete-phase disequilibrium for analysis of quantitative traits.

With the recent development of ESTs in cassava, sequence tagged sites (STSs), which are of a short DNA sequence generally, between 100 and 500 bp, that is easily recognisable and occurs only once in a chromosome or genome studied (Brown 1999), can be developed to further saturate the maps and facilitate fine mapping.

With the availability and enhancement of the JoinMap computer package future work on the cassava framework map TMS 30572×CM 2177-2 would involve developing a consensus map based on male- and female-derived framework maps. Following this, a composite map of cassava, based on all existing maps, could be used in planning experiments, to construct a genome database, to compare QTL identities in different genetic backgrounds, and for comparative mapping with other species. Composite maps have been reported in *Arabidopsis thaliana* and in maize (Beavis and Grant 1991; Hauge et al. 1993). Comparative mapping is a strategy that uses information obtained from the study of a species with a simple genome to make inferences about the map position and function of genes in species with a complex genome (Brown 1999). Comparative mapping has been used successfully in human and mouse genome mapping to increase the efficiency of mapping (Liu 1998). It has been suggested that comparative mapping has definite advantages in mapping plant genomes. For instance, genes in wheat with a genome size of 17,000 Mb would be mapped from information obtained from a small member of the Gramineae, rice, with a genome size of 400 Mb (Gale and Devos 1998; Liu 1998; Brown 1999).

For effective efficient map-based gene cloning and associating candidate genes with important biological or agronomic traits, there is a need for a physical map and an established correlation between the genetic maps and the physical map. The current efforts at fine mapping specific regions of the cassava framework map at CIAT, followed by contig mapping with BAC libraries, will be the start-off point for chromosome walking to genes of interest.

References

Aharoni A, Keizer LCP, Bouwmeester HJ, Sun Z, Alvarez-Huerta M, Verhoeven HA, Blaas J, van Houwelingen AMML, De Vos RCH, van der Voet H, Jansen RC, Guis M, Mol J, Davis RW, Schena M, van Tunen AJ, O'Connell AP (2000) Identification of the *SAAT* gene involved in strawberry flavor biogenesis by use of DNA microarrays. Plant Cell 12:647–661

Akano AO, Dixon AGO, Mba C, Barrera E, Fregene M (2002) Genetic mapping of a dominant gene conferring resistance to the cassava mosaic disease (CMD). Theor Appl Genet 105:521–525

Allen AC (1994) The origin of *Manihot Esculenta* Crantz (Euphorbiaceae). Genet Resour Crop Evol 41:133–150

Amma CSE, Sheela MN, Pillai PKT (1995) Combining ability heterosis and gene action for three major quality traits in cassava. J Root Crops 21(1):24–29

Anderson JV, Horvath DP (2001) Random sequencing of cDNAs and identification of mRNAs. Weed Sci 49:581–589

Anderson JV, Gedil M, Horvath DP, Dixon A (2001) Preliminary studies directed towards the development of *Euphorbiaceae*-specific microarrays. In: Taylor NJ, Ogbe F, Fauquet CM (eds) 5th Int Sci Meet Cassava Biotechnol Network: Abstr Book. Donald Danforth Plant Science Center, St. Louis, MO, pp S5-02

Anderson JV, Delseny M, Fregene MA, Jorge V, Mba C, Lopez C, Restrepo S, Soto M, Piegu B, Verdier V, Cooke R, Tohme J, Horvath DP (2005) An EST resource for cassava and other species of Euphorbiaceae. Plant Mol Biol 56(4):537–539

Angel F, Arias D, Tohme J, Iglesias C, Roca W (1993) Towards the construction of a molecular map of cassava (*Manihot esculenta* Crantz): comparison of restriction enzymes and probe source in detecting RFLPs. J Biotechnol 31:103–113

Awoleye F, van Duren M, Dolezel J, Novak FJ (1994) Nuclear DNA content in *in vitro* induced somatic polyploidisation cassava (*Manihot esculenta* Crantz) breeding. Euphytica 76:195–202

Beavis WD, Grant D (1991) A linkage map based on information from four F_2 populations in Maize (*Zea Mays*). Theor Appl Genet 82:636–644

Beckmann JS, Soller M (1983) Restriction fragment length polymorphisms in genetic improvement methodologies, mapping and costs. Theor Appl Genet 67:35–43

Beckmann JS, Soller M (1986) Restriction fragment length polymorphisms in plant genetic improvement. Oxford Surveys Plant Mol Cell Biol 3:196–250

Best R, Henry G (1994) Cassava Towards Year 2000. In: Intl Network for Cassava Genet Resour, Columbia, 18–23 August 1992. International Crop Network series, No. 10, International Plant Genetic Resource Institute, Rome, Italy, pp 12–15

Botstein D, White RL, Skolnick M, Davis RW (1980) Construction of a genetic map in man using restriction fragment length polymorphism. Am J Hum Genet 32:314–330

Brown TA (1999) Genomes. BIOS, New York, p 472

Bryne D (1984) Breeding cassava. Plant Breed Rev 2:73–134

Ceballos H, Fregene M, Lentini Z, Sanchez T, Puentes YI, Pérez JC, Rosero A, Tofino AP (2006) Development and identification of high-value cassava clones. Acta Hort (ISHS) 703:63–70 http://www.actahort.org/books/703/703_6.htm

Chavarriaga-Aguirre P, Maya MM, Bonierbale MW, Kresovich S, Fregene MA, Tohme J, Kochert G (1998) Microsatellites in cassava (*Manihot esculenta* Crantz): discovery, inheritance and variability. Theor Appl Genet 97:493–501

CIAT (1996) Development of Cassava Germplasm for the Drier Tropics and Subtropical Agroecosystems of Africa, Asia and Latin America. A Report by CIAT and CNMPF/EMBRAPA to the International Fund for Agricultural Development (IFAD). CIAT, Cali, Colombia

CIAT (1999) Programme for the Enhancement of Income-Generation Potential through the Development and Transfer of Improved Cassava Genetic Material and Adoption of Post-Harvest Technologies. A Report by CIAT and CNMPF/EMBRAPA to the International Fund for Agricultural Development (IFAD). CIAT, Cali, Colombia

CIAT (2003) Annual Report IP3. Improved Cassava for the Developing World. CIAT, Cali, Colombia, pp 8-1–8-3

Cock JH (1985) Cassava, New Potential for a Neglected Crop. West View, Boulder, CO, p 191

Cortes DF, Reily K, Okogbenin E, Beeching JR, Iglesias C, Tohme J (2002) Mapping wound response genes involved in post-harvest physiological deterioration (PPD) of cassava (*Manihot esculenta* Crantz). Euphytica 128:47–53

Edwards MD, Page NJ (1994) Evaluation of marker-assisted selection through computer simulation. Theor Appl Genet 88:376–382

Ekanayake IJ, Osiru DSO, Porto M (1997) Morphplogy of cassava. International Institute of Tropical Agriculture (IITA) Research Guide 61. Training Program, IITA, Ibadan, Nigeria

FAO (2000) http:// www.fao.org

FAO (2004) Food outlook No. 2. June 2004. http://www.fao.org/documents/show_cdr.asp

Fregene M (2000) Marking progress: collaboration to improve cassava. In: Kinley D (ed) Synergies in Science. CGIAR (Consultative Group on International Agricultural Research), Washington, DC, pp 6–7

Fregene MF, Angel F, Gomez R, Rodriguez P, Chavarriaga, Bonierbale M, Roca W, Tohme J (1997) A molecular genetic map of cassava (*Manihot esculenta* Crantz). Theor Appl Genet 95:431–441

Fregene M, Okogbenin E, Mba C, Angel F, Suarez MC, Gutiérez J, Chavarriaga P, Roca W, Bonierbale M, Tohme J (2001) Genome mapping in cassava improvement: challenges, achievements and opportunities. Euphytica 120: 159–165

Fregene M , Okogbenin E, Marin J, Moreno I, Ariyo O, Akinwale O, Barrera E, Ceballos H, Dixon A (2006) Molecular marker assisted selection (MAS) of resistance to the cassava mosaic disease (CMD). Euphytica (in press)

Fukuda WM, Saad N (2001) Participatory Research in Cassava Breeding with Farmers in North eastern Brazil. Working Document No. 99, CNPMF, Cruz das Almas-Bahia, Brazil, p 42

Gale MD, Devos KM (1998) Plant comparative genetics after 10 years. Science 282:656–659

Giovannoni JJ, Wing RA, Ganal MW, Tanksley SD (1991) Isolation of molecular markers from specific chromosomal

15 Yams

H. D. Mignouna[1,4], M. M. Abang[2], R. Asiedu[3]

[1] Virginia State University, Agricultural Research Station, Box 9061, Petersburg, VA 23806, USA, *e-mail*: jmignoun@vsu.edu
[2] International Center for Agricultural Research in the Dry Areas (ICARDA), P.O. Box 5466, Aleppo, Syria, and IRAD, Root and Tuber Program, B.P. 2123 Messa, Yaoundé, Cameroon
[3] International Institute of Tropical Agriculture (IITA), c/o L.W. Lambourn & Co., Carolyn House, 26 Dingwall Road, Croydon CR9 3EE, UK
[4] Current address: Technical Operations Manager, African Agricultural Technology Foundation (AATF), c/o ILRI, P.O. Box 30709, Nairobi, Kenya, e-mail: h.mignouna@aatf-africa.org

15.1 Introduction

15.1.1 Brief History of Crop

Yams are classified in the genus *Dioscorea*, family *Dioscoreaceae*, and Order *Dioscoreales*. More than 600 species are known to exist, the systematics of which remain uncertain (Miège and Lyonga 1982). For instance yams are considered as monocotyledons (Okonkwo 1985), although the occurrence of a vestigial element of a second cotyledon in a number of species, the shape and venation of the leaves, and the nature of the inflorescence provide strong evidence that *Dioscorea* spp. or their ancestors are closely related to the dicotyledons (Lawton and Lawton 1967; Degras 1993).

Despite initial disagreements concerning the domestication of yams in either discrete, closely circumscribed areas or over wide, diffuse areas in Sub-Saharan Africa and southeast Asia, most authors agree on the following three centers of origin for the most important cultivated edible yams:

- West Africa (*D. rotundata*, *D. cayenensis* and *D. dumetorum*)
- South-east Asia [Indo-China (Burkill 1951) or New Guinea (Lebot 1999)] (*D. alata*, *D. esculenta* and *D. opposita*)
- Tropical South America (*D. trifida*)

While certain varieties of *D. bulbifera* are generally acknowledged to be of Asian origin, some authors consider that this species appeared at the same time in Asia and in Africa (Coursey and Martin 1972; Onwueme 1978). *D. alata* is believed to have originated from spontaneous hybrids between *D. hamiltoni* and *D. persimilis* (Coursey 1967; Mignouna et al. 2002c). The species is believed to have been introduced to Africa by the Portuguese in the 16th century, although other accounts suggest that *D. alata* reached East Africa with the Malays about 300 to 1000 AD. Mignouna et al. (2004) investigated genetic relationships among wild and cultivated yams using PCR-based markers and showed that *D. rotundata* is a species separate from *D. cayenensis*. *D. rotundata* was found to be much more closely related to *D. praehensilis* and *D. liebrechtsiana* than it was to *D. cayenensis*.

15.1.2 Botanical Description

Dioscorea is an economically important, annual or perennial tuber-bearing, dioecious, climbing, tropical genus of monocots that looks like a dicot. Yams have a relatively small genome size (estimated at 550 Mbp/1C and 800 Mbp/1C for *D. alata* and *D. rotundata*, respectively), with a basic chromosome number $n = 10$, although polyploidy is widespread in the genus (Onwueme 1978; Mignouna et al. 2002 c, d).

The genus is fascinating from the genetic, developmental, physiological, and evolutionary points of view because of its distinctive biology and critical position in the phylogenetic tree of angiosperms. Yams exhibit distinct changes in shoot apical meristem (SAM) structure and phyllotaxy during phase transition from juvenile to adult (Burkill 1960). They have a dioecious habit with different

Genome Mapping and Molecular Breeding in Plants, Volume 3
Pulses, Sugar and Tuber Crops
C. Kole (Ed.)

morphologies of staminate and pistillate plants in some species, and a dicotlike leaf structure (net-veined, petiolate) with an early development intermediate between dicot and monocot modes (Bharathan 1996). The genus occupies a phylogenetic position that straddles the eudicots (containing most model species: *Arabidopsis*, snapdragon, tomato, *Petunia*) and the grasses (containing the rest: rice, maize). Indeed, *Dioscorea* occupies a position in relation to the grasses rather like that of the basal eudicot, *Aquilegia* (Ranunculaceae), in relation to the other eudicots (e.g. Kramer and Irish 1999).

The yam plant comprises the typical shoot and root portions with an underground storage tuber, while some species also produce bulbils (aerial tubers). Yam leaves exhibit great inter- and intraspecific variability (Miège 1982). The stems originate from the base of the emerging shoot (Coursey 1967) and may be winged, spiny or spineless, hairy or glabrous, and circular, rectangular, or polygonal in transverse section (Onwueme 1978). It is widely held that the tuber head, i.e., the cormlike structure attached to the proximal region of the tuber, is the organ or bud that produces the vine, roots, and tuber(s) during visible vegetative growth. Recent findings have shown, however, that the yam vine and feeder roots (and possibly the tuber) actually originate from the primary nodal complex, a small region of meristematic cells in a sprout that is known to occur on headless tubers as well as the tuber head of intact tubers (Ile et al. 2004).

Apart from temperate species such as *D. opposita*, the growth and development of yams is virtually arrested at temperatures below 20 °C. They grow optimally at about 30 °C (Kay 1973). The plant is relatively resistant to drought, but yields are considerably reduced when this coincides with the period of tuber initiation or early tuber development (Onwueme 1978). Yams exhibit photoperiodic responses, with long days favoring vegetative growth while short days tend to enhance tuber development (Kay 1973; Onwueme 1978). However, photoperiodic response varies with species, and it has been shown that the rate of change of photoperiod is an important determinant in the response of yams to light (Ile et al. 2004). Ideal conditions for yam production include at least 1000 mm of rain spread over five to six months and deep, fertile, friable, and well-drained soils to allow for tuber development (Degras 1993).

15.1.3 Economic Importance

The most important food species are *D. rotundata* (also called white Guinea yam) and *D. alata* (water or greater yam). *D. rotundata* is the most cultivated and consumed of the genus and also has the highest market value in West Africa owing to the superior suitability of its tubers to the preferred food use for the crop in the region. *D. alata* is considered to be the most widely distributed species of yam, and it is superior to *D. rotundata* in yield potential (especially under low to average soil fertility), ease of propagation (through production of bulbils and reliability of sprouting), early vigor for weed suppression, and storability of tubers.

Yams are produced throughout the tropical and subtropical regions of the world. The majority of yam production, however, is in Africa. West Africa accounts for about 95% of world production and 93% of the area (FAO 2003). More than 95% of the world's yams are currently grown in Sub-Saharan Africa, but the crop is also important and widely cultivated in the Caribbean and South Pacific islands, as well as in parts of Asia and South and Central America. Production of yams in Africa is largely confined to the "yam zone," comprising Cameroon, Nigeria, Benin, Togo, Ghana, and Ivory Coast, where ca. 90% of the world's production takes place. According to FAO statistics, 39.9 million tons of yam were produced worldwide in 2003, 96% of this in Sub-Saharan Africa. The leading producer was Nigeria with 27 million metric tons (mT), followed by Ghana (3.9 million mT), Ivory Coast (3.0 million mT), and Benin (1.7 million mT). In 2003, nearly 4.3 million ha of land was planted with yam throughout the world, 96% of which was in Sub-Saharan Africa. Nigeria alone accounted for 68% of the world total. The average yield was nearly 10 tons per hectare.

Yams constitute an important source of food and income and play a major role in sociocultural life for a wide range of smallholder households. They bring flexibility to the annual cycle of food availability through several species and cultivars, broad agroecological adaptation, early season drought tolerance, and diverse maturity periods. The tubers have organoleptic qualities that make them the preferred carbohydrate food where yams are grown, contributing up to 350 dietary calories per person each day for millions of people in the major producing countries.

Yam is important mainly for its high calorific value and, unlike cassava and sweet potato, is produced almost exclusively for human consumption. In west Africa, yam is one of the most appreciated and preferred staple foods. On a fresh-weight basis, the tubers contain about 70% water, 25% carbohydrates, 1 to 2% fat, and 1 to 2% protein (FAO 1988). It also tends to be rich in arginine, leucine, isoleucine, and valine, with substantial amounts of thiamine, riboflavin, niacin, and ascorbic acid. Yams are a good source of calcium (5 to 70 mg/100 g edible portion of tuber), phosphorus (5 to 60 mg/100 g), and iron (0.5 to 5.2 mg/100 g) (Eka 1985). The tubers are typically boiled and pounded for easy swallowing. Frying in oil and roasting are also common preparation methods. A few commercial products based on dry flakes or flours from the tuber are produced in Nigeria, Ghana, and Ivory Coast for export and sale in urban areas.

15.1.4 Breeding Objectives

Yam cultivation requires large amounts of labor for mounding, staking (especially in the humid forest zone), weeding, and harvesting (Nweke et al. 1991). As a result, labor input per hectare is roughly twice that for cassava and four times that for maize. The best land is often allocated to yams since the crop requires deep, fertile, well-drained soils. Cost of planting material is also high, as about one quarter of the total weight of each year's harvest is used to plant the next year's crop. The potential for increased production of yams seems highest in the savanna zone where higher intensity of solar radiation makes staking less important, and there are fewer limitations on the availability of fertile land. With a benefit:cost ratio of about 3:1, yam production is profitable (IITA 1999). However, the productivity of yam cultivation has been under intense pressure from reduction in soil fertility and pest buildup due to shortening fallow periods and increasing cost of production. Nematodes (*Scutellonema bradys* and *Meloidogyne* spp.), often interacting with fungal (e.g., *Botryodiplodia*, *Fusarium*) and bacterial (e.g., *Erwinia* sp.) pathogens, attack tubers of susceptible varieties in the field and continue their damage during storage, leading to loss of food quality and quantity as well as of planting materials. Yam production and marketing are also affected by a range of insect pests on the foliage, e.g., the leaf beetle (*Crioceris livida*) and the tuber, e.g., termites (*Amitermes* sp.), tuber moth (*Euzopherodes vapidella*), mealybug (*Planococcus* spp., *Phenacoccus* spp.), scale insect (*Aspidiella hartii*), and tuber beetles (*Heteroligus* spp., *Prionorcytes* spp.). Anthracnose disease (caused by *Colletotrichum gloeosporioides* Penz.) remains a major threat to the cultivation of *D. alata* in all yam-producing areas (Abang et al. 2003a,b). Yam cultivation is also severely constrained by the accumulation of viral infections, in particular by Yam mosaic virus (YMV), genus *Potyvirus*.

The International Institute of Tropical Agriculture (IITA) has a global mandate for research on yam within the Consultative Group on International Agricultural Research (CGIAR). The Institute's researchers develop and disseminate improved technologies targeted at increased productivity of yam-based systems in partnership with national programs and advanced laboratories. Specifics of this include strategies for integrated control of pests and diseases in the field and during storage, soil and crop management suited to intensified cultivation of yams, reduced labor input in yam-based systems, and manipulation of tuber dormancy to increase efficiency in propagation and flexibility in crop cycles. Expanded utilization opportunities through processing into various products and improvements in marketing channels would influence productivity and bring the benefits from the crop to a broader range of consumers. New and more productive varieties of yams are needed to increase and sustain the productivity of yam cultivation in the face of a deteriorating resource base. Farmers have been relying on natural variation for their selection of suitable varieties to cope with the situation, but the pace of this is not compatible with that of the changes in challenges in the physical and socioeconomic environment. Breeding and selection of yams are carried out at IITA and national research programs in the context of improving yam-based systems and with primary focus on *D. rotundata* and *D. alata*. The attributes desired in a good yam cultivar vary according to region and species involved. Hence, constraints to productivity of cultivation, and opportunities for improvement are identified through reviews of existing information, new surveys, and/or stakeholders' meetings. Generally, the principal objectives for most yam improvement programs include high and stable yield of marketable tubers, good tuber quality (e.g., in terms of dry-matter content, cooking quality/texture, taste, dormancy period, rate of enzymatic browning), resistance to biotic stresses in the field and during postharvest storage, tolerance to abiotic stresses (e.g., drought and low soil fertility), and suitability to prevalent cropping systems (e.g., plant architecture, vigor, and maturity period).

The main obstacles encountered in sexual hybridization of yams in the past have included sparse flowering, poor synchronization of male and female flowering phases, and lack of efficient pollination mechanisms. Advances have been made in studies of the reproductive biology of yams especially at the Central Tuber Crops Research Institute (CTCRI) in Trivandrum, India, and IITA in Ibadan, Nigeria (Asiedu et al. 1998). Many parental genotypes, especially of *D. rotundata*, have been identified over the years that combine good agronomic attributes with reliable flowering and high fertility. Techniques for manipulation of flowering periods to enhance synchronization and extend pollination periods have been established. Anthesis times, periods of pollen viability and stigma receptivity have also been determined for the relevant species. Hermetic cold storage of *D. rotundata* pollen without previous drying (wet–cold storage) in a viable condition at −20 °C or −80 °C for over 2 years has been demonstrated (Daniel et al. 2002). Hand pollination with pollen of *D. rotundata* frozen at −80 °C for 365 d gave 69.5% fruit set and 50% fruit set after 730 d in storage. This has the potential to iron out the nonoverlapping of male and female flowering phases. In spite of the foregoing, further work is in progress to develop protocols for inducing flowering in nonflowering but agronomically desirable varieties (especially of *D. alata*). Other technical challenges to breeding and selection of yams include the long breeding cycle, very low multiplication ratio of propagules, and the existence of a juvenile phase during the seminal and early clonal stages of selection. Moreover, the available methods for evaluating important traits such as tuber quality (based on sensory evaluation of food products) cannot be applied at early stages of the selection cycle when tubers per clone are few and small, and the populations are large.

The early clonal (observational) trials are constrained by the low multiplication ratio of the tubers, leading to a long period before multilocational yield trials can be established. As yams are a vegetatively propagated crop, the selection scheme involves repeated evaluation of clones selected from a seedling nursery or the germplasm collection over several years in clonal trials ending with on-farm testing. For yams, there is the natural break in the evaluation cycle each year due to dormancy in the tuber after harvest. A propagule storage period of up to 4 months is quite normal for most yam selection programs before the subsequent season. Severe losses are often incurred at this stage. Some of these losses are advantageous to the selection process as materials that are most susceptible to storage pests and pathogens are weeded out. Nonetheless, there are often proportions of the losses that most researchers would have liked to carry to the next season. Materials for genetic studies are particularly crucial in this category. It is, therefore, imperative for yam research programs to ensure good facilities for tuber storage at an appropriate location that would offer some measure of flexibility and security.

The breeding scheme begins with characterization and evaluation of germplasm received from farmers, National Agricultural Research Systems (NARS), and IITA's collection for field performance, tuber quality, morphology, and ploidy status leading to the selection of parents for hybridization that have desirable traits relevant to the objectives of the program. Botanic seeds are generated through biparental crosses and open pollination among selected clones planted in isolation from the main yam fields. Additional seeds are obtained through natural hybridization in clonal trials and on farmers' fields. Seedlings from these seeds are evaluated in nurseries from which selections go through a series of clonal trials toward the selection of superior genotypes.

The clonal trials start with unreplicated (observational) trials (clonal evaluation), with variable numbers of clones and stands per clone. At the preliminary (PYT), advanced (AYT), and uniform yield trial (UYT) stages, a randomized complete block design is used with 3 to 6 replications. By the fifth year of evaluation, the materials can be subjected to a series of tests for their cooking and processing attributes in addition to evaluation for yield and reaction to pests and diseases. Owing to the slow multiplication rate of yams, it is necessary to make a special effort after the first advanced yield trial in the sixth year to multiply planting materials before extensive multilocational testing of selected materials. Simultaneously, the Tissue Culture Unit at IITA, in collaboration with the Plant Quarantine Service in Nigeria, carries out the process of virus elimination, micropropagation, and certification. In addition to a broad-based population targeting the Guinea savanna zone of west Africa, specific populations, e.g., resistance to YMV in *D. rotundata* or anthracnose disease in *D. alata*, are being improved through recurrent selection.

15.1.5 Classical Breeding Achievements

Genetic improvement programs in Africa, India, and the Caribbean have been developing high-yielding *D. alata* and *D. rotundata* varieties with pest and disease resistance to meet farmers' requirements. To achieve breeding goals, these institutions and other yam improvement programs in various countries have collected and maintained more than 3000 germplasm accessions that are being used for yam improvement. Through classical breeding several clones and populations of white and water yams have been developed by IITA in partnership with national programs in Africa and disseminated for further evaluation and selection under local environmental conditions.

Subregional networking, collaborative research, and local selection from improved germplasm provided by IITA have increased in recent years. For instance, the West African Seed Development Unit (WASDU) initiated multinational farmer-managed on-farm adaptive trials on yams in west Africa in 1999 based on IITA-derived varieties that had previously been tested by NARS in on-station trials. These were conducted in Guinea, Togo, Benin, and Chad. In Burkina Faso 59 IITA-derived genotypes of *D. rotundata* toward selection for farmer-participatory trials were evaluated. Several other IITA-derived varieties of yams are evaluated on-station and on-farm annually. For example, 234 clones of *D. rotundata* were evaluated in various trials in Benin, 166 in Togo, 169 in Ghana, and 80 in Ivory Coast in 2000. Through collaborative evaluation of IITA-derived breeding lines with the National Root Crop Research Institute (NRCRI) in Umuahia, Nigeria, three and four varieties of white yam were officially released in Nigeria in 2001 and 2003, respectively. NRCRI is also testing over 100 IITA-derived varieties in six yam-growing states of Nigeria in a farmer-participatory selection scheme. More lines are in the pipeline to be released by root crop programs in other yam producing countries like Benin, Burkina Faso, Ivory Coast, Ghana, Sierra Leone, and Togo. These varieties have multiple pest and disease resistance, wide adaptability, and good organoleptic attributes.

15.1.6 Utility of Molecular Mapping

Progress in breeding for increased yield, adaptability to marginal environments, improved tuber quality, and pest and disease resistance has, however, been painfully slow due to biological constraints that impede the elucidation of the genetics of important traits in yam. These include a long growth cycle, dioecy, poor to nonflowering plants, polyploidy, vegetative propagation, a heterozygous genetic background, and poor knowledge of the organization of crop diversity. The long growth cycle of yam, lasting about 8 months or more, exposes the crop to a plethora of pests and diseases among which anthracnose (*C. gloeosporioides*) and YMV disease can be particularly damaging (Thouvenel and Dumont 1990; Abang et al. 2003 a, b). The long period to harvest hinders the flexibility of availability, an important trait for any industrial crop, and considerably lengthens the gestation period for new improved varieties. The vegetatively propagated nature of the crop means that securing clean and healthy planting materials can be an ordeal for poor farmers. Yam is cultivated in widely varying agroecological zones. In Nigeria, for instance, the crop is grown from the southern humid forest to the northern Guinea savanna. Performance of genotypes is very disparate across regions, thereby multiplying breeding goals.

Chromosome pairing in tetraploids can occur such that only homologs pair or such that any two homeologs may pair. These two types of pairing have very different consequences for segregation patterns so that these plants may, in the extreme, exhibit either diploid or tetraploid genetics. Intermediate types of behavior may also occur. Genes controlling important traits such as yield, tuber quality, and pest and disease resistance are usually distributed among several quantitative trait loci (QTLs), which may not be linked, thus making these traits difficult to manipulate using conventional breeding methods. Yams are polyploid, male and female plants are separate, and both are needed to obtain true botanic seeds. In addition, flowering of male and female plants is difficult to synchronize. The early testing of progenies from hybridization is complicated by yam's slow multiplication rate and the juvenile phase during which it is unreliable to use conventional screens for some economic traits. Molecular markers that are linked to genes controlling these economic traits would be useful in selection at an early stage of the plant's growth, thereby enhancing the speed and efficiency of selection. The utility of yam molecular mapping is best illustrated by the case of yam mosaic and anthracnose diseases, the most important biotic constraints to *D. rotundata* and *D. alata* production, respectively.

Genetic resistance is the only economically viable option of controlling yam mosaic, a potyvirus disease that is particularly damaging to *D. rotundata*. A study of the genetic control of YMV resistance in three *D. rotundata* cultivars to a Nigerian isolate of YMV showed that resistance is manifested differentially as the action of a single dominant gene in simplex condition or a major recessive gene in duplex condition (Mignouna et al. 2001 b). The dominant locus that contributes to YMV resistance has been tentatively named *Ymv*-1, pending tests of allelism. The availability of genotypic diversity for YMV resistance is quite interesting to breeding programs because both genes could be pyramided in the same genetic background or used separately against infection by this virus. However, the recessive nature of YMV resistance in some *D. rotundata* genotypes means that such resistance cannot be easily tracked at the phenotypic level, demanding refined diagnostic procedures such as molecular mapping for detailed genetic localization of specific genes (Mignouna et al. 2001 b). Screening by molecular markers linked to QTLs has the advantage of selecting pairs of parents with genes at different QTLs for the same trait (Solomon-Blackburn and Barker 2001 a, b). Using traditional cultivars/landraces and breeding lines, researchers at IITA thus initiated experiments to map the yam genome and identify QTLs controlling YMV resistance.

Anthracnose disease, caused by *Colletotrichum gloeosporioides* (Abang et al. 2002), is a serious constraint to the sustainable cultivation of *D. alata* worldwide (Abang et al. 2003 a, b). Resistance to the moderately virulent fast-growing salmon (FGS) strain of *C. gloeosporioides* in the tetraploid breeding line TDa 95/00328 was found to be strain-specific and appeared to be controlled by a single major dominant locus; however, results suggested the presence of an additional resistance gene or genes (Mignouna et al. 2001 a). Initial genetic inheritance studies showed that resistance to yam anthracnose in *D. alata* is dominantly but quantitatively inherited (Mignouna et al. 2001 a). A single major dominant locus controlling resistance in the breeding line TDa 95/00328 has been tentatively designated *Dcg*-1 pending development of an improved host differential series and investigation of allelism. The FGS strain is presently the predominant virulence phenotype in Nigeria and represents a genetically heterogeneous population (Thottappilly et al. 1999; Abang et al. 2002; Abang 2003). In view of the high genetic diversity of *C. gloeosporioides* from yam, an obvious strategy is to combine resistance to the FGS strain with resistance to strains such as the highly virulent slow-growing gray (SGG) strain, for which sources of resistance have been identified (Mignouna et al. 2001 a). Anthracnose resistance breeding is a considerably slow and cumbersome process, owing to the biological constraints of a heterozygous, vegetatively propagated crop (Tanksley and Nelson 1996; Asiedu et al. 1998). It is, therefore, important to develop varieties carrying as many different genes for resistance as possible in order to provide stable resistance against a broad spectrum of the fungal pathogen. The availability of a high-density genetic linkage map of *D. alata* could greatly improve the prospects for gene mapping and marker-assisted selection (MAS), as has been demonstrated in common bean (Adam-Blondon et al. 1994; Mendoza-Herrera et al. 1999; Geffroy et al. 1998, 2000), and potato (Hämäläinen et al. 1998; Van der Voort et al. 1999; Solomon-Blackburn and Barker 2001 a, b).

Thus, genome mapping and marker-assisted breeding now offer a practical solution to many of the intractable problems in yam breeding. Molecular genetic maps and marker-aided analysis of complex traits can be used to elucidate the genetic control of yield potential and tuber quality and locate genes of pest and disease resistance, nutrient-use efficiency, and flowering. For these reasons, a concerted effort to map the yam genome and dissect the inheritance of complex traits was initiated at IITA.

15.1.7 Brief History of Mapping

Within the genus *Dioscorea*, genetic linkage mapping using scored amplified fragment length polymorphism (AFLP) and microsatellite markers was first performed on the wild diploid species *D. tokoro* Makino (Terauchi and Kahl 1999). This wild species is widely distributed in east Asia and is classified into the botanical section *Stenophora*, which is phylogenetically distant from the section *Enantiophyllum* to which all the important cultivated yams belong within the genus *Dioscorea*. The *D. tokoro* map, therefore, has little immediate use for genetic improvement of cultivated yams.

Genome mapping of cultivated yams was initiated at IITA in the mid-1990s with funds provided by the Gatsby Charitable Foundation, UK and involved IITA, John Innes Centre, and the Natural Resources Institute, UK. However, the lack of mapping populations and adequate polymorphic molecular markers hampered rapid progress. The initial effort in yam genomics was devoted to the development of polymorphic DNA markers and assessment of their potential application in yams. Yams are monocots but very distantly related to the grasses, for example, banana and wheat are more closely related to each other than either is to yam. Thus there is no convenient model system for yam genomics. During the course of the yam mapping project, several attempts were made to exploit heterologous DNA sequences as a source of RFLP markers. This approach had limited success, and a set of yam gene sequences had to be isolated as a priority for anchoring yam genetic maps. However, the use of this type of DNA sequence for generating large data sets needed to populate genetic maps is not cost efficient, so the approach of using uncharacterized DNA sequences was adopted as a source of genetic markers. In this regard, random amplified polymorphic DNA (RAPD) (Mignouna and Asiedu 1999; Dansi et al. 2000) and AFLP were the molecular markers of choice (Mignouna et al. 1998).

It was anticipated that cultivated yams would have their origin from a cross between genetically distinct individuals, so the alleles derived from each parent would be different. Many cultivated yams are tetraploid ($2n=4x=40$), so each genetic locus is expected to be represented by two homeologous pairs. Thus for any gene a given individual may have up to four alleles. One general approach to mapping plants of this type is to examine the genotypes of selfed progeny; however, this is not feasible for dioecious yams, so the approach taken was to generate multiple F_1 individuals derived from crosses between the same parents. F_1 mapping populations of *D. alata* and *D. rotundata* were micropropagated in vitro based on techniques developed by Ng (1992).

A preliminary study on genetic mapping of *D. rotundata* revealed diploid segregation patterns for alleles at RAPD, isozyme, and a few AFLP marker loci (Mignouna and Asiedu 1999). In that study, the detection and estimation of linkage was carried out according to the pseudotestcross method using single-dose markers (SDMs) because the genome constitution of yam was not known (allo vs. autopolyploid). This approach resulted in a rather limited linkage map comprising four linkage groups (LGs) made of 12 markers each for the male and female parents (Mignouna and Asiedu 1999), mainly because only a small proportion of markers that were polymorphic between the two parents fit the condition of SDM. This situation explains the emphasis on, and use of, a more polymorphic source of markers such as AFLP. AFLP markers are an excellent source of polymorphisms in eukaryotic genomes and have been shown to be well suited for genotyping and map construction in several plant species (Peters et al. 2004).

15.2 Construction of Genetic Maps

Genetic mapping using AFLP led to the construction of the first, separate, comprehensive, molecular linkage maps of *D. rotundata* and *D. alata*. This effort also led to the identification of QTLs for YMV resistance in *D. rotundata* and anthracnose resistance in *D. alata* (Mignouna et al. 2002 c, d).

15.2.1 Mapping of *D. rotundata* Genome and Detection of QTLs for YMV Resistance

The ploidy levels of various flowering genotypes of *D. rotundata* were determined using flow cytometry (Dansi et al. 2001; Mignouna et al. 2002 d). Three levels of ploidy (tetraploid, hexaploid, and octoploid) were found in the cultivars analyzed based on previously reported chromosome numbers in yams (Zoundjihékpon et al. 1994). Two tetraploid genotypes with contrasting reactions to YMV were chosen and crossed for the development of a mapping population. The resistant female parent (TDr 93-1) and susceptible male parent (TDr 87/00211) were planted and cross-pollinated in the field. The susceptible parent TDr 87/00211 was a breeding line, while the resistant parent TDr 93-1 was a popular landrace cultivar that had consistently shown field resistance across locations. Both parents had previously been used to generate F_1 individuals segregating for resistance to a Nigerian isolate of YMV (Mignouna et al. 2001b). The F_1 population used in the present study consisted of 180 individuals. This population served as the source of individuals for YMV resistance screening and for marker segregation analyses. The virus isolate used, virus inoculation, and symptom evaluation were as reported by Mignouna et al. (2001b).

AFLP analysis was carried out as described by Vos et al. (1995), using the enzyme combinations *Eco*RI/*Mse*I (E/M) and *Pst*I/*Mse*I (P/M). A total of 10 E/M and 11 P/M primer combinations (PCs) were selected (Table 1) based on a prescreening of 64 E/M and 48 P/M PCs, respectively. The number, density, and complexity of polymorphic AFLP fragments were used as selection criteria. The adaptor and primer sequences employed were based on the core primer design as described by Vos et al. (1995). Gel images were electronically scanned and AFLP markers codominantly scored using the proprietary technology developed by Keygene N.V., Wageningen, The Netherlands (Vuylsteke et al. 1999; Park et al. 2000). Each polymorphic AFLP fragment was identified by a code referring to the PC, *Eco*RI/*Mse*I (E/M) and *Pst*I/*Mse*I (P/M), followed by the estimated size of the DNA fragment in nucleotides. The primer nomenclature of Keygene was used throughout and the primer sequences can be deduced from the marker designations (Table 1).

Linkage analyses were performed with the computer software package JoinMap 2.0 (Stam 1993; Stam and van Ooijen 1996), and the segregating population was treated as resulting from cross-pollination. Mapping analysis involved the assignment of markers to LGs based on logarithm of odds (LOD) scores (minimum LOD score of 3.0) and calculation of pairwise recombination frequencies and corresponding LOD scores for all pairs of markers that belonged to a certain LG (Stam 1993). Measures of the goodness-of-fit were expressed as chi-square values (χ^2). A chi-square value of <2 was considered to be reliable for this analysis. Marker segregation data were obtained by analyzing 180 individuals with the PC E/M, whereas 90 individuals were analyzed with the combination P/M.

Table 1. Primer combinations (PCs) used to screen *D. rotundata* mapping population, PC nomenclature, and number of scored markers per primer combination

Primer combination	Extension	Number of scored markers
P12/M15	AC/CA	16
P12/M19	AC/GA	13
P13/M16	AG/CC	13
P14/M15	AT/CA	14
P14/M20	AT/GC	10
P14/M22	AT/GT	8
P15/M20	CA/GC	7
P16/M15	CC/CA	10
P16/M16	CC/CC	9
P17/M15	CG/CA	10
P17/M22	CG/GT	10
E11/M55	AA/CGA	24
E11/M57	AA/CGG	23
E11/M58	AA/CGT	21
E12/M50	AC/CAT	21
E12/M59	AC/CTA	23
E12/M60	AC/CTC	26
E13/M51	AG/CCA	23
E13/M60	AG/CTC	21
E14/M49	AT/CAG	16
E14/M61	AT/CTG	23
Total		341

The data set of 341 markers was split into two, containing 199 (81 + 118) maternal and 223 (81 + 142) paternal markers, respectively. A separate genetic map was generated for each parent. The 199 markers in the maternal data set were allocated to 16 distinct LGs (LOD = 3.5). After a first mapping analysis it was found that some markers were responsible for a large increase of the chi-square value. Such markers were removed from the data set. In addition, some LGs that contained three or fewer markers were discarded. Finally, 155 markers were mapped in 12 LGs with a total map length of 891 cM. The size of the LGs ranged from 31 to 172 cM. The average distance between two adjacent markers was 5.7 cM. In the second round of JoinMap, 95% of all markers were mapped, indicating a

Table 2. Mapping data for *D. rotundata* population (TDr 93-1×TDr 87/00211) showing number of markers mapped after 1, 2, or 3 JoinMap rounds including mean chi-square value and distance (cM) per LG. Only Parent 1 [TDr 93-1; (<AB×AA>) and (<AB×AB>)] markers were used

Linkage group	JM round	Mean χ^2	No. of markers	Length (cM)
LG 1	Round 1	0.961	14/16	52.7
	Round 2	0.961	14/16	52.7
	Round 3	1.078	16/16	53.5
LG 2	Round 1	0.435	8/9	65.8
	Round 2	0.435	8/9	65.8
	Round 3	1.292	9/9	61.5
LG 3	Round 1	0.568	9/11	49.9
	Round 2	0.955	11/11	50.0
LG 4	Round 1	0.665	10/10	74.7
LG 5	Round 1	1.526	12/12	71.7
LG 6	Round 1	1.223	16/20	97.3
	Round 2	1.343	19/20	98.8
	Round 3	1.337	20/20	98.7
LG 7	Round 1	1.283	17/19	94.4
	Round 2	1.283	17/19	94.4
	Round 3	1.832	19/19	93.0
LG 8	Round 1	0.635	14/19	149.8
	Round 2	0.809	18/19	172.0
	Round 3	1.162	19/19	172.6
LG 9	Round 1	0.744	13/13	95.0
LG 10	Round 1	0.740	8/9	48.5
	Round 2	0.740	8/9	48.5
	Round 3	1.868	9/9	48.9
LG 11	Round 1	0.854	11/11	39.9
LG 12	Round 1	0.646	6/6	31.4
Total across LGs	Round 2	–	147 (95%)	894.9
	Round 3	–	155 (100%)	890.9

high reliability of the data. Furthermore, when all markers were mapped, the chi-square values remained low (Table 2). The genetic linkage map is shown in Fig. 1a. The 223 markers in the paternal data set were mapped to 16 distinct LGs (LOD = 7.5). A procedure similar to that described above was followed, resulting in a total of 157 markers that were mapped to 13 LGs with a total length of 852 cM. The 13 LGs ranged in size from 24 cM to 121 cM. In the second round of JoinMap 89% of all markers were mapped, which indicated the reliability of the data. Again, chi-square values remained low when all markers were mapped (Table 3). The genetic map is shown in Fig. 1b. The LGs did not contain enough <AB×AB> markers that were segregating in both parents to allow combination of the maternal and paternal maps. Three LGs consisted of maternal markers only, whereas two LGs had only paternal markers. These five LGs probably represent genomic regions unique to the respective parents. The total length of the parental genetic maps is similar (891 and 852 cM for the maternal and paternal maps, respectively). There are still some gaps in the genome to be covered, because 44 markers are still not mapped.

The markers segregated like a diploid cross-pollinator population, suggesting that the *D. rotundata* genome is an allotetraploid (2n = 4x = 40). However, the possibility that occasional homeologous exchanges take place cannot be ruled out (Mignouna et al. 2002 d). The total maternal and paternal map lengths of 891 cM and 852 cM, respectively, covered roughly 56% of the yam genome based on a known total diploid *D. tokoro* map length of 800 cM (Terauchi and Kahl 1999). If the yam genome is estimated to be 1000 to 2000 cM as inferred from the AFLP markers in this study, then the 155 loci of the female map covered 891/2000 (about 45% of

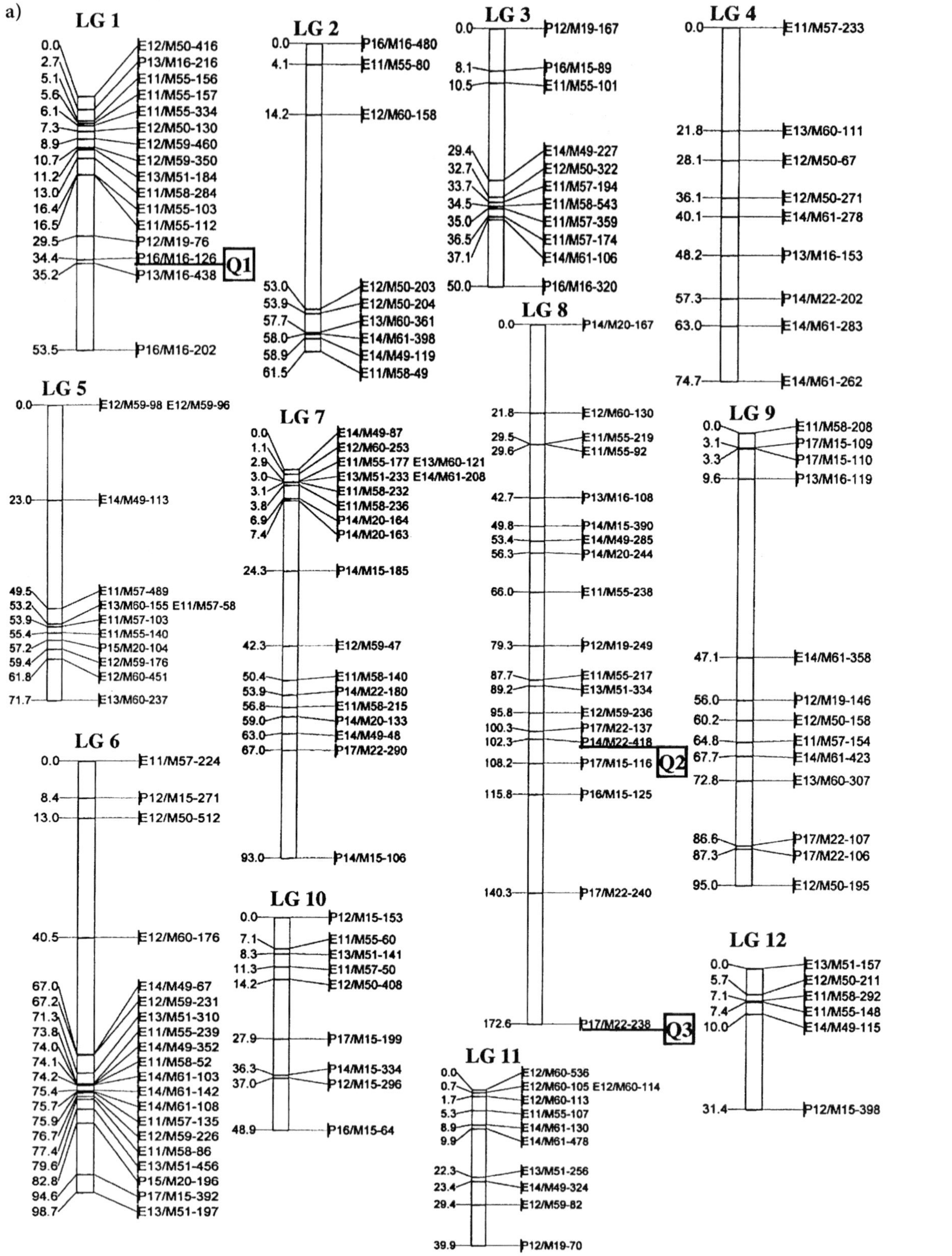

Fig. 1. Genetic linkage map of Guinea yam (*D. rotundata*) based on F_1 progeny from maternal line (TDr 93-1)×paternal line (TDr 87/00211). (**a**) Maternal and (**b**) paternal parent map. Map distances (to *left* of LGs) represent genetic distances in Kosambi centiMorgan (cM). Locus names are on *right side* of LGs. The loci are named with (1) the code referring to the corresponding

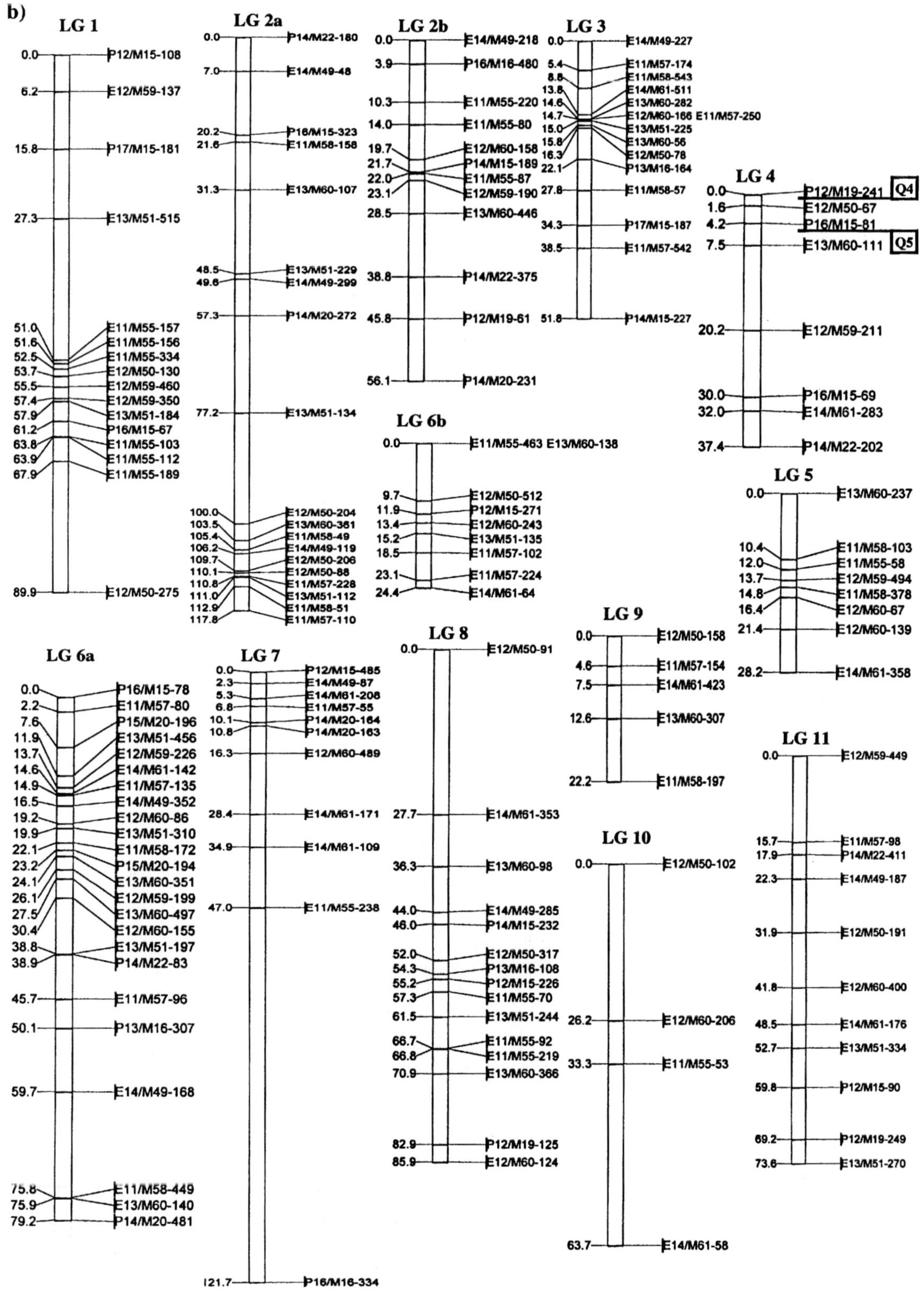

Fig. 1 (continued) primer combination followed by (2) the estimated size of the DNA fragment in nucleotides. Markers associated with QTLs for YMV resistance are marked Q1 to Q5 and explained 24, 22, 35, 13, and 16% of the phenotypic variance, respectively. The association with markers Q2 and Q3, and Q4 and Q5 may not be independent

Table 3. Mapping data for *D. rotundata* population (TDr 93-1×TDr 87/00211) showing number of markers mapped after 1, 2, or 3 JoinMap rounds including mean chi-square value and the cM distance per LG. Parent 2 (TDr 87/00211; (<AA×AB>) and (<AB×AB>)] markers were used

Linkage group	JM round	Mean χ^2	No. of markers	Length (cM)
LG 1	Round 1	0.518	15/16	96.5
	Round 2	0.518	15/16	96.5
	Round 3	0.932	15/16	89.9
LG 2 a	Round 1	1.103	15/19	85.9
	Round 2	1.103	15/19	85.9
	Round 3	1.554	19/19	117.8
LG 2 b	Round 1	1.526	9/12	51.9
	Round 2	1.526	9/12	51.9
	Round 3	2.593	12/12	56.1
LG 3	Round 1	0.481	14/15	48.1
	Round 2	0.481	14/15	48.1
	Round 3	0.808	15/15	51.8
LG 4	Round 1	1.643	8/8	37.4
LG 5	Round 1	1.390	8/8	28.2
LG 6 a	Round 1	1.345	19/25	79.7
	Round 2	1.490	21/25	78.1
	Round 3	1.818	24/25	79.2
LG 6 b	Round 1	0.915	8/9	27.5
	Round 2	0.915	8/9	27.5
	Round 3	1.953	9/9	24.4
LG 7	Round 1	1.306	13/15	85.7
	Round 2	1.372	14/15	85.9
	Round 3	1.819	15/15	85.9
LG 8	Round 1	0.952	10/11	42.2
	Round 2	0.952	10/11	42.2
	Round 3	0.817	11/11	121.7
LG 9	Round 1	0.431	5/5	22.2
LG 10	Round 1	0.100	4/4	63.7
LG 11	Round 1	1.385	9/11	58.5
	Round 2	1.385	9/11	58.5
	Round 3	2.325	11/11	73.6
Total across LGs	Round 2		140 (89%)	726.1
	Round 3		157 (99%)	851.9

the yam genome) while the 157 loci of the paternal map covered 852/2000 (43% of the genome). These estimates are the minimum genome coverage and, as more markers are added to the linkage maps, the unmapped markers from the present study could be fitted in the genetic linkage map. The genome size was estimated by Feulgen-stained root tip nuclei of tetraploid *D. rotundata* to be 0.8 pg per haploid nucleus, and thus is equivalent to that of species such as soybean, rice, and spinach (Conlan et al. 1995). Assuming the haploid nuclear DNA content of *D. rotundata* to be 800 Mbp/1C the physical distance per map unit could be estimated as 800 Mbp/2000 cM = 400 kb per cM, making map-based gene cloning feasible in white yam.

Detection of QTLs for YMV Resistance

The interval mapping procedure of the software package MapQTL version 4.0 (van Ooijen and Maliepaard 1996) was used to search for QTLs with effects on resistance to YMV. The symptom scores of all individuals were used as a phenotypic trait score (quantitative trait), and significant associations were searched for in both the maternal and the paternal linkage maps. An appropriate threshold value for declaring a significant QTL effect was sought through the permutation analysis of the MapQTL software. This was done for all mapped markers. The frequency distribution of the maximum LOD score was then determined. The LOD values at P = 0.05 and P = 0.01 were taken as the estimated critical values at which to declare the presence of a QTL.

The phenotypic response to YMV was characterized by continuous variation. Only mild symptoms (disease severity score of 2) were recorded for plants of the resistant landrace TDr 93-1, while plants of the susceptible breeding line TDr 87/00211 showed typical mosaic symptoms (disease severity score of 4). Progeny of the cross TDr 93-1 ×TDr 87/00211 segregated into 124 resistant: 56 susceptible. The observed segregation ratio fitted to a genetic ratio of 2.48:1 (χ^2 = 0.51; P = 0.47), which is to be expected when two simplex heterozygotes are crossed, indicating the possible modifying effect of the susceptible parent. A 3:1 ratio was also possible by the χ^2 test, suggesting the action of two dominant genes in simplex status, but there was only a weak fit to this ratio (χ^2 = 3.58; P = 0.06).

Since the population was considered to behave like a diploid cross-pollinator, the enhanced interval mapping method was applied. Three QTLs with effects on the resistance to YMV were identified on LGs 1 (1) and 8 (2) of the maternal parent TDr 93-1 at a LOD score minimum of 4.6 (P = 0.05). However, just one QTL (P17/M22-238; 35% of phenotypic variance) on LG 8 remained significant when the more stringent LOD value of 5.8 (P = 0.01) was applied. These three QTLs were associated with the marker P16/M16-126 on LG 1, which explained 24% of the phenotypic variance, and P14/M22-418 and P17/M22-238 on LG 8, which explained 22% and 35% of the phenotypic variance, respectively. Two QTLs for YMV were detected on the paternal LG 4 at a LOD score value of 4.5 (P = 0.05). At a higher LOD value of 5.4 (P = 0.01) only one of these two was retained. These QTLs were associated with the markers P12/M19-241 and P16/M15-81, which explained 13% and 16% of the phenotypic variation, respectively. This analysis showed that both parents contributed to the phenotypic resistance of the progeny.

15.2.2
Mapping of *D. alata* Genome and Detection of QTLs for Anthracnose Resistance

Two tetraploid *D. alata* genotypes that discriminate the FGS and SGG strains of *C. gloeosporioides* and carry different anthracnose resistance genes (Mignouna et al. 2001 a) were chosen for the development of a mapping population. The female parent (TDa 95/00328) and male parent (TDa 87/01091) were cross-pollinated in the field. Both parental accessions were breeding lines. TDa 95/00328 originated from a controlled cross between the susceptible landrace TDa 92-2 and the resistant breeding line TDa 85/00257, while TDa 87/01091 was obtained from open pollination of female line A19-165-445. The F_1 population used in the present study consisted of 176 individuals. This population served as the source of individuals for anthracnose resistance screening and for marker segregation analyses. The *C. gloeosporioides* isolate (Cg33), disease resistance screening and evaluation were as reported by Mignouna et al. (2001 a). The percentage leaf area affected by anthracnose was scored on a 0 to 6 scale. Genotypes with a mean leaf-area damage of 0 to 17.5% (corresponding to scores 0 to 4) were considered to be resistant, while those with a mean leaf-area damage of >17.5% (corresponding to scores 5 and 6) were considered susceptible (Mignouna et al. 2001 a). Chi-square tests for goodness-of-fit of the segregation ratios of resistant to susceptible (R:S) genotypes were carried out using the SAS statistics software (SAS Institute 1989). The test for the normal distribution of the frequency of the disease severity scores was performed by the Shapiro-Wilks W-statistic test using the program STATISTICA v. 6.0.

A total of 10 E/M and 11 P/M PCs were used in AFLP analysis, which was conducted following the same procedure used for the construction of the *D. rotundata* genetic map. Marker segregation data were obtained by analyzing the entire F_1 population with the PC E/M, whereas 90 individuals were analyzed with the combination P/M. Markers were tested against the expected segregation ratio using a χ^2 goodness-of-fit. A χ^2 value below 2.0 was considered to be reliable for this analysis. Linkage analysis was performed with the computer software package JoinMap 2.0 (Stam 1993; Stam and van Ooijen 1996), and the segregating population was treated as resulting from cross-pollination. Kosam-

Table 4. Primer combinations (PCs) used to screen *D. alata* mapping population, PC nomenclature, and number of scored markers per primer combination

Primer combination	Extension	Number of markers
P12/M15	AC/CA	23
P12/M19	AC/GA	13
P13/M16	AG/CC	18
P14/M15	AT/CA	9
P14/M20	AT/GC	14
P14/M21	AT/GG	8
P15/M20	CA/GC	10
P16/M16	CC/CC	15
P16/M20	CC/GC	16
P17/M15	CG/CA	14
P17/M19	CG/GA	6
E11/M55	AA/CGA	37
E11/M58	AA/CGT	45
E12/M48	AC/CAC	35
E12/M49	AC/CAG	31
E12/M50	AC/CAT	41
E13/M48	AG/CAC	30
E13/M52	AG/CCC	30
E14/M52	AT/CCC	43
E14/M54	AT/CCT	40
E14/M58	AT/CGT	30
Total		508

bi mapping function (Kosambi 1944), was used to convert the recombination frequency into map distance (cM). The pairwise analysis obtained from JoinMap was used to assign markers to LGs with a LOD score of ≥3.

The map was constructed using both parental markers. Of the total 508 markers (Table 4), 494 were assigned to 20 LGs (LOD score 4.5), while 14 markers could not be fitted into any of these LGs. Markers that caused a large increase in the χ^2 value were removed from the data set. Finally, 469 markers were mapped in 20 LGs with a total length of 1233 cM. Chi-square values remained low when all 469 markers were eventually mapped (Table 5). The genetic map is presented in Fig. 2. The average distance between markers was estimated at 2.62 cM.

Segregation of AFLP markers in this study supported a disomic inheritance that revealed an allotetraploid structure for *D. alata*. These results support the possible allotetraploid origin of water yam, although there are no known wild relatives of water yam. Using flow cytometry and chromosome counting, Gamiette et al. (1999) found three levels of ploidy in *D. alata* germplasm (4x, 6x, 8x) and concluded that these species are allotetraploid; however, they did not identify any diploid varieties. The allopolyploid nature of *D. alata* as revealed in this study is also supported by the observation of 20 bivalents during the meiosis of a tetraploid *D. alata* (Ramachandran 1968). The number of LGs reported in the present study is exactly the expected number of 20 (assuming disomic inheritance), which corresponds to the 20 gametic chromosomes of diploid (2n = 2x = 20) yam species such as *D. tokoro* and *D. gracillima*.

The total tetraploid map length of 1233 cM covered at least 77% of the yam genome based on a known total diploid *D. tokoro* map length of 800 cM (Terauchi and Kahl 1999). If the yam genome is estimated to be 1894 cM, as inferred from the AFLP markers from this study, the 494 loci of the current map covered 1233/1894 cM (ca. 65% of the yam genome). These estimates are the minimum genome coverage, and as more markers are added to the linkage map the unmapped markers from the present study could be fitted in the genetic linkage map. Assuming the haploid nuclear DNA content of *D. alata* to be 550 Mbp/1C the physical distance per map unit could be estimated as 550 Mpb/1894 cM = 290 kb per cM, making map-based gene cloning a feasible strategy in water yam.

Detection of QTLs for Anthracnose Resistance

A search for QTL effects for anthracnose resistance was carried out on the entire mapping population with all the segregating markers, using the nonparametric mapping procedure of the software package MapQTL v. 3.0 and 4.0 (van Ooijen and Maliepaard 1996). The mean leaf area infected (LAI) of the F_1 individuals was used as the phenotypic trait score and significant associations were searched on the *D. alata* linkage map. Because the Kruskal-Wallis rank-sum test that was performed in this nonparametric mapping method used both linked and unlinked markers, a more stringent significance level (P = 0.005) served as the threshold for declaring the likely presence of a QTL near a marker. In addition, association between molecular markers and anthracnose resistance was determined by simple linear regression of LAI data on marker genotype means using the computer package Q-GENE 3.06 running on a G3 Power Macintosh (Nelson 1997). The proportion of the phenotypic variance explained by the marker segregation was determined by the R^2 value. A putative QTL

Table 5. Mapping data for *D. alata* population (TDa 95/00328×TDa 87/01091) showing number of markers mapped after 1, 2, or 3 JoinMap rounds including mean chi-square value and distance (cM) per LG

Linkage group	JM round	Mean (X2)	No. of markers	Length (cM)
LG 1	Round 1	0.994	23/27	41.2
	Round 2	0.994	23/27	41.2
	Round 3	1.173	27/27	68.5
LG 2	Round 1	0.754	29/43	133.5
	Round 2	0.754	29/43	133.5
	Round 3	1.167	43/43	102.6
LG 3	Round 1	0.426	23/25	63.8
	Round 2	0.588	24/25	63.3
	Round 3	0.973	25/25	63.3
LG 4	Round 1	1.038	23/26	77.1
	Round 2	1.038	23/26	77.1
	Round 3	1.331	26/26	85.1
LG 5	Round 1	0.676	18/19	46.6
	Round 2	0.676	18/19	46.6
	Round 3	1.381	19/19	40.6
LG 6	Round 1	0.477	20/20	51.8
LG 7	Round 1	0.840	31/36	75.8
	Round 2	0.903	33/36	77.2
	Round 3	1.286	36/36	79.3
LG 8	Round 1	1.086	18/22	42.3
	Round 2	1.086	18/22	42.3
	Round 3	1.486	22/22	41.5
LG 9	Round 1	0.611	21/25	58.9
	Round 2	0.611	21/25	58.9
	Round 3	0.656	25/25	58.5
LG 10	Round 1	0.615	28/38	64.3
	Round 2	0.615	28/38	64.3
	Round 3	0.648	38/38	65.2
LG 11	Round 1	0.688	13/17	58.4
	Round 2	0.627	14/17	58.1
	Round 3	0.894	17/17	58.0
LG 12	Round 1	1.024	23/31	64.8
	Round 2	1.002	25/31	64.4
	Round 3	1.134	30/31	70.8
LG 13	Round 1	0.136	10/13	43.7
	Round 2	0.113	13/13	43.5
LG 14	Round 1	0.555	25/26	77.1
	Round 2	0.642	26/26	87.3
LG 15	Round 1	0.537	27/27	75.8
LG 16	Round 1	0.524	18/25	44.3
	Round 2	0.647	20/25	61.2
	Round 3	0.683	25/25	51.0
LG 17	Round 1	0.031	9/9	6.5
LG 18	Round 1	0.507	10/10	58.4
LG 19	Round 1	1.076	14/16	51.3
	Round 2	1.076	14/16	51.3
	Round 3	1.675	16/16	49.3
LG 20	Round 1	0.561	14/15	72.3
	Round 2	0.561	14/15	72.3
	Round 3	0.563	15/15	72.3
Total across LGs	Round 2		409 (87%)	1235.6
	Round 3		469 (100%)	1232.9

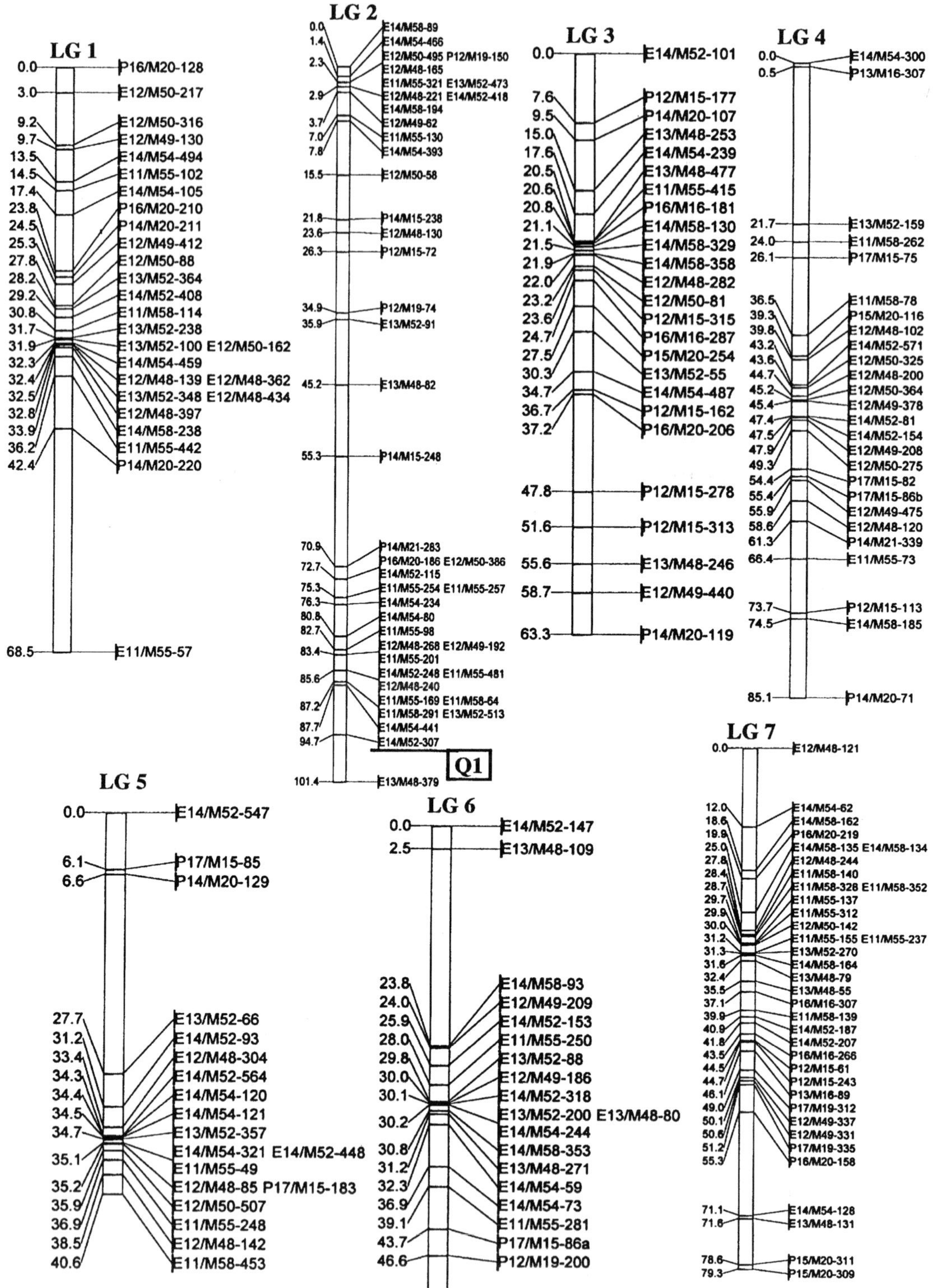

Fig. 2. Genetic linkage map of water yam (*D. alata*) based on F_1 progeny from maternal line (TDa 95/00328) × paternal line (TDa 87/01091). Distances (on *left side* of LGs) represent genetic distances in Kosambi centiMorgan (cM). Locus names are on *right side* of LGs. The loci are named with (1) the code referring to the corresponding primer combination (Table 1), followed by (2) the estimated size of the DNA fragment in nucleotides. One marker associated with a minor QTL for anthracnose resistance (explaining 10% of the phenotypic variance) is marked Q1

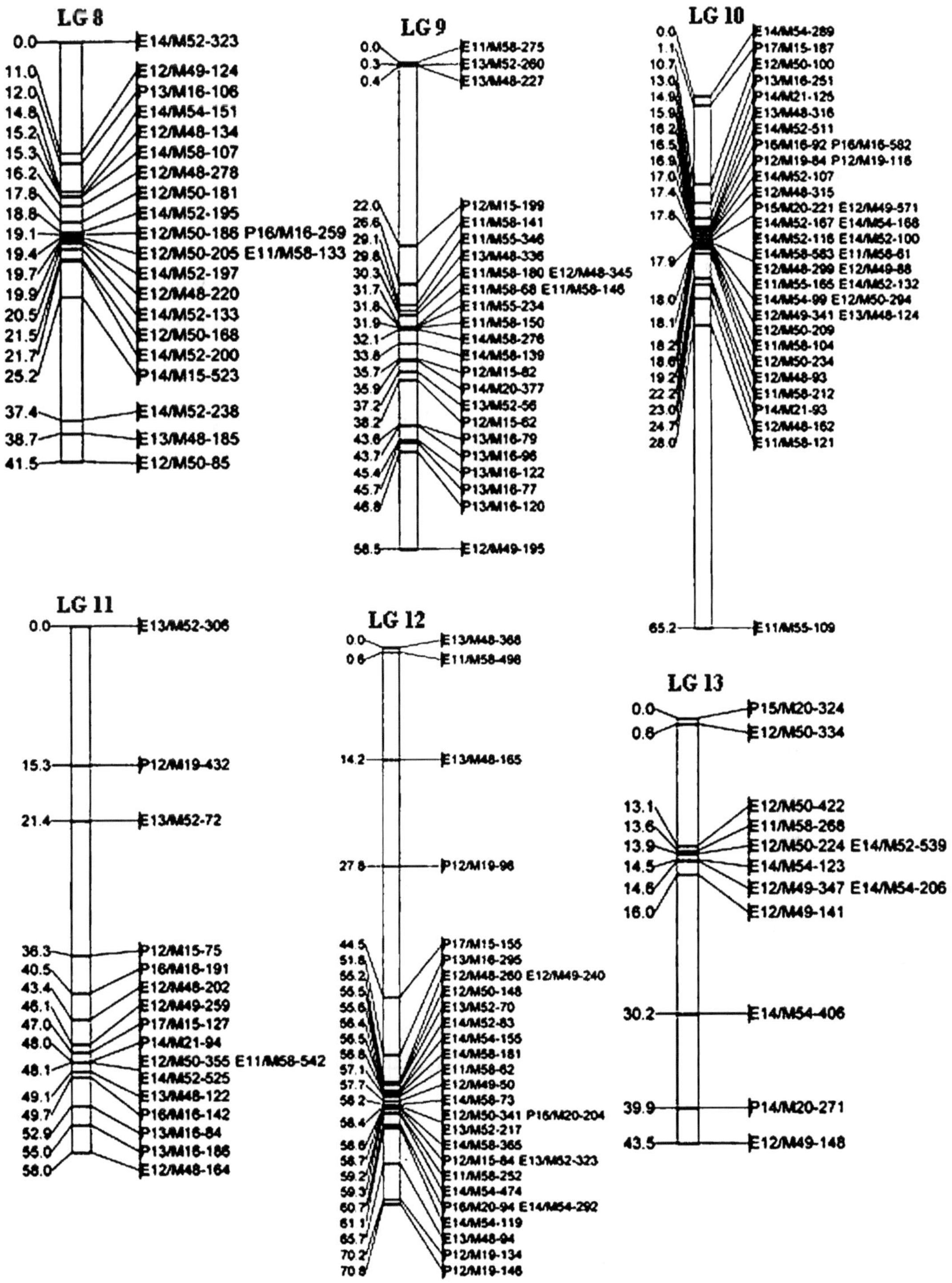

Fig. 2 (continued)

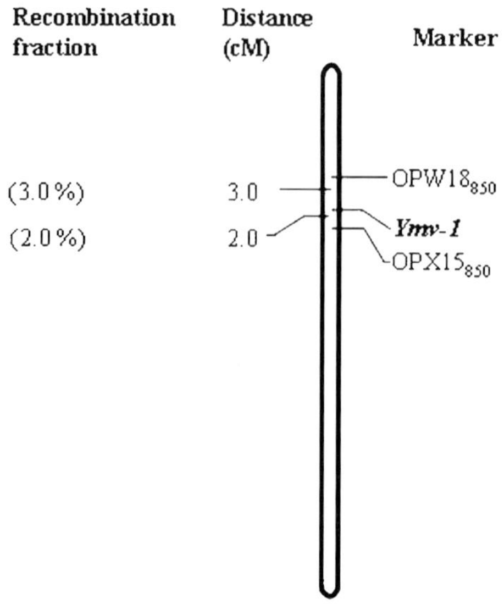

Fig. 3. Relative positions of RAPD markers linked to YMV resistance locus (*Ymv-1*) identified in a segregating population derived from cross TDr 87/00571×TDr 89/ 01444

Table 6. Segregation analyses of RAPD $OPW18_{850}$ and $OPX15_{850}$ markers and *Ymv-1* locus in cross TDr 87/00571× TDr 89/01444

Locus tested	Observed ratio	Expected ratio	χ^2	P	cM[a]
Ymv-1	45:55	1:1	1.00	P=0.32	
$OPWI8_{850}$[b]	48:52	1:1	0.16	P=0.69	3.0
$OPX15_{850}$	47:53	1:1	0.36	P=0.55	2.0

[a] Distance in centiMorgan in relation to YMV resistance locus tentatively named *Ymv-1*

[b] Markers $OPW18_{850}$ and $OPX15_{850}$ were both linked in coupling phase to *Ymv-1* locus

cultivar Gbongi. This population served as the source of individuals for forming DNA bulks and as a segregating population to confirm the presence and degree of putative linkages between RAPD markers and the *Ymv-1* locus. Eighty single-decamer primers were screened using bulks obtained from pooled DNA of individuals comprising each extreme of the disease phenotype distribution. Chi-squared analyses were carried out using the SAS statistical package to determine possible linkages between the RAPD marker and the resistance gene. Genetic distances between the markers and the resistance gene were determined with the aid of the program MAPMAKER v. 2.0 (Lander et al. 1987), using a log of the likelihood ratio (LOD) score minimum of 3.0.

Two RAPD markers, $OPW18_{850}$ and $OPX15_{850}$, closely linked in coupling phase with *Ymv-1* were identified, both of which were mapped on the same LG. The resistance locus *Ymv-1* was flanked on both sides by the RAPD markers (Fig. 3). A recombination value of 3.0 cM was obtained between OPW18850 and the *Ymv-1* locus, while OPX15850 had a recombination value of 2.0 cM (Table 6). Segregation ratios consistent with the presence of one allele (single dose) were observed for the two RAPD markers and the *Ymv-1* locus (Table 6). When used to screen 12 *D. rotundata* varieties with varying response to YMV, both markers successfully identified *Ymv-1* in the resistant genotypes. The selection efficiency for resistant and susceptible plants was 97% and 98% with markers $OPW18_{850}$ and $OPX15_{850}$, respectively. To facilitate MAS of *Ymv-1* across laboratories, the RAPD markers $OPW18_{850}$ and $OPX15_{850}$ will be converted into sequence characterized amplified regions (SCARs) (Miklas et al. 2000).

Although variation for YMV resistance is continuous, the working hypothesis assumed the existence of a single major gene, tentatively named *Ymv-1*. The markers $OPW18_{850}$ and $OPX15_{850}$ were each found to be closely linked to the *Ymv-1* locus; however, screening with both markers will be more effective in identifying phenotypes of resistant and susceptible plants. The markers will especially be useful in screening advanced generation lines that are nearly homozygous. The inability to detect linkage in the repulsion phase may be due to the large family size required to detect such linkages in autotetraploids (Wu et al. 1992). The RAPD markers showed good potential as genetic markers in white yam of West and Central African origin.

15.3.2 Molecular Markers for YAD Resistance

The identification of molecular markers linked to major dominant resistance in water yam (*D. alata*) to anthracnose (*C. gloeosporioides*) could be of considerable value for breeding programs since the incorporation of such resistance in susceptible genotypes would permit a direct increase of the resistance level in the improved genotypes. Also, resistance to the FGS strain needs to be combined with resistance to strains such as the highly virulent slow-growing gray (SGG) strain, for which sources of resistance were recently identified (Mignouna et al. 2001 a). RAPD markers can facilitate the pyramiding of resistance genes into elite yam cultivars, especially if they are tightly linked to the locus containing the resistance allele.

The tetraploid breeding line, TDa 95/00328, is a source of dominant genetic resistance to the moderately virulent fast-growing salmon (FGS) strain of *C. gloeosporioides*. BSA was used to search for RAPD markers linked to the anthracnose resistance locus (tentatively named *Dcg-1*) in F_1 progeny derived from a cross between TDa 95/00328 and the susceptible male parent, TDa 95-310. This population served as the source of individuals for forming DNA bulks and as a segregating population to confirm the presence and degree of putative linkages between RAPD markers and the *Dcg-1* locus. Two hundred and eighty decamer primers were screened using bulks obtained from pooled DNA of individuals comprising each extreme of the disease phenotype distribution. Chi-square analyses were carried out using the SAS statistical package to determine possible associations between the RAPD markers and the resistance gene. Recombination frequencies and LOD scores were calculated between all pairs of loci using the TetraMap software (C. Hackett, personal communication), which analyzes tetraploid data using the approach of Hackett et al. (1998) and Luo et al. (2001). The efficiency of putative linked RAPD markers to identify resistant or susceptible plants was determined in the entire segregating F_1 population. Two RAPD markers, $OPI7_{1700}$ and $OPE6_{950}$, closely linked to the dominant *Dcg-1* locus were screened across a sample of *D. alata* breeding lines with and without resistance to anthracnose to determine their potential application as genetic markers.

Segregation ratios consistent with the presence of two alleles (double dose) were observed for the two RAPD markers and the *Dcg-1* locus (Table 7). The resistance locus was flanked on both sides by $OPI7_{1700}$ and $OPE6_{950}$ (Fig. 4). A genetic distance of 2.3 cM was obtained between $OPI7_{1700}$ and the *Dcg-1* locus, while $OPE6_{950}$ was 6.8 cM distant from the resistance gene (Table 8). When used to screen 34 *D. alata* breeding lines with contrasting reactions to anthracnose, both markers successfully identified *Dcg-1* in the resistant genotypes. The selection efficiency for resistant and susceptible plants was 100% and 97% with markers $OPE6_{950}$ and $OPI7_{1700}$, respectively.

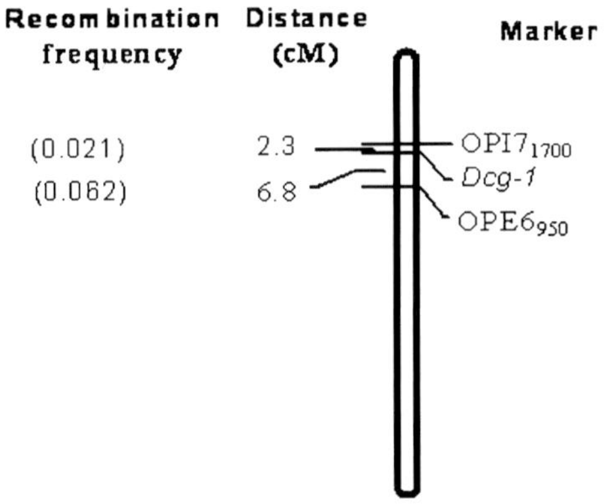

Fig. 4. Relative positions of RAPD markers linked to anthracnose resistance locus *Dcg-1* identified in a segregating population derived from cross TDa 95/00328×TDa 95-310

Table 7. Segregation analyses of RAPD $OPI7_{1700}$ and $OPE6_{950}$ markers and *Dcg-1* locus in F_1 population of cross TDa 95/00328×TDa 95-310

Locus tested	Observed ratio	Expected ratio	χ^2	P
Dcg-1	58:13	5:1	0.13	P=0.72
$OPI7_{1700}$	59:12	5:1	0.002	P=0.96
$OPE6_{950}$	57:14	5:1	0.465	P=0.49

Table 8. Recombination frequencies, LOD scores, and genetic distances (cM) between pairs of loci

Pair of loci[a]	Recombination frequency	LOD score	cM
$OPE6_{950}$ and $OPI7_{1700}$	0.085	6.27	9.1
$OPE6_{950}$ and *Dcg*-1	0.062	7.53	6.8
$OPI7_{1700}$ and *Dcg*-1	0.021	9.63	2.3

[a] The 1700-bp and 650-bp molecular markers were both linked in coupling phase to *Dcg-1* locus

The RAPD markers identified in this study will be made more reliable, specific, and easier to apply for indirect selection of *Dcg-1* in water yam cultivars by converting them into codominant PCR-based SCARs (Young et al. 1998; Ling et al. 2000). The use of molecular markers presents a valuable strategy for selection and pyramiding of anthracnose resistance genes in yam improvement. However, gene pyramiding is appropriate if the pathogen is exclusively asexual/clonal and if the potential for gene flow is low. But if the pathogen is recombining, then the recombination of virulence alleles may occur as quickly as breeders can recombine resistance genes, thus frustrating breeding efforts (McDonald and Linde 2002). This appears to be the case with *C. gloeosporioides* whose sexual stage *Glomerella cingulata* is present in yam fields (Abang 2003), and field populations have been shown to exhibit high genotypic diversity and gene flow (Thottappilly et al. 1999; Abang 2003). If the sexual stage plays an important role in yam anthracnose epidemics, then yam breeders will have to consider laying emphasis on race-nonspecific resistance and increasing diversity in host populations (McDonald and Linde 2002).

15.4 Future Scope of Works

Further AFLP mapping is planned to identify additional QTLs and strengthen existing marker-QTL linkages. Candidate-gene analyses are yet to be employed to investigate a variety of traits. To date, significant associations have been demonstrated for disease resistance in many crops, including tomato and potato. The yam breeding program at IITA is initiating the use of MAS for choosing parental lines for breeding purposes. It is likely that as QTL experiments are expanded, additional genes will be identified for use in breeding. At present, only disease resistance has been mapped on the *D. rotundata* and *D. alata* genetic linkage maps. Judging from the high heterozygosity levels in the parents used for the crossing experiment, many morphological and physiological trait loci should also be segregating among the progeny. Traits including resistance to other pathogen strains or diseases, and QTLs for flowering time and the period of dormancy, etc. could all be mapped using the same mapping populations. The ease of maintaining the individuals in tissue culture presents a unique advantage.

Conservative estimates put the genome coverage of the *D. rotundata* and *D. alata* maps at 56% and 65%, respectively. There are several reasons why the maps may not give complete coverage. The most obvious is that the two parents may have some common ancestry so that segments of the linkage maps may be devoid of polymorphisms and thus cannot be identified in genetic analysis. One approach to trying to gain some insight on this issue would be to attempt to align the *D. alata* and *D. rotundata* maps. This would give additional confidence in the general map structures and enable the development of suitable markers for genomic surveys of other populations. An attempt was made to derive gene-sequence-based markers, but unfortunately the cDNA library used for this analysis contained an unexpectedly high proportion of rRNA sequences. Nevertheless, this remains a sensible objective and would also permit the alignment of these maps with that presented for *D. tokoro* (Terauchi and Kahl 1999). Both maps provide useful tools for further genetic analysis of agronomically important traits in yam. While AFLPs continue to be identified and used for mapping the yam genome, efforts are geared toward saturating the map with microsatellites (simple sequence repeats, SSRs), and expressed sequence tags (ESTs) for greater ease of application in yam breeding.

Most of the currently available molecular markers for the yam genome are based on AFLP and RAPD. For a marker-assisted breeding to be feasible, it is important to develop user-friendly and high-throughput markers such as EST-SSR and SNP markers. In addition to the promise of gene discovery, a yam EST database would open new avenues to use candidate-gene-based approaches to understand the genetic control of phenotypic traits and to use this information in comparative studies and marker development for application in genetic, ecological, and evolutionary studies. Sequence data on expressed genes and on plant and crop genomes are rapidly accumulating and present powerful tools for plant science. The increasing availability of ESTs puts QTL cloning within reach (Zhu et al. 2003). EST collections also provide the basis for microarray technology that allows patterns of gene expression to be investigated in various physiological conditions, another potentially promising source of candidate genes. Combining information on mapped QTLs and ESTs provides a step toward identifying the genes that underlie QTLs. Although sequence data sets are, in themselves, imposing and cumbersome, increasingly powerful and friendly databases (e.g., Yuan et al. 2001), allow researchers to access genetic information and identify and exploit natural variation in ways previously not possible.

Candidate genes, genes known or suspected to be involved in conditioning the phenotype of interest, make it possible to localize desirable variants much more precisely. Credible candidate genes have been identified for many plant traits, including quantitative disease resistance in potato, *Solanum tuberosum* L. (Trognitz et al. 2002). A number of research approaches have converged to allow genes underlying QTLs to be cloned, the isolation of which will permit both the identification of potentially useful variants of agronomically important genes and the precise selection of the most useful alleles (Thornsberry et al. 2001). The availability of the isolated genes could allow natural molecular variation to be analyzed efficiently in a range of genotypes, enabling the identification of potentially useful variants for future use.

Advances in crop genomics have resulted in a more unified understanding of the biology of the entire plant kingdom, as well as a powerful set of molecular and bioinformatic tools and methods. Such advances provide an opportunity for efficient transfer of information systems from model species and major crops to orphan crops (Naylor et al. 2004). As a result, relatively small investments in the transfer of advanced science from major crops to orphan crops such as yam may potentially result in disproportionately high payoffs in terms of crop production, yield stability, and food security in developing countries. However, investment in genomics for any given species is only likely to be useful if a strong conventional breeding effort exists.

There are clearly reciprocal benefits of genomics research on orphan crops for improvement of major crops, derived from insights into the genetic bases for their distinctive attributes (Nelson et al. 2004). That is, some of the orphan crops can provide good models for traits not possessed by the model crops. For instance, the tuber is an important ecological (and economic) trait with few models: potato may serve for eudicots, but there is little basis to judge how suitable it might be as a model for monocots. Knowledge of gene expression at the appropriate stages in a tuberous monocot (e.g., *Dioscorea*, yam), matched with a candidate-gene approach, would address this issue. Phylogenetic morphological studies show that the "monocot" mode of leaf development typifies a nested group of some, but not all, monocots; others have either "dicot" or "intermediate" modes of development. The grass models may serve taxa with "monocot" modes, but other taxa (e.g., *Dioscorea*) may be needed to understand other developmental modes (Bharathan 1996).

With its dioecious habit and different morphologies of staminate and pistillate plants in some species, dicot-like leaf structure (net-veined, petiolate) with early development intermediate between "dicot" and "monocot" modes (Bharathan 1996); distinct changes in shoot apical meristem (SAM) structure and phyllotaxy during phase transition from juvenile to adult (Burkill 1960); tuber formation, small C-value, and widespread polyploidy (Dansi et al. 2001; Egesi et al. 2002), *Dioscorea* offers a system in which to raise general biological questions that cannot be addressed in many other species. *Dioscorea* thus holds great promise of yielding important clues to explain differences between eudicot and grass models (e.g., nonorthology of KNOX genes controlling SAM indeterminacy: Bharathan et al. 1999) and offering an example of biological phenomena such as dioecy and tuberization.

Acknowledgements

The authors thank the publishers of Annals of Applied Biology and Theoretical and Applied Genetics for permission to reproduce published tables and figures. The authors also thank the Gatsby Charitable Foundation for providing funding

References

Abang MM (2003) Genetic diversity of *Colletotrichum gloeosporioides* Penz. causing anthracnose disease of yam (*Dioscorea* spp.) in Nigeria. Bibliotheca Mycologica, vol 197. J. Cramer in der Gebr. Borntraeger, Berlin Stuttgart

Abang MM, Winter S, Green KR, Hoffmann P, Mignouna HD, Wolf GA (2002) Molecular identification of *Colletotrichum gloeosporioides* causing anthracnose of yam in Nigeria. Plant Pathol 51:63–71

Abang MM, Winter S, Mignouna HD, Green KR, Asiedu R (2003 a) Molecular taxonomic, epidemiological and population genetic approaches to understanding yam anthracnose disease. Afr J Biotechnol 2:486–496

Abang MM, Asiedu R, Mignouna HD, Green KR, Winter S, Wolf GA (2003 b) Anthracnose disease of yams in West Africa: recent advances in research and future perspectives. In: Akoroda MO (ed) Root Crops: The Small Processor and Development of Local Food Industries for Market Economy. Proc 8th Trienn Symp Int Soc Trop Root Crops–Africa Branch (ISTRC-AB), Nov 2001, Ibadan, Nigeria, pp 562–571

Adam-Blondon A-F, Sevignac M, Dron M, Bannerot H (1994) A genetic map of common bean to localize specific resistance genes against anthracnose. Genome 37:915–924

Asiedu R, Ng SYC, Bai KV, Ekanayake IJ, Wanyera NMW (1998) Genetic Improvement. In: Orkwor GC, Asiedu R,

Ekanayake IJ (eds) Food Yams: Advances in Research. IITA and NRCRI, Nigeria, pp 63–104

Asnaghi C, Roques D, Ruffel S, Kaye C, Hoarau J-Y, Télismart H, Girard JC, Raboin LM, Risterucci AM, Grivet L, D'Hont A (2004) Targeted mapping of a sugarcane rust resistance gene (*Bru1*) using bulked segregant analysis and AFLP markers. Theor Appl Genet 108:759–764

Bharathan G (1996) Does the monocot mode of leaf development characterize all monocots? Aliso 14:271–279

Bharathan G, Janssen B-J, Kellogg EA, Sinha N (1999) Phylogenetic relationships and evolution of the KNOTTED class of plant homeodomain proteins. Mol Biol Evol 16:553–563

Burkill IH (1951) The rise and decline of the greater yam in the service of man. Adv Sci 7:443–448

Burkill IH (1960) The organography and the evolution of the *Dioscoreaceae*, the family of the yams. J Linn Soc Bot London 56:319–412

Conlan SR, Griffiths LA, Napier JA, Shewry PR, Mantell S, Ainsworth C (1995) Isolation and characterization of cDNA clones representing the genes encoding the major tuber storage protein (dioscorin) of yam (*Dioscorea cayenensis* Lam.). Plant Mol Biol 28:369–380

Coursey DG (1967) Yams. Longmans, London

Coursey DG, Martin FW (1972) The past and future of the yams as crop plants. Plant Foods Hum Nutr 2:133–138

Crouzillat D, Phillips W, Fritz PJ, Pétiard V (2000) Quantitative trait loci analysis in *Theobroma cacao* using molecular markers. Inheritance of polygenic resistance in two related cacao populations. Euphytica 114:25–36

Daniel IO, Ng NQ, Tayo TO, Togun AO (2002) Wet-cold preservation of West African yam (*Dioscorea* spp.) pollen. J Agric Sci 138:57–62

Dansi A, Mignouna HD, Zoundjihékpon J, Sangaré A, Asiedu R, Ahoussou N (2000) Identification of some Benin Republic's Guinea yam (*Dioscorea cayenensis/Dioscorea rotundata*) cultivars using randomly amplified polymorphic DNA. Genet Resour Crop Evol 47:619–625

Dansi A, Mignouna HD, Pillay M, Zok S (2001) Ploidy variation in the cultivated yams (*Dioscorea cayenensis-D. rotundata* complex) from Cameroon as determined by flow cytometry. Euphytica 119:301–307

Degras LM (1993) The Yam: A Tropical Crop. Macmillan, London

Dudley JW (1993) Molecular markers in plant improvement: manipulation of genes affecting quantitative traits. Crop Sci 33:660–668

Egesi CN, Pillay M, Asiedu R, Egunjobi JK (2002) Ploidy analysis in water yam, *Dioscorea alata* L. germplasm. Euphytica 128:225–230

Eka OU (1985) The chemical composition of yam tubers. In: Osuji GO (ed) Advances in Yam Research. Biochem Soc, Nigeria/ASUTECH. Frontline Publishers, Enugu, Nigeria, pp 51–75

FAO (1988) Root and Tuber Crops, Plantain and Bananas in Developing Countries: Challenges and Opportunities. FAO Plant Production and Protection Paper, Rome, Italy. 58 pp

FAO (2003) FAOSTAT Agriculture Data. Food and Agriculture Organisation of the United Nations, Rome, Italy: *http://apps.fao.org/collections*

Gamiette F, Bakry F, Ano G (1999) Ploidy determination of some yam species (*Dioscorea* spp.) by flow cytometry and conventional chromosomes counting. Genet Resour Crop Evol 46:19–27

Geffroy V, Creusot F, Falquet J, Sévignac M, Adam-Blondon A-F, Bannerot H, Gepts P, Dron M (1998) A family of LRR sequences in the vicinity of the *Co-2* locus for anthracnose resistance in *Phaseolus vulgaris* and its potential use in marker-assisted selection. Theor Appl Genet 96:494–502

Geffroy V, Sévignac M, De Oliveira JCF, Fouilloux G, Skroch P, Thoquet P, Gepts P, Langin T, Dron M (2000) Inheritance of partial resistance against *Colletotrichum lindemuthianum* in *Phaseolus vulgaris* and co-localization of quantitative trait loci with genes involved in specific resistance. Mol Plant-Micr Interact 13:287–296

Hackett CA, Bradshaw JE, Meyer RC, McNicol JW, Milbourne D, Waugh R (1998) Linkage analysis in tetraploid potato: a simulation study. Genet Res 71:143–154

Haley SD, Afanador LK, Miklas PN, Stavely JR, Kelly JD (1994 a) Identification and application of a random amplified polymorphic DNA marker for the *I* gene (potyvirus resistance) in common bean. Phytopathology 84:157–160

Haley SD, Afanador LK, Kelly JD (1994 b) Selection for monogenic resistance traits with coupling- and repulsion-phase RAPD markers. Crop Sci 34:1061–1066

Hämäläinen JH, Sorri VA, Wanatabe KN, Gebhardt C, Valkonen JPT (1998) Molecular examination of a chromosome region that controls resistance to potato Y and A potyviruses in potato. Theor Appl Genet 96:1036–1043

IITA (1999) Annual Report of the International Institute of Tropical Agriculture, Ibadan, Nigeria

Ile E (2004) Control of tuber domancy and flowering in yam (*Dioscorea rotundata* Poir.). PhD thesis, University of Reading, Reading, UK

Kay DE (1973) Root Crops. Crop and Product Digest 2. Tropical Products Institute, London

Kelly JD, Anafador L, Haley SD (1995) Pyramiding genes for resistance to bean common mosaic virus. Euphytica 82:207–212

Kosambi D (1944) The estimation of map distances from recombination values. Ann Eugen 12:172–175

Kramer EM, Irish VF (1999) Evolution of genetic mechanisms controlling petal development. Nature 399:144–148

Lander ES, Green P, Abrahamson J, Barlow A, Daly MJ, Lincoln SE, Newburgh L (1987) MAPMAKER: An interactive computer package for constructing primary genetic linkage maps of experimental and natural populations. Genomics 1:174–181

Lawton JRS, Lawton JR (1967) The morphology of the dormant embryo and young seedling of five species of *Dioscorea* in Nigeria. Proc Linn Soc Lond 178:153–159

Lebot V (1999) Biomolecular evidence for plant domestication in Sahul. Genet Resour Crop Evol 46:619–628

Ling P, Duncan LW, Deng Z, Dunn D, Hu X, Huang S, Gmitter FG Jr (2000) Inheritance of citrus nematode resistance and its linkage to molecular markers. Theor Appl Genet 100:1010–1017

Luo ZW, Hackett CA, Bradshaw JE, McNicol JW, Milbourne D (2001) Construction of a genetic linkage map in tetraploid species using molecular markers. Genetics 157:1369–1385

McDonald BA, Linde C (2002) Pathogen population genetics, evolutionary potential, and durable resistance. Annu Rev Phytopathol 40:349–379

Mendoza-Herrera A, Hernandez F, Martinez O, Hernandez S, Acosta J, Simpson J (1999) Molecular mapping of one anthracnose resistant gene in the common bean. In: Plant and Animal Genome VII Conf, San Diego. *http://www.intlpag.org/pag/7/abstracts/*

Michelmore R (1995) Molecular approaches to manipulation of disease resistance genes. Annu Rev Phytopathol 15:393–427

Michelmore RW, Paran I, Kessell RV (1991) Identification of markers linked to disease resistance genes by bulked segregant analysis: a rapid method to detect markers in specific genomic regions using segregating populations. Proc Natl Acad Sci USA 88:9828–9832

Miège J (1982) Note sur les espèces *Dioscorea cayenensis* Lamk. et *Dioscorea rotundata* Poir. In: Miège J, Lyonga SN (eds) Yams. Ignames. Clarendon, Oxford, pp 367–375

Miège J, Lyonga SN (eds) (1982) Yams. Ignames. Clarendon, Oxford

Mignouna HD, Asiedu R (1999) A genetic linkage map of Guinea yam *(Dioscorea rotundata* Poir) based on isozyme, RAPD and AFLP markers. In: Plant and Animal Genome VII Conf, San Diego. *http://www.intl-pag.org*

Mignouna HD, Ellis NTH, Knox M, Asiedu R, Ng QN (1998) Analysis of genetic diversity in Guinea Yams (*Dioscorea* spp) using AFLP fingerprinting. Trop Agric (Trinidad) 75:224–229

Mignouna HD, Abang MM, Green KR, Asiedu R (2001 a) Inheritance of resistance in water yam (*Dioscorea alata*) to anthracnose (*Colletotrichum gloeosporioides*). Theor Appl Genet 103:52–55

Mignouna HD, Njukeng P, Abang MM, Asiedu R (2001 b) Inheritance of resistance to *Yam Mosaic Virus*, genus *Potyvirus*, in white yam (*Dioscorea rotundata*). Theor Appl Genet 103:1196–1200

Mignouna HD, Abang MM, Onasanya A, Asiedu R (2002 a) Identification and application of RAPD markers for anthracnose resistance in water yam (*Dioscorea alata*). Ann Appl Biol 141:61–66

Mignouna HD, Abang MM, Onasanya A, Agindotan B, Asiedu R (2002 b) Identification and potential use of RAPD markers linked to *Yam mosaic virus* resistance in white yam (*Dioscorea rotundata* Poir.). Ann Appl Biol 140:163–169

Mignouna HD, Mank RA, Ellis THN, van den Bosch N, Asiedu R, Abang MM, Peleman J (2002 c) A genetic linkage map of water yam (*Dioscorea alata* L.) based on AFLP markers and QTL analysis for anthracnose resistance. Theor Appl Genet 105:726–735

Mignouna HD, Mank RA, Ellis THN, van den Bosch N, Asiedu R, Ng SYC, Peleman J (2002 d) A genetic linkage map of Guinea yam (*Dioscorea rotundata* L.) based on AFLP markers. Theor Appl Genet 105:716–725

Mignouna HD, Abang MM, Wanyera NW, Chikaleke VA, Asiedu R, Thottappilly G (2005) PCR marker-based analysis of wild and cultivated yams (*Dioscorea* spp.) in Nigeria: genetic relationships and implications for *ex situ* conservation. Genet Resour Crop Evol 52:755–763

Miklas PN, Larsen RC, Riley R, Kelly JD (2000) Potential marker-assisted selection for *bc-1*^{2} resistance to bean common mosaic potyvirus in common bean. Euphytica 116:211–219

Naylor RN, Falcon WP, Goodman RM, Jahn MM, Sengooba T, Tefera H, Nelson RJ (2004) Biotechnology in the developing world: a case for increased investments in orphan crops. Food Policy 29:15–44

Nelson JC (1997) QGENE: software for marker-based genome analysis and breeding. Mol Breed 3:239–245

Nelson RJ, Naylor RL, Jahn MM (2004) The role of genomics research in improvement of "orphan" crops. Crop Sci 44:1901–1904

Ng SYC (1992) Micropropagation of white yam (*Dioscorea rotundata* Poir.). In: Bajai YPS (ed) Biotechnology in Agriculture and Forestry, High-tech and Micropropagation III, vol 19. Springer, Berlin Heidelberg New York, pp 135–159

Nweke FI, Ugwu BO, Asadu CLA, Ay P (1991) Production costs in the yam-based cropping systems of S.E. Nigeria. RCMP Research Monograph No. 6. RCMD, IITA, Ibadan, Nigeria

Okonkwo SNC (1985) The botany of the yam plant and its exploitation in enhanced productivity of the crop. In: Osuji GO (ed) Advances in Yam Research. Biochem Soc Nigeria/ASUTECH. Frontline, Enugu, Nigeria, pp 3–25

Onwueme IC (1978) The Tropical Tuber Crops. Wiley, Chichester, UK

Park YH, Sensoy S, Wye C, Antonise R, Peleman J, Havey MJ (2000) A genetic map of cucumber composed of RAPDs, RFLPs, AFLPs, and loci conditioning resistance to papaya ringspot and zucchini yellow mosaic viruses. Genome 43:1003–1010

Peters JL, Cnops G, Neyt P, Zethof J, Cornelis K, Van Lijsebettens M, Gerats T (2004) An AFLP-based genome-wide mapping strategy. Theor Appl Genet 108:321–327

Ramachandran K (1968) Cytological studies in *Dioscoreaceae*. Cytologia 33:401–410

Solomon-Blackburn RM, Barker H (2001 a) A review of host major-gene resistance to potato viruses X, Y, A and V in potato: genes, genetics and mapped locations. Heredity 86:8–16

Solomon-Blackburn RM, Barker H (2001 b) Breeding virus resistant potatoes (*Solanum tuberosum*): a review of traditional and molecular approaches. Heredity 86:17–35

Stam P (1993) Construction of integrated linkage maps by means of a new computer package: JOINMAP. Plant J 3:739–744

Stam P, van Ooijen JW (1996) JOINMAP^{TM} version 2.0: Software for the calculation of genetic linkage maps. CPRO-DLO, Wageningen, The Netherlands

Staub JE, Serquen FC, Gupta M (1996) Genetic markers, map construction, and their application in plant breeding. HortSci 31:729–741

Tanksley SD, Nelson JC (1996) Advanced backcross QTL analysis: a method for the simultaneous discovery and transfer of valuable QTLs from unadapted germplasm into elite breeding lines. Theor Appl Genet 92:191–203

Terauchi R, Kahl G (1999) Mapping of the *Dioscorea tokoro* genome: AFLP markers linked to sex. Genome 42:752–762

Thornsberry JM, Goodman MM, Doebley J, Kresovich S, Nielsen D, Buckler ES (2001) *Dwarf8* polymorphisms associate with variation in flowering time. Nat Genet 28:286–289

Thottappilly G, Mignouna HD, Onasanya A, Abang M, Oyelakin O, Singh NK (1999) Identification and differentiation of isolates of *Colletotrichum gloeosporioides* from yam by Random Amplified Polymorphic DNA markers. Afr Crop Sci J 7:195–205

Thouvenel JC, Dumont R (1990) Perte de rendement de l'igname infectée par le virus de la mosaique en Côte d'Ivoire. L'Agron Trop 45:125–129

Trognitz F, Manosalva P, Gysin R (2002) Plant defense genes associated with quantitative resistance to potato late blight in *Solanum phureja*×dihaploid *S. tuberosum* hybrids. Mol Plant-Micr Interact 15:587–597

Van der Voort JR, Kanyuka K, Van der Vossen E, Bendahmane A, Mooijman P, Klein-Lankhorst R, Stiekema W, Baulcombe D, Bakker J (1999) Tight physical linkage of the nematode resistance gene *Gpa2* and the virus resistance gene *Rx* on a single segment introgressed from the wild species *Solanum tuberosum* subsp. *andigena* CPC1673 into cultivated potato. Mol Plant-Micr Interact 12:197–206

Van Ooijen JW, Maliepaard C (1996) MapQTL^{TM} Version 3.0. Software for the calculation of QTL positions on genetic maps. In: Plant Genome IV Conf, San Diego. *http://probe.nalusda.gov:8000/otherdocs/pg/pg4/abstracts/p316.html*

Vos P, Hogers R, Bleeker M, Reijans M, van de Lee T, Hornes M, Frijters A, Pot J, Peleman J, Kuiper M, Zabeau M (1995) AFLP: a new technique for DNA fingerprinting. Nucleic Acids Res 23:4407–4414

Vuylsteke M, Mank R, Antonise R, Bastiaans E, Senior ML, Stuber CW, Melchinger AE, Lubberstedt T, Xia XC, Stam P, Zabeau M, Kuiper M (1999) Two high-density AFLP linkage maps of *Zea mays* L.: analysis of distribution of AFLP markers. Theor Appl Genet 99:921–935

Wingbermuehle WJ, Gustus C, Smith KP (2004) Exploiting selective genotyping to study genetic diversity of resistance to Fusarium head blight in barley. Theor Appl Genet 109:1160–1168

Wu KK, Burnquist W, Sorrells ME, Tew TL, Moore PH, Tanksley SD (1992) The detection and estimation of linkage in polyploids using single-dose restriction fragments. Theor Appl Genet 83:294–300

Young RA, Melotto M, Nodari RO, Kelly JD (1998) Marker-assisted dissection of the oligogenic anthracnose resistance in the common bean cultivar, 'G2333'. Theor Appl Genet 96:87–94

Yuan Q, Quackenbush J, Sultana R, Pertea M, Salzberg SL, Buell CR (2001) Rice bioinformatics: analysis of rice sequence data and leveraging the data to other plant species. Plant Physiol 125:1166–1174

Zhu W, Schlueter SD, Brendel V (2003) Refined annotation of the *Arabidopsis* genome by complete expressed sequence tag mapping. Plant Physiol 132:469–484

Zoundjihékpon J, Hamon S, Tio-Touré B, Hamon P (1994) First controlled progenies checked by isozymic markers in cultivated yams *Dioscorea cayenensis-rotundata*. Theor Appl Genet 88:1011–1016

Subject Index